Encyclopedia of
Plant Physiology

New Series   Volume 2   Part B

Editors

A. Pirson, Göttingen
M. H. Zimmermann, Harvard

Transport in Plants II

Part B Tissues and Organs

Edited by

U. Lüttge and M. G. Pitman

Contributors

W. P. Anderson E. Epstein A. E. Hill B. S. Hill
T. C. Hsiao W. D. Jeschke A. Läuchli U. Lüttge J. S. Pate
M. G. Pitman E. Schnepf R. M. Spanswick
R. F. M. Van Steveninck J. F. Sutcliffe

Springer-Verlag Berlin Heidelberg New York 1976

QK
871
.T73
v.2
pt. B

With 129 Figures

ISBN 3-540-07453-8 Springer-Verlag Berlin Heidelberg New York
ISBN 0-387-07453-8 Springer-Verlag New York Heidelberg Berlin

Library of Congress Cataloging in Publication Data. Main entry under title: Transport in plants. (Encyclopedia of plant physiology; v. Vol. 2, pt. A & B edited by U. Lüttge and M.G. Pitman. Bibliography: p. Includes indexes. Contents: v. 1. Phloem transport. — v. 2. pt. A. Cells. pt. B. Tissues and organs. 1. Plant translocation. I. Canny, M.J. II. Zimmermann, Martin Huldrych, 1926—   III. Lüttge, Ulrich. QK871.T73 582'.041 75-20178

© by Springer-Verlag Berlin · Heidelberg 1976
Printed in Germany.

Typesetting, printing and bookbinding: Universitätsdruckerei H. Stürtz AG, Würzburg.

Introduction

In the first part (Part A) of this volume on transport, there was an emphasis on the processes occurring at the membranes bounding the cells. It was convenient to distinguish active and passive processes of transport across the membranes, and to recognize that certain transport processes may be regulated by internal factors in the cells such as cytoplasmic pH, concentrations of ions, of malate or of sugar in the vacuoles, or the hydrostatic pressure.

Cells in tissues and organs show the same kinds of properties as individual cells, but in addition there can be cell to cell transport related to the organization of the tissue. Firstly cells within a tissue are separated from the external solutions by a diffusion path comprising parts of the cell walls and intercellular spaces; more generally this extra-cytoplasmic part of the tissue has been called the *apoplasm*. A similar term is "free space". Secondly, the anatomy of cells in tissues seems to allow some facilitated, local transport between cells in a *symplasm*. Entry into the symplast and subsequent transport in a symplasmic continuum seems to be privileged, in that ions may not have to mix with the bulk of the cytoplasm and can pass from cell to cell in particular cytoplasmic structures, plasmodesmata. In *Chara* plants, this kind of transport is found operating across the multi-cellular nodes as the main means of transport between the long internodal cells. In higher plants symplasmic transport appears important over distances of only a few cells. It has been proposed as acting in leaf parenchyma and in special situations such as glands (Chap. *5*); across the bundle sheaths of leaves, especially in C_4 plants; in transport of ions across the root (Chaps. *2* and *3*). As yet little is known of how this type of transport operates, nor how it is controlled, but it seems to be a non-metabolic process once ions have entered the symplast (Chap. *2*).

We have organized this Part (Part B) into three sections. The first two chapters in section I are concerned with the general properties of the apoplasm and the symplasm. In the following chapters in section II transport is studied in certain organized tissues that show symplasmic as well as cellular transport. Finally, in section III, aspects of regulation and integration of transport within organs and within the whole plant are considered.

We are grateful for help with Part B as with Part A, as acknowledged in the Introduction to Part A.

January 1976   ULRICH LÜTTGE   MICHAEL PITMAN
                           Darmstadt   Sydney

Instructions for the Reader

In using this book note that references to chapters in Part B are quoted as "Chap. *6*" whereas chapters in Part A are quoted as "Part A, Chap. *6*". Similarly parts of chapters are referred to as "*6*.4.2" if in Part B or as "Part A, *6*.4.2" if in Part A. In each case the italic number refers to the *Chapter* number and any roman numbers to the *part* within the chapter. The same conventions are used for Figures and Tables. "Part A, Fig. *5*.1" is Figure 1 in Chapter *5* in Part A; "Table *7*.2" refers to Table 2 in Chapter *7* of Part B. Chapters *3, 4* and *5* in Part B have been subdivided, and in this part (B) they are referred to as "Chap. *3.3*" and *3.3.4.1*", respectively, depending on whether reference is made to the whole or part of the subchapter. The index covers both Parts A and B; italic numbers refer to pages in Part A and ordinary (roman) script to pages in Part B.

Contents of Part B

Survey of Part A

List of Contributors (Part B)

W.P. ANDERSON
  Research School of Biological Sciences,
  Australian National University, Canberra
  ACT 2601/Australia

E. EPSTEIN
  Department of Land, Air, and Water
  Resources, Soils and Plant Nutrition
  Section, University of California, Davis,
  CA 95616/USA

A.E. HILL
  The Botany School, Downing Street,
  Cambridge CB4 3DQ/Great Britain

B.S. HILL
  The Botany School, Downing Street,
  Cambridge CB4 3DQ/Great Britain

T.C. HSIAO
  Department of Water Science and
  Engineering, Laboratory of Plant-Water
  Relations, University of California,
  Davis, CA 95616/USA

W.D. JESCHKE
  Botanisches Institut der Universität
  Würzburg, Mittlerer Dallenbergweg 64
  8700 Würzburg/Federal Republic
  of Germany

A. LÄUCHLI
  Botanisches Institut, Technische
  Hochschule, Schnittspahnstr. 3—5,
  6100 Darmstadt/Federal Republic
  of Germany

U. LÜTTGE
  Botanisches Institut, Technische
  Hochschule, Schnittspahnstr. 3—5,
  6100 Darmstadt/Federal Republic
  of Germany

J.S. PATE
  Department of Botany, University of
  Western Australia, Nedlands, W.A.
  6009/Australia

M.G. PITMAN
  School of Biological Sciences (A 12)
  University of Sydney,
  Sydney, N.S.W. 2006/Australia

E. SCHNEPF
  Lehrstuhl für Zellenlehre der
  Universität Heidelberg, Im Neuenheimer
  Feld 230, 6900 Heidelberg/Federal
  Republic of Germany

R.M. SPANSWICK
  Section of Genetics, Development
  and Physiology, Division of Biological
  Sciences, 255 Plant Science Building,
  Cornell University, Ithaca,
  New York 14853/USA

R.F.M. VAN STEVENINCK
  School of Agriculture, La Trobe University
  Bundoora, Victoria 3083, Australia

J.F. SUTCLIFFE
  School of Biological Sciences,
  University of Sussex, Falmer,
  Brighton, Sussex BN 1 9 Q6/Great Britain

I. Pathways of Transport in Tissues

1. Apoplasmic Transport in Tissues

A. LÄUCHLI

1. Introduction

Transport physiologists have recognized for many years that the cell walls of a plant tissue are continous and may therefore be looked upon as a continuum for transport of water and solutes. The term *apoplast* attributed to MÜNCH (1930) is used generally for this continuum. The term *symplasm* is used to describe the part of the tissue bound by membranes. There is a necessity for apoplasmic transport to the plasmalemma in cell walls of any type of plant, be it unicellular algae, more complex thallous organisms, or higher plants, whenever membrane-controlled transport occurs across this outer barrier of the symplasm.

The cell wall is considered to be readily permeable to water and ions. Hence, it is treated as *free space* for transport to the plasmalemma. Free space characteristics including water free space, Donnan free space, and exchange properties of the cell wall were covered in Part A, 5.3. It is attempted in this Chapter to look at the apoplast as a *transport pathway* in tissues, analogous to the following Chap. *2* dealing with the symplasm as a transport pathway. Long-distance transport in the apoplast, e.g. in the xylem, essentially is excluded from this discussion.

The apoplast has some important bearings on the overall transport characteristics of a tissue. Since movement in the apoplast is limited by the rate of diffusion, equilibration of tissues with the external environment *via* the apoplast is very much dependent on the size of the specimen, as for instance in leaf slices (SMITH and EPSTEIN, 1964; PITMAN et al., 1974; *4.2.2.2.1*). The same problem occurs with use of slices of storage tissues (Part A, *8.2.1*). The thickness of the specimen also affects the supply of oxygen in the tissue and may be particularly important when the intercellular spaces are injected with solution (Part A, *8.2.2*), as diffusion of oxygen in water ($D = 10^{-9}$ m^2 s^{-1}) is much slower than in air ($D = 10^{-5}$ m^2 s^{-1}). The apoplast of course restricts also the amount of material that can be transported through the tissue *in situ*. To a large extent the plant has overcome limitations of diffusion in the apoplast by evolution of the symplasm (Chap. *2*). In tissues of higher plants, where both pathways are developed, the apoplast could interfere with operation of an organ by acting as a wasteful "leak" pathway away from the direction of symplasmic transport. Certain chemical and structural changes of the cell wall can be interpreted as an evolution of blocks in the apoplasmic pathway (e.g. incrustation by lignin and suberin). The significance of such apoplasmic blocks for transport in tissues is stressed in the discussion of several anatomical examples (*1.4.2.2*).

Transport studies usually focus on the apoplast as the water-filled pathway for diffusion of solutes in the plant. However, in a similar sense one must regard the intercellular spaces when air-filled as a separate component of the apoplast. Indeed the intercellular spaces can be considered to be evolved as an adaptation for supply of gases to the cells to make use of the 10^4 times more rapid movement by diffusion of gases (e.g. O_2, CO_2, H_2O) in the gas phase than in the liquid phase.

2. Chemical and Cytological Properties of the Cell Wall in Relation to Apoplasmic Transport

2.1 Cell Wall Chemistry

2.1.1 Cellulose

The most characteristic constituent of the cell wall of higher plants is *cellulose*. The cellulose molecules are arranged to a large extent as cristallites in a chain lattice (cf. FREY-WYSSLING and MÜHLETHALER, 1965). These crystalline regions are called micelles. A variable number of micelles, often about six, form a microfibril (cf. ROBARDS, 1970). The crystalline core of the micelles is surrounded by paracrystal-line cellulose. The width of the whole microfibril is estimated to be about 10–25 nm, though it is extremely variable. A model of microfibrils in cross section is shown in Fig. *1*.1.

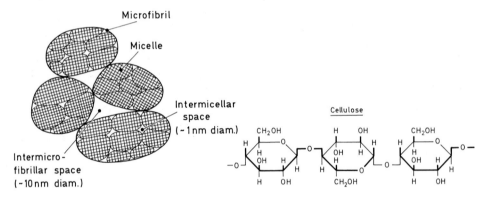

Fig. *1*.1. Model of microfibrils in cell walls. (Modified after ROBARDS, 1970)

Fig. *1*.2. Structure of *cellulose*

The chemical structure of cellulose (Fig. *1*.2) is such that H-bonds occur between the cellulose chains. However, cellulose does not have ion-binding capacities and hence does not have any direct effect on apoplasmic transport of ions. More important in transport are the intermicellar and intermicrofibrillar spaces (Fig. *1*.1) with approximate diameters of 1 and 10 nm, respectively. The diameter of an intermicrofi-brillar space is much greater than that of water and small solute particles, which hence apparently can pass through it and possibly to some extent even migrate within the intermicellar spaces. This is documented by Table *1*.1 which shows some particle diameters [nm]. In the case of ions, the hydrated ionic diameters are relevant in transport studies; unfortunately, data on hydrated ionic diameters are still incomplete and should be used with caution, as the values cited in the literature are very variable.

These properties of intermicrofibrillar and intermicellar spaces are most probably modified by matrix substances (*1*.2.1.2) which tend to decrease the size of these spaces. In spite of this apparent limitation in access of solutes to the spaces, ions do in fact migrate in cell walls whose water free space is certainly available for transport of anions.

Table *1*.1. Particle diameters [nm] of water, glucose, and some inorganic cations and anions

Particle	Molecule[a] or ionic crystal[b] diameter	Hydrated ionic diameter[d]	Particle	Molecule[a] or ionic crystal[b] diameter	Hydrated ionic diameter[d]
H_2O	0.39	—	Mg^{2+}	0.13	0.92
Glucose	0.89	—	Ca^{2+}	0.20	0.88
Na^+	0.19	0.60	Cl^-	0.36	0.50[e]
K^+	0.27	0.54	NO_3^-	0.41[c]	
NH_4^+	0.30				

[a] From BECK and SCHULTZ (1970). [b] From PAULING (1960). [c] Estimated from WELLS (1945).
[d] Estimated from MCFARLANE and BERRY (1974). [e] From ADAMSON (1973).

2.1.2 Matrix Substances (Hemicelluloses, Pectins, Glycoproteins)

Polysaccharides, which in contrast to cellulose are soluble in concentrated alkali, are known as *hemicelluloses* (FREY-WYSSLING and MÜHLETHALER, 1965). These matrix substances are made up of linearly oriented and also of branched polymers. Upon hydrolysis of hemicelluloses, several hexoses, pentoses, and uronic acids are obtained. Of particular interest to the transport physiologist are the *pectic substances* in the matrix of the cell wall. After extraction of the pectins from the wall with water or with aqueous solutions of chelating agents, various neutral (arabinogalactans) and acidic (rhamnogalacturonans) polysaccharides can be identified (NORTH-COTE, 1972). The acidic fraction contains α-D-polygalacturonic acid (Fig. *1*.3a). A possible structure of the rhamnogalacturonan proposed by TALMADGE et al. (1973) is presented in Fig. *1*.3b. According to these authors, the rhamnogalacturonan consists of an α-(1→4)-linked galac-

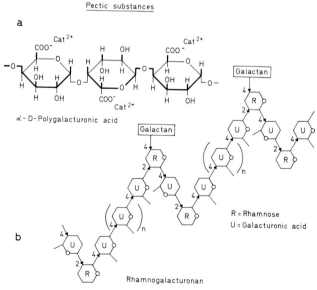

Fig. *1*.3a and b. *Pectic substances*. (a) α-D-polygalacturonic acid — the basic structure of the pectins. The free carboxyl groups can form salt with divalent and monovalent cations. (b) Possible structure of rhamnogalacturonan — the acidic fraction of the pectic polysaccharide. *n* an undetermined number, probably between 4 and 10. (From TALMADGE et al., 1973)

turonan chain which is interspersed with 2-linked rhamnosyl residues. The rhamnosyl residues possibly occur in units of rhamnosyl $(1 \rightarrow 4)$-galacturonosyl-$(1 \rightarrow 2)$-rhamnosyl. Approximately 50% of the rhamnosyl residues are branched, having a galactan substituent attached to carbon 4. A variable number of the carboxyl groups of α-D-polygalacturonic acid (Fig. *1*.3a) is esterified as methyl esters.

The free carboxyl groups give rise to fixed negative charges in the wall (Donnan free space) which can bind cations in preference to anions and divalent cations in preference to monovalent cations (see Part A, Tables 5.2 and 5.3). Thus, there are intimate interactions between the negatively charged groups of the pectins and cations, which by steric specificities in ion exchange may not only quantitatively but also qualitatively affect apoplasmic transport. In contrast to the more complex interactions with cations, transport of anions is possibly simply hindered by the negatively charged cell wall. Diffusion of anions in the water and solute accessible intermicellar and intermicrofibrillar spaces will, however, depend again on the way in which the negative charges in the cell wall are saturated by cations.

Isolated pectins form reversible gels and viscous solutions with water (NORTH-COTE, 1972). This property appears to be related to the role of pectins as filler substances within the matrix. This will influence the distribution of water in the cell wall. There is no evidence available that apoplasmic transport of water is also influenced by this property of the pectins. To what extent the cytological distribution of such matrix substances quantitatively affects apoplasmic transport by altering access of water and solutes is yet unknown, although quantitative data on exchange capacities of the free space are available (Part A, 5.3, Table 5.1).

In isolated cell walls, proteins are always detectable, apart from polysaccharides. There has been some controversy whether proteins are indeed a constituent of the wall or whether they occur in the wall fraction as contaminants from plasmalemma and plasmodesmata. The discovery in cell wall fractions of a protein rich in proline and hydroxyproline by DOUGALL and SHIMBAYASHI (1960) and LAMPORT and NORTHCOTE (1960) was therefore exciting in view of the fact that only minor amounts of hydroxyproline are found in the cytoplasm. Hence, it is now firmly accepted that cell walls contain endogenous proteins rich in hydroxyproline (Fig. *1*.4a). Additional proof of this assumption came from work by HEATH and NORTHCOTE (1971). They fractionated cell walls into cellulose, pectin, and hemicellulose and found that the bulk of the hydroxyproline-containing protein was associated with the cellulose fraction. More recent data suggest that the cell wall protein is a *glycoprotein* (reviews: LAMPORT, 1970; NORTHCOTE, 1972). According to LAMPORT (1973) the wall glycoproteins bear short arabinoside chains bound to hydroxyproline and galactose attached to the serine residues of the protein. A similar model of the structure of glycoprotein in cell walls was proposed independently by KEEGSTRA et al. (1973). The experiments of these authors indicate that the galactose attached to the serine residue of the protein belongs to an arabinogalactan. A possible structure of this protein-arabinogalactan complex is shown in Fig. *1*.4b. The arabinogalactan is a highly branched polymeric molecule containing predominantly 3,6-linked galactosyl residues as branch points with a single arabinosyl residue as the most prevalent side chain. Furthermore, a terminal rhamnosyl residue is present in the arabinogalactan polymer. TALMADGE et al. (1973) thought that all of the rhamnose in the cell wall is covalently linked in the rhamnogalacturonan of the pectic polysaccharide (cf. Fig. *1*.3b). This would lead to the hypothesis of a connection of the pectic rhamnogalacturonan to the hydroxyproline-rich protein through an arabinogalactan. This hypothesis forms an important part of a model of the macromolecular structure of primary cell walls discussed in *1*.2.1.3. KEEGSTRA et al. (1973) also presented evidence that the arabinoside chains bound to hydroxyproline of the glycoprotein (cf. LAMPORT, 1973) are arabinosyl tetrasaccharides which are glycosidically attached to hydroxyproline residues.

The cell wall glycoprotein in general is regarded as linked covalently to the carbohydrates (KEEGSTRA et al., 1973), but a significant portion of the protein can be isolated from the walls with salt solutions (BAILEY and KAUSS, 1974), suggest-

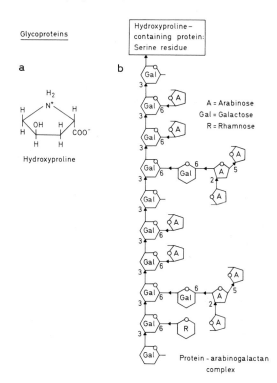

Fig. *1*.4a and b. *Glycoproteins.* (a) Hydroxyproline—the major amino acid of the cell wall glycoproteins (pK_a: 1.92, 9.73; pI: 5.83). (b) Possible structure of a glycoprotein: the glycoprotein is a protein-arabinogalactan complex. The linkage between the arabinogalactan and the hydroxyproline-containing protein is caused by a covalent bond between galactose and a serine residue of the protein. (From KEEGSTRA et al., 1973)

ing the occurrence of non-covalently bound glycoprotein. The latter protein fraction is able to agglutinate trypsinized red blood cells and can be classified as lectin or phytohemagglutinin (KAUSS and GLASER, 1974; for a review on lectins see LIS and SHARON, 1973). Lectins from seeds contain Ca^{2+} and Mn^{2+} as integral parts of their structure (GALBRAITH and GOLDSTEIN, 1970; LIS and SHARON, 1973). Since cell wall lectins rich in hydroxyproline can be extracted with complexing agents (e.g. EDTA), carbohydrate binding by the wall lectins is likely to involve cations of as yet unidentified nature (KAUSS and BOWLES, 1976). Thus it appears possible that the cell wall lectins may carry specific binding sites for divalent ions as well as for carbohydrate groups. Further studies have to show whether or not these properties are related to apoplasmic transport of ions or even sugars.

Proteins in the wall bear additional charged groups available for the binding of ions, the ion-binding capacity depending on pH. The possibility of binding anions to free amino groups is of particular interest in view of the fact that a small fraction of the apoplasmic Cl^- in barley roots is apparently bound to the cell wall. STELZER (1973), using the Ag^+-precipitation technique for localization of Cl^- (see *1*.4.1.2), found that the bulk of apoplasmic Cl^- could be washed out with 0.2 mM $CaSO_4$ in 1 min, but even after a wash-out period of 40 min there was still some Cl^- bound to the cell wall. In this context it seems pertinent to comment that a small Donnan free space for anions is measurable (Part A, *5*.3 and Table *5*.2), possibly due to the occurrence of glycoproteins in the cell wall matrix. It would be important to know the Cl^- binding capacity of hydroxyproline containing glycoproteins.

In fungal cell walls, the main constituent is chitin, a linear polysaccharide composed of N-acetyl glucosamine residues. The N in the N-acetyl group may carry a weak positive charge depending on pH. One might then speculate that fungal cell walls are positively charged and have anion exchange capacity.

2.1.3 Macromolecular Structure of Primary Cell Wall

A series of papers has been published by ALBERSHEIM and coworkers dealing with the macro-molecular structure of primary cell walls isolated from suspension-cultured sycamore (*Acer pseudoplatanus*) cells (TALMADGE et al., 1973; BAUER et al., 1973; KEEGSTRA et al., 1973). These studies were made by utilizing purified hydrolytic enzymes and methylation analyses and permitted the identification and quantitation of the macromolecular components of sycamore cell walls. TALMADGE et al. (1973) concluded that the sycamore primary wall is composed of:

23% cellulose
55% hemicellulose:
      34% pectins — 10% arabinan (neutral)
                       8% galactan (neutral)
                    16% rhamnogalacturonan (acidic)
21% xyloglucan

21% glycoprotein:
      10% protein (hydroxyproline-rich)
       9% oligo-arabinosides (attached to hydroxyproline)
       2% arabinogalactan (attached to serine).

On the basis of these analyses, KEEGSTRA et al. (1973) formulated a tentative model of the macromolecular structure of sycamore primary cell walls (Fig. *1*.5). In this model xyloglucan

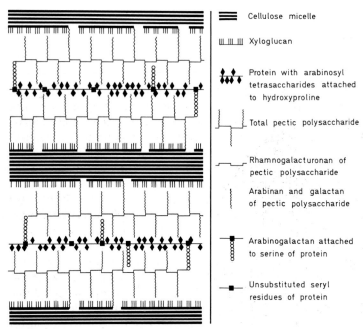

Fig. *1*.5. Tentative model of the macromolecular structure of a primary cell wall in suspension-cultured sycamore cells. The components are presented in approximately proper proportions except for the distance between cellulose micelles, which is expanded to allow room for the interconnecting structures. A detailed discussion of this model is given in the text. (From KEEGSTRA et al., 1973)

is considered to be non-covalently bound to cellulose through H-bonds and covalently linked to galactan chains. Galactan therefore appears to serve as a bridge between the xyloglucan-cellulose complex and the pectic rhamnogalacturonan. In addition, there is a covalent linkage between the pectic polysaccharides and serine residues of the protein through an arabinogalactan chain. Thus, this primary cell wall may be regarded as one giant complex macromolecule.

The studies of ALBERSHEIM's group are extremely stimulating and open the way for further endeavors of this kind. FRANKE et al. (1974) point out rightly that no similar study has been made of primary cell walls of other plant species or even of secondary walls. Only the hemicellulose xyloglucan in the primary wall of cells of *Phaseolus vulgaris* has been similarly characterized and found to have a basic structure almost identical to that in sycamore primary walls (WILDER and ALBERSHEIM, 1973).

What are the bearings of the cell wall model in Fig. *1*.5 to the problem of apoplasmic transport? Its primary message is that the intermicellar spaces (cf. Fig. *1*.1) cannot be considered as open spaces through which solutes migrate as in real pores. Rather, the intermicellar spaces contain a network of pectins and glycopro-teins, linked to each other and to the cellulose micelles, and giving rise to the Donnan free space of the apoplast. The negative charge of the Donnan free space is primarily due to the acidic rhamnogalacturonans, but the probable occurrence of non-covalently bound glycoprotein (BAILEY and KAUSS, 1974; KAUSS and BOWLES, 1976) may contribute additional charged groups to the Donnan free space (*1*.2.1.2). Furthermore, the fact that about one sixth of the primary wall consists of rhamno-galacturonan, suggests very strongly that the Donnan free space also encompasses the intermicrofibrillar spaces and extends into the cell wall matrix between the cellulose microfibrils.

2.2 Cell Wall Cytology

2.2.1 Incrustation (Lignin)

This discussion of primary cell walls has shown that some of their properties are similar to ion-exchange "membranes," but they contain spaces apparently large enough to allow water, ions, and small solutes to pass through them. When certain organic molecules penetrate the primary walls and form polymers, the cell wall undergoes incrustation. The consequences of incrustation for apoplasmic transport are debatable; they depend on the chemical properties of the incrusting substances and on the extent of incrustation in the wall (LÜTTGE, 1973).

Lignin is the most important incrusting substance. It may be described as the insoluble constituent of the cell wall that is aromatic, of high molecular weight, and derived by the enzymic dehydrogenation and subsequent polymerization of phenylpropane derivatives such as coniferyl, sinapyl, and coumaryl alcohols (Fig. *1*.6; FREUDENBERG, 1968). This author proposed a basic formula for the structure of lignin showing the types of linkages which could possibly arise between the aromatic alcohol molecules (cf. NORTHCOTE, 1972, Fig. 5). The formula shows that lignin contains some free hydroxyl groups; yet, it is considered as a hydrophobic filler material that replaces the water in the cell wall (NORTHCOTE, 1972). During incrustation of the wall with lignin, the more hydrophilic matrix shrinks and is compressed by displacement of water; this is very prominent in the region of the middle lamella (SITTE, 1965). One assumes that incrustation with lignin decreases the water permeability of the wall, although there is little experimental evidence

Lignin

H₂COH   H₂COH   H₂COH
 | | |
 CH   CH   CH
 ‖   ‖   ‖
 CH   CH   CH

H₃CO   H₃CO   OCH₃

OH   OH   OH

Coniferyl alcohol   Sinapyl alcohol   Coumaryl alcohol

Fig. *1.6. Lignin* is a complex polymer formed by condensation of precursors such as coniferyl, coumaryl, and sinapyl alcohols

in support of this point. CLARKSON (1974) quotes a value of the hydraulic conductivity L_p for lignified cells in *Pinus* of $6 \cdot 10^{-9}$ m s^{-1} bar^{-1} which is much lower than that considered for cellulose walls (cf. *1.4.2.1*, Table *1.2*). LÜTTGE (1973) discussed also the possibility of hindered transport and exchange of ions in lignified cell walls. In contrast, primary cell walls with cellulose and pectins as main constituents have long been known to be readily permeable to water and ions (e.g. STRUGGER, 1943). More experimental work on transport of water and solutes in lignified cell walls is needed, before any firm conclusions can be drawn.

2.2.2 Adcrustation (Suberin, Cutin)

The outer cell walls of the epidermis of leaves and other aerial organs, the walls of cork cells replacing the epidermis of roots and stems during secondary growth, and certain walls of inner sheaths in roots and shoots are covered or impregnated by *cutins* and *suberins* in a process called adcrustation (but note that suberins and cutins also occur as incrustations, e.g. in the Casparian strip of the root endodermis). Cutin is a complex mixture of polymeric, crossesterified fatty acids with hydroxy fatty acids predominating in the mixture; suberin has a similar composition. HEINEN and BRANDT (1963) were able to derive a general formula for cutin and suberin after identification of specific enzymes involved in synthesis of cutin. This formula is given in Fig. *1.7* and shows that the molecule contains subunits, each consisting

Suberin (cutin)

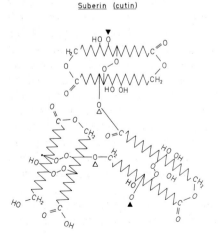

Fig. *1.7. Cutin* and *suberin* contain subunits with two chains each of hydroxy fatty acids. The figure shows three subunits connected with each other through ester and ether linkages, respectively (open triangles). More subunits can be linked at the sites indicated by closed triangles. (From SITTE, 1965; after HEINEN and BRANDT, 1963)

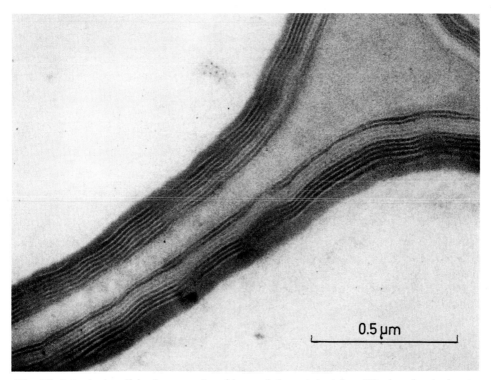

Fig. *1*.8. Suberized wall in the wound periderm of the potato tuber. The layering is due to alternation of suberin (dark-contrasted) and wax (light-contrasted). × 80,000. (By courtesy of H. FALK, Freiburg)

of two chains of hydroxy fatty acids. Suberized walls may also contain waxes and, furthermore, cutinized cell walls often are impregnated additionally with wax to form a cuticle (*1*.3). Waxes are a complex mixture of long-chain alkanes, alcohols, ketones, and fatty acids (NORTHCOTE, 1972).

An example of a suberized cell wall is presented in Fig. *1*.8 depicting an electron micrograph of the wall in the wound periderm of the potato tuber. In the cross-sectioned wall, alternating layers of suberin and wax are discernible. From such electron micrographs, SITTE (1962) constructed a model of the structure of the suberized wall in a cork cell (Fig. *1*.9). Over the primary wall, the secondary wall is adcrusted as a suberin layer consisting of alternating suberin and wax lamellae, and is covered by the tertiary cellulose wall. The suberin layer is traversed by pores which presumably originate from plasmodesmata but are occluded in the wall of mature cork cells. The lamellar structure of suberized cell walls is of significance for their permeability properties, as confirmed recently by SITTE (1975). He demonstrated that the lipophilic OsO_4 penetrated readily into the suberin layers but MnO_4^- penetrated only slowly and anisotropically. The anisotropic diffusion resistance of suberin layers appears to be a consequence of their lamellar fine structure.

While not much information is available on the effect of lignin on apoplasmic transport, there is no doubt that adcrustation of cell walls leads to a restriction

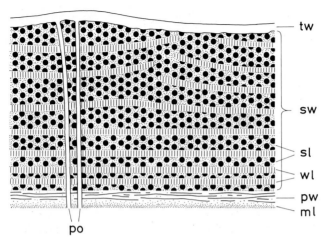

Fig. *1*.9. Schematic representation of the structure of the suberized wall in a cork cell. *ml* middle lamella; *pw* primary wall with cellulose; *sw* secondary wall, adcrusted (suberin layer); *sl* suberin lamellae; *wl* wax lamellae; *tw* tertiary wall containing cellulose; *po* pores, occluded, originating from plasmodesmata. The lines in *wl* indicate the orientation of wax molecules. (From SITTE, 1962)

on diffusion of water and solutes. Cutin is a lipophilic polymer, while suberin is considered semihydrophilic, the hydrophilic character depending on the ratio of unesterified to esterified hydroxyl and carboxyl groups in the molecule. The cuticles over the leaf surfaces protect the plant from excessive loss of water, particularly in xerophytes. Thus, cuticular transpiration is low as a consequence of the low water permeability of cutin. But cutin is not completely impermeable to water and solutes. STEINBRECHER and LÜTTGE (1969), for example, demonstrated that the cuticle on the outer surface of the onion epidermis is partially permeable to ions and impermeable to hexoses. The structure of cuticles and transport of substances across cuticles will be taken up in *1.3*.

Suberization also alters the permeability properties of cell walls. The Casparian strip, an incrustation in the root endodermis, composed of suberin and probably lignin (ESAU, 1965), is generally thought to render the wall almost impermeable to water and small solutes (*1.4.2.2.1*; *3.4.2.4*), though this statement is often disputed. However, when complete adcrusting suberization of the wall has taken place, as in barks and cork (cf. Figs. *1.8* and *1.9*) contact between the cells and the environment is prevented, leading eventually to cell death.

KOLATTUKUDY and DEAN (1974) identified the C_{18} fatty acid derivatives ω-hydroxy octadecenoic acid and octadec-9-enedioic acid as the major compounds of the suberin of potato tuber periderm. Using potato tuber slices, which were gradually regenerating new periderm and forming suberin adcrustations, these authors could correlate the resistance to water loss from the potato tuber tissue with suberization. The correlation was made possible by a quantitative gas chromatographic test of the C_{18} molecules for measurement of suberization, and a direct proportionality was observed. Similar experiments with other suberized tissues, including measurements of permeability to ions and neutral solutes, should be very helpful in elucidating the exact function of suberin in blocking apoplasmic transport.

3. Cytological and Chemical Properties of Cuticles in Relation to Apoplasmic Transport

We have seen above (1.2.2.2) that due to the presence of cutin a cuticle is lipophilic but not entirely impermeable to water and solutes. The question arises as to whether particular substructures in a cuticle facilitate transport across it or whether the permeability characteristics merely reflect the physicochemical properties of the substances composing the cuticle. SITTE and RENNIER (1963) have done an extensive study of the structure of cuticular walls of leaves using polarizing, interference and electron microscopy and also cytochemical techniques. They found the cuticular walls to be composed of four layers, the innermost cellulose wall (1), a cuticular layer with cellulose microfibrils (2), and the actual cuticle (3–4) which forms the outermost part of the wall, is double-layered and consists of cutin and wax. KOLAT-TUKUDY (1970) proposed a similar structure of leaf cuticle with the minor modification of a pectin layer between the cellulose wall and the cuticular layer. Furthermore, he emphasized the surface wax which may cover the cuticle as an extremely lipophilic surface film. Two electron micrographs depicting parts of the cuticle (a) and of the cuticular layer (b) in the outer epidermal cell wall of the leaf of *Ficus elastica* are shown in Fig. 1.10. The cuticle (Fig. 1.10a) exhibits a lamellar structure; neither pores nor cellulose microfibrils are present. By using standard electron microscopy techniques it thus appears that the cuticle is poreless; this was confirmed by LEDBET-TER and PORTER (1970). Between the cellulose wall and the cuticle there is a cuticular layer present which does contain cellulose microfibrils toward the cellulose wall (Fig. 1.10b). Cytochemical staining with Sudan-dyes revealed that the cutins of cuticular walls are stained but that the wax component cannot be stained (SITTE and RENNIER, 1963). Hence, cutins are less lipophilic than waxes, and the relative water impermeability of plant cuticles depends mostly on the impregnation and surface deposition with wax.

Long before the electron microscope was applied successfully by biologists, SCHUMACHER and HALBSGUTH (1939) demonstrated by light microscopy of plant material fixed with a sublimate fixative that cuticles apparently possess substructures that became known as "ectodesmata." These ectodesmata were, for some time, viewed as strands of protoplasm commonly extending into outer epidermal cell walls. Thus, in some respect ectodesmata were considered analogous to plasmodesmata (SCHUMACHER and LAMBERTZ, 1956) but, later on, FRANKE (1967) thought that they represented specific structures in the epidermal cell wall providing pathways of foliar penetration. This view no longer holds. SCHÖNHERR and BUKOVAC (1970) were able to demonstrate ectodesmata-like structures in enzymatically isolated cuticles using sublimate fixatives. The distribution of these structures was similar to that in the intact epidermis. This suggests that the characteristic properties of these structures are not caused by a continuity with the cytoplasm of the epidermal cells, but that they rather reflect areas in the cuticle which have a higher permeability to polar compounds than areas without them. By preparation using the freeze-etch technique these structures again were only detectable after sublimate fixation (LYON and MUELLER, 1974). This confirms that they are areas in the cuticle which are particularly easily accessible to the fixative. Thus, the sublimate fixative does not interact with distinct morphological structures of cytoplasmic origin.

Cuticular transpiration is a well-known phenomenon and therefore to a certain extent the cuticle must be permeable to water. How then does water permeate through cuticles? Are there polar pores through which water diffuses? Thus far not much quantitative work has been done to answer these questions. Polar pores in the cuticle, through which cuticular transpiration may proceed, have been postu-

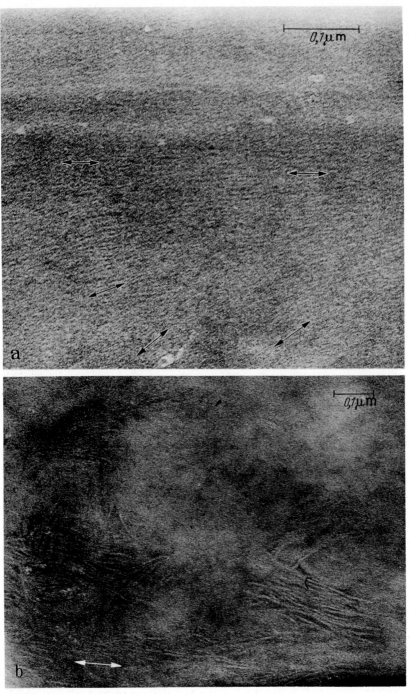

Fig. *1.*10. Legend see opposite page

lated occasionally, for instance by HÄRTEL (1951). Recently, SCHÖNHERR (1974) investigated in great detail the physicochemical properties of cuticles isolated enzymatically with pectinase and cellulase. A diffusive permeability coefficient (P_d) for water of about $5 \cdot 10^{-7}$ m s^{-1} was determined (SCHÖNHERR, 1974). On the other hand, P_d for glucose was very low (10^{-10} m s^{-1}), in agreement with data by STEINBRECHER and LÜTTGE (1969). Furthermore, the water permeability of the cuticle increased with increasing pH between pH 3 and 11; this was taken to mean that fixed dissociable carboxyl groups exist (cf. SCHÖNHERR and BUKOVAC, 1973). In contrast, McFARLANE and BERRY (1974) using cuticles isolated with $ZnCl_2$-HCl only observed this pH effect on the cuticle permeability with respect to K$^+$ but not to water. These authors arrived at the contrary conclusion, i.e. that the postulated cuticular pores are lined with a substance which has exposed positively charged sites, possibly due to the presence of proteins. Proteins indeed have been identified as minor components of cuticles, e.g. by HUELIN (1959). The differences in isolation procedures may well account for these conflicting models of cuticular pores; this problem awaits further experimental study. SCHÖNHERR (1974) also determined the diffusive (P_d) and osmotic (P_f) permeabilities of cuticles to various solutes. At pH 9 and 25° C the ratio of P_f/P_d was greater than or equal to 2. With the equation derived by NEVIS (1958) the average pore diameter d_p can be calculated:

$$d_p = 0.72 \sqrt{P_f/P_d} \qquad\qquad (1.1)$$

A value for $d_p = 0.82$ nm was estimated for the pore diameter. SCHÖNHERR (1974) suggests a smaller pore size for cuticles under natural conditions, where the pH is certainly well below 9, and hence water transport would be governed by diffusion. If SCHÖNHERR's measurements and calculations turn out to be correct and applicable to natural cuticles, their polar pores are indeed very much smaller than what was observed microscopically after sublimate fixation.

As already stated, cuticles are not only partially permeable to water but also to ions. The cuticle of isolated onion epidermis was about equally permeable to K$^+$ and Cl$^-$ (STEINBRECHER and LÜTTGE, 1969). Penetration of cations through apricot leaf cuticles was studied by McFARLANE and BERRY (1974). For monovalent cations, the rates of penetration followed a lyotropic series, i.e. Cs$^+$ > Rb$^+$ > K$^+$ > Na$^+$ > Li$^+$; divalent cations all permeated through the cuticle more slowly than monovalent cations. The latter results appear to be in accord with the ion exchange properties of cuticles measured by SCHÖNHERR and BUKOVAC (1973). According to this study cuticles behave like cation exchangers with a pronounced selectivity for Ca^{2+} over Na$^+$. One could then conclude that anions are

---

Fig. 1.10a and b. Cross section through the outer epidermal cell wall on the upper side of the leaf of *Ficus elastica*. The wall consists of 4 layers: The innermost layer (1) is the cellulose wall, followed by a cuticular layer (2) containing also cellulose microfibrils, and a double-layered cuticle (3–4) forming the outermost layer. Layers (3) and (4) are distinguishable only by different orientation of wax molecules and do not contain cellulose. (a) Electron micrograph depicting the lamellar structure of layer (4) with reorientation of the lamellae (arrows) toward layer (3) at the lower edge of the micrograph. ×205,000. (b) Electron micrograph of the cuticular layer (2). Cellulose microfibrils are present in the lower part of the micrograph (arrow) toward the cellulose wall (1) but absent in the upper part toward layer (3) of the cuticle. ×108,000. (From SITTE and RENNIER, 1963)

essentially excluded from the cuticles. This is not borne out by experimental evidence, as cations and anions do penetrate cuticles to a certain extent (YAMADA et al., 1964; STEINBRECHER and LÜTTGE, 1969). Binding of cations to cuticles, however, is far greater than that of anions (cf. FRANKE, 1967).

4. Apoplasmic Transport in Tissues

4.1 Demonstration of the Apoplasmic Pathway by Localization Methods

4.1.1 Water Pathway

Light microscopic demonstration of the apoplasmic water pathway in cell walls was first successful with the fluorescent dye K-fluorescein under conditions where the dye is carried along with a moving stream of H_2O molecules but does not penetrate the cytoplasm (STRUGGER, 1939). However, the distribution of fluorescein in the cell and also its fluorescence vary with pH, and hence doubts have been expressed concerning the localization of fluorescein (see *2.5.2.1*). STRUGGER and PEVELING (1961) using sols of noble metals as markers, were able to show electron microscopically that material carried along with transported water is located in fine capillaries (intermicrofibrillar spaces?) of the cell wall. Although the utilization of this method allows the conclusion that cell walls do constitute a water pathway, sols of metals are not ideal tracers for water. The metal particles are many times larger than the water molecules to be marked and, in addition, are water-insoluble; these properties may influence their distribution in a cell. One could envisage, for instance, that a symplasmic water pathway exists in a tissue whose apoplast is blocked by suberization. As sols of noble metals presumably cannot pass through the plasmalemma, a possible symplasmic water pathway would escape the investigator.

More recently, CROWDY and TANTON (1970) modified the heavy metal marking technique by using the water-soluble EDTA salt of Pb. Pb-EDTA has properties that make it more suitable as a marker for water than the insoluble metal particles used hitherto. The Pb-salt is not confined exclusively to label the apoplast but apparently penetrates the cytoplasm (TANTON and CROWDY, 1972a). The marker is made visible after treatment of the tissue with H_2S as PbS-precipitate, which appears black in the light microscope and is electron-dense in the electron microscope, respectively. The method was applied to study apoplasmic water transport in roots (TANTON and CROWDY, 1972a) and leaves (CROWDY and TANTON, 1970; TANTON and CROWDY, 1972b). As an example Fig. *1*.11 shows that in the epidermis of wheat leaves apoplasmic transport of water is confined to the cell walls.

4.1.2 Solute Pathway

Apoplasmic location of ions in the free space may be demonstrated indirectly by microautoradiography. KRICHBAUM et al. (1967) applied $^{45}Ca^{2+}$, $^{36}Cl^-$, and $^{35}SO_4^{2-}$ to maize roots and obtained microautoradiographs from unwashed roots and from roots washed for 120 min, respectively. By comparing the microautoradiographs

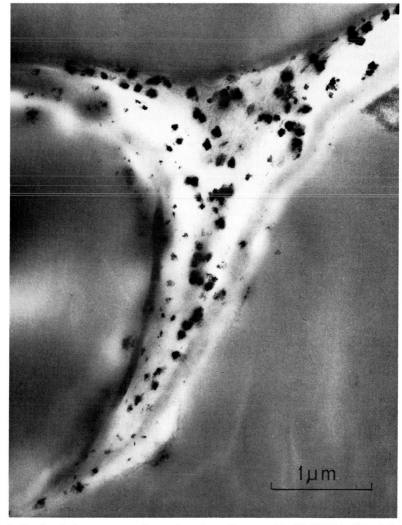

Fig. *1*.11. "Labeling" of the water pathway in wheat leaves with Pb-EDTA (25 mM). Precipitation as PbS. Electron micrograph showing the electron-dense deposits almost exclusively in the cell walls of an epidermis cell. ×27,300. (From CROWDY and TANTON, 1970)

with wash-out experiments in which the kinetics of tracer efflux from the roots were measured, the size of the apoplast could be calculated.

With electron microscopy direct evidence of the apoplasmic pathway of ions has only been obtained recently, either by "labeling" with electron-dense metal ions (UO_2^{2+}, La^{3+}) or after specific precipitation of certain ions *in situ* (e.g. Cl^-). ROBARDS and ROBB (1972, 1974) put barley roots in 0.1 mM solutions of uranyl acetate and subsequently processed them for electron microscopy. Electron-dense crystals, most probably due to the presence of uranium, were located mainly in the cell walls of the epidermis and cortex. The Casparian strip of the endodermis proved to be an effective barrier to the apoplasmic passage of UO_2^{2+} into the stele. Intracellular crystals were also observed in these investigations.

By administration of electron-dense La^{3+} tracer Gunning and Pate (1969) demonstrated that wall ingrowths of transfer cells are part of the apoplasmic pathway for ions (see *1.4.2.2.2*). Thomson and coworkers (Thomson et al., 1973; Nagahashi et al., 1974; Campbell et al., 1974) fed 1% $La(NO_3)_3$ (approx. 30 mM) to various plant tissues and observed the electron-dense deposits to occur only in cell walls and on the outer surface of the plasmalemma. The presence of these deposits in the apoplast of *Atriplex* leaves is borne out convincingly in Fig. *1.12*. However, the high concentration of La^{3+} used bears some problems. Robards and Robb (1974) applied only 0.01 mM La^{3+} in their experiments with barley roots and found the deposits also in the cytoplasm. By means of energy dispersive X-ray analysis the electron-dense deposits were confirmed by these authors to be due to La^{3+}. Furthermore, they showed that apoplasmic transport of La^{3+} appears impeded by the Casparian strip of the endodermis (Fig. *1.13*; cf. *1.4.2.2.1*; *3.4.2.4*).

The use of salts of heavy metals not normally present in plants to label the apoplasmic pathway for ions raises some questions. UO_2^{2+} is known as a metabolic inhibitor affecting membrane transport of solutes (Arisz, 1958; Rothstein, 1954). The high La^{3+} concentration used by Thomson and coworkers is inhibitory to salt excretion by salt glands (Campbell et al., 1974), and also to cell division in root meristems (Clarkson, 1965). One might argue that these experiments still serve to demonstrate the apoplasmic pathway. However, *in vivo* almost certainly there is intimate interaction between transport in the apoplasmic space and membrane transport absorbing from or feeding into the apoplasmic pathway. Thus, localizations using metabolic inhibitors represent doubtful approaches to depict the situation *in vivo*. Furthermore, one does not even know whether such metal ions in apoplasmic transport viewed *per se* behave similarly to nutrient ions. Their larger size and, in the case of La^{3+}, the high positive charge may affect transport of these tracers in a tissue.

Techniques of *in situ* precipitation specific for certain nutrient ions (Läuchli et al., 1974c) appear advantageous. One of the most widely-used examples is the localization of Cl^- in plant tissues after precipitation with organic Ag^+ salts and

---

Fig. *1.12*. "Labeling" of the apoplasmic pathway for ions in *Atriplex* leaves with $La(NO_3)_3$ (30 mM). The electron-dense deposits are abundant in the cell walls but absent in the cytoplasm. ×9,000. (From Thomson et al., 1973)

Fig. *1.13*. Part of an endodermal cell of a barley root in cross section. Labeling of the apoplasmic pathway for ions with $La(NO_3)_3$ (0.01 mM). The cell wall and intercellular space up to the Casparian strip appear heavily labeled with electron-dense deposits. ×15,000. (From Robards and Robb, 1974)

Fig. *1.14*. Part of an endodermal cell (*en*) of *Puccinellia peisonis* in the tertiary state showing AgCl-precipitates only in the primary wall (*pw*) toward the cortex (*co*), but not in the suberin lamella (*sl*) and in the thick, tertiary cellulose wall (*tw*). Plant collected from a salty, poorly aerated soil near the Neusiedlersee, Austria; cultivated in complete nutrient solution containing 200 mM NaCl, not aerated. Cross section less than 1 cm behind the tip. Fixation: 0.1% picric acid, 0.3% Ag-acetate, 0.5% Ag-lactate, 1% OsO_4 in 0.2 M cacodylate-acetate buffer pH 6.9 for $3^1/_2$ h. Section poststained with uranyl acetate. ×51,000. (Original micrograph by R. Stelzer, Darmstadt)

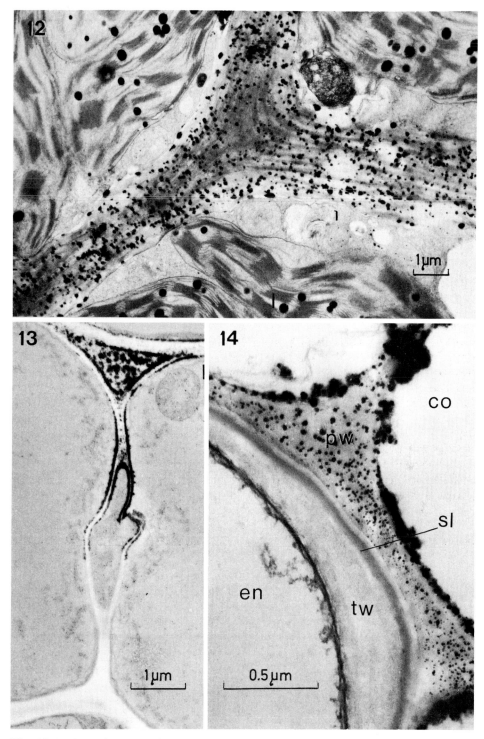

Figs. 12–14. Legend see opposite page

electron microscopic observation of the electron-dense deposits. (For a brief literature survey and a description and discussion of the method see LÄUCHLI et al., 1974c, and STELZER et al., 1975.) Important is the fact that the specificity for Cl^- of the Ag^+ precipitation technique has been confirmed by energy dispersive X-ray analysis (LÄUCHLI et al., 1974c). With this technique it has been demonstrated that the suberin lamella in a tertiary endodermal cell effectively blocks apoplasmic Cl^- transport in the root of *Puccinellia peisonis* (Fig. *1.*14) extending the result of the La^{3+} experiment shown in Fig. *1.*13 (cf. *1.*4.2.2.1).

It is now possible to determine the ionic composition of the apoplast with electron microprobe analysis using bulk deep-frozen, hydrated specimens on a cold stage of a scanning electron microscope fitted with an energy-dispersive detector (A. LÄUCHLI, J. GULLASCH and D. KRAMER, unpublished results). To the best of our judgment, artificial redistribution of ions due to the preparation procedure are possible only in dimensions that are below the resolution of the method. (For recent reviews of electron microprobe analysis of plant specimens see LÄUCHLI, 1972b, 1973, 1974.) Fig. *1.*15 exhibits the inorganic constituents of the apoplast in the cortex of a maize root grown in soil containing all the essential nutrient ions and also NaCl. (Salt was added to demonstrate that Na can be easily detected in spite of the limited sensitivity for Na of the energy-dispersive solid-state detector.) It can be seen that the apoplast of maize roots grown at moderate salinity contains Na^+ and Cl^- at high and about equimolar concentrations, and about 2–3 times less K^+; the elements Si, P, S, and Ca are also detectable. Si is possibly bound to the cell wall, and part of the Ca detectable may be bound in the Donnan free space. Note that under these conditions there is much more Cl^- in cell walls of roots as compared with leaf cell walls (see *1.*4.2.2.3).

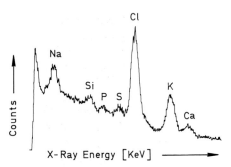

Fig. *1.*15. Energy dispersive X-ray spectrum from the wall of a cortex cell of a maize root about 5 cm behind the tip. Seedling grown in soil which was supplied periodically with a complete nutrient solution containing also 150 mM NaCl. Root rinsed with deionized water for about 2 s, excised, and frozen in liquid N_2. Frozen specimen analyzed on the cold stage (at about $-170°$ C) of the Autoscan scanning electron microscope fitted with the Kevex-ray energy dispersive X-ray analyzer. High voltage 10 kV, beam current 10^{-10} A, specimen uncoated. Note that the detection sensitivity in biological samples is about 2–3 mM except for Na with a sensitivity about 3 times less (self-absorption of Na K_α-radiation in the specimen is taken into account; cf. GULLASCH, 1974). Energy maxima: Na K_α (1.041 KeV); Si K_α (1.739 KeV); P K_{α_1} (2.013 KeV); S K_{α_1} (2.307 KeV); Cl K_{α_1} (2.621 KeV); K K_α (3.312 KeV); Ca K_α (3.690 KeV), the Ca peak is "contaminated" by K due to K K_β-radiation with an energy maximum at 3.589 KeV. (A. LÄUCHLI, unpublished)

4.2 Significance of the Apoplast for Transport in Tissues

4.2.1 Biophysical Measurements

Little quantitative information is available to evaluate the significance of apoplasmic transport in comparison with other possible transport pathways such as the symplasm. There are, however, some measurements and calculations of the hydraulic conductivity L_p of cell walls (Table *1.2*). BRIGGS (1967) estimated L_p of cellulose cell walls to be of the order of 10^{-7} m s^{-1} bar^{-1}. Measurements of L_p for cell walls of *Nitella* and *Chara* (KAMIYA et al., 1962; TAZAWA and KAMIYA, 1966; BARRY and HOPE, 1969; ZIMMERMANN and STEUDLE, 1975) gave values in reasonable agreement with that estimated by BRIGGS (1967). ZIMMERMANN and STEUDLE (1975) showed furthermore that L_p for cell walls is independent of cell turgor pressure, in contrast to L_p for whole cells (STEUDLE and ZIMMERMANN, 1974; Table *1.2*). The L_p values for whole cells approximately reflect those of cell membranes. Note that L_p for these cell walls is almost two orders of magnitude greater than that found usually for lipoprotein membranes (CLARKSON, 1974). However, there is a much smaller difference between the L_p values for cell walls and whole cells of *Nitella* (Table *1.2*).

Table *1.2*. Hydraulic conductivities of cell walls in comparison with cell membranes (f) and whole cells

Plant material	L_p [m s^{-1} bar^{-1}]	L_{pp} [m^2 s^{-1} bar^{-1}]	Ref.
Cellulose cell wall	$\sim 10^{-7}$ (a)	—	BRIGGS (1967)
Nitella cell walls	$5 \cdot 10^{-7}$ (b)	—	KAMIYA, TAZAWA and TAKATA (1962)
Nitella cell walls	$1.8 - 3.5 \cdot 10^{-7}$ (b)	—	TAZAWA and KAMIYA (1966)
Nitella cell walls	$6.9 \cdot 10^{-7}$ (b)	—	ZIMMERMANN and STEUDLE (1975)
Chara cell walls	$3.6 \cdot 10^{-7}$ (b)	—	BARRY and HOPE (1969)
Whole cells of *Nitella*	$1.8 - 2.8 \cdot 10^{-7}$ (c)	—	STEUDLE and ZIMMERMANN (1974)
Cell membranes	$1 - 6 \cdot 10^{-9}$ (d)	—	CLARKSON (1974)
Nitella cell walls	—	$1.4 \cdot 10^{-11}$ (b)	TYREE (1968)
Cell walls of maize roots	$1.4 \cdot 10^{-9}$ (e)	—	TYREE (1973)
Cortical cell membranes of maize roots	$4 \cdot 10^{-11}$ (e)	—	TYREE (1973)

(a) Estimated. (b) Measured. (c) Measured, range of values depending on cell turgor pressure. (d) Range of values. (e) Calculated. (f) See also Part A, *2.4.2* and *2.4.3*.

TYREE (1968) measured the hydraulic conductivity coefficient L_{pp}, which has the unit [m^2 s^{-1} bar^{-1}]. L_{pp} is thus a measure of the intrinsic conductivity of the material, whereas L_p depends on the thickness and dimensions of the material, too. On the basis of several assumptions, TYREE (1973) calculated L_p for cell walls and cortical cell membranes of maize roots from his measured L_{pp} values, since

$$L_p = \frac{\alpha_w \cdot L_{pp}}{l} \tag{1.2}$$

where α_w is the relative area occupied by the cell wall and l is the cell wall path length (TYREE, 1973). A value of L_{pp} is also shown in Table *1.2*. TYREE's calculated L_p values for cell walls and membranes are considerably lower than the other measured values. The point is, though, that he also arrived at an L_p for the cell membrane almost 100 times lower than that for cell walls. Hence cell walls (that are not altered by incrustation or adcrustation) are much more permeable to water than cell membranes. The difference in L_{pp}, of course, is even greater than that in L_p.

The difference in permeability to water between cell wall and cell membranes may be compared with the permeability to ions. Values of the diffusive permeability coefficient P_d for ions in the cell wall can be estimated to be about 10^{-3} to 10^{-4} m s^{-1} compared with the permeability coefficients P_K or P_{Cl} in membranes (10^{-9} to 10^{-10} m s^{-1}, see Table *3.11*).

Keeping this conclusion in mind we shall now proceed to look at the significance of apoplasmic transport in a few selected tissues of higher plants. The phenomenon is widespread and encompasses roots, stems, leaves and salt glands, to name just a few examples.

4.2.2 Special Systems

4.2.2.1 Roots

In nutrient solutions, ions and water molecules have ready access to the apoplast of a root and hence to the plasmalemma of the outer root cells. In soils, however, diffusion of ions through the dilute soil solution and the cell walls may become rate-limiting in ion uptake by the root (JENNY, 1966). Under such circumstances, contact exchange of adsorbed ions between soil particles and the root surface may gain importance. JENNY and OVERSTREET (1939) published evidence to show that cations absorbed to clay colloids can be exchanged with cations adsorbed to the root surface when their "oscillation volumes" overlap. These oscillation volumes originate from the vibration of ions around the adsorption site, as depicted diagrammatically in Fig. *1.16.* It shows furthermore that in such a contact exchange with clay colloids the mucigel (see Fig. *3.*1), which covers the root surface as a slime, plays an important role; particularly many negatively charged groups in the mucigel give rise to cation exchange properties (cf. *1.2.1.2*). Intimate contact between the mucigel layer of the root and soil particles was demonstrated by JENNY and GROSSENBACHER (1963) using electron microscopy. Naturally, contact exchange of ions can only be significant on the surface of the root.

In interior cell layers, particularly in the cortex, apoplasmic transport of ions proceeds in an aqueous medium which imbibes the cell walls (water free space, see *1.2.1.1*) or along the exchange sites of the Donnan free space (see *1.2.1.2*).

Whenever transport of water and ions from the cortex to the stele is considered, the role of the endodermis must be taken into account (see also *3.4.2.4*). A short distance behind the root apex, endodermal cells have reached their primary state

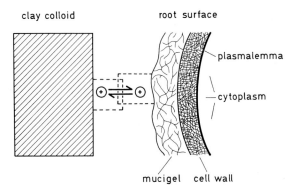

clay colloid

root surface

plasmalemma

cytoplasm

mucigel   cell wall

Fig. *1*.16. Contact exchange of cations between a clay colloid and the mucigel, the outermost zone of the root surface. Adsorbed cations vibrate around the site of adsorption occupying an "oscillation volume" (dashed lines). When the oscillation volume of a cation adsorbed to the clay colloid overlaps with that of a cation adsorbed to the mucigel, exchange of the two cations can occur. (Diagram drawn after results by JENNY and OVERSTREET, 1939; JENNY and GROSSENBACHER, 1963; JENNY, 1966)

of development with the Casparian strip forming a complete band in the radial walls. Due to incrustation with suberin, the Casparian strip decreases drastically the permeability of the radial walls of the endodermis. A typical Casparian strip is presented in Fig. *1*.17. The plasmalemma adheres intimately to the wall throughout the width of the Casparian strip (FALK and SITTE, 1960), even when the endodermis is plasmolyzed (e.g. BONNETT, 1968), not leaving any space for apoplasmic transport between the wall and the plasmalemma. Thus, transport of water and solutes across the endodermis into the stele must proceed in the symplasm, if the Casparian strip effectively blocks the cell wall apoplast. Evidence to this effect has been reviewed by LÄUCHLI (1972a). Localization experiments confirmed this role of the Casparian strip blocking the apoplasmic pathway for both transport of water (TANTON and CROWDY, 1972a; *1*.4.1.1) and of solutes (Fig. *1*.13; *1*.4.1.2). Only at the site of branch root emergence may a temporary apoplasmic pathway across the endodermis occur, as the formation of the Casparian strip appears to lag behind the division of the endodermal cells (DUMBROFF and PEIRSON, 1971).

Endodermal cells of monocotyledonous roots usually undergo further developmental changes. As is shown diagrammatically in Fig. *1*.18, a suberin lamella is deposited over the entire inner wall surface (secondary state), followed by the tertiary state where a thick and often lignified cellulose wall is layered onto the suberin lamella and covers mainly the inner tangential and the radial walls of the endodermis (for a recent review see CLARKSON and ROBARDS, 1975). The apoplasmic pathway across the secondary and tertiary endodermis is interrupted completely (see Fig. *1*.14), allowing the symplasm as the only possible pathway (see Chap. 2). CLARKSON et al. (1971) demonstrated the presence of numerous pits containing plasmodesmata in the inner tangential wall of the tertiary endodermis of barley roots and correlated this with fluxes of phosphate and water through the plasmodesmata. In the tertiary root of *Iris*, however, the thickened endodermal cells seem dead and without plasmodesmata, and symplasmic transport could only be demonstrated through the passage cells (ZIEGLER et al., 1963).

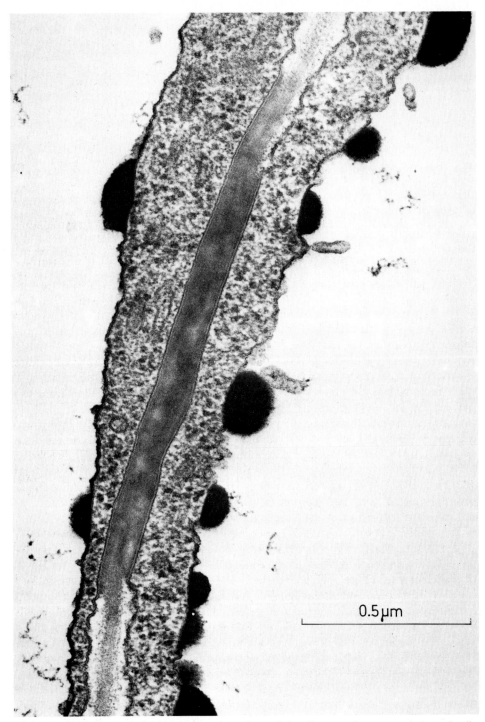

0.5 µm

Fig. *1.*17. Casparian strip in the cross section of a radial wall separating two endodermal cells in the root of *Limonium sinuatum*. Endodermis in the primary state. Note that the plasmalemma is intimately adherent to the cell wall throughout the width of the Casparian strip. ×91,500. (From LEDBETTER and PORTER, 1970)

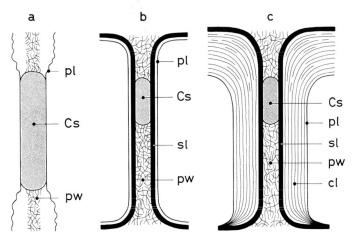

Fig. *1*.18a–c. Wall development of an endodermal cell in a monocotyledonous root depicted in schematic drawings of cross sections through the root. (a) Primary state: Casparian strip (*Cs*) in the radial wall, *pw* primary wall, *pl* plasmalemma. (b) Secondary state: a suberin lamella (*sl*) is deposited over the whole inner wall surface and separates the plasmalemma (*pl*) from the Casparian strip (*Cs*). (c) Tertiary state: a thick cellulose layer (*cl*) covers the inside of the wall, mainly the inner tangential and the radial walls. The entire cell wall may become lignified. Note that (b) and (c) are at a different scale than (a)

Ca^{2+} apparently is not transported in plasmodesmata and so there is no symplasmic Ca^{2+} transport (*2.5.3*). Ca^{2+} enters the stele of barley roots (ROBARDS et al., 1973) and of roots of *Cucurbita pepo* (HARRISON-MURRAY and CLARKSON, 1973) only in the region of the primary endodermis. ROBARDS et al. (1973) consider furthermore that the Casparian strip in the primary endodermis presents a high resistance to apoplasmic Ca^{2+} transport. Hence, the only way by which Ca^{2+} can get into the stele is by uptake through the plasmalemma of the endodermal cells at the outer tangential wall, transport through the cytoplasm, and release through the plasmalemma at the inner tangential wall into the apoplast of the stele. When the suberin lamella has covered the whole inner wall surface (secondary state), this pathway for Ca^{2+} transport across the endodermis is blocked.

Transport of water and solutes across the cortex up to the endodermis is probably predominant in the apoplast (*3.4.2.3*), although theoretical calculations led TYREE (1969) and MOLZ and IKENBERRY (1974) to assume that a significant portion of the water flux occurs through the cells. The ratio of water fluxes in the apoplast and symplasm of the root cortex cannot be constant and will depend on many parameters such as the water potential gradient. The ratio of ion fluxes in the two pathways is equally unknown and will again depend on various parameters. In studies of uptake of K^+ and Na^+ by barley roots, PITMAN (1965) showed that transport through the cells was more rapid than was possible by diffusion through the apoplast. That Cl^- is transported in the symplasm of the cortex was demonstrated by electron microscopic studies using the precipitation technique (*1.4.1.2*; STELZER et al., 1975). After water and ions have passed the endodermis, they can either be released into the apoplast just inside the endodermis or be transported in the symplasm up to the vessels. Symplasmic transport of Cl^- up to the vessels does occur (STELZER et al., 1975) and xylem parenchyma cells are considered to participate in the symplasmic ion transport pathway (LÄUCHLI et al., 1974a).

A much-neglected aspect which warrants more experimental studies is the transport of water and ions in suberized roots. When secondary root growth occurs, the epidermis, cortex, and endodermis all disappear. They are replaced by a suberized outer tissue, consisting of a peripheral bark layer and a layer of cork cambium, and situated over a layer of secondary phloem (SLATYER, 1967; KRAMER, 1969). SLATYER (1967) considers that the outer suberized layers have very low permeability but points out that breaks in the tissue and lenticels may provide areas of low resistance. Recent observations by FERGUSON (1974) indicate that suberization of the hypodermis in maize roots may be of greater significance in preventing PO_4^{3-} transport through the root than suberization of the endodermis. CLARKSON and ROBARDS (1975) suggest that suberization at the root periphery probably restricts the entry of water and ions into the symplasm.

4.2.2.2 Role of the Apoplast in Transfer Cells

Transfer cells have become known as highly specialized cells in which ingrowths of cell wall material increase the surface area of the plasmalemma (see Fig. *10*.3). PATE and GUNNING (1972) presented an extensive review of the structure and distribution of transfer cells and advanced the hypothesis that their specialized structures facilitate transmembrane flux of solutes. Transfer cells are found at sites where high rates of solute transport from one symplasmic system to another occur with a short apoplasmic route in between or just between apoplast and symplasm. The extensive wall ingrowths undoubtedly not only function by enlarging the surface of the plasmalemma but also by their other consequence, namely enlargement of the volume of the apoplasmic pathway.

Xylem parenchyma cells in the basal region of roots have been shown to be formed as transfer cells presumably functioning in selective re-absorption of ions from the xylem sap (LÄUCHLI et al., 1974b). Fig. *1*.19 shows a xylem parenchyma transfer cell in the root of *Phaseolus*. Extensive wall ingrowths are present on the side of the xylem parenchyma cell facing the vessel. The ingrowths protrude into the xylem parenchyma cell and are clearly distended from the plasmalemma. Some threads of probably microfibrillar nature (arrows in Fig. *1*.19) seem to connect the wall material of the ingrowths with the plasmalemma; this may indicate that growth of the protuberances occurs at the outer surface of the plasmalemma.

Plasmalemma distension from the cell wall had first been demonstrated in the sucking hairs of *Tillandsia* (DOLZMANN, 1964), where the space between the cell wall and the plasmalemma was found to swell upon water uptake indicating a function as an apoplasmic compartment for water storage. HILL and HILL (1973) advanced the hypothesis that these distensions give rise to a "solute-water coupling space." According to this hypothesis, in salt glands, the plasmalemma of the gland cells is pumping salt into the apoplasmic wall ingrowths at a high rate and water then follows osmotically. A pressure will develop and the plasmalemma distends itself from the ingrowths. This type of apoplast would therefore act as a coupling space for transport of water and solutes, the solution moving along the ingrowths by mass flow (see also *5*.2.2.2.3). In gland cells, belonging to the secretory transfer cells, the salt pump enriches the solute content of the ingrowths and the mass flow proceeds in the outward direction.

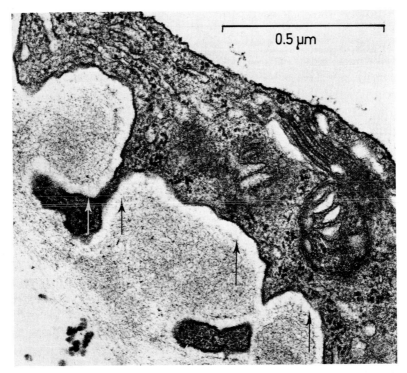

Fig. *1*.19. Part of a xylem parenchyma cell in the root of *Phaseolus vulgaris*, developed as transfer cell. Plant cultivated in complete nutrient solution containing also 100 mN Na_2SO_4. Cross section 15–20 cm behind the apex. The plasmalemma is distended from the extensive wall ingrowths but still seems to be connected with them by thin threads of probably microfibrillar material (arrows). Note also the dense structure of the cytoplasm with sheets of rough ER, a mitochondrion and a dictyosome present and the heavily contrasted tonoplast. × 85,000. (Original micrograph by D. KRAMER, Darmstadt)

In absorptive transfer cells, and the xylem parenchyma cell in Fig. *1*.19 presumably belongs to this type, the solute pumps may deplete the solute content of the ingrowths and the direction of the osmotic gradient would be such that mass flow passes inwards (PATE and GUNNING, 1972). If this were the case, the plasmalemma should not distend itself from the wall material; yet, distension occurs in water-absorbing (DOLZMANN, 1964) and in presumably ion-absorbing cells (Fig. *1*.19). In the latter case, it is possible that solute depletion in the ingrowths does not become manifest, because the rate of pumping at the plasmalemma may not be as high as in gland cells and the osmolarity in the wall material of the ingrowths may not be much different from that of the xylem sap. This conflict is unresolved and requires further experimental and theoretical studies.

4.2.2.3 Leaves

Water and ions are being conveyed through the xylem vessels into the leaves. The least resistance to water transport in the leaf is undoubtedly through the veins. Transpiration necessitates a water pathway from the leaf veins to the evaporative surfaces of the mesophyll and epidermal cells. One can distinguish between mesomorphic leaves, often having vein extensions and xeromorphic leaves without vein extensions. The vein extensions are in contact with usually both mesophyll and epidermis. SHERIFF and MEIDNER (1974) recently obtained evidence for the vein extensions maintaining a close hydraulic connection for water transport between the vascular tissue and the epidermis. In leaves without such vein extensions, however, the hydraulic connection between the two tissues was less apparent and the resistance to water transport into the epidermis appeared relatively high. Let us therefore examine whether the transpirational water pathway through the apoplast is hindered in leaves without direct vein access to the epidermis.

BOLLARD (1960) has already pointed out that the vessels at the vein endings are frequently surrounded by a sheath of parenchyma cells whose walls show staining reactions like those of endodermal cells. He did not know at the time whether and to what extent this would influence apoplasmic transport of water and ions to the mesophyll. O'BRIEN and CARR (1970) carried out a more detailed structural investigation of the walls of bundle sheath cells of grasses. In those leaves having two sheaths, the inner mestome sheath and the outer parenchymatous sheath, all walls of the mestome sheath contained suberized lamellae which were perforated by plasmodesmata. In grass leaves with just one parenchymatous bundle sheath, the cell walls of the sheath showed similar suberization and plasmodesmata occurred also. (See KUO and O'BRIEN, 1974, for the development of the suberized lamella in the mestome sheath of wheat leaves.) These findings suggest, though do not yet prove, that the apoplast is interrupted and that water and ions are transported to the mesophyll in the symplasm. Water labeled with Pb-EDTA was indeed retained in wheat leaves within the main veins by the mestome sheath (CROWDY and TANTON, 1970; Fig. *1.*11). However, using the same localization method, TANTON and CROWDY (1972b) obtained evidence for an apoplasmic water pathway to the mesophyll through the walls of the secondary veins, which are without the mestome sheath. They could not rule out the possibility of symplasmic water transport through the mestome sheath. Comparable studies with regard to apoplasmic ion transport to the mesophyll are unfortunately lacking.

In the mesophyll water appears to be transported in the apoplast, as shown with the Pb-EDTA method (CROWDY and TANTON, 1970; TANTON and CROWDY, 1972b; Fig. *1.*11). Using the La^{3+} technique for demonstration of the ion pathway, transport in the mesophyll was also confined to the apoplast (THOMSON et al., 1973; Fig. *1.*12; see, however, the critique against using high La^{3+} concentration in *1.*4.1.2). The Ag^+ precipitation technique for localization of Cl^- did in fact allow the demonstration of symplasmic Cl^- transport in leaves (ZIEGLER and LÜTTGE, 1967; VAN STEVENINCK and CHENOWETH, 1972). At the moment, one may have to settle this issue by assuming the occurrence of both apoplasmic and symplasmic ion transport through the mesophyll of leaves.

The extremely low concentration of Cl^- in cell walls of barley leaves (VAN STEVENINCK and CHENOWETH, 1972) is consistent with the strong Donnan phase in the cell wall apoplast (see *1.*2.1.2). A consequence should be that rates of uptake are limited either by low Cl^- levels when external concentration is low or by low K^+ levels when the concentration of Ca^{2+} is much higher than that of K^+. This effect has been observed as a depression of K^+ uptake in maize leaves by Ca^{2+} (LÜTTGE et al., 1974). It is interesting, though, that cell walls in barley roots do not seem to exclude Cl^- to the same extent as in leaves (see *1.*4.1.2).

The question now arises whether the significance of apoplasmic transport through the living leaf tissues can be evaluated by quantitative transport measurements. There are some results available, at least for water transport. The classical view was that water transport was from vacuole to vacuole along a water potential gradient. But there is now an increasing body of evidence that water flux in the cell walls is much greater than through the vacuoles. WEATHERLEY (1963), for instance, studied the time course of water uptake into the petioles of detached leaves in which transpiration was suddenly stopped by immersion in water and estimated the ratio of water flow in the apoplast compared with the vacuolar pathway to be about 50:1. The possibility of symplasmic water transport, however, was not taken into consideration. In conclusion, there is nonetheless no doubt that the mesophyll cell walls form the major pathway for water (cf. Chap. 2; WEATHERLEY, 1970).

The description of apoplasmic transport in leaves is rather sketchy in this Chapter and many aspects such as salt glands or the relation between the apoplast and stomatal function were neglected completely, partly because of lack of space and partly due to inconclusive evidence. I do not want to close without indicating the possibility for an apoplasmic step in the transport of sugars from the mesophyll to the sieve tube-companion cell complex which leads to long-distance transport in the phloem discussed in Volume 1 of this Encyclopedia. Suberizations blocking or controlling apoplasmic transport between mesophyll and bundle sheath cells may also be important in the co-operation of these cell layers in C_4 metabolism of photosynthesis (OSMOND, 1971).

5. Conclusions

The apoplast is generally looked upon as a much simpler structure than the symplast and, hence, apoplasmic transport as a less complex problem than for instance membranes and membrane transport (Part A, Chaps. *1, 3-5, 10*) or the cytoplasm, plasmodesmata and symplasmic transport (Chap. *2*). Yet, critical assessment of the role of the apoplasmic pathway in transport through plant tissues must base itself on cell wall chemistry, on cell wall cytology, and it must incorporate biophysical considerations. The present Chapter is an attempt to bring these three points of view together, a task which turns out to be quite difficult. Thus, it may not be too surprising that such a treatment brings up more open questions than solutions. Beyond the problem of exchange in the free space (Part A, *5.3*) the problem of transport in the apoplast becomes particularly significant in tissues and organs of higher plants. We are only at the beginning of understanding the role of apoplasmic transport in the overall regulation of transport in complex systems such as tissues, organs, or the intact higher plant (Chap. *10*).

References

ADAMSON, A.W.: A textbook of physical chemistry. New York-London: Academic Press 1973.

ARISZ, W.H.: Influence of inhibitors on the uptake and the transport of chloride ions in leaves of *Vallisneria spiralis*. Acta Botan. Neerl. **7**, 1–32 (1958).

BAILEY, R.W., KAUSS, H.: Extraction of hydroxyproline-containing proteins and pectic substances from cell walls and nongrowing mung bean hypocotyl segments. Planta **119**, 233–245 (1974).

BARRY, P.H., HOPE, A.B.: Electroosmosis in membranes: effects of unstirred layers and transport numbers. I. Theory. Biophys. J. **9**, 700–728 (1969).

BAUER, W.D., TALMADGE, K.W., KEEGSTRA, K., ALBERSHEIM, P.: The structure of plant cell walls. II. The hemicellulose of the walls of suspension-cultured sycamore cells. Plant Physiol. **51**, 174–187 (1973).

BECK, R.E., SCHULTZ, J.S.: Hindered diffusion in microporous membranes with known pore geometry. Science **170**, 1302–1305 (1970).

BOLLARD, E.G.: Transport in the xylem. Ann. Rev. Plant Physiol. **11**, 141–166 (1960).

BONNETT, H.T., JR.: The root endodermis: fine structure and function. J. Cell Biol. **37**, 199–205 (1968).

BRIGGS, G.E.: Movement of water in plants. Oxford: Blackwell 1967.

CAMPBELL, N., THOMSON, W.W., PLATT, K.: The apoplastic pathway of transport to salt glands. J. Exptl. Bot. **25**, 61–69 (1974).

CLARKSON, D.T.: The effect of aluminium and other trivalent metal cations on cell division in the root apices of *Allium cepa*. Ann. Bot. (London), N.S. **29**, 309–315 (1965).

CLARKSON, D.T.: Ion transport and cell structure in plants. London: McGraw-Hill 1974.

CLARKSON, D.T., ROBARDS, A.W.: The endodermis, its structural development and physiological role. In: The development and function of roots (J.G. TORREY, D.T. CLARKSON, eds.). London: Academic Press 1975.

CLARKSON, D.T., ROBARDS, A.W., SANDERSON, J.: The tertiary endodermis in barley roots: fine structure in relation to radial transport of ions and water. Planta **96**, 292–305 (1971).

CROWDY, S.H., TANTON, T.W.: Water pathways in higher plants. I. Free space in wheat leaves. J. Exptl. Bot. **21**, 102–111 (1970).

DOLZMANN, P.: Elektronenmikroskopische Untersuchungen an den Saughaaren von *Tillandsia usneoides* (Bromeliaceae). I. Feinstruktur der Kuppelzelle. Planta **60**, 461–472 (1964).

DOUGALL, D.K., SHIMBAYASHI, K.: Factors affecting growth of tobacco callus tissue and its incorporation of tyrosine. Plant Physiol. **35**, 396–404 (1960).

DUMBROFF, E.B., PEIRSON, D.R.: Probable sites for passive movement of ions across the endodermis. Canad. J. Bot. **49**, 35–38 (1971).

ESAU, K.: Plant anatomy. Second edition. New York-London-Sydney: John Wiley and Sons 1965.

FALK, H., SITTE, P.: Untersuchungen am Caspary-Streifen. Proc. Eur. Reg. Conf. on Electron Microscopy, Delft, vol. II, p. 1063–1066 (1960).

FERGUSON, I.B.: Ion uptake and translocation by the root system of maize. Agr. Res. Council Letcombe Lab. Ann. Rept. **1973**, 13–15 (1974).

FRANKE, W.: Mechanisms of foliar penetration of solutions. Ann. Rev. Plant Physiol. **18**, 281–300 (1967).

FRANKE, W.W., SCHEER, U., HERTH, W.: General and molecular cytology. Progress in botany (H. ELLENBERG, K. ESSER, H. MERXMÜLLER, E. SCHNEPF, H. ZIEGLER, eds.), vol. 36, p. 1–20. Berlin-Heidelberg-New York: Springer 1974.

FREUDENBERG, K.: Constitution and biosynthesis of lignin. In: Molecular biology, biochemistry and biophysics (A. KLEINZELLER et al., eds.), vol. 2, p. 45–122. Berlin-Heidelberg-New York: Springer 1968.

FREY-WYSSLING, A., MÜHLETHALER, K.: Ultrastructural plant cytology. Amsterdam: Elsevier 1965.

GALBRAITH, W., GOLDSTEIN, I.J.: Phytohemagglutinins: a new class of metalloproteins. Isolation, purification, and some properties of the lectin from *Phaseolus lunatus*. F.E.B.S. Letters **9**, 197–201 (1970).

GULLASCH, J.: Versuche zur Anwendung der Elektronenstrahlmikroanalyse beim Nachweis wasserlöslicher Substanzen in biologischem Weichgewebe. Dissertation, Universität Düsseldorf (1974).

GUNNING, B.E.S., PATE, J.S.: "Transfer cells". Plant cells with wall ingrowths, specialized in relation to short distance transport of solutes — their occurrence, structure, and development. Protoplasma **68**, 107–133 (1969).

HÄRTEL, O.: Ionenwirkung auf die Kutikulartranspiration von Blättern. Protoplasma **40**, 107–136 (1951).

HARRISON-MURRAY, R.S., CLARKSON, D.T.: Relationships between structural development and the absorption of ions by the root system of *Cucurbita pepo*. Planta **114**, 1–16 (1973).

HEATH, M.F., NORTHCOTE, D.H.: Glycoprotein of the wall of sycamore tissue-cultured cells. Biochem. J. **125**, 953–961 (1971).

HEINEN, W., BRANDT, I.V.D.: Enzymatische Aspekte zur Biosynthese des Blatt-Cutins bei *Gasteria verricuosa*-Blättern nach Verletzung. Z. Naturforsch. **18b**, 67–79 (1963).

HILL, A.E., HILL, B.S.: The *Limonium* salt gland: a biophysical and structural study. Intern. Rev. Cytol. **35**, 299–319 (1973).

HUELIN, F.E.: Studies in the natural coating of apples. IV. The nature of cutin. Australian J. Biol. Sci. **12**, 175–180 (1959).

JENNY, H.: Pathways of ions from soil into root according to diffusion models. Plant Soil **25**, 265–289 (1966).

JENNY, H., GROSSENBACHER, K.: Root-soil boundary zones as seen in the electron microscope. Soil Sci. Soc. Amer. Proc. **27**, 273–277 (1963).

JENNY, H., OVERSTREET, R.: Cation interchange between plant roots and soil colloids. Soil Sci. **47**, 257–272 (1939).

KAMIYA, N., TAZAWA, M., TAKATA, T.: Water permeability of the cell wall in *Nitella*. Plant Cell Physiol. **3**, 285–292 (1962).

KAUSS, H., BOWLES, D.J.: Some properties of carbohydrate-binding proteins (lectins) solubilized from cell walls of *Phaseolus aureus*. Planta in press (1976).

KAUSS, H., GLASER, C.: Carbohydrate-binding proteins from plant cell walls and their possible involvement in extension growth. F.E.B.S. Letters **45**, 304–307 (1974).

KEEGSTRA, K., TALMADGE, K.W., BAUER, W.D., ALBERSHEIM, P.: The structure of plant cell walls. III. A model of the walls of suspension-cultured sycamore cells based on the interconnections of the macromolecular components. Plant Physiol. **51**, 188–196 (1973).

KOLATTUKUDY, P.E.: Biosynthesis of cuticular lipids. Ann. Rev. Plant Physiol. **21**, 163–192 (1970).

KOLATTUKUDY, P.E., DEAN, B.B.: Structure, gas chromatographic measurement, and function of suberin synthesized by potato tuber tissue slices. Plant Physiol. **54**, 116–121 (1974).

KRAMER, P.J.: Plant and soil water relationships: A modern synthesis. New York: McGraw-Hill 1969.

KRICHBAUM, R., LÜTTGE, U., WEIGL, J.: Mikroautoradiographische Untersuchung der Auswaschung des „anscheinend freien Raumes" von Maiswurzeln. Ber. Deut. Botan. Ges. **80**, 167–176 (1967).

KUO, J., O'BRIEN, T.P.: Development of the suberized lamella in the mestome sheath of wheat leaves. In: 8th Intern. Congr. Electron Microscopy (J.V. SAUNDERS, D.J. GOODCHILD, eds.), vol. II, p. 604–605. Canberra: Australian Academy of Science 1974.

LAMPORT, D.T.A.: Cell wall metabolism. Ann. Rev. Plant Physiol. **21**, 235–270 (1970).

LAMPORT, D.T.A.: The glycopeptide linkages of extensin. O-D-Galactosyl serine and O-L-arabinosyl hydroxyproline. In: Biogenesis of plant cell wall polysaccharides (F. LOEWUS, ed.), p. 149–164. New York-London: Academic Press 1973.

LAMPORT, D.T.A., NORTHCOTE, D.H.: Hydroxyproline in primary walls of higher plants. Nature **188**, 665–666 (1960).

LÄUCHLI, A.: Translocation of inorganic solutes. Ann. Rev. Plant Physiol. **23**, 197–218 (1972a).

LÄUCHLI, A.: Electron probe analysis. In: Microautoradiography and electron probe analysis: their application to plant physiology (U. LÜTTGE, ed.), p. 191–236. Berlin-Heidelberg-New York: Springer 1972b.

LÄUCHLI, A.: Investigation of ion transport in plants by electron probe analysis: principles and perspectives. In: Ion transport in plants (W.P. ANDERSON, ed.), p. 1–10. London-New York: Academic Press 1973.

LÄUCHLI, A.: Localization of inorganic ions in plant cells and tissues by use of electron microprobe analysis. In: 8th Intern. Congr. Electron Microscopy (J.V. SAUNDERS, D.J. GOODCHILD, eds.), vol. II, p. 68–69. Canberra: Austral. Academy of Science 1974.

LÄUCHLI, A., GULLASCH, J., KRAMER, D.: Unpublished results.

LÄUCHLI, A., KRAMER, D., PITMAN, M.G., LÜTTGE, U.: Ultrastructure of xylem parenchyma cells of barley roots in relation to ion transport to the xylem. Planta **119**, 85–99 (1974a).

LÄUCHLI, A., KRAMER, D., STELZER, R.: Ultrastructure and ion localization in xylem parenchyma cells of roots. In: Membrane transport in plants (U. ZIMMERMANN, J. DAINTY, eds.), p. 363–371. Berlin-Heidelberg-New York: Springer 1974b.

LÄUCHLI, A., STELZER, R., GUGGENHEIM, R., HENNING, L.: Precipitation techniques as a means for intracellular ion localization by use of electron probe analysis. In: Microprobe analysis as applied to cells and tissues (T. HALL, P. ECHLIN, R. KAUFMANN, eds.), p. 107–118. London-New York: Academic Press 1974c.

LEDBETTER, M.C., PORTER, K.R.: Introduction to the fine structure of plant cells. Berlin-Heidelberg-New York: Springer 1970.

LIS, H., SHARON, N.: The biochemistry of plant lectins (phytohemagglutinins). Ann. Rev. Biochem. **42**, 541–574 (1973).

LÜTTGE, U.: Stofftransport der Pflanzen. Berlin-Heidelberg-New York: Springer 1973.

LÜTTGE, U., SCHÖCH, E.V., BALL, E.: Can externally applied ATP supply energy to active ion uptake mechanisms of intact plant cells? Australian J. Plant Physiol. **1**, 211–220 (1974).

LYON, N.C., MUELLER, W.C.: A freeze-etch study of plant cell walls for ectodesmata. Canad. J. Bot. **52**, 2033–2036 (1974).

McFARLANE, J.C., BERRY, W.L.: Cation penetration through isolated leaf cuticles. Plant Physiol. **53**, 723–727 (1974).

MOLZ, F.J., IKENBERRY, E.: Water transport through plant cells and cell walls: theoretical development. Soil Sci. Soc. Amer. Proc. **38**, 699–704 (1974).

MÜNCH, E.: Die Stoffbewegungen in der Pflanze. Jena: Gustav Fischer 1930.

NAGAHASHI, G., THOMSON, W.W., LEONARD, R.T.: The casparian strip as a barrier to the movement of lanthanum in corn roots. Science **183**, 670–671 (1974).

NEVIS, A.H.: Water transport in invertebrate peripheral nerve fibres. J. Gen. Physiol. **41**, 927–958 (1958).

NORTHCOTE, D.H.: Chemistry of the plant cell wall. Ann. Rev. Plant Physiol. **23**, 113–132 (1972).

O'BRIEN, T.P., CARR, D.J.: A suberized layer in the cell walls of the bundle sheath of grasses. Australian J. Biol. Sci. **23**, 275–287 (1970).

OSMOND, C.B.: Metabolite transport in C_4 photosynthesis. Australian J. Biol. Sci. **24**, 159–163 (1971).

PATE, J.S., GUNNING, B.E.S.: Transfer cells. Ann. Rev. Plant Physiol. **23**, 173–196 (1972).

PAULING, L.: Nature of chemical bond. Ithaca, New York: Cornell University Press 1960.

PITMAN, M.G.: Sodium and potassium uptake by seedlings of *Hordeum vulgare*. Australian J. Biol. Sci. **18**, 10–24 (1965).

PITMAN, M.G., LÜTTGE, U., KRAMER, D., BALL, E.: Free space characteristics of barley leaf slices. Australian J. Plant Physiol. **1**, 65–75 (1974).

ROBARDS, A.W.: Electron microscopy and plant ultrastructure. London: McGraw-Hill 1970.

ROBARDS, A.W., JACKSON, S.M., CLARKSON, D.T., SANDERSON, J.: The structure of barley roots in relation to the transport of ions into the stele. Protoplasma **77**, 291–311 (1973).

ROBARDS, A.W., ROBB, M.E.: Uptake and binding of uranyl ions by barley roots. Science **178**, 980–982 (1972).

ROBARDS, A.W., ROBB, M.E.: The entry of ions and molecules into roots: an investigation using electron-opaque tracers. Planta **120**, 1–12 (1974).

ROTHSTEIN, A.: The enzymology of the cell surface. Protoplasmatologia, Bd. II, E4. Wien: Springer 1954.

SCHÖNHERR, J.: The nature of the pH effect on water permeability of plant cuticles. Ber. Deut. Botan. Ges. **87**, 389–402 (1974).

SCHÖNHERR, J., BUKOVAC, M.J.: Preferential polar pathways in the cuticle and their relationship to ectodesmata. Planta **92**, 189–201 (1970).

SCHÖNHERR, J., BUKOVAC, M.J.: Ion exchange properties of isolated tomato fruit cuticular membrane: exchange capacity, nature of fixed charges and cation selectivity. Planta **109**, 73–93 (1973).

SCHUMACHER, W., HALBSGUTH, W.: Über den Anschluß einiger höherer Parasiten an die Siebröhren der Wirtspflanzen. Ein Beitrag zum Plasmodesmenproblem. Jahrb. Wiss. Botan. **87**, 324–355 (1939).

SCHUMACHER, W., LAMBERTZ, P.: Über die Beziehungen zwischen der Stoffaufnahme durch Blattepidermen und der Zahl der Plasmodesmen in den Außenwänden. Planta **47**, 47–52 (1956).

SHERIFF, D.W., MEIDNER, H.: Water pathways in leaves of *Hedera helix* L. and *Tradescantia virginiana* L. J. Exptl. Bot. **25**, 1147–1156 (1974).

SITTE, P.: Zum Feinbau der Suberinschichten im Flaschenkork. Protoplasma **54**, 555–559 (1962).

SITTE, P.: Bau und Feinbau der Pflanzenzelle. Jena: Gustav Fischer 1965.

SITTE, P.: Die Bedeutung der molekularen Lamellen-Bauweise von Kork-Zellwänden. Biochem. Physiol. Pflanzen **168**, 287–297 (1975).

SITTE, P., RENNIER, R.: Untersuchungen an cuticularen Zellwandschichten. Planta **60**, 19–40 (1963).

SLATYER, R.O.: Plant-water relationships. London-New York: Academic Press 1967.

SMITH, R.C., EPSTEIN, E.: Ion absorption by shoot tissue: technique and first findings with excised leaf tissue of corn. Plant Physiol. **39**, 338–341 (1964).

STEINBRECHER, W., LÜTTGE, U.: Sugar and ion transport in isolated onion epidermis. Australian J. Biol. Sci. **22**, 1137–1143 (1969).

STELZER, R.: Lokalisation von Chlorid in Gerstenwurzeln. Diplomarbeit, Technische Hochschule Darmstadt (1973).

STELZER, R., LÄUCHLI, A., KRAMER, D.: Interzelluläre Transportwege des Chlorids in Wurzeln intakter Gerstepflanzen. Cytobiologie **10**, 449–457 (1975).

STEUDLE, E., ZIMMERMANN, U.: Determination of the hydraulic conductivity and of reflection coefficients in *Nitella flexilis* by means of direct cell-turgor pressure measurements. Biochim. Biophys. Acta **332**, 399–412 (1974).

STEVENINCK, R.F.M. VAN, CHENOWETH, A.R.F.: Ultrastructural localization of ions. I. Effect of high external sodium chloride concentration on the apparent distribution of chloride in leaf parenchyma cells of barley seedlings. Australian J. Biol. Sci. **25**, 499–516 (1972).

STRUGGER, S.: Die lumineszenzmikroskopische Analyse des Transpirationsstromes in Parenchymen. I. Mitteilung: Die Methode und die ersten Beobachtungen. Flora (Jena) **133**, 56–68 (1939).

STRUGGER, S.: Der aufsteigende Saftstrom in der Pflanze. Naturwissenschaften **31**, 181–194 (1943).

STRUGGER, S., PEVELING, E.: Die elektronenmikroskopische Analyse der extrafaszikulären Komponente des Transpirationsstromes mit Hilfe von Edelmetallsuspensoiden adäquater Dispersität. Ber. Deut. Botan. Ges. **74**, 300–304 (1961).

TALMADGE, K.W., KEEGSTRA, K., BAUER, W.D., ALBERSHEIM, P.: The structure of plant cell walls. I. The macromolecular components of the walls of suspension-cultured sycamore cells with a detailed analysis of the pectic polysaccharides. Plant Physiol. **51**, 158–173 (1973).

TANTON, T.W., CROWDY, S.H.: Water pathways in higher plants. II. Water pathways in roots. J. Exptl. Bot. **23**, 600–618 (1972a).

TANTON, T.W., CROWDY, S.H.: Water pathways in higher plants. III. The transpiration stream within leaves. J. Exptl. Bot. **23**, 619–625 (1972b).

TAZAWA, M., KAMIYA, N.: Water permeability of a Characean internodal cell with special reference to its polarity. Australian J. Biol. Sci. **19**, 399–419 (1966).

THOMSON, W.W., PLATT, K.A., CAMPBELL, N.: The use of lanthanum to delineate the apoplastic continuum in plants. Cytobios **8**, 57–62 (1973).

TYREE, M.T.: Determination of transport constants of isolated *Nitella* cell walls. Canad. J. Bot. **46**, 317–327 (1968).

TYREE, M.T.: The thermodynamics of short-distance translocation in plants. J. Exptl. Bot. **20**, 341–349 (1969).

TYREE, M.T.: An alternative explanation for the apparently active water exudation in excised roots. J. Exptl. Bot. **24**, 33–37 (1973).

WEATHERLEY, P.E.: The pathway of water movement across the root cortex and leaf mesophyll of transpiring plants. In: The water relations of plants (A.J. RUTTER, F.H. WHITEHEAD, eds.), p. 85–100. Oxford: Blackwell 1963.

WEATHERLEY, P.E.: Some aspects of water relations. In: Advances in botanical research (R.D. PRESTON, ed.), vol. 3, p. 171–206. London-New York: Academic Press 1970.

WELLS, A.F.: Structural inorganic chemistry. New York: Oxford University Press 1945.
WILDER, B.M., ALBERSHEIM, P.: The structure of plant cell walls. IV. A structural comparison of the wall hemicellulose of cell suspension cultures of sycamore (*Acer pseudoplatanus*) and of red kidney bean (*Phaseolus vulgaris*). Plant Physiol. **51**, 889–893 (1973).
YAMADA, Y., WITTWER, S.H., BUKOVAC, M.J.: Penetration of ions through isolated cuticles. Plant Physiol. **39**, 28–32 (1964).
ZIEGLER, H., LÜTTGE, U.: Die Salzdrüsen von *Limonium vulgare*. II. Mitteilung. Die Lokalisierung des Chlorids. Planta **74**, 1–17 (1967).
ZIEGLER, H., WEIGL, J., LÜTTGE, U.: Mikroautoradiographischer Nachweis der Wanderung von $^{35}SO_4^{--}$ durch die Tertiärendodermis der *Iris*-Wurzel. Protoplasma **56**, 362–370 (1963).
ZIMMERMANN, U., STEUDLE, E.: The hydraulic conductivity and volumetric elastic modulus of cells and isolated cell walls of *Nitella* and *Chara* spp.: pressure and volume effects. Australian J. Plant Physiol. **2**, 1–12 (1975).

2. Symplasmic Transport in Tissues

R.M. SPANSWICK

1. Introduction

The discovery of plasmodesmata by TANGL in 1879 provided the structural basis for a direct intercellular pathway and TANGL himself suggested that plasmodesmata might facilitate intercellular diffusion. The system of protoplasts and connecting plasmodesmata is widely known as the *symplasm,* a term usually attributed to MÜNCH (1930). However, HABERLANDT (1904, p. 547) used the term "symplast" for this system, though in the context of stimulus transmission which was thought at that time to be its primary function.

The idea of symplasmic transport has not always received unanimous support. In the early part of the century, transport across cell membranes by diffusion was thought to be sufficiently rapid for intercellular transport. Later, there was a period when the cytoplasm was thought to form part of the free space (BRIGGS and ROBERTSON, 1957). Either of these ideas, had they proved to be well founded, would have rendered the symplasm superfluous as a transport pathway. However, it is now generally agreed that the plasmalemma is a significant permeability barrier, and hence symplasmic transport acquires a special significance for transport within plant tissues. For instance, in submerged aquatic plants, in which the apoplast is in direct contact with the medium, symplasmic transport would have advantages in terms of conservation of materials during transport from cell to cell.

Although the existence of symplasmic transport is now widely accepted, the idea is often presented without adequate justification. There follows an examination of the structural and physiological basis for the involvement of plasmodesmata in symplasmic transport. The older literature in this area has been reviewed by MEEUSE (1957) and ARISZ (1956).

2. The Ultrastructure of Plasmodesmata

Early work on plasmodesmata (see MEEUSE, 1957) was hampered by the limited resolving power of the optical microscope. Most recent measurements with the electron microscope show that the diameters of plasmodesmata are in the range 20–80 nm, which is at or beyond the limit of resolution of the optical microscope. Indeed, TAIZ and JONES (1973) suggest that it was not the plasmodesma itself that was observed but a structure in the surrounding cell wall which they call a "wall tube". The advent of electron microscopy has removed any doubts about the existence of plasmodesmata but has led to a controversy concerning their structure.

It is now clear that plasmodesmata exist in most tissues and have many configurations, some being branched. It is generally agreed that the plasmalemma is con-

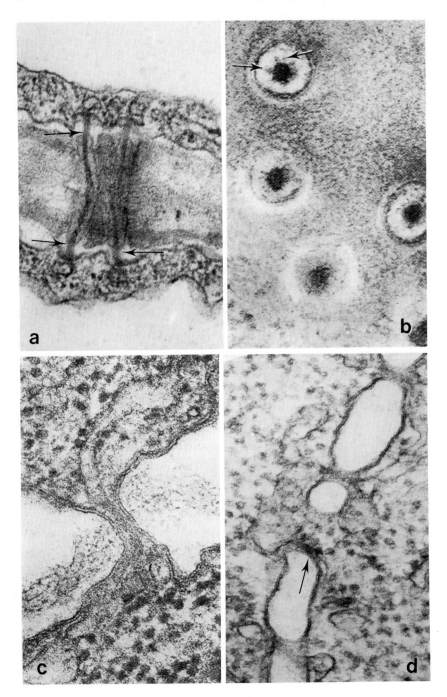

Fig. 2.1a–d. Legend see opposite page

tinuous from cell to cell. The main difficulty is interpretation of the structure of the plasmodesmatal core. There are technical problems both in preserving the structure and in interpreting the image from a biological structure of this size and shape (HELDER and BOERMA, 1969). However, several internal structures have been described (BURGESS, 1971; ROBARDS, 1971; Figs. 2.1a, b) and in many cases it does seem likely that the endoplasmic reticulum or a modified form of it, the "desmotubule" (ROBARDS, 1971), passes through the plasmodesma (Fig. 2.1c). This is observed most clearly in plasmodesmata that have expanded after the surrounding cell wall has been digested away (Fig. 2.1d; see also 3.1.2, Fig. 3.3).

Further clarification of the structure of the plasmodesma is highly desirable. In the meantime, physiological predictions should not be based on ultrastructural studies alone.

3. The Formation and Occurrence of Plasmodesmata

Plasmodesmata are laid down at the time of cell plate formation. However, the extent to which microtubules or endoplasmic reticulum are involved remains obscure (BURGESS, 1971). As the cell wall matures, the number of plasmodesmata per unit area may decrease due to expansion of the cell wall or loss of plasmodesmata. Mature guard cells represent the extreme case where most or all of the plasmodesmata have been lost (THOMSON and JOURNETT, 1970; ALLAWAY and SETTERFIELD, 1972; PALLAS and MOLLENHAUER, 1972). Nevertheless, it is probably more common for the number per cell to stay approximately constant (ROBARDS et al., 1973).

In *Zea mays* the number of plasmodesmata per unit area in the various cell walls of the root tip is in the range 4.5 to 22 per μm^2 (JUNIPER and BARLOW, 1969) but BURGESS (1971) has observed up to 140 per μm^2 in the root meristem of the fern *Dryopteris filix-mas* (cf. Table 2.1). CLOWES and JUNIPER (1968) have estimated that higher plant cells are each connected to their neighbors by 1,000 to 100,000 plasmodesmata. TYREE (1970) has compiled data for plasmodesmatal radius and frequency for a number of tissues (Table 2.1).

Plasmodesmata do not occur in the blue-green algae except during heterocyst formation. In the other groups of algae they appear to be associated with cytokinesis by cell-plate formation as they are absent in related species where division takes place by furrowing (FLOYD et al., 1971). It is clear that the genesis of plasmodesmata

---

Fig. 2.1a–d. Ultrastructure of the plasmodesma. (a) Plasmodesmata in a lateral wall of *Avena* coleoptile. The dark staining core approaches the plasmalemma at the constricted ends of the pore (arrows). × 80,000. (b) A transverse section through plasmodesmata of the type shown in Fig. 2.1a. The core assumes a tubular appearance when sectioned normally. Spokes appear to radiate between the core and the plasmalemma lining the pore (arrows). × 180,000. (c) Section through a plasmodesma in a transverse wall in the root meristem of *Colchicum*. The endoplasmic reticulum appears to be continuous with the dark staining core of the plasmodesma. × 137,000. (d) Oat root tip during enzyme digestion. The endoplasmic reticulum can be seen close to the plasmodesmata. A desmotubule is visible close to the plasmalemma on one side of the expanding plasmodesma (arrow). The desmotubule does not itself appear to expand. × 125,000. [(a), (b) and (c) from BURGESS (1971). (d) from WITHERS and COCKING (1972)]

Table 2.1. The pore radius and frequency of plasmodesmata in cell walls

Plant material	Radius of plasmodesma (nm)	Frequency of plasmodesmata (pores/m^2)
Laminaria digitata: trumpet cell cross wall[a]	30	$5\text{--}6 \cdot 10^{13}$
Tamarix aphylla: cross wall between collecting cell and salt gland cell[b]	30–40	$2 \cdot 10^{13}$
Allium cepa:		
(a) mature root cortical cell[c]	40–70	$1.5 \cdot 10^{12}$
(b) root meristem cell[d]	40–50	$6\text{--}7 \cdot 10^{12}$
Viscum album: mature root cortical cell[e]	20–30	$2 \cdot 10^{12}$
Avena sativa: mature cortical cell of coleoptile[f]	30–50	$3.6 \cdot 10^{12}$
Lycopersicum esculentum: pollen mother cells[g]	50	
Salix fragilis: cambial tissue[h]	10	
Nitella translucens: internodal cell wall[i]	35	$3 \cdot 10^{12}$
Zea mays: root cap cells at meristem[j]	—	$4.2 \cdot 10^{12}$

This Table has been adapted from TYREE (1970). The source references are: [a] ZIEGLER and RUCK (1967); [b] THOMPSON and LIU (1967); [c] STRUGGER (1957b); [d] STRUGGER (1957a); [e] KRULL (1960); [f] BÖHMER (1958); [g] WEILING (1965); [h] ROBARDS (1968); [i] SPANSWICK and COSTERTON (1967); [j] CLOWES and JUNIPER (1968).

was an important step in the evolution of complex forms of algae (eg. *Laminaria*, Table 2.1) and of all higher plants, as it permitted greater coordination in the development of the plant body.

4. Boundaries of Symplasmic Compartments in Plant Tissues

Cells of higher plants which do not have protoplasmic connections with their neighbors are the exception. However, this does not mean that the symplasm is continuous throughout the plant. Indeed, it may now be possible to consider defining the boundaries of tissues in terms of their symplasmic (or apoplastic) isolation. Emphasis has been lent to this approach by the recognition that cells lacking plasmodesmata in one or more walls may form wall ingrowths with a consequent proliferation of membrane area. It has been pointed out that these "transfer cells" occur at strategic sites in the plant where isolation is desirable for osmotic or developmental reasons yet where transport of materials is important (PATE and GUNNING, 1972; 1.4.2.2.2). Thus in *Pisum*, for example, the companion cells in the minor veins have the characteristics of transfer cells. They have elaborately branched connections (plasmodesmata) with the sieve cells, but few if any, connections with other cells. It might be postulated that loading of the phloem takes place at the surface of the companion cells and this is followed by "symplasmic" transport into the sieve cells. A similar situation occurs in embryos, other reproductive structures and at various sites in the plant that may be associated with transport. It should be noted, however, that transfer cells have not been observed in all plants or tissues and cannot, therefore, be used universally as a boundary marker for symplasmic compartments.

It might be expected that symplasmic transport would be especially important in situations where an osmotic gradient is set up between two regions of the apoplast which are separated by impermeable cell walls. The well-known examples are roots and most salt glands. In roots, the Casparian strip of the endodermis provides a barrier to solute diffusion (ROBARDS and ROBB, 1972; *1.4.1.2*; *1.4.2.2.1*; *3.4.2.4*). It is reasonable to assume that the plasmodesmata in the endodermal cell walls provide the main pathway for solute movement, particularly in the tertiary endodermis where the whole cell wall is suberized (CLARKSON et al., 1971). Glandular structures are often "insulated" by a cutinized wall which, again, is well endowed with plasmodesmata (*5.2.2.2.1*, *5.3.1.2.2.3*, and *5.3.1.3.3*).

5. Symplasmic Transport

Anatomical evidence, however good the correlation with the presumed function, does not prove that plasmodesmata do in fact provide an effective pathway for intercellular transport. The remainder of this Chapter will be devoted to an evaluation of the physiological evidence relating to this problem.

The earliest substantial work from which the existence of symplasmic transport could be deduced was the transport studies of ARISZ and his co-workers. However, a more direct method with greater spatial resolution uses electrical techniques and this will be dealt with first.

5.1 Demonstration of Symplasmic Continuity Using Electro-Physiological Techniques

Passage of electric current through a body will, in general, produce gradients of electrical potential proportional to the electrical resistance encountered from point to point. If a biological system is sufficiently simple, measurements of this type can be used to deduce the resistance of various parts of the system to movements of the endogenous ions. For plants, this was first done by LOU (1955) using external electrodes. The observations which can be most easily interpreted were made using a system consisting of a node and the two adjoining internodal cells of *Nitella*. He immersed the cells in oil, passed a current between the two ends and measured the variation of potential with distance along the surface of the cells (Fig. *2.2*). The system acts as a leaky cable, the cell walls being the outer conductor, the membranes the insulator (leaky) and the vacuoles the inner conductor. Fortunately, the resistance of the plasmalemma is much greater than that of the tonoplast and so the tonoplast can be ignored to a first approximation. Where current enters (or leaves) the cells at either end there is a sharp gradient of potential because most of the current is confined to the cell wall. With increasing distance, however,

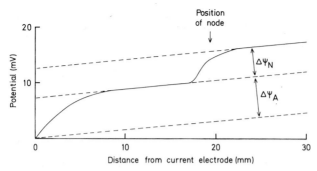

Fig. 2.2. Approximate representation of the potential gradient along a system consisting of a node and the two adjacent internodal cells of *Nitella* when a current is passed between external electrodes placed at either end of the preparation. External electrodes were also used to record the potential. (After LOU, 1955)

a greater proportion of the current enters the cell until the currents flowing in the cell wall and vacuole are inversely proportional to the resistance per unit length for the two phases. At this point no current flows across the membranes and the potential gradient along the cell is constant and has its minimum value (Fig. 2.2 dashed lines). If there were no direct connection between the internodal cells and the nodal cells, the current arriving at the node would have to leave the first internodal cell, pass through the cell walls and enter the second internodal cell across the plasmalemma. The extra change in potential in the nodal region, $\Delta\psi_N$, due to current crossing the membranes would then be approximately equal to twice the change in potential at the initial point of entry of the current into the internodal cell, $\Delta\psi_A$. LOU found that the change in potential at the node was significantly less than would have been the case if there were no direct connections between the cells (Fig. 2.2). It was not negligible, however.

LOU (1955) also made measurements on roots. However, in this case, the distance over which current continues to cross the cell membranes is greater than the lengths of the individual cells and the interpretation of the potential gradients is more difficult. He interpreted the greater spread of the initial potential gradient in living roots, as opposed to dead roots, as indicating penetration of current into the symplasm. However, the greater spread in the living roots could be due to impeded flow into the center of the root resulting from the presence of intact protoplasts and the endodermis. A better comparison would be between living roots with and without the plasmodesmata disrupted, but no method for selective disruption of the plasmodesmata is available at present.

Greater spatial resolution may be obtained by using microelectrodes to make intracellular measurements. The approach was first used for non-excitable animal cells by LOEWENSTEIN and KANNO (1964) and KUFFLER and POTTER (1964) and has since been used extensively in animal physiology. Current injected into a cell will cause a change in the membrane potential as it flows out across the membrane. If there are no direct intercellular connections, there will be no change in the membrane potential of the neighboring cell because the resistance of the extracellular space at the junction of the two cells provides a pathway to the external electrode with a much lower resistance than the alternative pathway through the membranes

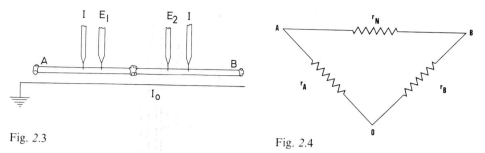

Fig. 2.3

Fig. 2.4

Fig. 2.3. Electrode arrangement for measurement of the electrical resistance of the node of *Nitella*, showing the relationship between the current microelectrodes, I, the potential microelectrodes, E_1 and E_2, the external current electrode, I_o, and the internodal cells, A and B. The external reference electrode is not shown. (From SPANSWICK and COSTERTON, 1967)

Fig. 2.4. A simplified equivalent circuit for a pair of internodal *Nitella* cells. The points A and B represent the cytoplasm of cells A and B, and O represents the external solution. r_A is the resistance of the plasmalemma of cell A, r_B is the resistance of the plasmalemma of cell B and r_N is the resistance of the node. (From SPANSWICK and COSTERTON, 1967)

of the neighboring cell. If there are direct connections between the cell interiors, however, current will pass into the neighboring cell and out across its membrane, producing a change in the membrane potential which may be detected using a separate microelectrode.

In plant cells the tonoplast is a complicating factor for such measurements. In *Nitella*, however, it was possible to measure the resistance of the plasmalemma and tonoplast separately by placing an additional microelectrode in the cytoplasm (SPANSWICK and COSTERTON, 1967). The resistance of the plasmalemma was about ten times that of the tonoplast under the prevailing experimental conditions. Using the electrode arrangement shown in Fig. 2.3 it was possible, after correction for the tonoplast resistances, to use the simple equivalent circuit shown in Fig. 2.4 to calculate all the resistances in the system from measurements of the potential changes when current was passed first from A to O and then from B to O. The specific resistance of the internodal cell plasmalemma was about 9 Ωm^2. However, the node (Fig. 2.5), which contains at least one layer of small nodal cells separating the internodal cells and hence four layers of plasmalemma, had a specific resistance of only 0.17 Ωm^2. This result clearly indicated that protoplasmic continuity existed between the internodal cells. However, a calculation of the nodal resistance based on measurements of the plasmodesmatal dimensions and frequency and the measured concentrations of the ions in the flowing cytoplasm (SPANSWICK and WILLIAMS, 1964), gave a value 330 times smaller than that observed. This provides an indication, therefore, that the plasmodesmata contain structures that hinder the free diffusion of ions or that most of them are blocked. SIBAOKA (1966) and SKIERCZYŃSKA (1968) have also demonstrated electrical coupling between characean internodal cells.

Application of this method to higher plants presents difficult technical problems, firstly in inserting more than one microelectrode into the preparation and, secondly, because generally it is not possible to insert a microelectrode directly into the cytoplasm. SPITZER (1970) was able to demonstrate electrical coupling between cells in the developing lily anther that were separated by distances of up to 600 μm.

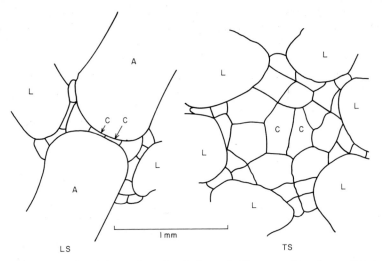

Fig. 2.5. Node of the main axis of plant of *Chara corallina* (*australis*), in longitudinal (*LS*) and transverse section (*TS*). *A* main axis internodes; *L* lateral internodes; *C* central cells of the node. In the longitudinal section, the thickness of the central cells *C* has been exaggerated for clarity. (From Walker and Bostrom, 1973)

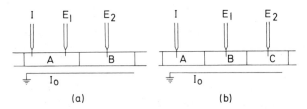

Fig. 2.6a and b. Microelectrode arrangements used in the demonstration of electrical coupling between cells (*A, B* and *C*) of *Elodea*. *I* is the electrode used to pass current between the vacuole of cell *A* and the electrode I_o in the external solution. E_1 and E_2 are the electrodes used to measure the electrical potential of the vacuoles relative to a reference electrode in the external solution (not shown). (a) Arrangement used initially to find the coupling ratio for cells *A* and *B*. (b) Arrangement used to find the coupling ratio for cells *B* and *C*. See Spanswick (1972) for experimental details

On the other hand, Goldsmith et al. (1972) were unable to detect coupling between parenchyma cells in the oat coleoptile. However, electrical coupling between cells has now been demonstrated in this tissue, between cortical cells of the maize root and between leaf cells of *Elodea canadensis* (Spanswick, 1972).

   Using the cells on the edge of the *Elodea* leaf, it was possible to circumvent some of the problems caused by the presence of the tonoplast. These problems arise because current injected into cell A (Fig. 2.6) must cross both the tonoplast and the plasmalemma as it leaves the cell and hence the recorded change in potential is the sum of that across both membranes. Current entering cell B through the plasmodesmata from cell A need not cross the tonoplast of cell B *en route* to the exterior. Therefore the change in portential recorded from cell B is that across the plasmalemma alone. As a result, the coupling ratio for the first two cells in the row (ψ_B/ψ_A, where ψ_A and ψ_B are potential changes recorded in cells A and B respectively), which was 0.29 for *Elodea,* does not give a measure of the attenuation across the plasmodesmata between cells A and B. However, by measuring in addition the coupling ratio for cells B and C (0.72 for *Elodea*) it was possible to calculate values for the resistance of the plasmalemma, the tonoplast and the plasmodesmata at the cell junction by using an approximate equivalent circuit for the system. Here too, the specific resistance of the junction between the cells (0.0051 Ωm^2) was much lower than the resistance of the plasmalemma (0.32 Ωm^2). But, although the precise anatomical information is not yet available, the intercellular resistance again appears to be higher than would be expected for unrestricted diffusion of ions through the plasmodesmata.

In certain specialized systems it is possible to use light-induced changes in membrane potential as an alternative to current injection for producing an electrical signal. Light-triggered transients of the membrane potential have been observed in the photosynthetic mesophyll cells of *Chenopodium* (LÜTTGE and PALLAGHY, 1969; *4.2.3.5*). Similar signals were also recorded from the bladder cells of the salt hairs when they were attached to the leaf but not from detached hairs. Since there are numerous plasmodesmata connecting the bladder, stem, epidermal and mesophyll cells, the simplest interpretation of these experiments would be that there is electrical coupling through the low-resistance pathway provided by the plasmodesmata. However, it is difficult to use these observations as a rigorous proof of electrical coupling because extensive cutinization of the cell walls (OSMOND et al., 1969) may mean that there is no low-resistance pathway from the free space at the junction of the bladder and stem cells to the external solution. In this case, it is conceivable that the electrical changes in the mesophyll cells would be detected in the bladder cells if the pathway with least resistance between the bladder cells and the reference electrode included the mesophyll cell membranes, even in the absence of symplasmic connections. More detailed knowledge of the resistance of the extracellular pathways is therefore required. However, it seems very probable that electrical coupling *via* the plasmodesmata does exist.

More recently, BRINCKMANN (1973, see also *4.2.3.5*) has used variegated mutants of *Oenothera* to perform similar experiments. In this system, the free space does provide a low resistance pathway to the external solution and therefore the reservations expressed about the experiments on *Chenopodium* do not apply. Light-induced potential changes were observed in the green cells but not in white cells when the tissue sample contained no green cells. However, if the tissue sample did contain green cells, the potential changes could also be recorded from the white cells. Since no time delay was observed even in widely separated cells, the white cells must be electrically coupled to the green cells and the change cannot be due to the transport of some product of photosynthesis. This is a unique demonstration of symplasmic continuity.

5.2 Demonstration of Symplasmic Transport Using Tracer Techniques or Net Transport

5.2.1 Transport of Dyes

Before the advent of radioisotopes, dyes were used extensively to demonstrate long-distance movement. However, their use to show direct movement between adjacent protoplasts is controversial except in cases, such as *Nitella* (BIERBERG, 1909), where the sites of loading and intercellular transport are widely separated and involvement of the apoplastic pathway would lead to losses to the external solution. In higher plants, it has been suggested that fluorescein, for example, is confined to the cytoplasm (Schumacher, 1967). However, doubts have been expressed concerning the apparent localization of fluorescein in the symplast which could be due to the variation of its fluorescence with pH (Bauer, 1953; see *1.4.1.1*). The fluorescence drops rapidly below pH 7 (Emmert, 1958) and, since vacuoles and cell walls are generally more acid than the cytoplasm, this could explain the apparent localization in the cytoplasm. It would also explain the apparent independence between the rate of protoplasmic streaming and movement of the dye, since streaming would have less effect if the whole cross section of the cell were available for diffusion.

5.2.2 Radioactive Tracers and Net Transport

With the aid of radioactive tracers and autoradiography, it is now a simple matter to demonstrate movement of substances on a gross scale in whole plants. Phloem and xylem translocation are responsible for the rapid long-distance transport and, since they are being treated separately (Vol. 1), the distinction from symplasmic transport will be carefully maintained here. (It should be noted, however, that Arisz (1969) considers the phloem to be part of the symplasm.)

Localization of molecules at the subcellular level using radioactive tracers and autoradiography presents many technical difficulties and the method has not been used successfully to demonstrate that substances are transported through, as opposed to being present in, plasmodesmata. However, at a grosser level, it is possible to demonstrate unambiguously that transport takes place through the symplasm. This may be done either using tracers or a system that will give significant net transport of an unlabeled substance. The requirements for such a demonstration are that the regions of uptake and transport can be physically separated, that the membranes have a low permeability to the substance being used so that intercellular transport does not take place *via* the membranes (active uptake is therefore desirable to achieve a high enough concentration in the symplasm), and that the apoplast should not be separated from the medium by a permeability barrier such as a cuticle (cf. *1.3*). The last point is important in demonstrating that the apoplast is not part of the transport pathway since, in the absence of a cuticle, solutes leaking from the symplasm can be detected in the external solution.

a) The Characeae. Submerged aquatic plants lacking cuticles are obviously suitable systems and the size of the internodal cells makes the Characeae particularly useful. Littlefield and Forsberg (1965) have used ^{32}P to confirm Bierberg's (1909) work on *Chara*. They showed that radioactive phosphate applied locally to one internodal cell was transported throughout the plant. Passage from one cell to another *via* the cell walls of the nodal complex would lead to large losses to the external solution. They demonstrated that such losses did not take place and the conclusion that transport took place through the plasmodesmata therefore appears inescapable.

It is, however, quite another matter when it comes to making accurate measurements of the fluxes across the node. This would require an accurate measurement or estimate of the specific activity of the ion or molecule in the streams of flowing cytoplasm approaching and leaving both sides of the node. In the absence of direct measurements, it is necessary to consider in great detail the effects of the vacuolar fluxes and compartmentation within the cytoplasm on the cytoplasmic specific activity. It is also necessary to estimate the back fluxes across the node. Preliminary estimates indicated that the fluxes were high (WALKER and BOSTROM, 1973). In more recent work (TYREE et al., 1974; BOSTROM and WALKER, 1975) the analysis has been refined and it is apparent that the fluxes are higher than expected on the basis of electrical resistance measurements (SPANSWICK and COSTERTON, 1967) and could be high enough to be consistent with the plasmodesmata containing only viscous cytoplasm. However, a great deal of work remains to be done in checking the assumptions on which these conclusions are based.

b) *Vallisneria*. For higher plants, the most conclusive evidence for symplasmic transport comes from the extensive investigations of ARISZ and his co-workers, particularly that on *Vallisneria spiralis*. While *Vallisneria* has the advantage of lacking xylem vessels and a cuticle, it does contain phloem. Attention here will be concentrated on those experiments relevant to symplasmic transport through parenchyma tissue and the equally important work on active transport and translocation will be dealt with elsewhere (*4.2.3.4.4*).

Uptake and transport in *Vallisneria* can be separated by placing a leaf section in a two or three compartment chamber in which the solution surrounding the absorbing part of the leaf is isolated from the remainder (Fig. *2.7*). The increase in the chloride content of the absorbing, middle and end sections, each 25 mm long, was measured 24 h after exposure of the absorbing section to a chloride solution (ARISZ, 1958). The amount of chloride moved to the middle and end sections during this period was usually only slightly less than that accumulated in the absorbing section. However, with intact material, no distinction can be made between movement through the phloem and the parenchyma. This problem was solved by ARISZ and SCHREUDER (1956) who removed sections of the leaf at the junction of the absorbing and transport zones, leaving tissue bridges containing either parenchyma tissue or a vascular bundle. After suitable pretreatment to allow the tissue to recover from the leakage induced by cutting, they were able to demonstrate that both parenchyma and bundle bridges were capable of transporting chloride. The transport through the parenchyma bridges provided clear evidence for symplas-

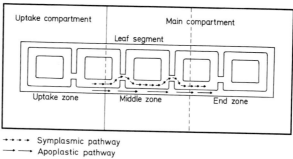

Fig. *2.7*. Schematic diagram of the apparatus used by ARISZ to study transport in leaf sections of *Vallisneria spiralis*. The main compartment could be subdivided to divide the leaf into three zones. The alternative symplasmic and apoplastic pathways in the leaf are also illustrated diagrammatically. The vascular bundles are not shown

mic transport. The possible objection that transport could have occurred by diffusion through the free space of the tissue held in the divider separating the uptake and main compartments (Fig. 2.7), followed by uptake into the tissue, has been dealt with in two ways. Firstly, separate sections of leaf tissue placed in the main compartment were shown not to take up any chloride (ARISZ, 1960), indicating that no chloride leaked into the main compartment. Secondly, extensive studies with inhibitors (ARISZ, 1958) have shown that inhibition of uptake (at the plasmalemma) by uranyl nitrate or cyanide practically abolishes movement through the parenchyma bridges to the end zone. This reduction of chloride in the end zone would not be expected if it were due to an independent uptake of chloride that had leaked into the main compartment. Furthermore, application of uranyl nitrate or cyanide to the middle zone caused no inhibition of the accumulation there (mainly due to transport across the tonoplast into the vacuoles) or of transport to the end zone. Again, this suggests that the initial uptake did indeed take place in the loading zone.

The effect of azide is also of interest, particularly with regard to the differentiation between phloem and symplasmic transport. ARISZ (1958) found that 10^{-5} M azide slightly inhibited accumulation in the loading zone but did not affect transport to the middle and end zones through the parenchyma bridges. However, in mature tissue, transport through the bundle bridges was severely inhibited. As mentioned earlier, the definition of the symplasm used here excludes the phloem and this experiment suggests that it can in fact form a compartment separate from the phloem, as do the vacuoles where most of the accumulation takes place but where accumulation can be inhibited without affecting transport through the symplasm.

ARISZ's work on *Vallisneria* has thus provided strong evidence for symplasmic transport and has made it possible to demonstrate differential effects of inhibitors on transport at the plasmalemma and tonoplast.

Investigation of the effect of the length of the bridges on the amount of chloride transported showed that transport through the parenchyma bridges was reduced when the length of the bridge was greater than 4 mm but transport through the bundle bridges was only slightly reduced when the length was 16 mm. This would be consistent with diffusion playing an important role in symplasmic transport as opposed to some special mechanism operating in the phloem.

The importance of phloem in transport over long distances was emphasized by autoradiographic studies of transport through parenchyma bridges. In Fig. 2.8 from ARISZ (1960) it is clear that, once the parenchyma bridges have been passed, most of the accumulation takes place in the vascular bundles and accumulation in the parenchyma decreases more with distance from the vascular tissue than with distance from the loading zone. Thus phloem transport appears to be important for rapid transport over long distances in this system. However, there can be no doubt that the symplasm plays a vital role in short-distance transport. This is also evident from the autoradiograms of transport through bundle bridges where it can be seen that after passage throughout the vascular system considerable amounts of chloride move into the parenchyma (Fig. 2.8).

c) *Drosera*. Another system used by ARISZ is the tentacle of the carnivorous plant *Drosera capensis* (see HELDER, 1967b). The tentacle stalk consists of a xylem vessel surrounded by two layers of parenchyma cells and a cuticle. At the head there is a gland that secretes mucilage. It also secretes enzymes to digest trapped insects and absorbs the products. The parenchyma

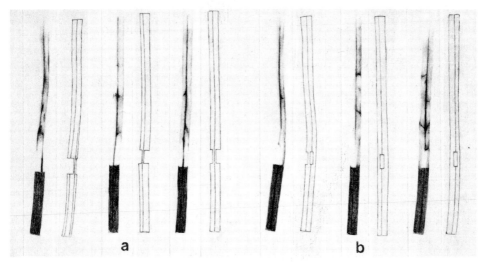

Fig. 2.8a and b. Movement of chloride through tissue bridges in *Vallisneria* leaves. Comparison of some dissected leaf lengths (7.5 cm) with their autoradiograms: (a) bundle bridges, (b) parenchyma bridges. 24 h uptake from a 1.6 mM potassium chloride solution with labelled chloride. (From ARISZ, 1960)

cells are the obvious pathway for the transport of these products to the leaf blade since there is no phloem. While the presence of a cuticle and therefore the possibility of apoplastic diffusion makes it difficult to demonstrate this rigorously, inhibition of transport by low temperatures, inhibitors or lack of oxygen suggests that entry into the symplasm is a prerequisite for transport. A complication not evident at the time is the action potentials that travel down the tentacle stalks (WILLIAMS and PICKARD, 1972). These would have been triggered by the method of applying solutes to the tentacles and are important because they could involve changes in the permeability of the cell membrane to the substances under study. However, the evidence does appear to be consistent with the symplasm theory.

d) Roots. The large body of evidence on the lateral transport of salts across roots (ARISZ, 1956; LÄUCHLI, 1972; *1*.4.2.2.1; Chap. *3.4*) has not provided an independent proof of the existence of symplasmic transport as opposed, for example, to transport across a single layer of cells. However, the evidence can be interpreted most easily in terms of symplasmic transport and it is difficult to interpret the recent work on barley roots (CLARKSON et al., 1971; ROBARDS and ROBB, 1972) in any other way.

5.3 Solutes Transported

From the evidence presented above it is clear that salts can penetrate the plasmodesmata. If it is accepted that transport from the phloem to the parenchyma in *Vallisneria* without leakage to the external solution must involve symplasmic transport, then the work of ARISZ and WIERSEMA (1966) provides evidence for the movement of a variety of salts and organic molecules within the symplasm. Rubidium, chloride and phosphate moved rapidly in the plant but Ca^{2+} was not translocated from the loading zone. This could be due to a combination of a low rate of uptake of Ca^{2+} (Spanswick and Williams, 1965) and binding of the Ca^{2+} which does enter the cells. They also demonstrated transport of several sugars, amino acids, malic acid and IAA. However, it should be noted that polar movement in coleoptiles is probably due to transport across cell membranes (Goldsmith and Ray, 1973); there is as yet no evidence that the plasmodesmata themselves are responsible for directed movements, though this possibility has been suggested (Ziegler, 1973).

Rapid symplasmic transport of organic acids appears to be especially important in plants with C_4 photosynthesis. CO_2 is fixed into malate and aspartate in the mesophyll cells and the organic acids then move into the bundle sheath where they are metabolized to sugars. Osmond (1971) has shown that this transport is very rapid (65 nmol g_{FW}^{-1} s^{-1}) and that the fluxes could be accounted for by symplasmic transport, assuming reasonable values for the area of the cell walls occupied by plasmodesmata, but not by the usual values for the fluxes (0.3–0.6 nmol g_{FW}^{-1} s^{-1}) of these substances across the cell membranes (cf. Black, 1973; Lüttge, 1974; Smith, 1971).

There is no evidence for the passage of macromolecules through plasmodesmata. The appearance of certain viruses in plasmodesmata suggests that such movement should be possible (Esau et al., 1967). However, the increased diameter of plasmodesmata containing viruses suggests that penetration of the virus may follow modification of the plasmodesmatal structure (Kitijama and Lauritis, 1969). The electrical evidence for restricted diffusion of ions through plasmodesmata indicates that diffusion of macromolecules may well be severely impeded.

5.4 Water Transport

Since plasmodesmata permit the passage of solutes, they undoubtedly permit the passage of water also. The question of importance here, then, is not whether they can transport water but whether the symplasmic pathway carries a significant fraction of the flow through a tissue subjected to a water potential gradient. The factors which differentiate water from solute transport in this context are the high permeability of cell membranes to water and the possibility of bulk flow through the cell walls. Quantitative analyses of this problem (Briggs, 1967; Tyree, 1969; Weatherley, 1970) have been concerned only with the importance of the cell wall pathway *versus* the vacuole pathway. A paucity of accurate values for the permeability of higher plant cell membranes and cell walls lends considerable latitude to the calculations. However, it is generally agreed that the cell walls provide the major pathway, the estimates of the ratio of volume flows in the two pathways being of the order of 50:1 for leaf mesophyll tissue (*1.4.2.2.3*) but lower for the root cortex.

None of these calculations takes into account the possibility of transport through the plasmodesmata giving rise to a parallel symplasmic pathway or at least a reduction in the vacuole to vacuole resistance. For a short file of cells it would make little difference because the water must in any case enter and leave the file across a membrane. An effect would only be obvious in long files of cells, where entry and exit would be relatively less important, or in a system where transport is confined to the symplasm at some point. It is extremely unlikely that a reasonable estimate of symplasmic water transport can be made for a file of cells because, in addition to the uncertainties in membrane and cell wall permeabilities referred to above, there are problems posed by our sketchy knowledge about the structure and composition of plasmodesmata. The importance of this may be judged by the calculations of TYREE (1970) and the attempt of CLARKSON et al. (1971) to reconcile measurements of the flux of water across fully suberized regions of the endodermis of barley roots, where it is presumably confined to the symplasm, with measurements of the size and frequency of the plasmodesmata in the endodermal cell walls. Both sets of calculations, which are based on Poiseuille's law, are critically dependent on the value for the viscosity of the fluid within the plasmodesma and of the cross-sectional area actually available for flow. We do not in fact know that bulk flow can take place at all; it is possible that the content of the plasmodesma is a gel, the viscosity infinite and mass flow zero. Alternatively, mass flow might quickly be reduced by closure of the pore by organelles or endoplasmic reticulum on the high pressure side. It should also be pointed out that, although the plasmodesmata may greatly facilitate the flow of water between adjacent protoplasts, most of the volume flow in highly vacuolated cells will probably take place through the vacuole. This means that much of the water must cross the tonoplast and this will further reduce the effect of the symplasm on transport through a file of cells. TYREE (1970) appears to ignore the tonoplast and also assumes that measured permeability coefficients give the permeability of the plasmalemma to water. In fact they give the permeability of the whole plasmalemma-cytoplasm-tonoplast complex and, as yet, there are no values for the individual membranes.

Thus, although it is difficult to make a definitive statement about the movement of water through the symplasm, it seems probable that for movement over intermediate distances in most tissues it will not affect the conclusion that the cell walls form the major pathway (see 1.4.2.1).

6. Conclusions

It may reasonably be concluded that the principle of symplasmic transport is well established. However, the information required for a quantitative treatment of the system is largely lacking. More and better information concerning the structure of the plasmodesma is urgently required, particularly its relationship to the endoplasmic reticulum. CLARKSON et al. (1971) have made the interesting suggestion that the lumen of the endoplasmic reticulum may provide a pathway for transport in one direction while the space outside the endoplasmic reticulum could provide a pathway for transport in the opposite direction. In this way it might be easier to account for the opposing flows of salts and carbohydrate across the root.

Quantitative comparisons have been made between the electrical resistance of plasmodesmata and the resistance predicted from measurements of plasmodesmatal size and frequency (SPANSWICK and COSTERTON, 1967; SPANSWICK, 1972) and progress is now being made with measurements of intercellular chloride fluxes in Chara (TYREE et al., 1974). However, the problems involved in measuring cytoplasmic concentrations and specific activities may mean that similar measurements will not be possible for higher plants.

The relationship between cytoplasmic streaming and symplasmic transport is another neglected area. However, in tissues with small cells, TYREE (1970) has shown theoretically that transport is probably not rate-limited by cyclosis. This may account for repeated failures to show a dependence on streaming. For instance, Müller and Bräutigam (1973) showed that colchicine inhibited transport in *Vallisneria*. However, it did not inhibit streaming, though it is possible that it inhibited phloem transport. Conversely, Cande et al. (1973) showed that cytochalasin B inhibited cytoplasmic streaming in maize coleoptiles but had little effect on the rate of auxin movement. Circumstantial evidence for the minimal importance of streaming comes from the remark of Helder (1967a) that streaming is least active in the most active material. Indeed, it has long been known that wounding may initiate or accelerate streaming in *Vallisneria* or *Elodea* and that cyclosis is absent in many plants (Pfeffer, 1906, p. 358).

Finally, it should be pointed out that, while transport of nutrients is probably the most important function of the symplasm, it also plays an important role in the co-ordination of the activities of the plant. For instance, the existence of action potentials in plants is receiving renewed attention (Pickard, 1973) and transmission from cell to cell undoubtedly involves flow of current through the plasmodesmata (Spanswick and Costerton, 1967; Spanswick, 1974).

It may also be expected that the distribution of plasmodesmata will prove to be an important factor in morphogenesis, controlling to some extent the direction and strength of growth substance gradients. The relationship between the distribution of plasmodesmata and cell differentiation in the root apex has already received explicit consideration (Juniper and Barlow, 1969; Barlow, 1971).

It is also possible that electric fields may be as important in determining polarity in higher plants as they are in *Acetabularia* (Novak and Bentrup, 1972) and *Fucus* (Jaffe, 1968), in which case the symplasm would provide an internal pathway completing a circuit that would enable the field to act at the tissue level (Scott, 1967).

References

Allaway, W.G., Setterfield, G.: Ultrastructural observations on guard cells of *Vicia faba* and *Allium porrum*. Canad. J. Bot. **50**, 1405–1413 (1972).

Arisz, W.H.: Significance of the symplasm theory for transport across the root. Protoplasma **46**, 5–62 (1956).

Arisz, W.H.: Influence of inhibitors on the uptake and the transport of chloride ions in leaves of *Vallisneria spiralis*. Acta Botan. Neerl. **7**, 1–32 (1958).

Arisz, W.H.: Symplasmatischer Salztransport in *Vallisneria*-Blättern. Protoplasma **52**, 309–343 (1960).

Arisz, W.H.: Intercellular polar transport and the role of the plasmodesmata in coleoptiles and *Vallisneria* leaves. Acta Botan. Neerl. **18**, 14–38 (1969).

Arisz, W.H., Schreuder, M.J.: The path of salt transport in *Vallisneria* leaves. Proc. Koninkl. Ned. Akad. Wetenschap. Ser. C **59**, 454–460 (1956).

Arisz, W.H., Wiersema, E.P.: Symplasmatic long-distance transport in *Vallisneria* plants investigated by means of autoradiograms. Proc. Koninkl. Ned. Akad. Wetenschap., Ser. C **69**, 223–241 (1966).

Barlow, P.W.: Properties of the cells in the root apex. Riv. Fac. Agron. Univ. Nac. La Plata **47**, 275–301 (1971).

Bauer, L.: Zur Frage der Stoffbewegungen in den Pflanzen mit besonderer Berücksichtigung der Wanderung von Fluorochromen. Planta **42**, 367–451 (1953).

BIERBERG, W.: Die Bedeutung der Protoplasmarotation für den Stofftransport in den Pflanzen. Flora (Jena) **99**, 52–80 (1909).

BLACK, C.C.: Photosynthetic carbon fixation in relation to net CO_2 uptake. Ann. Rev. Plant Physiol. **24**, 253–286 (1973).

BÖHMER, H.: Untersuchungen über das Wachstum und den Feinbau der Zellwände in der *Avena*-Koleoptile. Planta **50**, 461–497 (1958).

BOSTROM, T.E., WALKER, N.A.: Intercellular transport in plants. I. The flux of chloride and the electric resistance in *Chara*. J. Exptl. Bot. **26**, 767–782 (1975).

BRIGGS, G.E.: Movement of water in plants. Oxford: Blackwell 1967.

BRIGGS, G.E., ROBERTSON, R.N.: Apparent free space. Ann. Rev. Plant Physiol. **8**, 11–30 (1957).

BRINCKMANN, E.: Zur Messung des Membranpotentials und dessen lichtabhängigen Änderungen an Blattzellen höherer Landpflanzen. Dissertation Darmstadt (1973).

BURGESS, J.: Observations on the structure and differentiation of plasmodesmata. Protoplasma **73**, 83–95 (1971).

CANDE, W.Z., GOLDSMITH, M.H.M., RAY, P.M.: Polar auxin transport and auxin-induced elongation in the absence of cytoplasmic streaming. Planta **111**, 279–296 (1973).

CLARKSON, D.T., ROBARDS, A.W., SANDERSON, J.: The tertiary endodermis of barley roots: fine structure in relation to radial transport of ions and water. Planta **96**, 292–305 (1971).

CLOWES, F.A.L., JUNIPER, B.E.: Plant cells. Oxford: Blackwell 1968.

EMMERT, E.W.: Observations on the absorption spectra of fluorescein, fluorescein derivatives and conjugates. Arch. Biochem. **73**, 1–8 (1958).

ESAU, K., CRONSHAW, J., HOEFERT, L.L.: Relation of beet yellows virus to the phloem and to movement in the sieve tube. J. Cell Biol. **32**, 71–87 (1967).

FLOYD, G.L., STEWART, K.D., MATTOX, K.R.: Cytokinesis and plasmodesmata in *Ulothrix*. J. Phycol. **7**, 306–309 (1971).

GOLDSMITH, M.H.M., FERNÁNDEZ, H.R., GOLDSMITH, T.H.: Electrical properties of parenchymal cell membranes in the oat coleoptile. Planta **102**, 302–323 (1972).

GOLDSMITH, M.H.M., RAY, P.M.: Intracellular localization of the active process in polar transport of auxin. Planta **111**, 297–314 (1973).

HABERLANDT, G.: Physiologische Pflanzenanatomie. Leipzig: Engelmann 1904.

HELDER, R.J.: Translocation in *Vallisneria spiralis*. Handbuch Pflanzenphysiol. **13**, 20–43 (1967a).

HELDER, R.J.: Transport of substances through the tentacles of *Drosera capensis*. Handbuch Pflanzenphysiol. **13**, 44–54 (1967b).

HELDER, R.J., BOERMA, J.: An electron-microscopical study of the plasmodesmata in the roots of young barley seedlings. Acta Botan. Neerl. **18**, 99–107 (1969).

JAFFE, L.: Localization in the developing *Fucus* egg and the general role of localizing currents. Advan. Morphogenesis **7**, 295–328 (1968).

JUNIPER, B.E., BARLOW, P.W.: The distribution of plasmodesmata in the root tip of maize. Planta **89**, 352–360 (1960).

KITIJAMA, E.W., LAURITIS, J.A.: Plant virions in plasmodesmata. Virology **37**, 681–684 (1969).

KRULL, R.: Untersuchungen über den Bau und die Entwicklung der Plasmodesmen im Rindenparenchym von *Viscum album*. Planta **55**, 598–629 (1960).

KUFFLER, S.W., POTTER, D.D.: Glia in the leech central nervous system: physiological properties and neuron-glia relationship. J. Neurophysiol. **27**, 290–320 (1964).

LÄUCHLI, A.: Translocation of inorganic solutes. Ann. Rev. Plant Physiol. **23**, 197–218 (1972).

LITTLEFIELD, L., FORSBERG, C.: Absorption and translocation of phosphorous-32 by *Chara globularis* Thuill. Physiol. Plantarum **18**, 291–296 (1965).

LOEWENSTEIN, W.R., KANNO, Y.: Studies on an epithelial (gland) cell junction. I. Modifications of surface membrane permeability. J. Cell Biol. **22**, 565–586 (1964).

LOU, C.H.: Protoplasmic continuity in plants. Acta Botan. Sinica **4**, 183–222 (1955).

LÜTTGE, U.: Co-operation of organs in intact higher plants: a review. In: Membrane transport in plants (U. ZIMMERMANN, J. DAINTY, eds.), p. 353–362. Berlin-Heidelberg-New York: Springer 1974.

LÜTTGE, U., PALLAGHY, C.K.: Light-triggered transient changes of membrane potentials in green cells in relation to photosynthetic electron transport. Z. Pflanzenphysiol. **61**, 58–67 (1969).

MEEUSE, A.D.J.: Plasmodesmata (Vegetable Kingdom). Protoplasmatologia **2**, Pt. A1c, 1–43 (1957).

MÜLLER, E., BRÄUTIGAM, E.: Symplasmic translocation of α-aminoisobutyric acid in *Vallisneria* leaves and the action of kinetin and colchicine. In: Ion transport in plants (W.P. ANDERSON, ed.). London-New York: Academic Press 1973.

MÜNCH, E.: Die Stoffbewegungen in der Pflanze. Jena: Fischer 1930.

NOVAK, B., BENTRUP, F.W.: An electrophysiological study of regeneration in *Acetabularia mediterranea*. Planta **108**, 227–244 (1972).

OSMOND, C.B.: Metabolite transport in C_4 photosynthesis. Australian J. Biol. Sci. **24**, 159–163 (1971).

OSMOND, C.B., LÜTTGE, U., WEST, K.R., PALLAGHY, C.K., SCHACHAR-HILL, B.: Ion absorption in *Atriplex* leaf tissue. II. Secretion of ions to epidermal bladders. Australian J. Biol. Sci. **22**, 797–814 (1969).

PALLAS, J.E., MOLLENHAUER, H.H.: Physiological implications of *Vicia faba* and *Nicotiana tabacum* guard-cell ultrastructure. Amer. J. Bot. **59**, 504–514 (1972).

PATE, J.S., GUNNING, B.E.S.: Transfer cells. Ann. Rev. Plant Physiol. **23**, 173–196 (1972).

PFEFFER, W.: The physiology of plants, vol. III, 2nd ed., transl. by EWART, A.J. London: Oxford 1906.

PICKARD, B.G.: Action potentials in higher plants. Botan. Rev. **39**, 172–201 (1973).

ROBARDS, A.W.: On the ultrastructure of differentiating secondary xylem in willow. Protoplasma **65**, 449–464 (1968).

ROBARDS, A.W.: The ultrastructure of plasmodesmata. Protoplasma **72**, 315–323 (1971).

ROBARDS, A.W., JACKSON, S.M., CLARKSON, D.T., SANDERSON, J.: The structure of barley roots in relation to the transport of ions into the stele. Protoplasma **77**, 291–311 (1973).

ROBARDS, A.W., ROBB, M.E.: Uptake and binding of uranyl ions by barley roots. Science **178**, 980–982 (1972).

SCHUMACHER, W.: Der Transport von Fluorescein in Haarzellen. Handbuch Pflanzenphysiol. **13**, 17–19 (1967).

SCOTT, B.I.H.: Electric fields in plants. Ann. Rev. Plant Physiol. **18**, 409–418 (1967).

SIBAOKA, T.: Action potentials in plants. Symp. Soc. Exptl. Biol. **20**, 49–73 (1966).

SKIERCZYNSKA, J.: Some of the electrical characteristics of the cell membrane of *Chara australis*. J. Exptl. Bot. **19**, 389–406 (1968).

SMITH, F.A.: Transport of solutes during C_4 photosynthesis: assessment. In: Photosynthesis and photorespiration (M.D. HATCH, C.B. OSMOND, R.O. SLATYER, eds.), p. 302–306. New York-London-Sydney-Toronto: Wiley-Interscience 1971.

SPANSWICK, R.M.: Electrical coupling between cells of higher plants: a direct demonstration of intercellular communication. Planta **102**, 215–227 (1972).

SPANSWICK, R.M.: Symplasmic transport in plants. Symp. Soc. Exptl. Biol. **28**, 125–135 (1974).

SPANSWICK, R.M., COSTERTON, J.W.F.: Plasmodesmata in *Nitella translucens:* structure and electrical resistance. J. Cell Sci. **2**, 451–464 (1967).

SPANSWICK, R.M., WILLIAMS, E.J.: Electrical potentials and Na^+, K^+, and Cl^- concentrations in the vacuole and cytoplasm of *Nitella translucens*. J. Exptl. Bot. **15**, 193–200 (1964).

SPANSWICK, R.M., WILLIAMS, E.J.: Ca fluxes and membrane potentials in *Nitella translucens*. J. Exptl. Bot. **16**, 463–473 (1965).

SPITZER, N.C.: Low-resistance connections between cells in the developing anther of the lily. J. Cell Biol. **45**, 565–575 (1970).

STRUGGER, S.: Der elektronenmikroskopische Nachweis von Plasmodesmen mit Hilfe der Uranylimprägnierung an Wurzelmeristem. Protoplasma **48**, 231–236 (1957a).

STRUGGER, S.: Elektronenmikroskopische Beobachtungen an den Plasmodesmen des Urmeristems der Wurzelspitze von *Allium cepa;* ein Beitrag zur Kritik der Fixation und zur Beurteilung elektronenmikroskopischer Größenangaben. Protoplasma **48**, 365–367 (1957b).

TAIZ, L., JONES, R.L.: Plasmodesmata and an associated cell wall component in barley aleurone tissue. Amer. J. Bot. **60**, 67–75 (1973).

TANGL, E.: Über offene Kommunikationen zwischen den Zellen des Endosperms einiger Samen. Jahrb. Wiss. Botan. **12**, 170–190 (1879).

THOMSON, W.W., JOURNETT, R.DE: Studies on the ultrastructure of guard cells of *Opuntia*. Amer. J. Bot. **57**, 309–316 (1970).

THOMSON, W.W., LIU, L.L.: Ultrastructural features of the salt gland of *Tamarix aphylla* L. Planta **73**, 201–220 (1967).

TYREE, M.T.: The thermodynamics of short-distance transport. J. Exptl. Bot. **20**, 341–349 (1969).

TYREE, M.T.: The symplast concept. A general theory of symplastic transport according to the thermodynamics of irreversible processes. J. Theoret. Biol. **26**, 181–214 (1970).

TYREE, M.T., FISCHER, R.A., DAINTY, J.: A quantitative investigation of symplasmic transport in *Chara corallina*. II. The symplasmic transport of chloride. Canad. J. Bot. **52**, 1325–1334 (1974)

WALKER, N.A., BOSTROM, T.E.: Intercellular movement of chloride in *Chara*—a test of models for chloride influx. In: Ion transport in plants (W.P. ANDERSON, ed.). London-New York: Academic Press 1973.

WEATHERLEY, P.E.: Some aspects of water relations. Advan. Botan. Res. **3**, 171–206 (1970).

WEILING, F.: Zur Feinstruktur der Plasmodesmen und Plasmakanäle bei Pollenmutterzellen. Planta **64**, 97–118 (1965).

WILLIAMS, S.E., PICKARD, B.G.: Receptor potentials and action potentials in *Drosera* tentacles. Planta **103**, 193–221 (1972).

WITHERS, L.A., COCKING, E.C.: Fine-structural studies on spontaneous and induced fusion of higher plant protoplasts. J. Cell Sci. **11**, 59–75 (1972).

ZIEGLER, H.: Bericht über die Arbeiten zum Wuchsstofftransport im Parenchym. Fortschr. Botan. **35**, 67–75 (1973).

ZIEGLER, H., RUCK, I.: Untersuchungen über die Feinstruktur des Phloems. III. Die „Trompetenzellen" von *Laminaria*-Arten. Planta **73**, 62–73 (1967).

II. Particular Tissue Systems

3. Transport Processes in Roots

3.1 General Introduction

M.G. Pitman, W.P. Anderson, and U. Lüttge

1. Introduction

In relation to ions, plant roots have two main functions. At the surface of the root ions are absorbed from the soil or solution; from within the root, ions are transported to other parts of the plant. Active processes maintain concentrations of ions in the vacuole and cytoplasm of cortical cells as well as participating in the general movement across the root to the stele.

Absorption from the external environment, and supply of ions to the shoot are interdependent and often it is not possible to separate them experimentally from accumulation in the cells of the cortex. Changes in tracer and in ion content of the cortical cells may be affected by both processes as well as by fluxes into the vacuole. The complexity of interactions was illustrated in Part A, Fig. 5.17 which showed how fluxes into the root, into the vacuoles of cortical cells and into the xylem may be related. This system will be used for discussion of uptake in the following Chapters. During passage through the root, certain ions may be metabolized to other compounds (e.g. NO_3^- to amino acids; Part A, 13.2.3).

Plant roots have been used to study the mechanism of transport into root cells using "kinetic analysis" of rates of uptake (Chap. 3.2) and also (Chap. 3.3) following approaches similar to the studies with algal cells already discussed. The function of the root as an organ supplying ions to the shoot is discussed in Chap. 3.4. In this introductory Chap. 3.1 some general features of roots are discussed.

2. General Features of Roots

2.1 Organization of the Root in Relation to Solute Uptake

The structure of the root is covered in standard reference books of plant anatomy (e.g. ESAU, 1965) but some aspects need to be reviewed here briefly as the organization of the root is intimately related to the absorption of ions from the soil and the physiological function of transport from root to shoot. Different cell types perform different transport functions in such a way that the integrated working of all the cells in the root constitutes the overall capability of the organ.

Mobilization of solutes from the soil and uptake by the roots is related to properties of the root surface and those parts of the cortex readily accessible to the outside solution through the free space. The *root surface* varies greatly with changing developmental stages along the length of the root. Starting with the root tip, the decomposing and budding-off of root cap cells provides a slime cylinder, in which the soft root can migrate even in hard soils. This slime also provides a mucigel across which the root can establish intimate contact with soil particles and crystals (see also *1.4.2.2.1*). Adsorption exchange between soil and mucigel may play a large role in ion uptake (Fig. *3.1*). Bacteria growing in this mucigel may be important in root soil interactions (NISSEN, 1973; cf. *9.3.3.5*).

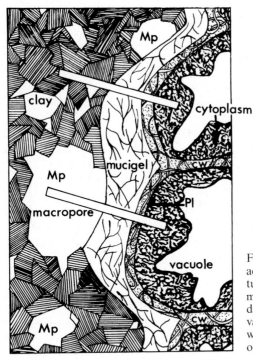

Fig. *3.*1. Boundary region of root and soil according to electron micrographs. Major features are: clay gel with macro-pores (*Mp*), mucigel, cell wall (*CW*), plasma membrane (*Pl,* dark, thin), cytoplasm with mitochondria, vacuole. The two white bars indicate ion pathways. (From JENNY, 1966, Fig. 1; the pathways of ion diffusion are discussed in his paper)

In more basal parts of the root special epidermal cells grow out to form root hairs, whose cell walls largely consist of pectic acid through which again close adsorption exchange with soil particles is possible. (Interestingly in this context, roots often develop fewer root hairs in water culture than in the soil or in a humid atmosphere.) Further from the apex the root-hair walls and all epidermal cell walls, plus part of the anticlinal walls of the outermost cortex cell layer will be suberized and cutinized, so that uptake of solutes from the medium may be impaired (see Chap. *1*). In young hydroponically grown roots there is essentially no cuticle on the epidermis. Even in young roots, though, the surface may act as a barrier to ion diffusion, especially for divalent cations (LEGGETT and GILBERT, 1969). BANGE (1973) has suggested, too, that uptake of univalent ions is limited by transport across the epidermis; such an effect will be most evident at low external concentrations (*3.4.2.3*).

Special structures like formation of an exodermis from outer cortical cells or of a *velamen radicum* in aerial roots promoting water (and ion?) uptake in tropical rain forest epiphytes may further complicate the situation at some distance from the root tip.

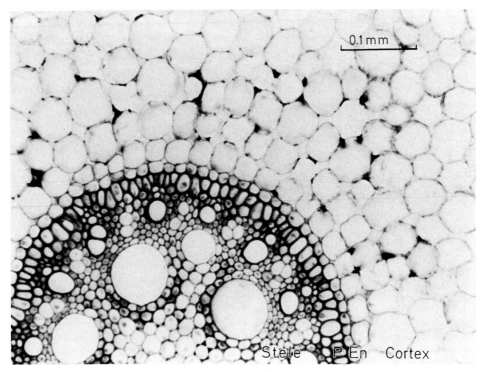

Stele   P En   Cortex

Fig. *3*.2. Structure of plant root as represented by a transverse section of a *Zea mays* primary root. *En* endodermis; *P* pericycle, the bar represents 0.1 mm. (From a slide prepared by Dr. M. McCully)

Young roots of most species, and certainly of those species most commonly used in transport studies, contain two morphologically distinct zones, the *cortex* and the *stele* (see Fig. *3*.2). At the outer surface, the cortex is bounded by a sheath of epidermal cells which may develop root hairs depending upon culture conditions. At the inner surface, the cortex is delimited from the stele by a single cell layer, the endodermis. These features can be distinguished at an early stage in root onto-geny, usually within a millimetre or so from the meristem, but the exact distance depends on species and on rate of root growth. Also at this stage the Casparian band forms in the endodermis, a suberized strip in the endodermal cell wall which completely surrounds the cell and is in contact from one cell to the next so as to appear to form a continuous barrier in the apoplasm, separating the cortex from the stele (see *1*.4.2.2.1). The function of the Casparian band has been much debated (see *3*.4.2.4), but it is pertinent to point out that similar structures are found surrounding other secretory organs, e.g. salt glands (ZIEGLER and LÜTTGE, 1966; see also Fig. *5*.2) and the nitrogen fixing root nodules of the legumes (GUNNING et al., 1974; see also Figs. *6*.7 and *6*.8). This coincidence may be taken as *prima facie* evidence that the Casparian bands in roots and the suberized bands in other secretory tissues are impermeable barriers which isolate one region of extracellular space from another, the role which has been classically assumed (CRAFTS and BROYER, 1938; VAN FLEET, 1961). In roots there is also evidence (HELDER and BOERMA,

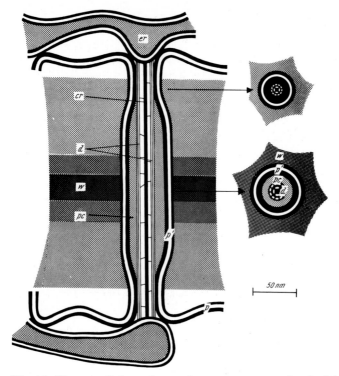

Fig. *3.3.* Diagram of plasmodesmatal structure. *cr* central rod; *d* desmotubule; *er* endoplasmic reticulum; *p* plasmalemma; *p′* plasmalemma through plasmodesmatal canal; *pc* plasmodesmatal canal; *w* cell wall. Reproduced from ROBARDS (1971)

1969) that both the outer and inner tangential walls of the endodermis are penetrated by plasmodesmata so that the symplasm is continuous from cortex to stele.

The epidermal and cortical cells of the root are interconnected by cytoplasmic bridges, plasmodesmata, to form the cortical symplasm. Perhaps the most complete study of plasmodesmatal structure is ROBARDS'(1971) from which Fig. *3.3* has been produced. It is probably fair to say that there is general agreement that plasmodesmatal structure in a large variety of species is similar to that shown here, although there may be dissent as to whether the tubular inclusion (the desmotubule) is or is not simply a strand of endoplasmic reticulum. All are agreed that there is continuity of both plasmalemma and cytoplasm from one cell to the next through the plasmodesmata (see *2.2*).

The significance of plasmodesmata for transport through the root will be discussed later. In this connexion, the frequency of plasmodesmata is obviously of equal importance as the pore size of a single plasmodesma, and TYREE (1970) has performed a useful service in searching the literature and gathering together the relevant data. His Table has been reproduced as Table *2.1*, and should be studied in this respect. The fine study of plasmodesmata in the endodermis of barley by CLARKSON et al. (1971) will be referred to in detail later (*3.4.2.4*).

In contrast to the cortex which contains only a single cell type, the stele has several distinct cell types within it. There are the pericycle cells, xylem and phloem parenchyma, xylem elements, phloem elements and in many species a central core of pith. In the present context of correlating structure and function it is the differentiation of the xylem which must be focused upon. The opinion of most anatomists is that the early metaxylem, through which the exudation stream can be seen to flow in roots excised from young seedlings by simply looking at the cut end under a microscope, is fully mature within a few millimetres of the meristem. The implications of full maturity in the metaxylem are that the vessels have completed secondary wall deposition, have perforated end walls and no longer contain viable cytoplasm; in other words the present consensus is that xylem translocation is conducted along open, dead vessels which play no part in that translocation other than to provide an open conduit of relatively low hydraulic resistance. However, there does seem to be marked variation in xylem development between species as discussed in *3.4.2.5*.

The parenchyma cells of the stele are usually well vacuolated, containing similar concentrations of K^+ as the cortical cells (Fig. *3.4*). The cytoplasm contains the normal complement of mitochondria and often the endoplasmic reticulum is well developed particularly adjacent to pits in the secondary wall (LÄUCHLI et al., 1974 a). The xylem parenchyma may also show infolding of the cell wall associated with "transfer cells" (PATE and GUNNING, 1972). Both development of ER and cell-wall proliferation have been suggested to be involved in secretion of ions from the stele to the xylem (LÄUCHLI et al., 1974a, b).

In general the parenchyma cells of the stele are inter-connected by plasmodesmata in much the same manner as are the cortical cells; the cortical and stelar symplasms are continuous through the endodermis (see Chap. *1*).

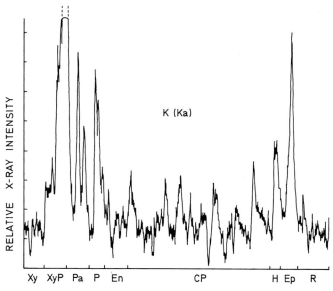

Fig. *3.4*. Line scan of K (*K*α) radiation across a corn root section. Electron probe X-ray analyzer operated at 15 KeV accelerating voltage, 0.5 μm beam diameter, and 0.023 μA sample current on Be. *R* epoxy resin outside of tissue; *Ep* epidermis; *H* hypodermis; *CP* cortical parenchyma; *En* endodermis; *P* pericycle; *Pa* unspecified stelar parenchymatous tissue; *XyP* xylem parenchyma; *Xy* xylem vessel. (From LÄUCHLI et al., 1971)

Table 3.1. Percentage of roots showing endomycorrhiza in pasture plants. (Data from J.R. Crush, unpublished)

Trifolium ambiguum	96%	*Lolium perenne*	15–79%
T. hybridum	85–96%	*Dactylis glomerata*	72–83%
T. pratense	71–96%	*Phleum pratense*	46%
T. repens	74–94%	*Festuca arundinacea*	82–95%

The root systems of plants in the soil are commonly found to be infected with endomycorrhizal fungi (Table 3.1). The hyphae of the mycorrhiza may extend several cm from the root surface into the soil. Within the cortex the hyphae spread between and into the cortical cells right up to the endodermis and have arbuscular tufts of haustoria and vesicular storage organs in the cells of the plant. The fungal system increases the surface area in contact with the soil and can act as a pathway for transfer of nutrients from the soil to the root.

2.2 Roots in Culture Solution and Soil

For many physiological experiments it is convenient to use roots grown in culture solution, since the nutrient content of the solution may be controlled and adsorption of ions on soil particles is not a problem. Roots may be transferred easily from one solution to another without damage. Use of culture solutions is discussed in more detail by Epstein (1972), and representative concentrations are given in 3.1.4.

However, at some stage, most physiologists are interested in how their experimental plants and solutions compare with the "real" world of the soil.

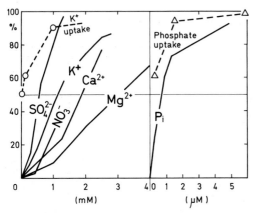

Fig. 3.5. Reisenauer (1966) measured the content of nutrients in solution in a number of soils. His data are plotted here as cumulative frequency against concentration, so that (for example) 44% of soils had less than 1 mM K$^+$ (and 56% therefore had more than 1 mM); (18–44)% of soils are between 0.5 and 1.0 mM K. Intersections with the 50% line give the modal concentrations (i.e. 0.7 μM phosphate and 1.1 mM K$^+$). The symbols and dotted lines (○ = K; △ = phosphate; P$_i$) are rates of uptake taken from Fig. 3.23 plotted on the same concentration scale to show the efficiency of the process of uptake compared with the modal concentration in the soil solution. In each case 100% was taken as the maximum value in Fig. 3.23. (Reisenauer's data taken from Epstein, 1972)

Reference to standard books on soil analysis shows that determination of concentrations in soil solution around the roots is not unambiguous. General figures are about 0.5 to 5 mM for NO_3^-, 0.1 to 1 mM for Na^+ and 0.001 to 0.07 mM for phosphate. Further data is given in Fig. 3.5, but see also EPSTEIN, 1972. The resistance to diffusion of ions to the root due to unstirred layers can reduce uptake of ions and oxygen at very low concentrations. POLLE and JENNY (1971) found that stirring only affected uptake from solutions below about 10 µM and clearly, the rate of adsorption also affects the limitation due to low external concentrations. Interpretation of diffusion resistance to roots in soils is complicated by adsorption of ions on soil particles and the capacity of the soil to maintain supply of nutrients

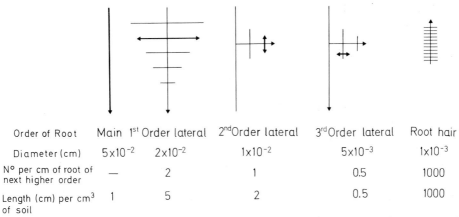

Order of Root	Main	1st Order lateral	2nd Order lateral	3rd Order lateral	Root hair
Diameter (cm)	5×10^{-2}	2×10^{-2}	1×10^{-2}	5×10^{-3}	1×10^{-3}
N° per cm of root of next higher order	—	2	1	0.5	1000
Length (cm) per cm³ of soil	1	5	2	0.5	1000

Fig. 3.6. The constituent parts of the root system. The geometrical data pertain to roots of cereals in topsoils. The arrows show order of root in relation to the root system. (From BARLEY, 1970, Fig. 1)

Table 3.2. Comparative data for a few common experimental plants grown under various conditions

	$mg_{DW}\,g_{FW}^{-1}$	$m\,g_{FW}^{-1}$	$m\,g_{DW}^{-1}$	$m^2_{surface}\,g_{FW}^{-1}$	$m^2_{cells}\,g_{FW}^{-1}$
Hordeum vulgare					
(5 days old,[a] in solution)	70	3	43	0.008	0.1
(soil)[b] (0–15 cm)	—	—	194	—	—
Avena sativa					
soil (0–15 cm)[b]	—	—	174	—	—
Triticum sativum					
soil (0–15 cm)[b]	—	—	148	—	—
Zea mays					
solution[c]	75	0.9	12	—	—
dark grown[d], vermiculite					
4 days	66	—	—	—	—
6 days	58	—	—	—	—
Pisum sativum					
7 days old, solution[e]	65	1.5	240	0.004	—
(13.5 cm of radicle)					

[a] PITMAN, M.G., unpublished. [b] WELLBANK, P., unpublished. [c] BAKER, D., unpublished. [d] BARLEY, K.P., unpublished. [e] GRASMANIS and BARLEY, 1969.

to the root. For example, phosphate extracted from certain soils may be restricted to that within the volume of soil occupied by the root and its hairs (or mycorrhiza), due to adsorption of phosphate on soil particles.

BARLEY (1970) has discussed extensively the configuration of the root system in the soil in relation to nutrient uptake, and gives examples of l_v, the length of the root per volume of soil (Fig. *3.6*). For herbaceous plants l_v is about 10–50 cm^{-2} in the top 25 cm of soil and decreases with depth to about 0.1 to 1.0 cm^{-2}; under woody plants l_v in the topsoil is much smaller (about 2–4 cm^{-2}) than under crops or pasture. The length of root under 1 cm^2 of the soil surface, l_A, is perhaps simpler to relate to whole plants and BARLEY suggested values ranged from 50 to 500 cm^{-1} for crops but quoted NEWMAN (1969) as finding 3,000 cm^{-1} for perennial grasses.

Relating root length to physiological systems based on root fresh weight has many problems such as variation in root diameter and in uptake capacity along the root. Some examples are given in Table *3.2*.

2.3 Comparison of Excised Roots and Roots of Intact Plants

Most workers with excised roots have at some time tested the extent to which their experimental system behaved like roots of an intact plant. The answer is not simple and depends to some extent on what is being measured and on the plants used. For each system there is the need to test the extent to which excision affects ion transport; even if results are clearly different, then reasons for the difference may be valuable in understanding the general processes of transport in the root.

In short-term experiments (up to 1 h) rates of uptake to barley roots of varied salt status were found to be the same in excised and intact roots (JOHANSEN et al., 1970). Excision was also found to have a negligible effect on rate of transport of ions out of the stele (but see *3.4.2.1*). In barley seedlings K$^+$ transport was unaffected for up to 3 h (PITMAN, 1971) and in maize up to 24 h (ANDERSON and HOUSE, 1970). However the response of the plant depends very much on the conditions of growth. In barley there can be large reductions in uptake and transport from the stele within a few hours from cutting, particularly when plants are grown in the light (PITMAN et al., 1974), and in cell potential (ψ_{vo}) over 4–6 h following excision (PITMAN et al., 1971). HAY and ANDERSON (1972) reported *development* of ion transport in onion roots following excision.

Over longer periods, clearly there will be reduction in transport as reserves and respiration decrease and eventually the excised roots will die. For many purposes, though, excised roots are more convenient than intact plants, especially when the emphasis is on transport at cortical cell membranes.

2.4 Variability within the Root

Measurements have been made of distribution of ions in the root using electron probe analysis (Fig. *3.4*) or K$^+$-sensitive microelectrodes (DUNLOP and BOWLING, 1971). It is clear that cells across the cortex of roots grown on K$^+$-nutrient solutions usually have high and often uniform concentrations of K$^+$. However, some experiments show that cells at the outside and inside of the cortex may have different ability to take up ions (CROSSETT, 1967; PALLAGHY and SCOTT, 1969).

The stele and cortex may differ in their ability to discriminate between K^+ and Na^+. In barley roots, the ratio of K^+ to Na^+ in the cortex was 2.1 compared with 4.1 in the stele (PITMAN, 1965). However, in maize and in bean plants RICHTER and MARSCHNER (1974) found selectivity for K^+ greater in the cortex than in the stele as the following ratios show:

		K^+/Na^+	Ca^{2+}/K^+
Maize	cortex	35–45	0.8–1.6
	stele	18–24	0.4–0.5
Bean	cortex	7.0	1.2
	stele	2.7	1.0

Since growth takes place from the tip of the root, development of cells occurs along the length of the root and at later stages there may be production of laterals. Hence it is not surprising to find variation in capacity for ion uptake along the length of some roots (BOWEN and ROVIRA, 1967; ESHEL and WAISEL, 1973) (but see also tabulation below, 3.3.4.1). Differences between cells seem less extreme in short lengths (5 cm) of roots of young seedlings, though this possible variability is a factor that may need to be taken into account in interpreting results. As roots age, there may also be different proportions of young laterals to older main roots, which may also introduce variation.

2.5 Low-Salt Roots and Salt Saturation

Early investigators of ion transport in plants were restricted by techniques available to measurements of net changes in content. Such changes are negligible in roots growing in culture solutions. It was found convenient, for example, to use tissues of storage organs which develop net uptake on slicing (Part A, Chap. 8). A major discovery made by HOAGLAND and colleagues was that roots of barley seedlings grown on dilute $CaSO_4$ solution contained a low content of K^+, but when transferred to solutions of univalent salts accumulated the ions in amounts large enough to permit ready measurement. Such roots have been called "low-salt" roots. Using this material, HOAGLAND made considerable progress in the study of ion transport in plant roots. More recently low-salt roots have been used extensively for kinetic studies (Chap. 3.2).

With the advent of radioactive isotopes it has become possible to measure isotopic fluxes and to study transport processes in cells at steady state (zero net flux), as shown in Part A, Chap. 6 for giant algal cells. Roots grown on culture solution also come to a steady state in which vacuolar concentration is steady, i.e. the cells are "salt-saturated". (The term "high-salt" is often used but can be misleading as the level is not pathologically "high" but rather normal for the root.) Unlike storage tissues and algal cells, these roots maintain transport *across* the root so that the net flux across the outer cell membrane of the cells need not be zero though $dQ_v/dt=0$ (see Part A, 5.5.3).

Both low-salt roots and salt-saturated roots have been used to study the processes of transport at the *cellular* level, and salt-saturated roots have been used to study transport across the root, as well as uptake to the cells of the cortex. Results from both kinds of root preparation should be mutually compatible, but there are some differences that may affect interpretation.

Table *3.3* gives the content of barley roots grown on $CaSO_4$ and on an aerated solution containing $(2.5 \text{ mM } K^+ + 7.5 \text{ mM } Na^+)$ in addition to other nutrients. Low-salt roots show little preference for K^+ over Na^+ in the range of Mechanism II (see *3.2*.3.3.), but in contrast fluxes into salt-saturated barley roots show strong preference for K^+ (PITMAN et al., 1968). This difference in selectivity is correlated with H^+ efflux from the roots. Efflux of H^+ from low-salt roots in a 10 mM KCl solution is about 4 μmol g_{FW}^{-1} h^{-1} but from salt-saturated roots is almost zero.

In practice, roots may be intermediate between these extremes (e.g., a "pretreatment of 6 h" would bring low-salt roots near, but not to, salt saturation). The activity of the roots will also be affected by the temperature, type of nutrient solution, degree of aeration and no doubt by other factors too.

3. Conclusion

The particular point we wish to make here is that results of experiments with roots are very rarely generalizations applicable to *all* roots or in *all* conditions. Often the results apply for a limited range of environmental conditions. However, if the experimenter is aware of this limitation he may find useful correlations between growth conditions and ion uptake that help understand how roots adapt to environmental conditions or how uptake is affected by cell development.

4. Appendix: Culture Solutions

Most physiologists are faced with the problems of growing plants in culture solution at some stage. Each species may have particular requirements for best growth but

Table *3.3*. Comparison of roots of barley grown on aerated 0.5 mM $CaSO_4$ solution (low-salt) and of roots of plants grown on aerated full nutrient solution containing (2.5 K + 7.5 Na). (Data from PITMAN et al., 1968, PITMAN, 1969)

	Amounts in roots (μmol g_{FW}^{-1})	
	Low-salt roots ($CaSO_4$)	Salt-saturated roots (Nutrient solution)
K^+	18	90
Na^+	2	28
Cl^-	2	2
Ca^{2+}	7	2
Mg^{2+}	5	8
Glucose + fructose	70	20
Sucrose	nil	< 2

a good starting point is the standard solutions which are discussed in detail by
HEWITT (1963). Content of solutions can be considered in terms of major nutrients;
trace nutrients; pH.

Major Nutrients

ARNON and HOAGLAND (1940) recommended the following solution containing:

	g liter^{-1}
KNO_3	1.011
$Ca(NO_3)_2$	0.492
$MgSO_4 \cdot 7H_2O$	0.490
$NH_4H_2PO_4$	0.230

giving concentrations of: $K^+ = 10$; $Ca^{2+} = 3$; $Mg^{2+} = 2$; $NH_4^+ = 2$; $NO_3^- = 16$; SO_4^{2-}
$= 3$ and $H_2PO_4^- = 2$, all mM. This solution is relatively highly concentrated, and
there is the problem that phosphate may be toxic, or else it may be precipitated
with Ca^{2+}. Commonly, the solution is diluted two or more times. HEWITT (1963)
gives a "Long Ashton" formula in which $K^+ = 5$, $Ca^{2+} = 5$, $Mg^{2+} = 3$; $NO_3^- = 15$;
$SO_4^{2-} = 3$; and $H_2PO_4^- = 1.33$ mM. GREENWAY (1962) suggested a solution for barley
in which $NH_4H_2PO_4$ was 0.8 mM, but otherwise major ions were as in "HOAG-
LAND's" solution.

Trace Nutrients and Iron

ARNON and HOAGLAND (1940) recommended adding to the major nutrients:

	mg liter^{-1}
H_3BO_3	2.86
$MnCl_2 \cdot 4H_2O$	1.81
$CuSO_4 \cdot 5H_2O$	0.08
$ZnSO_4 \cdot 7H_2O$	0.22
$H_2MoO_4 \cdot H_2O$	0.09

It is convenient to make these nutrients into a stock solution that can be diluted
1:2,000, when making up the culture solution.
   The "Long Ashton" solution (HEWITT, 1963) contained

	mg liter^{-1}
$MnSO_4$	2.23
$CuSO_4 \cdot 5H_2O$	0.24
$ZnSO_4 \cdot 7H_2O$	0.296
H_3BO_3	1.86
$(NH_4)_6Mo_7O_{24} \cdot 4H_2O$	0.035
$CoSO_4 \cdot 7H_2O$	0.028
$NaCl$	5.85

Note the inclusion of Co^{2+} and the higher level of Cl^-.

ARNON and HOAGLAND (1940) recommended adding 0.6 ml 3 times a week of 0.5% $FeSO_4 \cdot 7H_2O$ plus 0.4% tartaric acid. A convenient alternative is to add Fe as a chelate of EDTA, diluting the following solution 1:1,000 in the culture media to give a final Fe concentration of 90 μM.

Dissolve 26.1 g EDTA (the acid, not a salt) in 260 cm^3 1 M NaOH. Add 24.9 g $FeSO_4 \cdot 7H_2O$ and dilute to 1 liter. Aerate for 12 h (overnight) and adjust the pH to 5.5 with NaOH or H_2SO_4.

Alternatively, Ferric citrate can be added at 24.5 mg $liter^{-1}$, again to give a final Fe concentration of about 100 μM.

pH

Culture solutions are generally adjusted to about pH 6.5–6.0, but again plant species differ in their tolerance. At high pH there may be symptons of Fe deficiency, especially in cereals. At low pH there may be enhanced uptake of trace elements to toxic levels, but this is more often a problem in soils where the pH determines the content of the soil solution as well as rates of uptake. If odd symptoms appear it is worth checking pH around the plants as a first test.

References

ANDERSON, W.P., HOUSE, C.R.: A correlation between structure and function in the root of *Zea mays.* J. Exptl. Bot. **18**, 544–555 (1967).

ARNON, D.I., HOAGLAND, D.R.: Crop production in artificial solutions and in soils with special reference to factors affecting yields and absorption of inorganic nutrients. Soil Sci. **50**, 463–484 (1940).

BANGE, G.G.J.: Diffusion and absorption of ions in plant tissue 111. The role of the root cortex cells in ion absorption. Acta Botan. Neerl. **22**, 529–542 (1973).

BARLEY, K.P.: The configuration of the root system in relation to nutrient uptake. Advan. Agron. **22**, 159–207 (1970).

BOWEN, G.D., ROVIRA, A.D.: Phosphate uptake along attached and excised wheat roots measured by an automatic scanning method. Australian J. Biol. Sci. **20**, 369–378 (1967).

CLARKSON, D.T., ROBARDS, A.W., SANDERSON, J.: The tertiary endodermis in barley roots; fine structure in relation to transport of ions and water. Planta **96**, 292–305 (1971).

CRAFTS, A.S., BROYER, T.C.: Migration of salts and water into xylem of the roots of higher plants. Amer. J. Bot. **24**, 415–431 (1938).

CROSSETT, R.N.: Autoradiography of ^{32}P in maize roots. Nature **213**, 312–313 (1967).

DUNLOP, J., BOWLING, D.J.F.: The movement of ions to the xylem exudate of maize roots 1. Profiles of membrane potential and vacuolar potassium activity across the root. J. Exptl. Bot. **22**, 434–444 (1971).

EPSTEIN, E.: Mineral nutrition of plants: principles and perspectives. New York: Wiley and Sons 1972.

ESAU, K.: Plant anatomy, 2nd ed. New York: Wiley and Sons 1965.

ESHEL, A., WAISEL, Y.: Variations in uptake of sodium and rubidium along barley roots. Physiol. Plantarum **28**, 557–560 (1973).

FLEET, D.S. VAN: Histochemistry and function of the endodermis. Botan. Rev. **27**, 165–220 (1961).

GRASMANIS, V.O., BARLEY, K.P.: The uptake of nitrate and ammonium by successive zones of the pea radicle. Australian J. Biol. Sci. **22**, 1313–1320 (1969).

GREENWAY, H.: Plant response to saline substrates. I. Growth and ion uptake of several varieties of *Hordeum* during and after sodium chloride treatment. Australian J. Biol. Sci. **15**, 16–38 (1962).

GUNNING, B.E.S., PATE, J.S., MINCHIN, F.R., MARKS, I.: Quantitative aspects of transfer cell structure in relation to vein loading in leaves and solute transport in legume nodules. In: Transport at the Cellular Level. Soc. Exptl. Biol. Symp., vol. 28, p. 87–126. Cambridge: Cambridge University Press 1974.

HAY, R.K.M., ANDERSON, W.P.: Characterisation of exudation from excised roots of onion, *Allium cepa* l. Water flux. J. Exptl. Bot. **23**, 577–584 (1972).

HELDER, R.J., BOERMA, J.: An electron microscopical study of the plasmodesmata in the roots of young barley seedlings. Acta Botan. Neerl. **18**, 99–107 (1969).

HEWITT, E.J.: Mineral nutrition of plants in culture media. In: Plant physiology, vol. III (F.C. STEWART, ed.), p. 97–133. New York-London: Academic Press 1963.

JENNY, H.: Pathways of ions from soil into root according to diffusion models. Plant Soil **25**, 255–285 (1966).

JOHANSEN, C., EDWARDS, D.G., LONERAGAN, J.F.: Potassium fluxes during potassium absorption by intact barley plants of increasing potassium content. Plant Physiol. **45**, 601–603 (1970).

LÄUCHLI, A., KRAMER, D., PITMAN, M.G., LÜTTGE, U.: Ultrastructure of xylem parenchyma cells of barley roots in relation to ion transport to the xylem. Planta **119**, 85–99 (1974a).

LÄUCHLI, A., KRAMER, D., STELZER, R.: Ultrastructure and ion localization in xylem parenchyma cells of roots. In: Membrane transport in plants (U. ZIMMERMANN and J. DAINTY, eds.), p. 363–371. Berlin-Heidelberg-New York: Springer 1974b.

LÄUCHLI, A., SPURR, A.R., EPSTEIN, E.: Lateral transport of ions into the xylem of corn roots. 11. Evaluation of a stelar pump. Plant Physiol. **48**, 118–124 (1971).

LEGGETT, J.E., GILBERT, W.A.: Magnesium uptake by soybeans. Plant Physiol. **44**, 1182–1186 (1969).

NEWMAN, E.I.: Resistance to water flow in soil and plant 1. Soil resistance in relation to amounts of root: theoretical estimates. J. Appl. Ecol. **6**, 1–12 (1969).

NISSEN, P.: Bacteria-mediated uptake of choline sulfate by plants. Sci. Rept. Agr. Univ. Norway **52**, 1–53 (1973).

PALLAGHY, C.K., SCOTT, B.I.H.: The electrochemical state of cells of broad bean roots. Australian J. Biol. Sci. **22**, 585–600 (1969).

PATE, J., GUNNING, B.E.S.: Transfer cells. Ann. Rev. Plant Physiol. **23**, 173–196 (1972).

PITMAN, M.G.: Sodium and potassium uptake by seedlings of *Hordeum vulgare*. Australian J. Biol. Sci. **18**, 10–24 (1965).

PITMAN, M.G.: Adaptation of barley roots to low oxygen supply and its relation to potassium and sodium uptake. Plant Physiol. **44**, 1233–1240 (1969).

PITMAN, M.G.: Uptake and transport of ions in barley seedlings 1. Estimation of chloride fluxes in cells of excised roots. Australian J. Biol. Sci. **24**, 407–421 (1971).

PITMAN, M.G., COURTICE, A.C., LEE, BARBARA: Comparison of potassium and sodium uptake by barley roots at high and low salt status. Australian J. Biol. Sci. **21**, 871–881 (1968).

PITMAN, M.G., LÜTTGE, U., LÄUCHLI, A., BALL, E.: Action of abscisic acid on ion transport as affected by root temperature and nutrient status. J. Exptl. Bot. **25**, 147–155 (1974).

PITMAN, M.G., MERTZ, S.M. JR., GRAVES, J.S., PIERCE, W.S., HIGINBOTHAM, N.: Electrical potential differences in cells of barley roots and their relation to ion uptake. Plant Physiol. **47**, 76–80 (1971).

POLLE, E.O., JENNY, H.: Boundary layer effects in ion absorption by roots and storage organs of plants. Physiol. Plantarum **25**, 219–224 (1971).

REISENAUER, H.M.: Mineral nutrients in soil solution. In: Environmental biology (P.L. ALTMAN and D.S. DITTMER, eds.), p. 507–508. Bethesda: Federation of American Societies for Experimental Biology 1966.

RICHTER, C.H., MARSCHNER, H.: Distribution of K^+, Na^+ and Ca^2 between root cortex and stele. Z. Pflanzenphysiol. **71**, 95–100 (1974).

ROBARDS, A.W.: The ultrastructure of plasmodesmata. Protoplasma **72**, 315–323 (1971).

TYREE, M.T.: The symplast concept: a general theory of symplastic transport according to the thermodynamics of irreversible processes. J. Theoret. Biol. **26**, 181–214 (1970).

ZIEGLER, H., LÜTTGE, U.: Die Salzdrüsen von *Limonium vulgare*. I. Die Feinstruktur. Planta **70**, 193–206 (1966).

3.2 Kinetics of Ion Transport and the Carrier Concept

E. Epstein

1. Introduction

Cells characteristically differ from the solutions bathing them in mineral composition. These differences are nowhere more dramatic than in the cells of plants, and especially, the roots of plants. Roots in many of the world's soils are bathed by soil solutions in which the concentrations of at least some of the essential mineral nutrients are low, on the order of 1 mM or less (Asher and Ozanne, 1967; Reisenauer, 1966; see Fig. 3.5). For example, the concentrations of K^+ in over 40 per cent of 155 samples of soil solutions surveyed by Reisenauer (1966) were 1 mM or less, and in 7.7 per cent of the samples, they were 0.256 mM or less (Fig. 3.5). Within the cells of roots and other plant organs, the concentration of K^+ is typically two orders of magnitude higher than the latter value.

The actual discrepancy between the concentration of many nutrient ions in the solution bathing the roots and their intracellular concentrations will often be much greater than the figures cited above would suggest. The reason for that conclusion is that the values for the ionic concentrations in soil solutions represent the results of determinations made on bulk samples of solution extracted from soil. But in the immediate vicinity of roots, the concentration of some ions in the soil solution may be much lower, as a result of depletion through absorption by the roots (Barber, 1975). Reactions between dissolved ions and the solid phase of the soil complicate the picture but do not change the above conclusions.

The knowledge that intracellular ionic concentrations are much higher than those of the ambient medium suggests a high degree of impermeability of the cellular membranes. But inasmuch as the intracellular ions are acquired from the supply in the medium the suggestion arises that mechanisms of transport must exist within the membranes to ferry the ions across the (diffusively) impermeable membranes into the cell.

The phenomenon of selectivity lends force to this idea. Most terrestrial plants accumulate K^+ in preference to Na^+, even when the medium contains more Na^+ than K^+, and numerous other instances of selectivity between members of pairs of ions might be cited (see 3.2.2.1; 3.3.3.1; and Part A, Tables 6.3; 7.2). It is tempting to assign to the postulated transport agents the function of selective binding and hence, selective transport of ions.

On the basis of such considerations, Hoagland (1944) proposed the idea of an ion "pump." "The mechanism of the pump is not revealed, but there occurs apparently some preliminary combination of protoplasmic constituents with the solute..." The process of ion transport or ion pumping was found to depend on concomitant metabolic activity of the tissue, and the term "active transport" was used by Hoagland to describe this process of trans-membrane pumping of ions contingent on active cell metabolism. The term "carriers" is now commonly used to refer to the agents responsible for pumping ions across cellular membranes of plants.

Still earlier, VAN DEN HONERT (1937) had arrived at somewhat similar ideas on transport, on the basis of experiments on the absorption of $H_2PO_4^-$ by sugarcane plants, in which he found a hyperbolic relationship between the concentration of $H_2PO_4^-$ in the solution and the rate of its absorption by the plants: "Recent investigations proved that the intake of salts by plant roots... is by no means due only to passive permeability of the protoplasm, but to an active mechanism, operated by aerobic respiration... It is assumed that the phosphate is adsorbed by the surface layer of the protoplasm of the root cells and subsequently carried on by a mechanism resembling a constantly rotating belt conveyer, which moves its charge from the surface, depositing it inside and returning empty to be charged again."

VAN DEN HONERT's ideas, published in a Netherlands East Indies journal, received no wide currency. OSTERHOUT (1952) expressed very similar ideas, on the basis of experiments with algae. He listed a number of questions that might be asked about ion transport and set forth the utility of a "carrier" concept in providing answers to them. For an account of current work on the biochemistry of carriers, see Part A, Chap. *10.*

However, the bulk of work, and the most important work, on ion transport in plants during the 1930's and 40's was devoted to the connection between ion transport by plant cells and active aerobic metabolism (HOAGLAND and BROYER, 1936, 1942, HOAGLAND, 1944; LUNDEGÅRDH, 1946; ROBERTSON and WILKINS, 1948; STEWARD and HARRISON, 1939). Both LUNDEGÅRDH and ROBERTSON and his collaborators developed rather detailed ideas on the nature of the linkage between respiratory metabolism and the transport of ions.

LUNDEGÅRDH's hypothesis was contested by other workers in the field (HOAGLAND and STEWARD, 1939, 1940), and the nature of the linkage between respiration and ion transport still represents an unsolved problem. It is surprising, in retrospect, that detailed hypotheses on energetics of ion transport should have been advanced before major aspects of the phenomenology of the process, including its kinetics, had been worked out.

2. The Enzyme-Kinetic Formulation

2.1 Michaelis-Menten Kinetics Interpreted in Terms of a Carrier Concept

Against the background outlined above, EPSTEIN and HAGEN (1952) and EPSTEIN (1953) formulated the enzyme-kinetic hypothesis of ion transport and carrier function. These researches became the point of departure for numerous investigations, and a large literature on the kinetics of membrane transport and the function of trans-membrane carriers has since come into being. The central ideas will be but briefly outlined here, because the enzyme-kinetic hypothesis of membrane transport has been extensively reviewed in recent years (BÖSZÖRMÉNYI et al., 1972; CLARKSON, 1974; EPSTEIN, 1972, 1973; HIGINBOTHAM, 1973, 1974; HODGES, 1973; LATIES, 1975; NEAME and RICHARDS, 1972; NISSEN, 1974).

EPSTEIN and HAGEN (1952) and EPSTEIN (1953) likened the process of carrier-mediated transport of an ion across the cell membrane to the process of enzyme-mediated catalysis of a substrate. They concluded that kinetically, the two kinds of processes should be alike, the basis for this similarity being transitory occupancy of finite numbers of active sites of agents (carrier or enzyme) by their substrates (ions or organic substrates, respectively).

Ion transport, like enzymic catalysis, would be expected to follow Michaelis-Menten kinetics. That is, the rate of transport, v, would be given by

$$v = \frac{V_{max} \cdot c_s}{K_m + c_s} \qquad (3.1)$$

where V_{max} is the maximal rate of transport, c_s the concentration of the substrate ion in the medium, and K_m the Michaelis constant equal to the substrate ion concentration giving half the maximal rate of transport. Formally, this is equivalent to the adsorption (LANGMUIR) equation; see EPSTEIN (1972) and NEAME and RICHARDS (1972).

Ion transport would be expected to exhibit selectivity or specificity based on differential affinities of active carrier sites for different ions. Competitive inhibition would result if two or more different ions had appreciable affinity for identical carrier sites.

Since net change in the medium-tissue system is most pronounced when the initial level of the ion in question within the tissue is low, EPSTEIN and co-workers used a modification of the technique of HOAGLAND and BROYER (1936) to grow low-salt barley plants whose excised roots they used in short-term absorption experiments. The approach was in keeping with what HALDANE (1930) had written about enzymes many years before: "The key to a knowledge of enzymes is the study of reaction velocities, not of equilibria." (For comparisons of absorption by low-salt and high-salt tissues see *3.1*.2.5 and Part A, 5.5.3.)

The initial investigations bore out the expectations that had been entertained. Adherence to Michaelis-Menten kinetics and mutual competition between similar ions such as K^+ and Rb^+ and Cl^- and Br^- were demonstrated, and so was failure of competition between more dissimilar ions such as K^+ and Na^+, and Br^- and NO_3^- (EPSTEIN and HAGEN, 1952, for the cations; EPSTEIN, 1953, for the anions). The hypothesis of transmembrane carriers operating by a transitory complexation of the ions transported was considered to be borne out by the evidence and the general conclusion drawn "that all active transport of ions must needs implicate their combination with protoplasmic constituents, and that this forms the chemical basis for selectivity" (EPSTEIN, 1953).

At the same time, these and subsequent investigations provided evidence suggesting kinetics more complex than the simple Michaelis-Menten formulation. Specifically, if the concentration of Rb^+ did not exceed 1 mM, Na^+ failed to interfere with absorption of Rb^+ whereas at high Rb^+ concentrations Na^+ competed (EPSTEIN and HAGEN, 1952). Potassium and Rb^+ were competitive at all concentrations and ratios tested. The conclusion was that K^+ and Rb^+ occupy identical carrier sites, but that "in the presence of Na^+, two types of such sites can be distinguished, one site having little affinity for Na^+". Further evidence of dual kinetics was described by EPSTEIN and LEGGETT (1954) for absorption of Sr^{2+}, by LEGGETT and EPSTEIN

(1956) for absorption of SO_4^{2-}, and by FRIED and NOGGLE (1958) for absorption of several cations, barley roots being used in all these investigations.

2.2 Technical Improvements

Two developments subsequently made possible a more definitive delineation of ion absorption kinetics by plant tissue.

First, it was recognized by several groups that diffusion and cation exchange unrelated to active ion transport might account for an appreciable fraction of the total ions taken up in brief experimental periods. This subject has been repeatedly reviewed; see especially BRIGGS et al. (1961) and EPSTEIN (1956, 1972). Although BRIGGS et al. (1961) considered that part at least of the cytoplasm might be included in the volume invaded by ions in this passive and readily reversible fashion, the more general conclusion was that this space, the "outer" or "free" space, is in the cell wall and intercellular space and does not extend across the plasmalemma (cf. CONWAY and DOWNEY, 1950; see also 1.1 and Part A, Chap. 5.3).

In experiments on active ion transport, it is therefore mandatory to take into account those ions that have become associated with the tissue in this relatively superficial fashion, without being transported into the "inner space" (the cytoplasm and its inclusions). The most common technique of dealing with this problem is the one devised by EPSTEIN and LEGGETT (1954) in a study of the absorption of Sr^{2+} by barley roots. After a period of absorption of Sr^{2+} labeled with ^{89}Sr the tissue was exposed to a solution of unlabeled Sr^{2+} for 30 min. This stripped the readily exchangeable labeled Sr^{2+} from the tissue, leaving another, nonexchangeable or much more slowly exchangeable fraction of labeled Sr^{2+} within the tissue. This fraction was deemed to be the one that had negotiated the plasmalemmas of epidermal and cortical cells and had entered the cytoplasm. Uptake and desorption of ^{86}Rb/Rb are shown in Fig. 3.7.

The same technique of stripping the labile, "outer space" fraction after a period of absorption was used in studies on the transport of SO_4^{2-} (EPSTEIN, 1955; LEGGETT and EPSTEIN, 1956), and it has since been used to good advantage in numerous investigations. In short-term experiments, the fraction of the total ions taken up

Fig. 3.7. Absorption and desorption of rubidium labeled with ^{86}Rb by excised barley roots as a function of time. Solid circles: tissue in solutions of 5 mM labeled RbCl, 0.5 mM $CaCl_2$, 30° C, rinsed with water at the end of the absorption period. Open circles: tissue in solutions of 5 mM KCl, 0.5 mM $CaCl_2$, 5° C. (Potassium and rubidium are close analogs in ion transport by plants. For the sake of economy, KCl was used in the desorbing solution instead of RbCl. The results were identical whether KCl or RbCl was used.) (After EPSTEIN et al., 1963b)

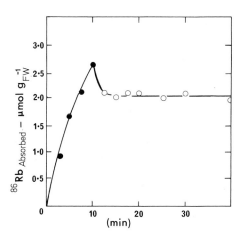

by tissue that is in its "outer" or "free" space can be large, and therefore, if not eliminated or accounted for, can confound measurements of absorption. The technique described above, by virtue of leaving for final measurement only that fraction of the ions that had been transported, made possible precise measurement of rates of transport in short and very short experimental periods. For a recent discussion, see Methods in EPSTEIN (1973).

A second major advance, like the one discussed above indispensable for the refined development of ion transport kinetics that was to follow, occurred in the early 1960's in three laboratories. HANSON (1960) showed that removal of calcium from root tissue by treatment with ethylenediamine tetraacetic acid (EDTA) impaired its ability to absorb and retain ions and caused other physiological derangements. JACOBSON et al. (1961) found Ca^{2+} to have marked effects on the selectivity of cation absorption by roots. Finally, EPSTEIN (1961) showed Ca^{2+} to be indispensable for unimpaired cation absorption by barley roots; specifically, the selectivity of the K^+-Rb^+ transport mechanism vis-à-vis Na^+ was shown to depend on the presence of Ca^{2+}, whose absence caused a marked breakdown in selectivity within minutes. He concluded that Ca^{2+} was "essential" for selective cation transport and henceforth included Ca^{2+} routinely in experimental solutions, believing a solution containing Ca^{2+} to represent a minimal "physiological saline" for plant tissue (EPSTEIN, 1965; see also HANSON, 1965; LÄUCHLI and EPSTEIN, 1970). Under certain conditions and with certain plant materials, however, Ca^{2+} may inhibit the absorption of monovalent cations, although it may be better to say that the presence of Ca^{2+} prevents abnormally high rates of absorption of the monovalent cations, appreciable concentrations of Ca^{2+} being the normal physiological condition. For instances of diminished rates of absorption of monovalent cations as a function of the Ca^{2+} concentration, see ELZAM and HODGES (1967), FALADE (1973), GAUCH (1972), JOHANSEN and LONERAGAN (1975a) and MINCHIN and BAKER (1973).

Two important advances emerged from the investigations described above—the technique of stripping off the readily exchangeable fraction of labeled ions present in the cell walls of plant tissues at the end of a period of absorption of a radioion, and inclusion of Ca^{2+} in experimental solutions. EPSTEIN et al. (1963b) described a technique for short-term experiments on solute absorption by plant tissue incorporating these features. The technique made possible precise determinations of absorption rates in very short experiments. Periods of 10 or 20 min became routine in their laboratory.

In addition, SMITH and EPSTEIN (1964) devised a method for exposing cells within leaf tissue of terrestrial plants to experimental solutions, thereby extending the range of plant tissues with which kinetic experiments on ion transport can be done (see 4.2.2.2.1). For such experiments the techniques described above are also appropriate, as is the case for experiments with storage tissues (see Part A, 5.4.4, 5.5.1 and Chap. 8). Since 1963, experiments using the methods described above have led to a redefinition of ion absorption kinetics and revealed a complexity not suspected before. The findings are outlined in 3.2.3.

3. The Dual Pattern of Ion Transport

3.1 General Outline

If the rate of absorption of K^+ and ions of many other elements is examined as a function of the external concentration of that ion, a pattern like that shown in Fig. 3.8 has been observed time and again. It is essentially a dual pattern: two ranges of concentration (low and high) at which the rate of absorption of the ion increases markedly with increasing external concentrations are separated by an intermediate range over which there is little or no change in the rate of absorption. Numerous experiments in which this dual pattern has been observed are listed in Table 3.4. They include plant materials ranging from excised roots to unicellular algae and from below-ground storage organs to leaf tissue. As for different elements, they include ions of the majority of mineral elements known to be essential nutrients for plants. It is thus a pattern of impressive generality. The two ranges of concentration separated by an intermediate plateau in the rate of absorption, and the features of ion absorption characteristic of each, will be discussed in 3.2.3.2 and 3.2.3.3. The examples will deal mainly with the absorption of K^+ and its analog, Rb^+, because absorption of these two elements has been most closely examined.

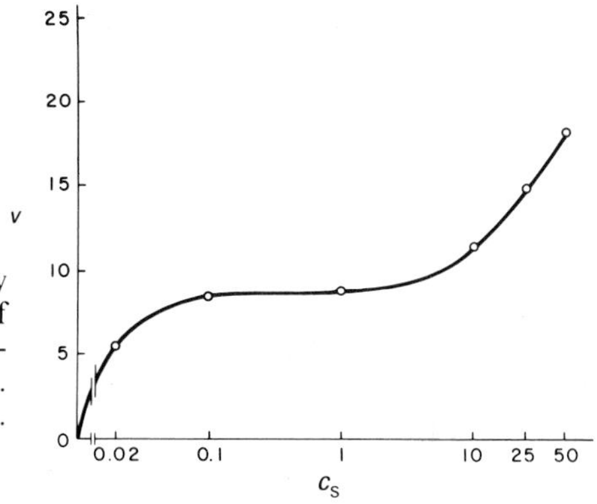

Fig. 3.8. Rate of absorption, v, by excised barley roots of K^+ labeled with ^{86}Rb as a function of the concentration of KCl, c_s, plotted logarithmically, where v is in $\mu mol \ g_{FW}^{-1} \ h^{-1}$ and c_s in mM. The roots were grown under sterile conditions. After EPSTEIN (1968)

Table 3.4. Dual pattern of ion absorption by plants[a]

Substrate ion	Plant species	Plant organ	Remarks	Ref.
K^+	Barley, *Hordeum vulgare*	Roots	–	EPSTEIN et al. (1963a)
K^+	Barley	Roots	Fig. 1: semi-log	HIATT (1967)
K^+	Barley	Roots	Figs. 1, 2: semi-log	HIATT (1968)
K^+	Barley	Roots	Roots grown under sterile conditions; Fig. 1: semi-log	EPSTEIN (1968)
K^+	Corn, *Zea mays*	Roots	Entire plants	LÜTTGE and LATIES (1966)

[a] Specific figures listed in the column headed "Remarks" refer to plots in which no break is introduced in the concentration scale over the entire range of concentrations of the external solution used. These plots all serve to demonstrate the basically dual pattern of absorption, without any risk of "optical illusions" being introduced by a change of scale in the concentration axis. See p. 85.

Table *3.4* (continued)

Substrate ion	Plant species	Plant organ	Remarks	Ref.
K$^+$	Corn	Roots	—	Leigh et al. (1973)
K$^+$	Oat, *Avena sativa*	Roots	Fig. 13: Hofstee	Leonard and Hodges (1973)
K$^+$	*Avicennia marina* (a mangrove)	Leaf tissue	—	Rains and Epstein (1967c)
K$^+$	Beet, *Beta vulgaris*	Discs	Freshly cut and aged	Osmond and Laties (1968)
K$^+$	Beet	Discs	—	Osmond and Laties (1970)
K$^+$	Wheatgrass, *Agropyron* spp.	Roots	—	Elzam and Epstein (1969)
K$^+$	Bean, *Phaseolus vulgaris*	Stem tissue	Fig. 10: semi-log	Rains (1969)
K$^+$	Waterweed, *Elodea densa*	Leaves	—	Jeschke (1970a)
K$^+$	Waterweed	Leaves	—	Jeschke (1970b)
Rb$^+$	Barley	Roots	—	Epstein et al. (1963a)
Rb$^+$	Perennial ryegrass, *Lolium perenne*	Roots	—	Jackman (1965)
Rb$^+$	Mung beans, *Phaseolus aureus*	Roots	—	Jackman (1965)
Rb$^+$	Subterranean clover, *Trifolium subterraneum*	Roots	—	Jackman (1965)
Rb$^+$	*Chlorella pyrenoidosa* (a green alga)	Cells	—	Kannan (1971a)
Rb$^+$	Waterweed	Leaves	—	Jeschke (1970b)
Rb$^+$	Corn	Roots	Low and high salt tissue	Leigh and Wyn Jones (1973)
NH$_4^+$	Corn	Roots	Fig. 4: Hofstee	Berlier et al. (1969)
Na$^+$	Barley	Roots	Fig. 1: semi-log	Rains and Epstein (1965)
Na$^+$	Barley	Roots	Fig. 6: semi-log	Rains and Epstein (1967a)
Na$^+$	Barley	Roots	Fig. 2c: semi-log	Thellier (1973)
Na$^+$	Wheatgrass	Roots	—	Elzam and Epstein (1969)
Na$^+$	Bean	Stem tissue	Fig. 10: semi-log	Rains (1969)
Na$^+$	Tomato, *Lycopersicon esculentum*	Roots	Fig. 4: Hofstee	Picciurro and Brunetti (1969)
Na$^+$	Corn	Roots	Low and high salt tissue	Leigh and Wyn Jones (1973)
Na$^+$	Beet	Roots	Entire plants; long-term experiments	El-Sheikh and Ulrich (1971)
Ca^{2+}	Corn	Roots	—	Maas (1969)
Ca^{2+}	Cotton, *Gossypium hirsutum*	Roots	Tissue culture; long-term experiments	Johanson and Joham (1971)

Table 3.4 (continued)

Substrate ion	Plant species	Plant organ	Remarks	Ref.
Ca^{2+}	Horse bean, *Vicia faba*	Roots	–	SALSAC (1973)
Ca^{2+}	Yellow lupine, *Lupinus luteus*	Roots	–	SALSAC (1973)
Mg^{2+}	Barley	Roots	–	KAPLAN (1969)
Fe^{2+}	Rice, *Oryza sativa*	Roots	–	KANNAN (1971b)
Cl^-	Barley	Roots	–	ELZAM et al. (1964)
Cl^-	Barley	Roots	Fig. 1: semi-log	HIATT (1967)
Cl^-	Barley	Roots	Figs. 1, 3: semi-log	HIATT (1968)
Cl^-	Barley	Roots	Fig. 1: semi-log	BÖSZÖRMÉNYI (1966)
Cl^-	*Mnium cuspidatum* (a moss)	Branches	–	LÜTTGE and BAUER (1968)
Cl^-	Beet	Discs	Freshly cut and aged	OSMOND and LATIES (1968)
Cl^-	Wheatgrass	Roots	–	ELZAM and EPSTEIN (1969)
Cl^-	Corn	Roots	–	MAAS (1969)
Cl^-	Corn	Roots	Figs. 2a, b: semi-log	THELLIER (1973)
Cl^-	Bean	Roots, leaf tissue	–	JACOBY and PLESSNER (1970)
Br^-	Barley	Roots	Fig. 1: semi-log	BÖSZÖRMÉNYI (1966)
Br^-	Wheat	Roots	Fig. 2: log-log	BÖSZÖRMÉNYI et al. (1969)
$H_2PO_4^-$	Waterweed	Leaves	Fig. 3: Hofstee	JESCHKE and SIMONIS (1965)
$H_2PO_4^-$	Barley	Roots	Fig. 4: Hofstee	LEGGETT et al. (1965)
$H_2PO_4^-$	Barley	Roots	Roots grown under sterile conditions	BARBER (1972)
$H_2PO_4^-$	Corn	Roots	Fig. 2B: Hofstee	CARTER and LATHWELL (1967)
$H_2PO_4^-$	Barley	Roots	Roots grown under sterile conditions, Figs. 2, 3: Hofstee	CARTWRIGHT (1972)
$H_2PO_4^-$	Lettuce, *Lactuca sativa*, and cabbage, *Brassica oleracea*	Roots	Roots grown under sterile conditions, Fig. 3: Hofstee	TEMPLE-SMITH and MENARY (1974)
$H_2PO_4^-$	Several species	Roots	Figs. 3–7: Hofstee	ANDREW (1966)
SO_4^{2-}	Waterweed	Leaves	Fig. 4: Hofstee	JESCHKE and SIMONIS (1965)
SO_4^{2-}	*Mnium cuspidatum*	Branches	–	LÜTTGE and BAUER (1968)
SO_4^{2-}	*Lemna minor*	Plants	Figs. 3, 4: semi-log	THELLIER et al. (1971)
SO_4^{2-}	Corn	Roots	Fig. 5: Hofstee	BERLIER et al. (1969)
SO_4^{2-}	Barley	Roots	Fig. 1: log-log	NISSEN (1971)
$H_2BO_3^-$	Sugarcane, *Saccharum officinarum*	Leaf tissue	Fig. 3: Hofstee	BOWEN (1968)

3.2 The Range of Low Concentrations

Over the range of external K^+ concentrations extending from low values (0.002 or 0.005 mM) to about 0.2 mM, the rate of absorption of the ion shows the characteristic features of Michaelis-Menten kinetics. The most detailed experiment for this range seems to be one on absorption of Rb^+ by barley roots (Epstein, 1965). To test conformance of the data to Michaelis-Menten kinetics, the original experimental data were entered into a computer-plotter programmed to analyze the data statistically, calculate the kinetic parameters, and plot the calculated straight line for v vs. v/c_S, where v is the rate of absorption and c_S the external concentration (Hofstee, 1952). The results, computer-generated and computer-plotted from the experimental data inputs, are shown in Fig. 3.9. A conventional (v vs. c_S) plot of the same data is presented in Fig. 3.10.

(The Hofstee plot used in Fig. 3.9 represents one of three linear transformations of the Michaelis-Menten equation. Two others are the Lineweaver-Burk plot in which $1/v$ is plotted as a function of $1/c_S$, and a plot of c_S/v as a function of c_S. The Hofstee plot is probably the best in that it clearly displays deviant experimental points, while the Lineweaver-Burk plot tends to give the appearance of a good fit even for unreliable data, and gives rise to other "optical illusions". For discussions, see Dowd and Riggs (1965), Hofstee (1952, 1959), and Neame and Richards (1972).)

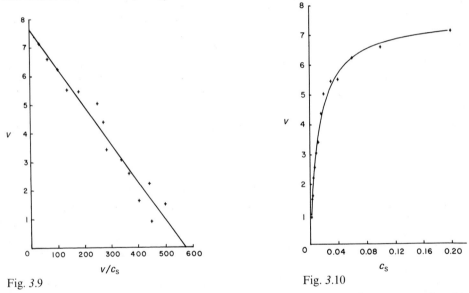

Fig. 3.9   Fig. 3.10

Fig. 3.9. Hofstee plot of the rate of absorption, v, of Rb^+ by excised barley roots as a function of the ratio, v/c_S, where v is in μmol g_{FW}^{-1} h^{-1} and c_S is the external concentration of Rb^+, ranging from 0.002 to 0.2 mM. Calculation of the kinetic parameters and plotting were done by computerized plotting board from the experimental data inputs. The calculated kinetic parameters are as follows: V_{max}, 7.63 μmol g_{FW}^{-1} h^{-1} and K_m, 0.0133 mM. The square of the correlation coefficient, r^2, is 0.9537. For discussions of the advantages of the Hofstee plot, see Dowd and Riggs (1965), Hofstee (1952, 1959), and Neame and Richards (1972)

Fig. 3.10. Conventional plot of the experiment shown in Fig. 3.9. The rate of absorption, v, of Rb^+, in μmol g_{FW}^{-1} h^{-1}, is plotted as a function of its external concentration, c_S, in mM. The line is a plot of the Michaelis-Menten equation, computer-plotted on the basis of the same kinetic parameters as those of Fig. 3.9

Table 3.5. Michaelis constants from experiments on ion absorption[a, b]

Substrate ion	Plant species	Plant organ	Michaelis constant (K_m) (mM)
K^+	Barley, *Hordeum vulgare*	Roots	0.021
K^+	Corn, *Zea mays*	Leaf tissue	0.034
K^+	Tall wheatgrass, *Agropyron elongatum*	Roots	0.008
Rb^+	Barley	Roots	0.017
Rb^+	Corn	Leaf tissue	0.015
Rb^+	Ryegrass, *Lolium perenne*	Roots	0.012
Rb^+	Mung bean, *Phaseolus aureus*	Roots	0.012
Rb^+	Subterranean clover, *Trifolium subterraneum*	Roots	0.008
Cl^-	Barley	Roots	0.014
Cl^-	Tall wheatgrass	Roots	0.013

[a] From Table 6-1, EPSTEIN (1972). [b] Rapidly absorbed, monovalent ions only, for the range of low concentrations.

For K^+ absorption by roots of barley, *Hordeum vulgare*, the Michaelis constant, K_m, is on the order of 0.02 mM and the maximal rate, V_{max}, approximately 10 μmol g_{FW}^{-1} h^{-1} at 30° C. Rates of absorption of Rb^+ are slightly lower, as are the Michaelis constants. Representative values for the Michaelis constants obtained in experiments on the absorption of several ions are given in Table 3.5.

Absorption of K^+ has the following characteristics in barley roots, in addition to the quantitative parameters just mentioned: it is competitively inhibited by Rb^+, it is hardly affected by Na^+ even if the latter is present in considerable excess, it is little affected by substitution of SO_4^{2-} for Cl^- as the counterion, and it is accelerated by Ca^{2+} present at high concentrations, on the order of several millimolar. (A low level of Ca^{2+}, routinely 0.5 mM $CaSO_4$, is always present in the solutions; see 3.2.2.2.)

The failure of Na^+ to compete with K^+-Rb^+ in their absorption by barley roots is in line with the earlier findings of EPSTEIN and HAGEN (1952) for Rb^+ concentrations not exceeding 1 mM. The indifference to the identity of the anion is interesting in that Cl^- is itself absorbed at an appreciable rate, though not equaling that of K^+-Rb^+ uptake (ELZAM et al., 1964; ELZAM and EPSTEIN, 1965), while the rate of SO_4^{2-} absorption is very low by comparison (LEGGETT and EPSTEIN, 1956).

The plateau of the absorption rate at the higher concentrations of this range is wide. Often, there is little or no change in the rate of absorption of alkali cations and halides when their concentrations are varied over a factor of 4, 5, or even 10. For example, in experiments with barley roots (grown under sterile conditions) EPSTEIN (1968) found no response in the rate of K^+ absorption between 0.1 and 1 mM (Fig. 3.8). The identical observation was made in experiments on the absorption of K^+ by *Lemna minor* (YOUNG and SIMS, 1972). A plateau over the same range of concentrations was shown for absorption of Cl^- by corn roots in experiments of MAAS (1969), and for absorption of K^+ by corn roots in work by LEIGH et al. (1973, Fig. 3, low salt isotherm). For K^+ absorption by bean stem tissue, the plateau was wider yet (RAINS, 1969). (See also Part A, Fig. 6.7.)

3.3 The Range of High Concentrations

When absorption rates are measured at concentrations much beyond those just discussed, ranging up to 50 mM and even more in some experiments, the rate of absorption by low-salt tissue rises to levels much beyond the plateau referred to. Originally this rise was thought to be a second Michaelis-Menten hyperbola (Epstein et al., 1963a) but later results showed the dependence of the rate of absorption on concentration in the higher range to be complex as shown in Fig. *3.*11. This was first shown for absorption of Cl^- by Elzam et al. (1964) and for absorption of K^+ and Na^+ by Epstein and Rains (1965).

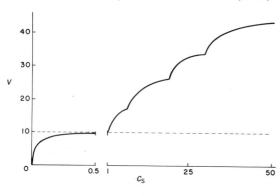

Fig. *3.*11. Generalized diagram of the rate of absorption, v, by plant tissue of an alkali cation or halide as a function of its concentration, c_S, in the external solution, in mM. The units for the rate of absorption, v, are arbitrary

    This complexity has been demonstrated in numerous experiments with many different plant materials and ions of several elements; see on this Epstein (1972, 1973). Although undoubtedly not every inflection in every published isotherm is real, the generality of the complex pattern is such that the validity of these findings seems established. Among current authors only Cram (1973) seems to believe that the "bumps" in these isotherms may be due to random variation.

    Apart from the complex kinetics of ion absorption characteristic of the range of high concentrations, features of absorption in this range differ diametrically from those that characterize absorption in the range of low concentrations (see *3.2.3.2*). Potassium absorption by low-salt barley roots over the range of high concentrations is severely inhibited by Na^+, substitution of SO_4^{2-} for Cl^- as the counter-ion depresses K^+ absorption, and Ca^{2+} at high concentrations, far from accelerating the rate of K^+ absorption, diminishes it (Rains and Epstein, 1967a, b; Welch and Epstein, 1968; see also *3.3.1*).

    During uptake of K^+ and Na^+ from solutions of their chlorides it has frequently been observed that there is net H^+ efflux from the tissue (*3.3.1*; Part A, *12.5.1*). This H^+ efflux accompanies formation of organic acids which are accumulated in the cell vacuole (Part A, *13.2.5*).

4. Interpretations of the Experiments on Ion Absorption Kinetics

Enzyme kinetics began with the straightforward application of the Michaelis-Menten equation to enzyme-mediated catalysis, but since then, far more complex observations and hence, more varied and sophisticated interpretations have given to the field a novel and often more controversial aspect (PLOWMAN 1972). We have seen that the experimental work on carrier-mediated ion transport in plant tissues has also revealed, over a period of more than two decades, increasingly complex kinetic patterns. Inevitably, these discoveries have given rise to a variety of hypotheses to account for the facts. In what follows, these interpretations will be presented and discussed.

4.1 Dual Mechanisms

4.1.1 Dual Mechanisms in the Plasmalemma

EPSTEIN et al. (1963a) considered, in keeping with earlier conclusions (EPSTEIN and HAGEN, 1952), that the duality apparent in studies of K^+-Rb^+ absorption kinetics reflected the operation of two mechanisms with greatly different affinities for the ions. At low concentrations (0.2 mM and less) only a single, high-affinity "mechanism I" is in evidence. Over the range of high concentrations the rate of absorption consists of two terms: the contribution from mechanism I operating at virtually maximal velocity, and an additional contribution from the low-affinity "mechanism II," the magnitude of this increment of the absorption rate depending on the external concentration. In other words, two mechanisms operating in parallel were envisioned. The implication was, furthermore, that these mechanisms lay in the membrane normally bathed by the soil solution, i.e., the outer cell membrane. The two "mechanisms" were defined in operational terms, on the basis of the kinetics observed, and interpreted as reflecting the existence of "two sets of carrier sites" (EPSTEIN, 1966), not of two different carrier molecules as implied by LEONARD and HODGES (1973); (see also 3.2.2.1).

Much work has confirmed and extended the principal findings that had led to the conclusion outlined above. The basic duality of the absorption kinetics, with marked response of the absorption rate over two concentration ranges separated by an intermediate range of little or no response, and the complexity of the absorption isotherm over the range of high concentrations have been observed by many investigators, using diverse plant materials and studying the absorption of ions of many elements. The conclusions themselves, however, in regard to the operation of two ion transport mechanisms operating in parallel across the plasmalemma have been challenged by several investigators who offer interpretations that differ from the one outlined above, and from each other.

4.1.2 Dual Mechanisms in the Plasmalemma and Tonoplast, Respectively

In the experiments in LATIES' laboratory, the duality of the absorption kinetics discovered by EPSTEIN and his coworkers was confirmed for roots of corn, *Zea mays* (TORII and LATIES, 1966; LÜTTGE and LATIES, 1966, 1967a, b), but a different interpretation of the findings was offered. These workers agreed with the original conclusion that the low concentration isotherm reflects the operation of a transport mechanism at the plasmalemma but argued that the mechanism that becomes evident at high external concentrations (mechanism II) resides in the tonoplast and transports ions into the vacuole. According to this interpretation, the two mechanisms do not operate in parallel but in series, mechanism II lying behind mechanism I.

This scheme cannot account for the facts without additional assumptions. If mechanism I in the plasmalemma were the sole supplier of ions to the cytoplasm, mechanism II, if located in the tonoplast, could not supply ions to the vacuole at a higher rate than the maximal rate of operation of mechanism I. Mechanism II could not be rate-limiting and should be kinetically "invisible" in experiments on the rate of ion absorption from an external solution. These workers therefore assumed that in the range of high concentrations, ions cross the plasmalemma by a process of "diffusive penetration" at a rate "rapid enough... not to be rate-limiting" (Torii and Laties, 1966). Mechanism II, considered by them to reside in the tonoplast, then becomes the rate-limiting factor.

The initial evidence claimed in support of this scheme was that root tips of corn showed a typical Michaelis-Menten isotherm over the low (mechanism I) range of concentrations for Cl^- and Rb^+, but a more nearly linear isotherm in the high (mechanism II) range of concentrations, whereas proximal root sections exhibited the typical patterns of a Michaelis-Menten isotherm at the low, and a complex isotherm at the high concentrations (Torii and Laties, 1966). Available evidence to the contrary notwithstanding (Whaley et al., 1960), tip cells were considered "predominantly nonvacuolate", hence lacking tonoplast and mechanism II. The linear absorption isotherm over the range of high concentrations was therefore taken to reflect diffusion across the plasmalemma.

Additional evidence from experiments on ion translocation into the xylem was considered to support this interpretation (Lüttge and Laties, 1967a, b; see also Laties, 1967, 1969, and Lüttge, 1969, 1973). The tonoplast not lying in the main pathway of transport into the xylem, this transport (into the exudate of excised roots or the tops of entire seedlings) ought not to exhibit the characteristic inflections of the absorption isotherm in the range of high concentrations, if mechanism II lay in the tonoplast and hence resulted in transport into the vacuole. The evidence, and further evidence from ion flux measurements in beet tissue, *Beta vulgaris,* (Osmond and Laties, 1968, 1970) was interpreted as favoring the series model of the disposition of the two transport mechanisms in the plasmalemma and tonoplast, respectively.

Both the actual experiments leading to this conclusion and their interpretation have been questioned (Weigl, 1969; Hiatt, 1967; Welch and Epstein, 1968; Welch, 1969; Läuchli and Epstein, 1971; Anderson, 1972; Epstein, 1972, 1973; Gerson and Poole, 1972; Läuchli, 1972; Hodges, 1973; Nissen, 1973a, b; Johansen and Loneragan, 1975b). In particular, the conclusion of rapid diffusive penetration of inorganic ions from the medium into the cytoplasm across the plasmalemma was rejected on the basis of independent evidence by Kannan (1971a), Kannan and Ramani (1974), Venrick and Smith (1967), Welch (1969), and Welch and Epstein (1968, 1969); see also the discussions by Epstein (1972, pp. 138–139, 170–172), Epstein (1973), Leigh and Wyn Jones (1973), and Leigh et al. (1974).

In a recent review, Laties (1975) has himself abandoned the concept of vacuolar kinetic control of ion absorption by barley roots at concentrations but slightly more than 1 mM, concluding instead vacuolar uptake to assume kinetic control only at concentrations in excess of 10 mM. Furthermore, while diffusion of ions across the plasmalemma was the very cornerstone upon which Laties' theoretical edifice was built, he now concludes that a passive (diffusive) element in transport across the plasmalemma need not be large.

Clearly, the point at issue is not whether there is some kind of transport at the tonoplast, but whether diffusion of ions across the plasmalemma can account for the difference between mechanism I and mechanism II. The argument is thus largely a quantitative one, depending on the degree of permeability of the plasma membrane.

Using computer simulation, PITMAN (1969) tested the effect of various values for permeability of the plasmalemma on the extent of diffusion of K^+ and Cl^-. For Cl^-, the required permeability coefficient was 100 times larger than that estimated from flux values (about $1 \cdot 10^{-10}$ m s^{-1}) and it was concluded that mechanism I and an additional, diffusional component at the plasmalemma in the concentration range of mechanism II were inadequate to account for observed rates of uptake at the plasmalemma in that range. For K^+ no such clear distinction could be made by this approach. GERSON and POOLE (1972) measured Cl^- accumulation and electrochemical activity in mung bean roots and showed that at least up to about 30 mM Cl^- in the external solution, i.e., well into the range of mechanism II, there was evidence for active Cl^- transport at the plasmalemma; see also KANNAN and RAMANI (1974) and LÄUCHLI and EPSTEIN (1971).

There is much prior evidence supporting the conclusion that the unimpaired plasmalemma is largely impermeable to inorganic ions in terms of diffusive penetration; see ARISZ (1963, 1964), BROUWER (1965), COLLANDER (1959), EPSTEIN (1972, 1973), and GREENHAM (1966). Although experiments on efflux of ions from plant tissue have sometimes been interpreted in terms of the LATIES hypothesis, i.e., in terms of an ion-permeable plasmalemma (RICHTER and MARSCHNER, 1973), such findings are more likely a reflection of carrier-mediated transport (POOLE, 1969; PIERCE and HIGINBOTHAM, 1970; JESCHKE, 1972, 1973), impairment of the tissue, or other factors (EPSTEIN, 1973, pp. 152–153).

An independent kind of support for the view that transport over the entire physiological range of concentrations is governed at the plasmalemma comes from

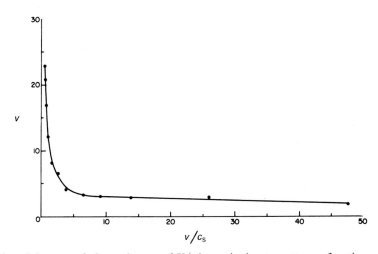

Fig. 3.12. Hofstee plot of the rate of absorption, v, of K^+ by excised oat roots as a function of the ratio, v/c_S, where v is in µmol g_{FW}^{-1} h^{-1} and c_S is the external concentration of K^+, ranging from 0.05 to 50 mM. After LEONARD and HODGES (1973)

an important investigation by Leonard and Hodges (1973); see also Part A, Chap. *10*. They isolated an adenosine triphosphatase (ATPase) from plasma membranes of oat roots, *Avena sativa,* and showed its activity to depend on the concentration of KCl in a manner strikingly parallel to the rate of K^+ absorption by these roots. In each case, the response showed the familiar dual pattern (see *3.2.4.2.1* and Fig. *3.12* and Part A, Fig. *10*.9). On the basis of this and other evidence, the authors concluded that this ATPase mediates K^+ transport. Since on the evidence the enzyme is located in the plasmalemma and shows the dual pattern of activity when the K^+ concentration is varied over the entire concentration range of mechanisms I and II the conclusion emerges that the energy-dependent transport occurs across the plasmalemma at both low and high concentrations.

4.2 Unitary Mechanisms

4.2.1 A Single, Multiphasic Mechanism

Nissen (1971) performed experiments on the absorption of SO_4^{2-} by excised roots and leaf slices of barley, using the techniques of Epstein et al. (1963b) and Smith and Epstein (1964). His findings were much as described before for absorption of many ions including SO_4^{2-}: at low external concentrations, the absorption isotherm followed Michaelis-Menten kinetics (cf. Leggett and Epstein, 1956); at intermediate concentrations, the isotherm flattened out, to rise again at high concentrations and exhibiting in this range a number of inflections (cf. Penth and Weigl, 1969).

Nissen (1971) proposed a "single, multiphasic mechanism" to account for these findings and suggested a similar explanation for the kinetic features of the absorption of other inorganic ions as well. In later publications, he analyzed published work on the kinetics of ion uptake in terms of this hypothesis (Nissen, 1973a, b, c, d), extended his own earlier work (Nissen, 1971) on absorption of sulfate by barley roots (Vange et al., 1974; Holmern et al., 1974), and comprehensively reviewed his hypothesis (Nissen, 1974).

Nissen's analysis is based on plotting a series of Lineweaver-Burk plots and determining the kinetic constants for each straight-line segment representing one of the partial isotherms between two inflections in the conventional v vs. c_S plot (Fig. *3.13*). He concludes that there is a single entity which changes characteristics at certain discrete external concentrations of the ion, *viz.,* those concentrations at which the inflections in the conventional isotherm occur (or the changes in slope of the Lineweaver-Burk plot).

Nissen denies that there is any duality in the kinetic pattern of ion absorption. He also believes that, except for Na^+ absorption, the isotherm in the range of low concentrations does not follow simple Michaelis-Menten kinetics but is biphasic. Thus in this interpretation, the transition from "mechanism I" to "mechanism II" is merely one of numerous changes in "phase," no different in kind from all the other inflections along Nissen's isotherm—a total of as many as eight all-or-nothing transitions in "phase."

The basically dual nature of the absorption isotherm does indeed tend to be obscured by Nissen's technique of breaking up the isotherms into numerous short

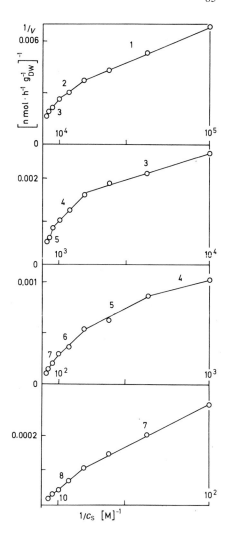

Fig. *3*.13. Data on SO_4^{2-} uptake by barley roots obtained over a wide range of external concentrations and plotted as numerous short segments in the Lineweaver-Burk form (after Fig. 9B from NISSEN, 1971)

segments and by other features of his presentation of the data. But whenever the absorption isotherm over a wide range of concentrations is displayed in a single graph without break or change of scale, then, no matter what particular form of graph is chosen, it almost invariably shows a pronounced dual pattern. Such graphs are listed in Table *3*.4 (see its footnote), and additional ones have been presented by THELLIER (1970). An example of a semi-log plot is shown in Fig. *3*.8, and of a Hofstee plot, in Fig. *3*.12.

Reinforcing the evidence for a basic duality of the absorption process are the contrasting features of absorption in the low and high concentration ranges already discussed under *3*.2.3.2 and *3*.2.3.3, and evidence that the rate of absorption of an ion, if its external concentration is high, is experimentally divisible into two increments (WELCH, 1969; WELCH and EPSTEIN, 1968; RAINS and EPSTEIN, 1967a, b).

4.2.2 Electrokinetic Formalism

In a series of papers, Thellier and coworkers have applied an electrokinetic formulation to the kinetics of ion transport (Thellier, 1968, 1970, 1971, 1973; Thellier et al., 1971; Ayadi et al., 1971). Thellier prefers not to consider ion transport in terms of a "mechanistic" model, that is, a model invoking molecular mechanisms of carriers and their operation, in analogy to catalysis effected by enzymes. Rather, he favors a formulation based on analogy to electrical currents and expressed in terms of conductance, driving force, and current (ion) flow. He shows that such a description yields a linear relation between flow and the logarithm of the external concentration over the range of low concentrations that Epstein and many others have associated with the high-affinity mechanism I.

Over the range of high concentrations (the low-affinity, mechanism II range) the experimental data do not fit a straight line in the semi-log plot. This is accounted for by an additional term including an exponential function of the driving force. The sum of two such terms based on the electrical analogy yields curves which describe a basically dual relationship between concentration and flow (absorption rate). By suitable choice of parameters, data of many experiments can be shown to be described by Thellier's electrokinetic formalism (Thellier, 1970), though no account is taken of the fine structure of the absorption isotherm in the range of high concentrations (see *3.2.3.3*).

Thellier's formulation shows that a fundamentally biphasic response of the rate of absorption of an ion to its concentration does not necessarily implicate dual mechanisms, as indeed had been pointed out before on the basis of the carrier model (Epstein et al., 1963a, p. 691). In fact, a number of different models can yield dual kinetics, for both transport and catalysis (Bange and Meijer, 1966; Borst-Pauwels, 1973; Harper, 1971, 1973; Koch, 1972; Neame and Richards, 1972; Oertli, 1967; Smith, 1973; Teipel and Koshland, 1969; see also *3.2.4.2.3*). Thellier's formalism is distinguished by its abandonment of the mechanistic model of carriers. That model has had the heuristic advantage of prompting a biochemical quest for carrier molecules. It seems vindicated by the apparent success of this search and has provided diagnostic criteria for recognition of carrier molecules (see Part A, Chap. *10*).

4.2.3 Michaelis-Menten Kinetics Modified by Membrane Potential

Gerson and Poole (1971) considered a carrier model operating according to Michaelis-Menten kinetics but including the stipulation that the carrier does not neutralize the charge of the ion. The kinetics of transport, specifically the term for the maximal velocity of transport, is accordingly modified by progressive changes in the membrane potential as the external concentration of the ion rises.

Values for rates of anion absorption calculated on this basis show a dual pattern like that which has been observed. At the same time, the model has severe limitations, as recognized by the authors. Above all, it is restricted to anion transport, while many experiments have demonstrated the dual pattern of transport kinetics for cations (Table *3.4*). Also, like the other single-mechanism schemes described above, it does not make provision for the inflections of the absorption isotherm in the range of high concentrations.

4.2.4 Ion-Induced Changes in Carrier Conformation

HODGES (1973) has proposed a model to account for the complex kinetics of ion transport. It is based on the analysis of enzyme kinetics advanced by KOSHLAND (1970), LEVITZKI and KOSHLAND (1969), and TEIPEL and KOSHLAND (1969). KOSHLAND (1970) has drawn attention to two principal modifications of classical Michaelis-Menten kinetics. One is positive cooperative kinetics, the other, negative cooperative kinetics. In positive cooperativity, the plot of rate, v, as a function of substrate concentration, c_S, yields a sigmoidal curve and the $1/v$ vs. $1/c_S$ plot is concave upward. In negative cooperativity the v vs. c_S plot appears qualitatively like a Michaelis-Menten (saturation) type curve, but the $1/v$ vs. $1/c_S$ plot is concave downward.

These contrasting behaviors can be accounted for in a variety of ways of which the preferred one is the assumption of a multisubunit enzyme. Binding of substrate (or other ligand) by one of the subunits induces conformational changes in neighboring subunits. These changes cause either an increase in their affinity for the substrate (positive cooperativity) or a decrease in the affinity (negative cooperativity). A combination of (mainly) negative cooperativity with some positive component might account for complex, multi-step isotherms that have been obtained in studies on enzymes. A similar explanation might be invoked to explain carrier-mediated ion transport and its complex kinetics (HODGES, 1973). BAKER and HALL (1973) also have proposed a model of ion uptake that involves ion-induced conformational changes in membrane proteins, as well as pinocytotic vesiculation at the plasmalemma. Transport into the vacuole may be effected by similar mechanisms. Thus LEIGH and WYN JONES (1974) have suggested that an ion-stimulated ATPase they have isolated is located at the tonoplast. They found a kinetic response for this enzyme similar to that described by HODGES for the plasmalemma ATPase (see Part A, Chap. *10*.5).

4.3 Concluding Comments on Ion Transport Kinetics

It is clear from the preceding discussion that the basic observations on the complexity of ion transport kinetics that emerged from the work of EPSTEIN and his collaborators have been widely confirmed and extended, but that there is no unanimity as to their interpretation. It is useful to classify the various interpretations, as follows.

With the exception of THELLIER's (*3.2*.4.2.2), all the models incorporate the original concepts of carriers possessing active sites and obeying enzyme kinetics. Only one, that of LATIES (*3.2*.4.1.2) of necessity requires two separate carrier molecules for the rate-limiting step in the absorption of a given ionic species, because he assigns two mechanisms to two different membranes, the plasmalemma and tonoplast, respectively. (In analogy to this model NISSEN (1973a, 1974) proposed that multiphasic systems may reside both in the plasmalemma and in the tonoplast so that at each of these membranes, transport is governed by a single entity of multiphasic nature, i.e., the multiphasic plasmalemma system and the multiphasic tonoplast system respectively, the latter playing a minor role in the overall absorption, and coming into play only at very high external concentrations.) Only one hypothesis among those invoking carriers and their binding sites, that of GERSON and POOLE (*3.2*.4.2.3), implies a single type of site, the duality of the kinetics being accounted for by changes in membrane potential at increasing external concentrations of the ion.

Nissen's (*3*.2.4.2.1) discussions of "multiphasic" absorption are not specific as to their meaning in terms of carriers. He speaks of "different configurations of the same rate-limiting structure" (Nissen, 1971, p. 323), but elsewhere invokes "sudden alterations in membrane resistance caused by changes in ionic strength" (Nissen, 1973b, p. 118). He finds it "difficult to envisage a protein undergoing a series of all-or-none transitions", but yet postulates "all-or-none phase transitions in a common protein or lipid moiety" (Nissen, 1973d, pp. 38–39).

Only two of the schemes unequivocally postulate multi-site carriers in the plasmalemma and account for the complexity of the absorption isotherm on the basis of different affinities of carrier sites for the ions. These are the hypotheses of Epstein (*3*.2.4.1.1) and Hodges (*3*.2.4.2.4). Epstein (1966) recognized "two sets of carrier sites" (those referred to as mechanisms I and II, respectively), with a "spectrum of carrier sites differing slightly in their affinity for the ion" in the second (mechanism II) set. Hodges' (1973) innovation lies in his application of Koshland's (1970) concept of negative and positive cooperativity as the mechanism whereby different ionic affinities of binding sites for a given ion are brought about, bearing in that respect some resemblance to Nissen's "phase" transitions.

Finally, although the above discussion and classification of hypotheses has been in terms of "dual" *vs.* "unitary" mechanisms this distinction may well represent a semantic exaggeration. None of the schemes with the exception of Laties' requires the assumption of two discrete carrier molecules for the rate-limiting step in the absorption of a given ionic species. On the other hand, even the "unitary" mechanisms must take into account the marked basic duality of the absorption isotherm documented in the numerous experiments listed in Table *3*.4 and illustrated in Figs. *3*.8 and *3*.12, and reflected as well in the contrasting features of ion absorption in the ranges of low and high concentrations referred to in *3*.2.3.2 and *3*.2.3.3. Also, the "unitary" schemes must come to terms with and account for the experimental results showing that absorption rates of ions at high concentrations can be shown to be summations of two fractions quantitatively and qualitatively corresponding to the contributions to the absorption rates identified on other evidence as those assigned to "mechanism I" and "mechanism II," respectively (see the last paragraph of *3*.2.4.2.1, and for a fuller discussion, Epstein, 1973). In terms of binding of ions to sites with differential affinities, the wide transition observed as the concentration of an ion is increased from the low ("mechanism I") to the high ("mechanism II") range must reflect differences of molecular conformation much more far-reaching than those responsible for the fine structure of the isotherm within the high ("mechanism II") range.

Acknowledgments

Principal support for the research by the author and his associates resulting in the discovery of the complex ion transport kinetics described here came from the National Science Foundation and Office of Saline Water, United States Department of the Interior. Current support is from the National Science Foundation and National Sea Grant Program, United States Department of Commerce. I am indebted to my colleague, Dr. R.G. Burau, for his help in the computer plotting of Figs. *3*.9 and *3*.10. The manuscript of this Chapter was extensively revised while I was a Senior Fulbright Research Scholar at the Department of Scientific and Industrial Research, Palmerston North, New Zealand.

References

ANDERSON, W.P.: Ion transport in the cells of higher plant tissues. Ann. Rev. Plant Physiol. **23**, 51–72 (1972).

ANDREW, C.S.: A kinetic study of phosphate absorption by excised roots of *Stylosanthes humilis, Phaseolus lathyroides, Desmodium uncinatum, Medicago sativa,* and *Hordeum vulgare.* Australian J. Agr. Sci. **17**, 611–624 (1966).

ARISZ, W.H.: Influx and efflux by leaves of *Vallisneria spiralis.* I. Active uptake and permeability. Protoplasma **57**, 5–26 (1963).

ARISZ, W.H.: Influx and efflux of electrolytes. Part II. Leakage out of cells and tissues. Acta Botan. Neer. **13**, 1–58 (1964).

ASHER, C.J., OZANNE, P.G.: Growth and potassium content of plants in solution cultures maintained at constant potassium concentrations. Soil Sci. **103**, 155–161 (1967).

AYADI, A., DEMUYTER, P., THELLIER, M.: Interprétation électrocinétique des interactions compétitives réciproques K$^+$/Rb$^+$ lors de l'absorption de ces ions par la *Lemna minor.* Compt. Rend. **273** D, 67–70 (1971).

BAKER, D.A., HALL, J.L.: Pinocytosis, ATP-ase and ion uptake by plant cells. New Phytologist **72**, 1281–1291 (1973).

BANGE, G.G.J., MEIJER, C.L.C.: The alkali cation carrier of barley roots: a macromolecular structure? Acta Botan. Neerl. **15**, 434–450 (1966).

BARBER, D.A.: "Dual isotherms" for the absorption of ions by plant tissues. New Phytologist **71**, 255–262 (1972).

BARBER, S.A.: Nutrients in the soil and their flow to plant roots. In: The belowground ecosystem (J.K. MARSHALL, ed.). In press, 1975.

BERLIER, Y., GUIRAUD, G., SAUVAIRE, Y.: Etude avec l'azote 15 de l'absorption et du métabolisme de l'ammonium fourni à concentration croissante à des racines excisées de maïs. Agrochimica **13**, 250–260 (1969).

BÖSZÖRMÉNYI, Z.: The ion uptake of excised barley roots with special reference to the low concentration process. Advan. Frontiers Plant Sci. **16**, 11–29 (1966).

BÖSZÖRMÉNYI, Z., CSEH, E., GÁRDOS, G., KERTAI, P.: Transport processes in living organisms. Budapest: Akadémiai Kiadó 1972.

BÖSZÖRMÉNYI, Z., CSEH, E., MESZES, G.: Characterization of some parameters of ion transport and translocation. I. Effects of isolation and preloading on bromide transport. Acta Agron. Acad. Sci. Hung. **18**, 195–207 (1969).

BORST-PAUWELS, G.W.F.H.: Two site-single carrier transport kinetics. J. Theoret. Biol. **40**, 19–31 (1973).

BOWEN, J.E.: Borate absorption in excised sugarcane leaves. Plant Cell Physiol. **9**, 467–478 (1968).

BRIGGS, G.E., HOPE, A.B., ROBERTSON, R.N.: Electrolytes and plant cells. Oxford: Blackwell 1961.

BROUWER, R.: Ion absorption and transport in plants. Ann. Rev. Plant Physiol. **16**, 241–266 (1965).

CARTER, O.G., LATHWELL, D.J.: Effects of temperature on orthophosphate absorption by excised corn roots. Plant Physiol. **42**, 1407–1412 (1967).

CARTWRIGHT, B.: The effect of phosphate deficiency on the kinetics of phosphate absorption by sterile excised barley roots, and some factors affecting the ion uptake efficiency of roots. Soil Sci. Plant Anal. **3**, 313–322 (1972).

CLARKSON, D.T.: Ion transport and cell structure in plants. London: McGraw-Hill 1974.

COLLANDER, R.: Cell membranes: their resistance to penetration and their capacity for transport. In: Plant Physiology—a treatise (F.C. STEWARD ed.), vol. II, p. 3–102. New York-London: Academic Press 1959.

CONWAY, E.J., DOWNEY, M.: An outer metabolic region of the yeast cell. Biochem. J. **47**, 347–355 (1950).

CRAM, W.J.: Chloride fluxes in cells of the isolated root cortex of *Zea mays.* Australian J. Biol. Sci. **26**, 757–779 (1973).

DOWD, J.E., RIGGS, D.S.: A comparison of estimates of Michaelis-Menten kinetic constants from various linear transformations. J. Biol. Chem. **240**, 863–869 (1965).

El-Sheikh, A.M., Ulrich, A.: Sodium absorption by intact sugar beet plants. Plant Physiol. **48**, 747–751 (1971).

Elzam, O.E., Epstein, E.: Absorption of chloride by barley roots: kinetics and selectivity. Plant Physiol. **40**, 620–624 (1965).

Elzam, O.E., Epstein, E.: Salt relations of two grass species differing in salt tolerance. II. Kinetics of the absorption of K, Na, and Cl by their excised roots. Agrochimica **13**, 196–206 (1969).

Elzam, O.E., Hodges, T.K.: Calcium inhibition of potassium absorption in corn roots. Plant Physiol. **42**, 1483–1488 (1967).

Elzam, O.E., Rains, D.W., Epstein, E.: Ion transport kinetics in plant tissue: complexity of the chloride absorption isotherm. Biochem. Biophys. Res. Commun. **15**, 273–276 (1964).

Epstein, E.: Mechanism of ion absorption by roots. Nature **171**, 83–84 (1953).

Epstein, E.: Passive permeation and active transport of ions in plant roots. Plant Physiol. **30**, 529–535 (1955).

Epstein, E.: Mineral nutrition of plants: mechanisms of uptake and transport. Ann. Rev. Plant Physiol. **7**, 1–24 (1956).

Epstein, E.: The essential role of calcium in selective cation transport by plant cells. Plant Physiol. **36**, 437–444 (1961).

Epstein, E.: Mineral metabolism. In: Plant biochemistry (J. Bonner and J.E. Varner, eds.), p. 438–466. New York: Academic Press 1965.

Epstein, E.: Dual pattern of ion absorption by plant cells and by plants. Nature **212**, 1324–1327 (1966).

Epstein, E.: Microorganisms and ion absorption by roots. Experientia **24**, 616–617 (1968).

Epstein, E.: Mineral nutrition of plants: principles and perspectives. New York: Wiley and Sons 1972.

Epstein, E.: Mechanisms of ion transport through plant cell membranes. Intern. Rev. Cytol. **34**, 123–168 (1973).

Epstein, E., Hagen, C.E.: A kinetic study of the absorption of alkali cations by barley roots. Plant Physiol. **27**, 457–474 (1952).

Epstein, E., Leggett, J.E.: The absorption of alkaline earth cations by barley roots: kinetics and mechanism. Amer. J. Bot. **41**, 785–791 (1954).

Epstein, E., Rains, D.W.: Carrier-mediated cation transport in barley roots: kinetic evidence for a spectrum of active sites. Proc. Natl. Acad. Sci. U.S. **53**, 1320–1324 (1965).

Epstein, E., Rains, D.W., Elzam, O.E.: Resolution of dual mechanisms of potassium absorption by barley roots. Proc. Natl. Acad. Sci. U.S. **49**, 684–692 (1963a).

Epstein, E., Schmid, W.E., Rains, D.W.: Significance and technique of short-term experiments on solute absorption by plant tissue. Plant Cell Physiol. **4**, 79–84 (1963b).

Falade, J.A.: Interrelationships between potassium, calcium, and magnesium nutrition of *Zea mays* L. Ann. Bot. (London), N.S. **37**, 345–353 (1973).

Fried, M., Noggle, J.C.: Multiple-site uptake of individual cations by roots as affected by hydrogen ion. Plant Physiol. **33**, 139–144 (1958).

Gauch, H.G.: Inorganic plant nutrition. Stroudsberg: Dowden-Hutchinson and Ross 1972.

Gerson, D.F., Poole, R.J.: Anion absorption by plants: a unary interpretation of "dual mechanisms." Plant Physiol. **48**, 509–511 (1971).

Gerson, D.F., Poole, R.J.: Chloride accumulation by mung bean root tips: a low-affinity active transport system at the plasmalemma. Plant Physiol. **50**, 603–607 (1972).

Greenham, C.G.: The relative electrical resistances of the plasmalemma and tonoplast in higher plants. Planta **69**, 150–157 (1966).

Haldane, J.B.S.: Enzymes. London: Longmans, Green 1930.

Hanson, J.B.: Impairment of respiration, ion accumulation, and ion retention in root tissue treated with ribonuclease and ethylenediamine tetraacetic acid. Plant Physiol. **35**, 372–379 (1960).

Hanson, J.B.: Metabolic aspects of ion transport. In: Genes to genus (F.A. Greer, A.T. Army, eds.), p. 63–74. Skokie: International Minerals and Chemical Corporation 1965.

Harper, E.T.: Kinetics of the two-sited enzyme. I. Activation and inhibition by substrate. J. Theoret. Biol. **32**, 405–414 (1971).

Harper, E.T.: Kinetics of the two-sited enzyme. II. A method of distinguishing between anticooperative and independent active sites based on competitive inhibition. J. Theoret. Biol. **39**, 91–102 (1973).

HIATT, A.J.: Relationship of cell sap pH to organic acid change during ion uptake. Plant Physiol. **42**, 294–298 (1967).

HIATT, A.J.: Electrostatic association and Donnan phenomena as mechanisms of ion accumulation. Plant Physiol. **43**, 893–901 (1968).

HIGINBOTHAM, N.: The mineral absorption process in plants. Botan. Rev. **39**, 15–69 (1973).

HIGINBOTHAM, N.: Conceptual developments in membrane transport, 1924–1974. Plant Physiol. **54**, 454–462 (1974).

HOAGLAND, D.R.: Lectures on the inorganic nutrition of plants. Waltham: Chronica Botanica 1944.

HOAGLAND, D.R., BROYER, T.C.: General nature of the process of salt accumulation by roots with description of experimental methods. Plant Physiol. **11**, 471–507 (1936).

HOAGLAND, D.R., BROYER, T.C.: Accumulation of salt and permeability in plant cells. J. Gen. Physiol. **25**, 865–880 (1942).

HOAGLAND, D.R., STEWARD, F.C.: Metabolism and salt absorption by plants. Nature **143**, 1031–1032 (1939).

HOAGLAND, D.R., STEWARD, F.C.: Salt absorption of plants. Nature **145**, 116–117 (1940).

HODGES, T.K.: Ion absorption by plant roots. Advan. Agron. **25**, 163–207 (1973).

HOFSTEE, B.H.J.: On the evaluation of the constants V_m and K_m in enzyme reactions. Science **116**, 329–331 (1952).

HOFSTEE, B.H.J.: Non-inverted *versus* inverted plots in enzyme kinetics. Nature **184**, 1296–1298 (1959).

HOLMERN, K., VANGE, M.S., NISSEN, P.: Multiphasic uptake of sulfate by barley roots. II. Effects of washing, divalent cations, inhibitors, and temperature. Physiol. Plantarum **31**, 302–310 (1974).

JACKMAN, R.H.: The uptake of rubidium by the roots of some graminaceous and leguminous plants. New Zealand J. Agr. Res. **8**, 763–777 (1965).

JACOBSON, L., HANNAPEL, R.J., MOORE, D.P., SCHAEDLE, M.: Influence of calcium on selectivity of ion absorption process. Plant Physiol. **36**, 58–61 (1961).

JACOBY, B., PLESSNER, O.E.: Some aspects of chloride absorption by bean leaf tissue. Ann. Botany **34**, 177–182 (1970).

JESCHKE, W.D.: Der Influx von Kaliumionen bei Blättern von *Elodea densa*, Abhängigkeit vom Licht, von der Kaliumkonzentration und von der Temperatur. Planta **91**, 111–128 (1970a).

JESCHKE, W.D.: Über die Verwendung von ^{86}Rb als Indikator für Kalium, Untersuchungen am lichtgeförderten ^{42}K/K- und ^{86}Rb/Rb-Influx bei *Elodea densa*. Z. Naturforsch. **25b**, 624–630 (1970b).

JESCHKE, W.D.: Wirkung von K^+ auf die Fluxe und den Transport von Na^+ in Gerstenwurzeln, K^+-stimulierter Na^+-Efflux in der Wurzelrinde. Planta **106**, 73–90 (1972).

JESCHKE, W.D.: K^+-stimulated Na^+ efflux and selective transport in barley roots. In: Ion transport in plants (W.P. ANDERSON, ed.), p. 285–296. London-New York: Academic Press 1973.

JESCHKE, W.D., SIMONIS, W.: Über die Aufnahme von Phosphat- und Sulfationen durch Blätter von *Elodea densa* und ihre Beeinflussung durch Licht, Temperatur und Außenkonzentration. Planta **67**, 6–32 (1965).

JOHANSEN, C., LONERAGAN, J.F.: Effects of anions and cations on potassium absorption by plants of high potassium chloride content. Australian J. Plant Physiol. **2**, 75–83 (1975a).

JOHANSEN, C., LONERAGAN, J.F.: Effect of anions on potassium transport to shoots in plants of high potassium chloride content. Z. Pflanzenphysiol. **75**, 415–421 (1975b).

JOHANSON, L., JOHAM, H.E.: The influence of calcium absorption and accumulation on the growth of excised cotton roots. Plant and Soil **34**, 331–339 (1971).

KANNAN, S.: Plasmalemma: the seat of dual mechanisms of ion absorption in *Chlorella pyrenoidosa*. Science **173**, 927–929 (1971a).

KANNAN, S.: Kinetics of iron absorption by excised rice roots. Planta **96**, 262–270 (1971b).

KANNAN, S., RAMANI, S.: Mechanisms of ion absorption by bean leaf slices and transport in intact plants. Z. Pflanzenphysiol. **71**, 220–227 (1974).

KAPLAN, O.B.: Kinetics of magnesium absorption by excised barley roots. Ph.D. Thesis, University of California, Davis (1969).

KOCH, A.L.: Deviations from hyperbolic dependency of transport processes. J. Theoret. Biol. **36**, 23–40 (1972).

Koshland, D.E., Jr.: The molecular basis for enzyme regulation. In: The enzymes (P.D. Boyer, ed.), 3rd ed., vol. 1, p. 341–396. New York: Academic Press 1970.

Läuchli, A.: Translocation of inorganic solutes. Ann. Rev. Plant Physiol. **23**, 197–218 (1972).

Läuchli, A., Epstein, E.: Transport of potassium and rubidium in plant roots: the significance of calcium. Plant Physiol. **45**, 639–641 (1970).

Läuchli, A., Epstein, E.: Lateral transport of ions into the xylem of corn roots. I. Kinetics and energetics. Plant Physiol. **48**, 111–117 (1971).

Laties, G.G.: Metabolic and physiological development in plant tissues. Australian J. Sci. **30**, 193–203 (1967).

Laties, G.G.: Dual mechanisms of salt uptake in relation to compartmentation and long-distance transport. Ann. Rev. Plant Physiol. **20**, 89–116 (1969).

Laties, G.G.: Solute transport in relation to metabolism and membrane permeability in plant tissues. In: Historical and current aspects of plant physiology: A symposium honoring Professor F.C. Steward. New York State College of Agriculture and Life Sciences. p. 98–151. New York: Cornell University 1975.

Leggett, J.E., Epstein, E.: Kinetics of sulfate absorption by barley roots. Plant Physiol. **31**, 222–226 (1956).

Leggett, J.E., Galloway, R.A., Gauch, H.G.: Calcium activation of orthophosphate absorption by barley roots. Plant Physiol. **40**, 897–902 (1965).

Leigh, R.A., Wyn Jones, R.G.: The effect of increased internal ion concentration upon the ion uptake isotherms of excised maize root segments. J. Exptl. Bot. **24**, 787–795 (1973).

Leigh, R.A., Wyn Jones, R.G., Williamson, F.A.: The possible role of vesicles and ATPases in ion uptake. In: Ion transport in plants (W.P. Anderson, ed.), p. 407–418. London-New York: Academic Press 1973.

Leigh, R.A., Wyn Jones, R.G., Williamson, F.A.: Ion fluxes and ion-stimulated ATPase activities. In: Membrane transport in plants (U. Zimmermann and J. Dainty, eds.), p. 307–316. Berlin-Heidelberg-New York: Springer 1974.

Leonard, R.T., Hodges, T.K.: Characterization of plasma membrane-associated adenosine triphosphatase activity of oat roots. Plant Physiol. **52**, 6–12 (1973).

Levitzki, A., Koshland, D.E., Jr.: Negative cooperativity in regulatory enzymes. Proc. Natl. Acad. Sci. U.S. **62**, 1121–1128 (1969).

Lüttge, U.: Aktiver Transport (Kurzstreckentransport bei Pflanzen). Protoplasmatologia **VIII**, 7.b, 1–146 (1969).

Lüttge, U.: Stofftransport der Pflanzen. Berlin-Heidelberg-New York: Springer 1973.

Lüttge, U., Bauer, K.: Die Kinetik der Ionenaufnahme durch junge und alte Sprosse von *Mnium cuspidatum*. Planta **78**, 310–320 (1968).

Lüttge, U., Laties, G.G.: Dual mechanisms of ion absorption in relation to long distance transport in plants. Plant Physiol. **41**, 1531–1539 (1966).

Lüttge, U., Laties, G.G.: Selective inhibition of absorption and long distance transport in relation to the dual mechanisms of ion absorption in maize seedlings. Plant Physiol. **42**, 181–185 (1967a).

Lüttge, U., Laties, G.G.: Absorption and long distance transport by isolated stele of maize roots in relation to the dual mechanisms of ion absorption. Planta **74**, 173–187 (1967b).

Lundegårdh, H.: Transport of water and salts through plant tissues. Nature **157**, 575–577 (1946).

Maas, E.V.: Calcium uptake by excised maize roots and interactions with alkali cations. Plant Physiol. **44**, 985–989 (1969).

Minchin, F.R., Baker, D.A.: The influence of calcium on potassium fluxes across the root of *Ricinus communis*. Planta **113**, 97–104 (1973).

Neame, K.D., Richards, T.G.: Elementary kinetics of membrane carrier transport. New York: Wiley and Sons 1972.

Nissen, P.: Uptake of sulfate by roots and leaf slices of barley: mediated by single, multiphasic mechanisms. Physiol. Plantarum **24**, 315–324 (1971).

Nissen, P.: Multiphasic ion uptake in roots. In: Ion transport in plants (W.P. Anderson, ed.), p. 539–553. London-New York: Academic Press 1973a.

Nissen, P.: Kinetics of ion uptake in higher plants. Physiol. Plantarum **28**, 113–120 (1973b).

Nissen, P.: Multiphasic uptake in plants. I. Phosphate and sulfate. Physiol. Plantarum **28**, 304–316 (1973c).

NISSEN, P.: Multiphasic uptake in plants. II. Mineral cations, chloride, and boric acid. Physiol. Plantarum **29**, 298–354 (1973d).

NISSEN, P.: Uptake mechanisms: inorganic and organic. Ann. Rev. Plant Physiol. **25**, 53–79 (1974).

OERTLI, J.J.: The salt absorption isotherm. Physiol. Plantarum **20**, 1014–1026 (1967).

OSMOND, C.B., LATIES, G.G.: Interpretation of the dual isotherm for ion absorption in beet tissue. Plant Physiol. **43**, 747–755 (1968).

OSMOND, C.B., LATIES, G.G.: Effect of poly-L-lysine on potassium fluxes in red beet tissue. J. Membrane Biol. **2**, 85–94 (1970).

OSTERHOUT, W.J.V.: The mechanism of accumulation in living cells. J. Gen. Physiol. **35**, 579–594 (1952).

PENTH, B., WEIGL, J.: Unterschiedliche Wirkung von Licht auf die Aufnahme von Chlorid und Sulfat in *Limnophilia*blätter. Abhängigkeit der Lichtwirkung von der Konzentration der Anionen. Z. Naturforsch. **24b**, 342–348 (1969).

PICCIURRO, G., BRUNETTI, N.: Assorbimento del sodio (Na^{22}) in radici escisse di alcune varietà di *Lycopersicum esculentum*. Agrochimica **13**, 347–357 (1969).

PIERCE, W.S., HIGINBOTHAM, N.: Compartments and fluxes of K^+, Na^+ and Cl^- in *Avena* coleoptile cells. Plant Physiol. **46**, 666–673 (1970).

PITMAN, M.G.: Simulation of Cl^- uptake by low-salt barley roots as a test of models of salt uptake. Plant Physiol. **44**, 1417–1427 (1969).

PLOWMAN, K.M.: Enzyme kinetics. New York: McGraw-Hill 1972.

POOLE, R.J.: Carrier-mediated potassium efflux across the cell membrane of red beet. Plant Physiol. **44**, 485–490 (1969).

RAINS, D.W.: Sodium and potassium absorption by bean stem tissue. Plant Physiol. **44**, 547–554 (1969).

RAINS, D.W., EPSTEIN, E.: Transport of sodium in plant tissue. Science **148**, 1611 (1965).

RAINS, D.W., EPSTEIN, E.: Sodium absorption by barley roots: role of the dual mechanisms of alkali cation transport. Plant Physiol. **42**, 314–318 (1967a).

RAINS, D.W., EPSTEIN, E.: Sodium absorption by barley roots: its mediation by mechanism 2 of alkali cation transport. Plant Physiol. **42**, 319–323 (1967b).

RAINS, D.W., EPSTEIN, E.: Preferential absorption of potassium by leaf tissue of the mangrove, *Avicennia marina:* an aspect of halophytic competence in coping with alt. Australian J. Biol. Sci. **20**, 847–857 (1967c).

REISENAUER, H.M.: Mineral nutrients in soil solution. In: Environmental biology (P.L. ALTMAN and D.S. DITTMER, eds.), p. 507–508. Bethesda: Federation of American Societies for Experimental Biology 1966.

RICHTER, C., MARSCHNER, H.: Umtausch von Kalium in verschiedenen Wurzelzonen von Maiskeimpflanzen. Z. Pflanzenphysiol. **70**, 211–221 (1973).

ROBERTSON, R.N., WILKINS, M.J.: Studies in the metabolism of plant cells. VII. The quantitative relation between salt accumulation and salt respiration. Australian J. Sci. Res. **B 1**, 17–37 (1948).

SALSAC, L.: Absorption du calcium par les racines de Féverole (calcicole) et de Lupin jaune (calcifuge). Physiol. Vég. **11**, 95–119 (1973).

SMITH, P.G.: Dependence of ion flux across a membrane on ionic concentration. J. Theoret. Biol. **41**, 269–286 (1973).

SMITH, R.C., EPSTEIN, E.: Ion absorption by shoot tissue: technique and first findings with excised leaf tissue of corn. Plant Physiol. **39**, 338–341 (1964).

STEWARD, F.C., HARRISON, J.A.: The absorption and accumulation of salts by living plant cells. IX. The absorption of rubidium bromide by potato discs. Ann. Bot. (London), N.S. **3**, 427–453 (1939).

TEIPEL, J., KOSHLAND, D.E., JR.: The significance of intermediary plateau regions in enzyme saturation curves. Biochemistry **8**, 4656–4663 (1969).

TEMPLE-SMITH, M.G., MENARY, R.C.: Phosphate uptake by sterile and non-sterile excised roots of cabbage and lettuce. In: Mechanisms of regulation of plant growth (R.L. BIELESKI, A.R. FERGUSON, M.M. CRESSWELL, eds.), p. 139–144. Bulletin 12. Wellington: The Royal Society of New Zealand, 1974.

THELLIER, M.: Essai d'interpretation électrocinétique du fonctionnement de mécanismes transporteurs ou enzymatiques, spécialement dans le cas de l'oxydoréduction. Compt. Rend. **266 D**, 826–829 (1968).

Thellier, M.: An electrokinetic interpretation of the functioning of biological systems and its application to the study of mineral salts absorption. Ann. Botany **34**, 983–1009 (1970).

Thellier, M.: Non-equilibrium thermodynamics and electrokinetic interpretation of biological systems. J. Theoret. Biol. **31**, 389–393 (1971).

Thellier, M.: Electrokinetic formulation of ionic absorption by plant samples. In: Ion transport in plants (W.P. Anderson, ed.), p. 47–63. London-New York: Academic Press 1973.

Thellier, M., Thoiron, B., Thoiron, A.: Electrokinetic formulation of overall kinetics of *in vivo* processes. Physiol. Vég. **9**, 65–82 (1971).

Torii, K., Laties, G.G.: Dual mechanisms of ion uptake in relation to vacuolation in corn roots. Plant Physiol. **41**, 863–870 (1966).

Van den Honert: Over Eigenschappen van Plantenwortels, welke een Rol spelen bij de Opname van Voedingszouten. Natuurk. Tijdschr. Nederl. Ind. **97**, 150–162 (1937).

Vange, M.S., Holmern, K., Nissen, P.: Multiphasic uptake of sulfate by barley roots. I. Effect of analogues, phosphate, and pH. Physiol. Plantarum **31**, 292–301 (1974).

Venrick, D.M., Smith, R.C.: The influence of initial salt status on absorption of rubidium by corn root segments of two stages of development. ·Bull. Torrey Bot. Club **94**, 501–510 (1967).

Weigl, J.: Efflux und Transport von Cl^- und Rb^+ in Maiswurzeln. Wirkung von Außenkonzentration, Ca^{++}, EDTA und IES. Planta **84**, 311–323 (1969).

Welch, R.M.: The plasmalemma as the seat of the dual ion carrier mechanisms of plant cells. Ph.D. Thesis, University of California, Davis (1969).

Welch, R.M., Epstein, E.: The dual mechanisms of alkali cation absorption by plant cells: their parallel operation across the plasmalemma. Proc. Natl. Acad. Sci. (U.S.) **61**, 447–453 (1968).

Welch, R.M., Epstein, E.: The plasmalemma: seat of the type 2 mechanisms of ion absorption. Plant Physiol. **44**, 301–304 (1969).

Whaley, W.G., Mollenhauer, H.H., Leech, J.H.: The ultrastructure of the meristematic cell. Amer. J. Bot. **47**, 401–449 (1960).

Young, M., Sims, A.P.: The potassium relations of *Lemna minor* L. I. Potassium uptake and plant growth. J. Exptl. Bot. **23**, 958–969 (1972).

3.3 Ion Uptake by Plant Roots

M.G. PITMAN

1. Processes Involved in Ion Uptake

Chap. *3.2* describes results from a particular approach to study of the mechanism of uptake to plant roots, in which the rate of uptake has been used to characterize the carrier systems involved in transport. There is an interesting correlation between these rates of uptake to the roots and the kinetics of stimulation of an ATPase associated with plasmalemma membrane fractions (Part A, *10*.5; Fig. *10*.10), implying that the carriers are ATPases.

Other approaches such as measurement of fluxes or cell PD and the judicious use of inhibitors give different types of information, with an emphasis on the systems of transport and related processes in the cell. This type of work was the basis for much of Part A, Chap. *6* (algal cells), Part A, Chap. *7* (fungal cells) and Part A, Chap. *8* (storage tissues). This Chapter (*3.3*) is concerned with applications of these techniques to root cells. A particular problem is to bring together such results with the findings of Chap. *3.2*. There are an incredible number of studies of transport to roots, many of which have had to be omitted for lack of space and because the aim here has been to extract some consistent generalizations. EPSTEIN (1972) gives a good account of much of this work. See also LÜTTGE (1973); CLARKSON (1974); BAKER and HALL (1975).

Operation of the carriers defined by kinetic analysis is dependent on the cells' metabolism. For example, uncoupling oxidative phosphorylation with CCCP or DNP brings uptake to a dramatic stop (e.g. LÄUCHLI and EPSTEIN, 1971). Table *3*.6 gives further examples of the effects of various inhibitors and antibiotics on transport. A difficulty with such studies is that inhibition is not evidence that a *particular* flux is metabolically active. In low salt roots tracer influx is due to large net uptake of both cation and anion, so that inhibition of one will lead to inhibition of the other, even if it is passive.

The difficulty of deciding which flux is active has a bearing on the problem of locating ion transport in root cells, i.e. of explaining tracer uptake in terms of fluxes at the plasmalemma and tonoplast. As discussed in *3.2.4.1* there seems to be a need for active transport at both tonoplast and plasmalemma. The difficulty is in deciding the relative contributions of these processes to net tracer uptake. This topic will be examined further, using salt-saturated roots (*3.1.2.5*; Table *3.3*) for measurement of cell PD's (*3.3.2*) and fluxes (*3.3.3*).

Transfer of low-salt roots to KCl or NaCl solutions initiates rapid net uptakes of K^+, Na^+ and Cl^-, but may also stimulate organic acid synthesis and H^+ ion influx. This synthesis makes use of reducing sugars stored in the vacuoles and the sugar concentration falls rapidly as salt is accumulated (PITMAN et al., 1971 b). The amount of organic acid formed is equivalent to the difference between cation

Table *3.6*. Effects of inhibitors on uptake of K^+, Na^+ and Cl^- by plant roots. All measurements are Φ_{ov} except where shown otherwise. Low-salt roots are shown as LS

Inhibitor	Plant	Inhibitor Concn (μM)	Ion	Ion Concn (mM)	Inhibition (%)	Ref.
DNP	*Phaseolus*	100	K^+	3.0	92	Drew and Biddulph
		1000	K^+	3.0	89	(1971)
	Barley (LS)	100	K^+	0.01	100	Jackson and Adams
		100	Na^+	1.0	90	(1963)
	Barley (LS)	100	Na^+	0.1	84	Rains and Epstein
		100	Na^+	5 (+ 1 mM K^+)	82	(1967a)
CCCP	Maize cortex (LS)	1	K^+	0.2	99	Baker (1973)
	Maize cortex	3	Cl^-	0.1	78	Cram (1973a)
		3	Cl^-	50.0	48	
		3	Cl^-	50.0	18	
	Barley (LS)	5	Cl^-	0.1	100	Pitman (1970)
		5	Cl^-	10.0	95–100	
	Maize ϕ_{oc} (5 min)	1	Cl^-	0.5	60	Läuchli and Epstein
		1	Cl^-	5.0	36	(1971)
	Maize	1	Cl^-	0.5	92	
		1	Cl^-	5.0	82	
	Maize Φ_{vo}	10	Cl^-	0.1 (KCl)	20	Weigl (1970)
	Barley (LS)	5	Cl^-	0.1	100	Pitman (1970)
				10.0	95–100	
	Barley	6	K^+	2.5 KCl +7.5 NaCl	85	Pitman and Saddler (1967)
		6	Na^+	2.5 KCl +7.5 NaCl	0–50	M.G. Pitman (unpublished)
		6	Cl^-	2.5 KCl +7.5 NaCl	90	
		6	Na^+	10 NaCl	75	
		6	Cl^-		85	
	Barley (ϕ_{oc})	4	Cl^-	10 KCl	60–90	Pitman (1972a)
	Barley (ϕ_{oc})	10	Cl^-	10 KCl	80–90	Pitman (1972a)
	Barley (Φ_{vo})	10	Cl^-	10 KCl	$< \pm 5$	Pitman (1972a)
KCN	*Phaseolus*	1000	K^+	0.3	58	Drew and Biddulph
		1000		3.0	62	(1971)
Azide	Barley (LS)	100	Na^+	0.1	66	Rains and Epstein
		100		5 (+1 mM K^+)	48	(1967a)
N_2	Barley (LS)	—	Na^+	0.1	50	Rains and Epstein
				10 (+1 mM K^+)	65	(1967a)

Table 3.6. (continued)

Inhibitor	Plant	Inhibitor Conc^n (μM)	Ion	Ion Conc^n (mM)	Inhibition (%)	Ref.
		(g m^{-3})				
Oligo-mycin	Maize ϕ_{oc} (10 min)	2	Cl$^-$	0.5	61	LÄUCHLI and EPSTEIN
		2		5.0	45	(1971)
		2		0.5	85	
		2		5.0	69	
	Barley (LS)	10	K$^+$	0.1	50	JACOBY and PLESSNER
		5	Na$^+$	0.1	40	(1970)
		5	Cl$^-$	0.1	80	
		10	K$^+$	5.0	50	
		5	Na$^+$	5.0	40	
		5	Cl$^-$	5.0	60	
		(μM)				
Arsenite	Barley (LS)	100	Cl$^-$	10	85–90	PITMAN (1970)
	Phaseolus	1000	K$^+$	0.3	56	DREW and BIDDULPH
		1000	K$^+$	3	64	(1971)
Ouabain	Phaseolus	1000	K$^+$		+25%	DREW and BIDDULPH (1971)
	Barley	100	K$^+$, Na$^+$		No effect	PITMAN and SADDLER (1967); JESCHKE (1973)
	Barley (ϕ_{cx})	500	Na$^+$		+20%	NASSERY and BAKER (1972b)
Mannose		10 mM	K$^+$	0.01	56	JACKSON and ADAMS (1963)
UO$_2^{2+}$	Maize (Φ_{vo})	500	^{86}Rb/K$^+$		33	WEIGL (1970)
		(g m^{-3})				
Valino-mycin	Oat (LS)	1	K$^+$	0.2	+10	HODGES et al. (1971)
Nonactin		1	K$^+$	0.2	+20	
Nigericin		1	K$^+$	0.2	+100	
Gramici-din-D		1	K$^+$	0.2	+430	

and anion uptake and to the net H$^+$ efflux (ULRICH, 1941, 1942; HIATT, 1967; Part A, Chaps. *12*; *13*). The H$^+$ efflux appears to be an active process and it increases with [KCl] (Fig. *3*.14).

Initiation of synthesis of organic acids has been suggested to be due to a decrease in cytoplasmic pH brought about by the H$^+$ extrusion (HIATT, 1967; HIATT and HENDRICKS, 1967). The continuing synthesis presumably depends on removal of organic acid from the cytoplasm to the vacuole, and hence on the rate of entry of K$^+$ or Na$^+$ across the plasmalemma, in excess of the flux of other anions to the vacuole (Cl$^-$, NO$_3^-$).

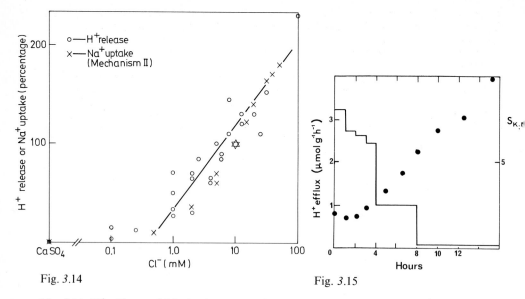

Fig. *3.14*

Fig. *3.15*

Fig. *3.14*. H^+ efflux and Mechanism II uptake. Rate of H^+ release ($\circ$) relative to that into 10 mM KCl plotted against $[Cl^-]_o$ on a logarithmic scale; $100\% \simeq 3\ \mu mol\ g_{FW}^{-1}\ h^{-1}$. Values of Mechanism II uptake ($\times$) estimated from data of RAINS and EPSTEIN (1967a) plotted on the same scale; $100\% \simeq 7\ \mu mol\ g_{FW}^{-1}\ h^{-1}$. (Fig. from PITMAN, 1970)

Fig. *3.15*. Low-salt barley roots were put into a solution of [2.5 mM KCl+7.5 mM NaCl+0.5 mM CaSO$_4$]. Loss of H^+ and uptake (Φ_{ov}) of K^+ and Na^+ was measured. Note that the selectivity ($\bullet$; $S_{K, Na}$) rises as H^+ efflux falls; $S_{K, Na}$ is defined in Table *3.7*. (Data from PITMAN et al., 1968; PITMAN, 1969a)

On the basis of measurements of H^+ fluxes and cation/anion balance during salt accumulation JACKSON and ADAMS (1963) suggested that H^+ efflux and OH^- efflux could be the driving forces for cation and anion uptake respectively. HIGIN-BOTHAM in various publications (e.g. 1973) has also pointed out the possibility of K^+ influx being coupled electrogenically with another process, possibly Cl^- influx or H^+ efflux. Clearly H^+ and OH^- fluxes provide a means of balancing cation and anion transports and organic acid synthesis (Part A, Chap. *12*). There has been a recurrence of interest in active H^+ efflux and its relation to K^+ and Na^+ transport following demonstrations of its importance in *Chara* (Part A, Chaps. *6, 9.4*) and in *Neurospora* (see Part A, Chap. *7*).

A difficulty in attempting to compare data for different root preparations is that salt status appears to affect the permeability of the cell membranes. For example, low-salt roots of barley and maize have lower selectivity for K^+ relative to Na^+ than roots at higher salt status (Table *3.7*). The change in selectivity can be detected about 4 to 6 h after transferring low-salt roots to a salt solution (barley, PITMAN et al., 1968; maize, LEIGH and WYN JONES, 1973). The change in selectivity in barley roots is associated with a decrease in net H^+ efflux (Fig. *3.15*) and with a change in P_{Na}/P_K as determined both by measurements of ψ_{vo} (*3.3.2*) and of fluxes (*3.3.3.3* and Table *3.7*).

Table 3.7. Effect of external concentration and salt status on $S_{K, Na}$ for maize and barley. Low = Low-salt status; High = plants grown on salt or culture solution. $S_{K, Na}$ measured either from tracer influx or from total content.[f]

Solution [K$^+$ + Na$^+$] (mM)	maize (Low) influx[a]	maize (High) influx[a]	barley (Low) influx[b]	barley (Low) influx[c]	barley (High) influx[d]	barley (High) Content[e]
0.01	32	57	–	–		
0.1	13	24	–	–	–	
0.2	–	–	7	–	–	
0.3	7	25	–	–	–	
1	37	36	0.8	3	–	
2	–	–	–	–	7	
5	19	38	–	–	10	
10	18	32	1.6	1.6	11	8
20	–	–	–	1.0	10	10
30	15	25	–	1.0	–	9
40	–	–	–	1.0	11	8
60	–	–	–	1.2	–	9
80	–	–	–	1.4	–	9
100	–	–	–	1.7	–	8

[a] LEIGH and WYN JONES (1973). [b] PITMAN (1969a). [c] RAINS and EPSTEIN (1967a, b). [d] PITMAN et al. (1968). [e] PITMAN (1965a).

[f] It is convenient in describing preferential uptake to use the term $S_{K, Na}$ for selectivity of K$^+$ with respect to Na$^+$ defined as:

$$S_{K, Na} = \frac{\text{Uptake of K}^+}{[K^+]_o} : \frac{\text{Uptake of Na}^+}{[Na^+]_o}$$

so that

$$\frac{\text{Uptake of K}^+}{\text{Uptake of Na}^+} = S_{K, Na} \cdot \frac{[K^+]_o}{[Na^+]_o}.$$

A further factor affecting transport into low-salt roots (and other preparations) is the availability of Ca^{2+}, which can also affect K$^+$ and Na$^+$ selectivity (EPSTEIN, 1961; 3.2.3.2).

MARSCHNER (1964) and MARSCHNER and MENGEL (1966) have commented on the parallels between Ca^{2+} and H$^+$ effects on ion uptake and exchange in roots and the dependence of membrane permeability on the local concentrations of these ions. A relationship between H$^+$ efflux and reduced selectivity for K$^+$ relative to Na$^+$ was found for barley roots grown under different levels of aeration (PITMAN, 1969a), and in this case too, H$^+$ concentration near the membrane was suggested to account in part for the differences in K$^+$ selectivity.

The possibility cannot be ignored too that such factors as H$^+$ level, Ca^{2+} and salt status affect the relationship of rates of ion uptake to external concentrations. For example KOMOR and TANNER (1974) determined the kinetics of hexose uptake at different pH. Fig. 3.16 shows that there was a shift from a process with $K_m = 0.3$ mM to one with $K_m = 30$ mM glucose as pH was made alkaline, and at neutral pH the process showed the duality in kinetics discussed in Chap. 3.2 for ions. The total amount of carrier seemed to be the same at each pH, but there was a shift in affinity with external pH.

The intention here (*3.3.1*) has been to introduce the components of transport in root cells that need to be considered together, that is: organic acid synthesis; H^+ ion efflux; levels of H^+ and Ca^{2+} (and perhaps Mg^{2+}) near the cell membranes; energy metabolism; and membrane permeability; as well as the ion transport itself. In addition the differences between low-salt and salt-saturated roots have been developed from the preliminary data in *3.1.2.5*. Later, in *3.3.9*, the integration of these processes will be discussed further to bring in additional data from studies with salt-saturated roots.

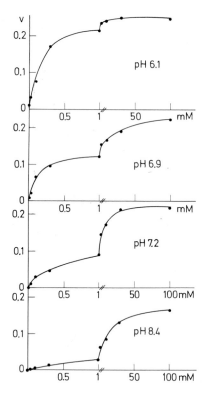

Fig. *3.16*. Kinetics of hexose uptake by *Chlorella vulgaris* cells at different pH. Cells (36 µl packed cells) induced for hexose uptake (see *9.3.3.2*) were incubated in 2.2 ml 40 mM Na-phosphate buffer of the pH as indicated. (^{3}H)-6-deoxyglucose was then added at different concentrations (0.01–100 mM). Samples were taken in $^1/_2$-min intervals, filtered and the amount of 6-deoxyglucose determined. v is given as mmoles h^{-1} cm^{-3} packed cells. 0.2 mmoles h^{-1} cm^{-3} correspond to 11.2 pmoles cm^{-2} s^{-1}. (From Komor and Tanner, 1975)

2. Electrical Measurements

Measurements of electrical potential differences between vacuoles of root cells and the external solution (ψ_{vo}) have been made for several plant species. Values are collected in Table *3.8*. In general ψ_{vo} is large and negative (vacuole relative to solution). Fig. *3.17* shows that ψ_{vo} decreases with increasing concentration as expected if diffusive permeability to K^+ is much greater than that to Cl^-. The change in potential was about 32 mV for each tenfold change in concentration above 1 mM. Similar results were found for low-salt barley roots (Pitman et al., 1971a).

There are few measurements of the effect of Na^+ or H^+ on cell potentials for roots. For low-salt barley roots variation of external pH between 4 and 7 was found to have no effect on ψ_{vo} and transferring from 10 mM KCl to 10 mM NaCl

Table *3*.8. Measurement of potential difference between vacuole and solution (ψ_{vo}) and estimation of equilibrium concentration of K^+ and Na^+. Comparison of predicted and observed concentrations suggests active efflux (e) or active influx (i) of K^+ or Na^+

Plant	Solution		PD (ψ_{vo})	Predicted		Observed	
	$[K^+]_o$ (mM)	$[Na^+]_o$ (mM)	(mV)	$[K^+]_v$ (mM)	$[Na^+]_v$ (mM)	$[K^+]_v$ (mM)	$[Na^+]_v$ (mM)
Barley[1]	2.8	8.1	− 87	86	250	80	23[e]
	0.22	9.9	−110	17	750	34	68[e]
Bean[2] (*Vicia*)	1.0	1.0	−112	80	80	83	22[e]
Sunflower[3]	7.0	0.25	− 28	21	0.8	111[i]	19[i]
	7.0	1.0	− 30	23	3.2	82[i]	25[i]
	7.0	10.0	− 32	25	35	83[i]	39
	0.7	0.25	− 72	12	4.2	59[i]	43[i]
Pea[4]	10	10	−108	240	240	122[e]	−
Maize[5]	1.0	−	− 75	19	−	110[i]	−
Ranunculus flammula[6]	0.03	0.69	− 28	0.09	2.1	43[i]	28[i]
Equisetum palustre	0.03	0.69	− 33	0.11	2.5	36[i]	24[i]
Potamogeton natans	0.055	0.89	− 33	0.20	3.3	84[i]	23[i]
Menyanthes trifoliata	0.055	0.89	− 30	0.18	3.0	50[i]	23[i]
Equisetum fluviatile	0.055	0.89	− 38	0.25	4.0	35[i]	29[i]
Triglochin maritima[7]	2	100	− 85	56	2,800	70[i]	130
	6	300	− 92	190	9,600	70	200[e]
	10	500	− 72	168	8,400	78	270[e]

[1] Pitman and Saddler (1967). [2] Pallaghy and Scott (1969). [3] Bowling and Ansari (1971). [4] Etherton (1968). [5] Dunlop and Bowling (1971). [6] Shepherd and Bowling (1973). [7] Jefferies (1973).

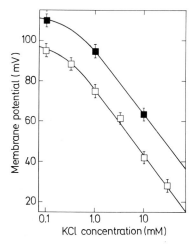

Fig. *3*.17. Dependence of the membrane potential (ψ_{vo}) of maize root epidermal cells on the KCl concentration of the bathing solution. Open symbols: tissue grown in $CaCl_2$ solution; closed symbols: tissue grown in a nutrient medium. Vertical bars indicate 95% confidence limits (Dunlop and Bowling, 1971)

changed ψ_{vo} by only -8 mV, from which it was inferred that P_{Na}/P_K was about 0.8 to 0.9 (PITMAN et al., 1971). In roots at salt-saturation however, there can be large changes in ψ_{vo} on transfer from KCl to NaCl (PITMAN et al., 1975b) showing that P_{Na}/P_K has become much smaller (about 0.3 to 0.03), as expected from measurements of K^+ and Na^+ influx (*3.3.3.1* and Table *3.9*).

Measurements of potential differences between cytoplasm and vacuole in higher plant cells are few, but it seems likely that ψ_{vc} is small and that ψ_{vo} is mainly due to ψ_{co}, as found for other higher plant cells (ETHERTON and HIGINBOTHAM, 1960; DENNY and WEEKS, 1968; ETHERTON, 1970).

In many plant tissues it has been shown that ψ_{vo} is more negative than expected for ψ_K, i.e. for electro-chemical equilibrium between K^+ inside the cell and out. See for example the discussion in Part A, *4.2.6* and the data collected in Part A, Table *4.2*. As evidence for an electrogenic pump in higher plant tissues HIGIN-BOTHAM et al. (1970) have shown that ψ_{vo} in pea stem and oat coleoptile can be depolarized (made less negative) by about 70 mV by the inhibitors KCN and DNP. Further evidence for electrogenic pumps in higher plant tissues is discussed by HIGINBOTHAM and ANDERSON (1974). Transport into root cells also seems to be electrogenic. The values of ψ_{vo} measured for low-salt barley roots (PITMAN et al., 1971a) were more negative by about 35 mV over the range 0.01 to 25 mM KCl than expected for ψ_K. In barley roots at salt saturation ψ_{vo} is close to ψ_K in 5 mM KCl but addition of fusicoccin (see *7.7.3*) hyperpolarizes the cell by about 20 mV, at the same time stimulating H^+ efflux. The addition of fusicoccin seems to be increasing an electrogenic H^+ efflux (MARRÈ et al., 1974; PITMAN et al., 1975a).

The value of ψ_{vo} in low-salt barley roots was found to drift to a more negative value within a few hours of excising the roots (PITMAN et al., 1971a). This change could not be accounted for by salt accumulation. It was suggested that the drift in potential was due to diffusion of growth substances within the root. LIN and HANSON (1974) have shown that the change in potential of maize roots following washing was due to increase in an electrogenic component. Whatever the explanation, though, it raises a practical problem in interpretation of potential measurements of root cells.

ANDERSON et al. (1974) have used a single electrode technique to determine the resistance of pea root cortical cell membranes. The resistivity was 0.21 Ω m^2 and ψ_{vo} was -138 mV (see also Part A, *4.1.3*).

3. Flux Measurements Using Roots

Part A, Chap. *5* discussed measurement of fluxes in plant cells. The particular problems of using roots were mentioned in Part A, *5.5.3* and Part A, Fig. *5.17* was suggested as a system relating various fluxes in the root to each other, which could be used as a basis for calculation of fluxes from efflux data. There are advantages in using cells near vacuolar flux equilibrium as the calculations are then simplified and less sensitive to error in efflux measurements. The simpler system of Part A, Fig. *5.4* can be used with stripped cortical cells (CRAM, 1973a).

Measurements have more usually been made of net tracer uptake over short or long periods. See Part A, *5.4.4* and *5.5.3.2* for discussion of this method.

The problems to be considered in particular here are (a), can the kinetic data of Chap. *3.2* be interpreted in terms of individual fluxes and (b), where are active fluxes located?

3.1 Selectivity of K$^+$ Relative to Na$^+$

Estimation of fluxes of K$^+$ and Na$^+$ by efflux analysis was made by PITMAN and SADDLER (1967) for barley roots at 5° C (Table *3.9*). At the tonoplast $\phi_{cv,K}$ was much greater than $\phi_{cv,Na}$, which could be explained by the lower concentration of Na$^+$ in the cytoplasm relative to K$^+$. *Selective uptake to the vacuole appeared to be due to the operation of an active Na$^+$ efflux at the plasmalemma,* and not to low permeability of the tonoplast membrane to Na$^+$, since Na$^+$ entered the vacuole readily when K$^+$ was absent from solution. Thus in 10 mM solutions the net rates of tracer uptake of Na$^+$ and Cl$^-$ (Φ_{ov}) were nearly equal and little different from uptake of ^{86}Rb (tracer for K$^+$) from (2.5 KCl+7.5 NaCl) mM, e.g. (PITMAN et al., 1968):

Solution (mM)	Influx (Φ_{ov}^*)		
	^{86}Rb	^{22}Na	^{36}Cl
	(μmol g$_{FW}^{-1}$ h^{-1})		
2.5 KCl+7.5 NaCl	4.7	0.5	4.7
10 NaCl	—	4.5	4.2

An elegant demonstration of the Na$^+$ efflux and its stimulation by K$^+$ in the solution was made by JESCHKE (1970, 1972, 1973). His values of Na$^+$ fluxes at 25° C (also given in Table *3.9*) confirm that the content of Na$^+$ in the cytoplasm is reduced disproportionately by K$^+$ in solution, and that the change in K$^+$/Na$^+$ in the cytoplasm affects uptake to the vacuole. JESCHKE's data also show that when Na$^+$ replaces K$^+$ in the solution the flux of Na$^+$ into the vacuole is increased due to the higher Na$^+$ content of the cytoplasm (see also JEFFERIES, 1973).

Transport of K$^+$ into or Na$^+$ out of root cortical cells has not been found to be affected by ouabain (Table *3.6*), the specific inhibitor of K$^+$/Na$^+$ coupled transport in certain algal cells (Part A, Chap. *6*), though Na$^+$ efflux from excised barley roots has been shown to be slightly inhibited and ϕ_{cx} increased (NASSERY and BAKER, 1972a, b).

Table *3.9*. K$^+$ and Na$^+$ fluxes in barley roots

Solution (mM)		Temp (°C)	Fluxes (μmol g$_{FW}^{-1}$ h^{-1})				Cytoplasmic content (μmol g$_{FW}^{-1}$)	
K$^+$	Na$^+$		$\phi_{oc,K}$	$\phi_{oc,Na}$	$\phi_{cv,K}$	$\phi_{cv,Na}$	K$^+$	Na$^+$
2.5[a]	7.5	5	0.44	0.36	0.50	0.02	4.6	0.4
0[b]	1.0	22	—	6.0	—	1.4	—	6.0
0.2[b]	1.0	22	—	1.5	—	0.35	—	1.0

[a] PITMAN and SADDLER, 1967.   [b] JESCHKE, 1973.

3.2 Effects of External Concentration on Fluxes

The effect of external concentration on Cl^- fluxes in cortical cells of maize roots was determined by Cram (1973a) using isolated half-cortices to eliminate transport from the stele and reduce variability in cell type (as in Part A, 5.5.3). He found that ϕ_{oc} increased with concentration and that ϕ_{cv} was almost independent of increasing concentration (Fig. 3.18). Net flux was about 90% of ϕ_{cv} so that ϕ_{co} increased almost in parallel with ϕ_{oc} while ϕ_{vc} was small and little affected by external concentration. These results are generally similar to earlier estimates of influx at plasmalemma and tonoplast based on short- and long-term tracer uptake (Cram and Laties 1971). Similar patterns of response of flux to concentration was also found for K^+ and Cl^- in storage tissue (Part A, Fig. 8.3).

Measurements of fluxes of Cl^- in barley roots in 0.5 and 10.0 mM KCl were made by Pitman (1971) who found less increase in ϕ_{oc} with external concentration than shown in Fig. 3.17 (Table 3.10).

There are no data for K^+ and Na^+ fluxes like those of Fig. 3.17 showing the response to external concentration. Estimates of K^+ fluxes for barley in 10 mM K^+-culture solution show that ϕ_{oc} is much larger than ϕ_{cv} (Table 3.15 qv). There are some measurements of net tracer uptake to salt-saturated roots which are consistent with the general pattern of rising ϕ_{oc} and near-constant ϕ_{cv} (barley, Pitman et al., 1968; maize, Leigh and Wyn Jones, 1973).

These measurements of Φ_{ov} also showed that K^+ enters these roots preferentially to Na^+ in up to 30 to 50 mM external concentration and that low-salt roots showed less preference for K^+ relative to Na^+ than roots that had accumulated salt (Table 3.7).

Jefferies (1973) gives data for fluxes of K^+ and Na^+ from mixtures of KCl and NaCl of varied total concentration using roots of *Triglochin maritima*. The results support these generalizations about the effects of external concentration at the plasmalemma and tonoplast, but there is the added complexity that discrimination between K^+ and Na^+ increased at higher concentrations.

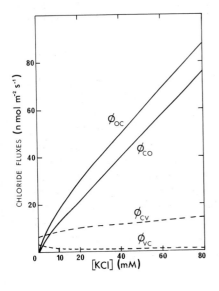

Fig. 3.18. Chloride fluxes at the plasmalemma and tonoplast in maize root cortical cells over a range of KCl concentrations. Calculated from measurements of plasmalemma influx; Φ_{ov}; and net influx. (From Cram, 1973a)

Table 3.10. Fluxes of Cl^- in barley roots at low and high concentration (Pitman, 1941)

$[KCl]_o$ (mM)	Fluxes ($\mu mol\ g_{FW}^{-1}\ h^{-1}$)				
	ϕ_{oc}	ϕ_{co}	ϕ_{cv}	ϕ_{vc}	ϕ_{cx}
0.5	3.3	0.8	1.6	1.6	2.5
10.0	4.7	1.1	1.6	1.6	3.6

3.3 Active and Passive Fluxes of K^+, Na^+ and Cl^-

Table 3.8 gave values for $[K^+]_v$ and $[Na^+]_v$ as well as ψ_{vo}, from which the difference in electrochemical activity, $\bar{\mu}_{vo}$, can be estimated. Where the cells were growing in the solutions listed it can be assumed that the vacuoles were at or near flux equilibrium and that $\bar{\mu}_{vo}$ is a guide to active transport. Generally, $\bar{\mu}_{vo,K}$ is negligible at high external concentration, as found for other higher plant tissues (HIGINBOTHAM, 1973). At lower external concentrations there is evidence from $\bar{\mu}_{vo}$ for inward active transport of K^+ and also of Na^+ (e.g. *Helianthus* and the 5 aquatic spp.). PITMAN and SADDLER (1967) suggested that there was active K^+ transport at the higher K^+ concentrations too, but it did not contribute to ψ_{vo} due to the large diffusive flux of K^+ into the cell.

Calculation of $\bar{\mu}_{vo,Na}$ gives evidence for active efflux of Na^+ from cells of bean and barley roots, as found widely for algal cells (Part A, 6.5.4). Such an efflux pump was suggested from this type of evidence for barley (PITMAN and SADDLER, 1967), for *Vicia faba* (SCOTT et al., 1968), and for pea roots (ETHERTON, 1967). An active efflux is consistent with the flux data of 3.3.3.1. JEFFERIES (1973) found evidence for Na^+ efflux from the halophyte *Triglochin maritima* (Table 3.8), and the K^+ influx was strongly stimulated by 10 to 100 mM NaCl, but not by LiCl, $MgCl_2$ or $CaCl_2$.

Inhibitors of energy metabolism reduce accumulation and tracer influx of K^+ and Na^+ very strongly (Table 3.6). Either ϕ_{oc} or ϕ_{cv} (or both) appears to be drawing on metabolic energy, possibly in the form of ATP. Efflux of K^+ and Na^+ from the vacuoles of barley roots (ϕ_{vc}) was not affected by CCCP and appears to be a passive process (PITMAN, 1972a).

There is evidence that both $\phi_{oc,Cl}$ and $\phi_{cv,Cl}$ contain large active components. Firstly, the plasmalemma seems to be the site of the major difference in electrochemical activity for Cl^-, and $\phi_{oc,Cl}$ can be inhibited substantially by CCCP at lower external concentrations of Cl^- (0.1 mM in maize, CRAM, 1973a; to 10 mM in barley, PITMAN, 1972a). Secondly, tracer uptake to the vacuole of maize roots from 50 mM Cl^- can be inhibited by CCCP, even though at this concentration the plasmalemma flux $\phi_{oc,Cl}$ was inhibited less than 20% (CRAM, 1973a) and it appears that transport at the tonoplast must be active too.

Part of the Cl^- flux at the plasmalemma may be exchange diffusion, particularly at $[Cl^-]_o$ above 3 mM (CRAM, 1973a), as found for carrot tissue (Part A, 8.4.2; Fig. 8.4). The effluxes of Cl^- from vacuole and cytoplasm (Φ_{vo} and ϕ_{co}) in barley are not inhibited by CCCP and appear to be passive (PITMAN 1972a; see Fig. 3.27).

Decreased ion uptake in the presence of inhibitors of protein synthesis has been reported (UHLER and RUSSELL, 1963, SUTCLIFFE, 1973) but the experiments were of 20 to 24 h duration. In short-term experiments no effect has been found for CHM (LÄUCHLI et al., 1973; LÜTTGE et al., 1974) or p-flouro phenyl alanine (SCHAEFER et al., 1975). However these inhibitors have a strong effect on transport across the root, as pointed out by UHLER and RUSSELL (1963) for chloramphenicol, and it seems likely that their long-term action on ϕ_{oc} and ϕ_{cv} may be indirectly due to changed ATP levels or build-up of ion concentration in the symplast due to inhibition of ϕ_{cx}. The difference in rate of action on uptake to the root and transport to the xylem may also reflect different rates of turnover of the carrier enzymes (see also 3.4.2.5).

Table 3.11. Permeability coefficients estimated for cells of roots at 25° C. Calculated from Goldman equation (Part A, 3.5.3.1)

Plant	Permeability coefficients (m s^{-1}) $\times 10^{10}$						
	Plasmalemma				Tonoplast		
	K^+	Na^+	Cl^-	Ca^{2+}	K^+	Na^+	Cl^-
Vicia faba[a]	22	6	–	0.04	–	–	–
Hordeum vulgare[b]	33	10	1.2	–	1.1	0.23	1.3
Triglochin maritima[c]	48	0.3	–	–	2.9	0.07	–

[a] Scott et al. (1968).   [b] Pitman (1969b).   [c] Jefferies (1973).

Estimates of permeability coefficients (P) have been made assuming certain fluxes were not active, i.e. $\phi_{co,K}$; $\phi_{vc,K}$; $\phi_{vc,Na}$; $\phi_{oc,Na}$ (at high $[Na^+]_o$); $\phi_{co,Cl}$; $\phi_{vc,Cl}$. The permeability coefficient is defined by the equation used to relate fluxes, concentrations and potentials, in this case the Goldman equation (Part A, 3.5.3.1). There is good agreement between values of P_K in roots (Table 3.11) and those determined for giant algal cells but the ratio P_{Na}/P_K is much higher in root cells than in many algal cells (Part A, 6.6.5). See too the values for P in storage tissues in Part A, Table 8.5. Cram (1973a) estimated P_{Cl} for the plasmalemma to be $0.2 \cdot 10^{10}$ m s^{-1} when $[Cl^-]_o$ was 0.1 mM rising to 4.4 and $8.2 \cdot 10^{10}$ m s^{-1} in 20 mM and 50 mM $[Cl^-]_o$ respectively. Part of this increase he attributed to exchange diffusion and Cram suggested P_{Cl} was "about $1 \cdot 10^{10}$ m s^{-1}".

These estimates of permeability coefficients can be combined with measurements of cell PD to calculate the passive components of K^+ and Cl^- fluxes into the cell (ϕ_{oc}) as a function of external concentration. This has been done in Fig. 3.19 for low-salt barley roots. The values of $\Phi_{ov,K}$ and $\Phi_{ov,Cl}$ are lines of best fit to actual results at 25° C, and are equivalent to data given by Epstein in various publications for roots at 30° C (e.g. 1966). The calculations show firstly that the

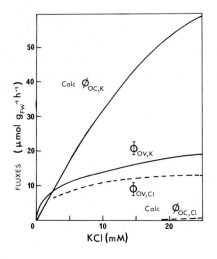

Fig. 3.19. Calculation of passive fluxes of K^+ and Cl^- into low-salt roots of barley ($\phi_{oc,K}$; $\phi_{oc,Cl}$) using values of P_K and P_{Cl} given in Table 3.11; values of ψ_{co} (taken as ψ_{vo}) from Pitman et al., 1971; and the Goldman equation (Part A, Chap. 3.5.3.1). $\Phi_{ov,Cl}$ is the line of best fit for rates of Cl^- uptake to low-salt roots at 25° C (Pitman, 1969b). $\Phi_{ov,K}$ is calculated from $\Phi_{ov,Cl}$ by adding net H^+ efflux (Pitman, 1971a). See text for discussion

passive component of $\phi_{oc,Cl}$ is negligible up to $[Cl^-]_o = 25$ mM, implying that the balance of this flux (and of $\Phi_{ov,Cl}$) needs to be an active process. Similar calculations made by CRAM (1973a) for maize cortical cells showed that the passive components of $\phi_{oc,Cl}$ were negligible up to 3 mM and only 13% of the total influx at 20 mM.

Secondly, the diffusive fluxes of K^+ are lower than $\Phi_{ov,K}$ up to about 1 mM, supporting the view that mechanism I (3.2.3.2) for K^+ is an active process. At higher concentrations the diffusive flux of K^+ becomes as large, or larger than $\Phi_{ov,K}$ and $\phi_{oc,K}$ could have large passive as well as active components. There is of course the possibility of error in the value used for P_K, which would alter $\phi_{oc,K}$ but probably not by more than a factor of 2; the particular point is that at higher concentrations of $[KCl]_o$, $\Phi_{ov,K}$ may be limited by ϕ_{cv} and not by ϕ_{oc}. Hence it becomes difficult to estimate the magnitude of active K^+ flux at the plasmalemma and so determine real kinetics for carrier operation. This point has been discussed by PITMAN (1969b) on theoretical grounds and by CRAM and LATIES (1971) in relation to short and long term measurements of Φ_{ov}.

In conclusion then, it seems there is justification for active transport of K^+ and Cl^- inward at the plasmalemma and tonoplast, though K^+ influx may be limited to "mechanism I" of Chap. 3.2. Influx of K^+ (and Na^+) also appears to be related to H^+ efflux over the range of mechanism II at least and so determined by P_K and P_{Na} as well as by metabolism. Sodium appears to be extruded as an active efflux, which may be in competition with H^+ efflux as in *Neurospora* (SLAYMAN, 1970).

4. Ion Content of Plant Roots

The content of ions in plant roots is very much affected by the content of the soil or solution in which the plants grow. Some generalizations can be made though about the levels of various ions.

Ions such as K^+ and Na^+ in the vacuoles of root cells may be in flux equilibrium with the external solution. Their levels are related to π for the vacuoles and may be controlled by the cell (Part A, Chap. 11). Other ions such as SO_4^{2-}, NO_3^- and phosphate which are metabolically incorporated (Part A, Chap. 13) and predominantly in the cytoplasm, are determined largely by the amounts of such compounds as protein and nucleotides present. Other ions may be at indeterminate levels related to the particular history of ion availability and such factors as water flow through the root.

4.1 K^+ and Na^+ Content

In the mature parts of the root, average concentrations of K^+ and Na^+ are found to vary little with time, provided the culture solution does not become depleted, despite the large transport of these ions through this part of the root to the shoot. The following data from BOWLING and ANSARI (1972) for *Helianthus* roots illustrate this point. The segments represent cells of increasing age at increasing distance from the tip with probably about 4 days between youngest and oldest:

Distance from tip (cm)	K⁺	Na⁺
	(μmol g_{FW}^{-1})	
0–1	120	18
1–2	111	19
2–3	123	18
3–4	113	18
4–5	108	18
5–6	113	17

Similar data are given for *Vicia faba* by Pallaghy and Scott (1969). Greenway, in various publications (Greenway et al., 1965, 1966), showed levels of K⁺ and Na⁺ in barley roots were relatively constant over several months. Constant concentrations over shorter periods for barley and mustard are also well established (Pitman, 1965a, 1966).

Thus for the major univalent cations, cells of roots of plants grown in culture solution are in a quasi-steady state. Net uptake to the vacuoles of the root is low but there may be large net fluxes across the plasmalemma, due to supply of ions to the shoot.

Fig. *3.20* gives measurements of K⁺ (or K⁺ + Na⁺) in plant roots when concentrations in the solution were varied. Above about 0.5 mM, the concentrations of univalent cation in the root increased to only a small extent compared with

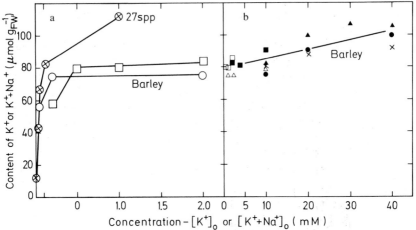

Fig. *3.20a* and b. K⁺ or K⁺ + Na⁺ content of plant roots. (a) (○): K⁺ in barley roots, data from Johansen et al. (1968b). (□): (K⁺ + Na⁺) in barley roots, data from Pitman et al. (1968). (×): K⁺ as mean content of 27 different species (spp), data from Asher and Ozanne (1967). (b) All barley roots. (▲): (K⁺ + Na⁺) when $K_o^+/Na_o^+ = 1/19$; (△, ●, □): (K⁺ + Na⁺) when $K_o^+/Na_o^+ = 1/3$; (×): K⁺ from solutions containing no Na⁺ (data from Pitman, 1965a, b; Pitman et al., 1968). In calculating content relative to fresh weight it was assumed the dry weight was 70 mg g_{FW}^{-1}

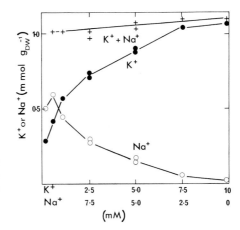

Fig. 3.21. Effect of variations of K$^+$/Na$^+$ in the solution on the total K$^+$ + Na$^+$ content (+) and the individual content of K$^+$ (●) and Na$^+$ (○), (From PITMAN, 1965a)

Table 3.12. Effect on relative content of the roots of variation in total concentration of $([K^+]_o+[Na^+]_o)$; $[K^+]_o/[Na_o^+]=1/3$. $(1.0\ \mu mol\ mg_{DW}^{-1}\simeq 70\ \mu mol\ g_{FW}^{-1})$

Concentrations [K$^+$ + Na$^+$]$_o$ (mM)	Barley[a]				Mustard[b]			
	K$^+$	Na$^+$	Total	$S_{K, Na}$	K$^+$	Na$^+$	Total	$S_{K, Na}$
	($\mu mol\ mg_{DW}^{-1}$)				($\mu mol\ mg_{DW}^{-1}$)			
5	—	—	—	—	1.72	0.47	2.19	11
10	1.03	0.29	1.32	10.8	1.23	0.52	1.75	7.1
20	1.07	0.33	1.40	9.6	1.43	0.66	2.09	6.5
40	1.49	0.35	1.84	12.6	1.38	0.79	2.17	5.2
60	1.33	0.38	1.71	10.5	1.42	0.98	2.40	4.3

[a] PITMAN (1965a). [b] PITMAN (1966).

that in solution. These values were apparently in steady-state equilibrium with the external solution.

In most roots there is preferential uptake of K$^+$ with respect to Na$^+$, providing there is adequate supply of these ions. Thus Fig. 3.21 shows the amounts of K$^+$ and Na$^+$ in barley roots grown in a culture solution containing varied proportions of K$^+$ and Na$^+$ in addition to other essential elements. The total external concentration $([K^+]_o+[Na^+]_o)$ was kept constant at 10 mM. Uptake was reduced and growth retarded when K$^+$ was completely absent but with this exception the total concentration of (K$^+$ + Na$^+$) in the root was constant despite variation in ratio of $[K^+]_o/[Na^+]_o$ from 0.05 to 3.0. However the ratio of K$^+$/Na$^+$ in the roots varied with $[K^+]_o/[Na^+]_o$. Similar results have been found for *Sinapis* and barley seedlings over a wide range of $([K^+]_o+[Na^+]_o)$ and $[K^+]_o/[Na^+]_o$ (PITMAN, 1965a, b, 1966) and for the tropical grass *Chloris gayana* (SMITH, 1974).

Compare the above results with those of Table 3.12 which shows that when the ratio $[K^+]_o/[Na^+]_o$ was kept constant, there was also little effect on $S_{K,Na}$ of changing total $([K^+]_o+[Na^+]_o)$ concentration for barley roots, but for mustard there was some reduction in $S_{K,Na}$ with increased concentration due to increased Na$^+$ uptake possibly at the expense of K$^+$.

The levels of K$^+$ and Na$^+$ in roots (and other tissues) can be seen as fulfilling two roles. Certain concentrations of K$^+$ (10 to 20 mM) are needed for maximum operation of enzymes (Evans and Sorger, 1966). (This is the amount found in low-salt roots, where though low, it seems adequate for operation of the K^+-dependent pyruvic kinase (Wildes and Pitman, 1975.) The other role is maintenance of π in the vacuoles when either K^+ or Na^+ is suitable. This (osmotic) concentration may be controlled by some property of the root cells (as in barley) or may rise with external concentration (as in the halophyte *Triglochin maritima,* Table *3.8*).

The content of root cells can be related to the flux measurements described above. The selectivity of K^+ relative to Na^+ can be due both to greater P_K than P_{Na} but particularly to operation of a Na^+ efflux. The level of $(K^+ + Na^+)$ in the vacuole is determined mainly by ϕ_{cv} and by the permeability of the tonoplast membrane (since it represents the concentration in which ϕ_{vc} becomes equal to ϕ_{cv}). As seen in Fig. *3.18*, ϕ_{cv} was virtually independent of external concentration, as found for the tissue content.

The K^+ and Na^+ in the roots of most naturally growing plants will be balanced by an equivalent concentration of organic acids or partly by Cl^- if this is abundant in the soil (Part A, *13.2.2*).

4.2 Divalent Cations

Loneragan and Snowball (1969a, b) give determinations of Ca^{2+} content of 30 different plant species growing in flowing nutrient solution in which the level of Ca^{2+} was varied between 0.3 and 1000 µM. Fig. *3.22* gives mean values of Ca^{2+} in roots of 16 legumes and 11 cereals and grasses. The values are about 2 µmol g_{FW}^{-1} and very much less than the $(K^+ + Na^+)$ content (Fig. *3.20*).

The relationship between Ca^{2+} content and $[Ca^{2+}]_o$ followed the same trend for each species but there were differences in absolute content. Thus roots of maize contained nearly twice as much Ca^{2+} as other species.

At higher $[Ca^{2+}]_o$ the content of barley roots increased regularly from about 4 µmol g_{FW}^{-1} at 2.5 to 50 µmol g_{FW}^{-1} at 50 mM (Lazaroff and Pitman, 1966). There was little interaction between $([K^+ + Na^+])$ content and Ca^{2+} content as $[K^+ + Na^+]_o$ was varied. Johansen et al. (1968a), however, found that Ca^{2+} content of barley roots increased by about 20% as $[K^+]_o$ was reduced from 2 mM to 20 µM.

There are fewer data for Mg^{2+} in roots, and nothing so extensive as that in Fig. *3.22*.

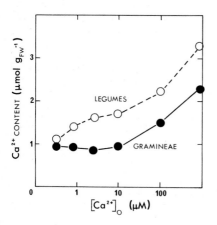

Fig. *3.22*. Mean Ca^{2+} content of 16 legume species (○----○) and 11 grass and cereal species (●——●). Data calculated from Loneragan and Snowball (1969a, b) using percentage dry weights from Asher and Ozanne (1967). Note the values are lower than those quoted for maize and soybean (*3.3.4.2*)

COLLANDER (1941) made a survey of content of various species, varying $[Ca^{2+}]_o/$ $[Mg^{2+}]_o$, but his results were for whole plants. In barley Mg^{2+} content was about 50% of Ca^{2+} in the roots (LAZAROFF and PITMAN, 1966), but in soybean Mg^{2+} was much higher (25 µmol g_{FW}^{-1}) than Ca^{2+} (8 µmol g_{FW}^{-1}) (LEGGETT and FRERE, 1971). MAAS and OGATA (1971) found Mg^{2+} in maize was taken up to about 10 to 17 µmol g_{FW}^{-1}. This difference will be returned to below (3.3.5).

Amounts of other divalent cations are normally much smaller, and some data are given in 3.3.5 in relation to the mechanisms of divalent cation uptake.

Zinc content of 8 species in relation to growth was studied by CARROLL and LONERAGAN (1968, 1969). The effective concentration range in solution for good growth was between about 0.25 and 5 µM, and the average content of the roots at maximum growth was about 0.1 µmol g_{FW}^{-1}.

4.3 Anion Content

In general it can be expected that total anion content of root cells will follow the same trends and reach the same concentrations as found for total cation content, due to the balance of charge in the vacuole. Analyses of anions in plant roots have been made for a number of cases; examples are shown in Table 3.13. Relative amounts of each ion will vary with external concentrations, but the total content should be nearly equal to the amount of diffusible cation in the root. Chloride is taken up less strongly than NO_3^- and is usually present in only small amounts in roots, except in very saline soils or when grown in chloride solutions.

Measurements of phosphate content in 8 species of plants growing in culture solutions of between 0.04 and 25 µM were made by ASHER and LONERAGAN (1967). With increasing $[P]_o$, the average content of the roots increased from 4 to 25 µmol g_{FW}^{-1} reaching a maximum at about 1 µM $[P]_o$. This maximum level is close to that given in Table 3.13. BOUMA (1967c) measured S content of roots of *Trifolium subterraneum* and found 15 µmol g_{FW}^{-1} for plants grown on 125 µM SO_4^{2-}, but only about 4 µmol g_{FW}^{-1} in roots growing at levels where growth was retarded.

Nitrogen in roots will be present as protein (about 140 to 280 µmol g_{FW}^{-1}) and also as NO_3^- (Table 3.13). The abundance of NO_3^- is determined by its availability

Table 3.13. Anions in plant roots grown on culture solutions. Data from KIRKBY and MENGEL (1967) for tomato in solution containing $PO_4 = 2$; $SO_4^{2-} = 0.75$ and $NO_3^- = 5$ mM and from HIGINBOTHAM et al. (1967) for oat and pea roots in solution containing $(1\times)$ $Cl^- = 1.0$: $PO_4 = 0.95$; $SO_4^{2-} = 0.25$; $NO_3^- = 2$ mM, $(10\times) = 10$ times these concentrations; (assuming $1 g_{FW} = 0.08\ g_{DW}$ for tomato roots)

Plant	Anion content in root (µmol g_{FW}^{-1})					Diffusible cation content (µmol g_{FW}^{-1})	
	SO_4^{2-}	PO_4	Cl^-	NO_3^-	Organic	$K^+ + Na^+$	$Ca^{2+} + Mg^{2+}$
Tomato	4	22	10	28	23	87	10
Pea 1×	10	21	7	28	–	83	2
10×	26	26	6	35	–	96	7
Oat 1×	2	17	3	56	–	69	10
10×	5	14	4	38	–	76	12

from the solution but also by the rate at which NO_3^- is reduced or transported as NO_3^- out of the root to the shoot. Uptake of NO_3^- is discussed further in *3.3.6.3*, Chaps. *6, 9* and Part A, Chaps. *11, 13*.

Both phosphate and SO_4^{2-} are involved in metabolism as well as forming part of the total ion balance. Jackson and Hagen (1960) found at least 95% of PO_4 entering the root from 10 µM solution was incorporated into organic compounds during uptake. Their data and that of Loughman and Russell (1957) show that inorganic phosphate ranged from 10 to 60% of total P as external concentration (and internal content) increased. Sulfur incorporated in protein should be about 0.07 of the protein nitrogen content (Holford, 1971), and so 10 to 20 µmol g_{FW}^{-1}.

In a number of species, phosphate may be stored as polyphosphate during part of the year and used later during a growing period (e.g. *Banksia ornata*, Specht and Groves, 1966; Jeffrey, 1968). Plants of *Scabiosa* grow on soils normally low in P but it may become locally abundant. *Scabiosa* is able to absorb P to levels well above those normally found in roots, when P is available and then use it for growth of the plant at later times (Rorison, 1969).

This particular problem points out too some of the difficulties in extrapolating from the constant environment of culture solutions to soils where abundance can fluctuate with depth in the soil and at different times of the year. To some extent, local variations are overcome by differential root development in areas where nutrients are abundant.

5. Mechanisms of Divalent Cation Absorption

Most divalent cations in roots are far from electrochemical equilibrium and $\bar{\mu}_{vo}$ is of little use in detecting active transport. Various responses have been obtained from use of inhibitors of energy metabolism (Table *3.14*), and there are also data showing competition between pairs of divalent cations. Generally, uptake of divalent cations seems to involve metabolically dependent transport.

5.1 Calcium and Magnesium

Various authors have commented on the difference in rate of uptake of Ca^{2+} by excised roots of maize and wheat on the one hand, and barley, *Phaseolus*, soybean, pinto bean, mung bean, and pea on the other (e.g. Maas, 1969). Maas showed that the rate of Ca^{2+} uptake from 10 mM Ca^{2+} solution to low-salt maize roots was about 2 µmol g_{FW}^{-1} h^{-1} and that it could be inhibited at least 75% by 10 µM DNP, or by reducing the temperature to 2° C. In contrast Moore et al. (1961) found little or no Ca^{2+} uptake to barley, apart from the initial non-metabolic absorption, which seems odd when compared with determinations of Ca^{2+} content (*3.3.4.2*).

Leggett and Gilbert (1969) suggested for soybean that the low rate of Ca^{2+} accumulation in the excised roots was due to rapid equilibration of a small component (e.g. symplast) and exclusion of Ca^{2+} from the cell vacuoles. They pointed out

Table 3.14. Effects of inhibitors on uptake of divalent cations, and SO_4^{2-} by plant roots. All measurements are of Φ_{ov} except where shown

Inhibitor	Plant	Inhibitor Concn (μM)	Ion	Ion Concn (mM)	Inhibition (%)	Ref.
DNP	Maize (Φ_{ox})	50	Ca^{2+}	2.5	91	HUTCHIN and VAUGHAN
			Sr^{2+}	0.5	98	(1967)
	Phaseolus	100	Ca^{2+}	2.5	15	DREW and BIDDULPH
		1,000	Ca^{2+}	2.5	+7%	(1971)
	Rice	100	Fe^{2+}	0.2	35	KANNAN (1971)
		100	Fe^{2+}	10.0	28	
	Barley (LS)	10	Mn^{2+}	0.025 (pH 5)	100	MAAS et al. (1968)
		10	SO_4^{2-}	100	15–33	NISSEN (1971)
		10	SO_4^{2-}	1	42–52	
		10	SO_4^{2-}	10 μM	72–67	
		10	SO_4^{2-}	0.1 μM	78	
	Helianthus	10	SO_4^{2-}	0.5	30–40	PETTERSSON (1966)
		100	SO_4^{2-}	0.5	77	
CCCP	Rice	2	Fe^{2+}	0.2	20	KANNAN (1971)
Arsenite	Phaseolus	1,000	Ca^{2+}	2.5	24	DREW and BIDDULPH
		1,000	Ca^{2+}	0.25	22	(1971)
N_3^-	Barley	100	Mn^{2+}	0.025 (pH 5)	85	MAAS et al. (1968)
	Helianthus	10	SO_4^{2-}	0.5	30	PETTERSSON (1966)
		100	SO_4^{2-}	0.5		
		500	SO_4^{2-}	0.5	85–90	
	Rice	100	Fe^{2+}	0.2	49	KANNAN (1971)
		100	Fe^{2+}	10.0	69	
KCN	Helianthus	10	SO_4^{2-}	0.5	65	PETTERSSON (1966)
		100	SO_4^{2-}	0.5	70	
		500	SO_4^{2-}	0.5	80	
	Phaseolus	1,000	Ca^{2+}	2.5	31	DREW and BIDDULPH
		1,000	Ca^{2+}	0.25	15	(1971)

that the results of MOORE et al. (1965) for barley showed increased concentration of Ca^{2+} in the xylem exudate compared with solution even though there was very little accumulation in the root. Such an explanation certainly rationalizes both whole-plant and excised-root studies, and shows how there could be active Ca^{2+} influx but little accumulation.

Magnesium uptake appears to be strongly inhibited when energy metabolism is blocked, and its rate of uptake may be affected by the level of Ca^{2+} in the solution. In barley (MOORE et al., 1961) and maize, the content and rate of uptake are greater from solutions of Mg^{2+} alone than when Ca^{2+} is present. The reduction is not always accompanied by an equivalent Ca^{2+} uptake and MAAS and OGATA (1971) suggested Ca^{2+} was having a non-competitive effect at a site other than that involved in uptake. This effect of Ca^{2+} is not general; Leggett and Gilbert (1969) found little reduction in Mg^{2+} uptake by excised soybean roots from culture solutions with and without Ca^{2+} provided K^+ was present. When K^+ was absent from solution Mg^{2+} increased in the vacuole of the roots from 6 to about 28 µmol g_{FW}^{-1} and in this case seems to be replacing K^+. (In the shoots however, *calcium* accumulated to nearly twice its normal level and Mg^{2+} was unaffected.)

5.2 Manganese

Manganese uptake shows some similarity to Ca^{2+} and Mg^{2+}. There is non-competitive inhibition by other divalent cations and uptake appears to be active since it is reduced by inhibitors of energy metabolism (Table *3.14*; Maas et al., 1968, 1969). It has been suggested that there are two different types of Mn^{2+} uptake, one independent of Ca^{2+} and the other affected by it (van Diest and Schuffelen, 1967). Certainly there is marked interaction between Ca^{2+} and H^+ ions. Robson and Loneragan (1970) found Ca^{2+} inhibited Mn^{2+} uptake by *Trifolium* and *Medicago* spp. but the degree of inhibition depended on pH. At pH 7 there was less inhibition of Mn^{2+} uptake by Ca^{2+} (74%) than at pH 5.4 (96%). The effect of pH has been investigated by Munns et al. (1963), who showed that in one genotype Mn^{2+} uptake was strongly dependent on pH, but not dependent on pH in another. This interaction of Mn^{2+} uptake with Ca^{2+} and pH is important in development of Mn^{2+}-toxicity, especially in water-logged soils. The rates of uptake to *Trifolium* spp. (Robson and Loneragan, 1970) ranged from 1 to 30 nmol g_{FW}^{-1} h^{-1} and were much less than the 1 µmol g_{FW}^{-1} h^{-1} found by Maas et al., 1968; this rate of uptake would have probably proved toxic if in whole plants and over a long time period.

5.3 Zinc

Rates of Zn^{2+} absorption are given in Fig. *3.23*. The effective concentration range appears to be between 0.25 and 5 µM and as with Mn^+, high levels can be toxic. Zinc uptake also appears to be active and is inhibited strongly by Ca^{2+}. As with Mn^{2+}, there appears to be a Ca^{2+} sensitive and a Ca^{2+} insensitive mechanism which was inhibited by H^+ (Chaudhry and Loneragan, 1972).

5.4 Iron

Iron uptake to plant roots has been extensively studied for its importance in plant nutrition. Uptake of Fe^{2+} has features in common with other divalent cations, (i.e. an active process which can be inhibited by other divalent cations) but in addition, there may be reduction of Fe^{3+} to Fe^{2+} at the cell surface. Transport of Fe^{2+} across the root to the xylem involves complex formation with citrate ions (see also *3.4.2.8*).

Reduction of Fe^{3+} has been shown very neatly by production of blue precipitate from potassium ferricyanide at the surface of the root (Ambler et al., 1970). Chaney et al. (1972) showed that uptake of Fe^{2+} from Fe^{3+}-chelates involved reduction of Fe^{3+}-chelate and release of Fe^{2+} from the chelate prior to uptake, by trapping the Fe^{2+} as a complex with BPDS (bathophenanthropinedisulphonate).

Uptake of iron from Fe^{3+}-chelate was inhibited by Zn^{2+} at concentrations above 5 µM. Zinc was shown to be acting by preventing reduction of Fe^{3+} to Fe^{2+}, since little precipitate was formed with potassium ferricyanide when Zn^{2+} was present. There was a decrease too in the reducing capacity of root sap and of the used nutrient solution (AMBLER et al., 1970). Release to the solution of factors affecting iron uptake was found for genotypes of soybean inefficient in iron uptake (ELMSTROM and HOWARD, 1970).

Iron-deficient plants have been shown to lower the pH of the solution around the roots (RAJU et al., 1972). This could be due to H^+ release and the roots would then be analogous to yeast where reduction or H^+ transport can take place at the cell surface (Part A, 7.2).

5.5 Summary

There seems to be clear evidence for metabolically dependent transport of divalent cations into plant roots. The actual rates of uptake may be low (e.g. Zn^{2+}) but the concentrations of such ions in soil solutions are low too. Apart from the ions mentioned here, others such as Ni^{2+} and Cu^{2+} can be taken up actively. Due to slow diffusion of divalent cations in the free space, the effective site of transport may be at the plasmalemma of cells at the surface of the root (LEGGETT and GILBERT, 1967).

Uptake of Fe shows that there is a powerful reducing activity at the root surface and it may be expected that this can act on ions of other transition elements (e.g. V; WELCH, 1973). This reducing power, and the interactions of uptake with H^+ suggest that proton (or electron) movement across the membranes may be important in divalent cation uptake, as seems to be the case with uptake of univalent cations (3.3.1). Transport could then be *via* a "general" divalent cation transport as well as *via* "specific" carriers. Though there are interactions between univalent and divalent cations, the content of many species seems independent of wide variation in proportions of uni- to divalent cations in the culture solutions (LAZAROFF and PITMAN, 1966), suggesting that divalent cation uptake is a separate process (or group of processes) from K^+ and Na^+ uptake.

6. Mechanisms of Anion Absorption

6.1 Phosphate

Uptake of phosphate has long been recognized to be complex, due to the number of ionic species available for transport. Suggestions have been made too for existence of sites differing in sensitivity to external concentration and inhibitors. The various results are summarized by CARTER and LATHWELL (1967a). They suggested one site (or reaction) dominating at very low concentrations ($K_m = 6$ µM for maize) and another at higher concentrations ($K_m = 140$ µM). Collecting together results of various studies, it appeared that the low-concentration site was activated by Ca^{2+}, and involved HPO_4^{2-} ions and NADH. The high-concentration site was unaffected by Ca^{2+} and involved transport of $H_2PO_4^-$ (see also Hagen and Hopkins, 1955; Hagen et al., 1957). However, the estimate of $K_m = 6\,\mu M$ for the low concentration process seems high when compared with the concentrations found in soils (Fig. *3.5*) and needed to produce maximum uptake (1 to $5\,\mu M$; Fig. *3.23* qv).

Evidence that phosphate and Cl^- are taken up by different processes comes from the demonstration by Carter and Lathwell (1967b) that phosphate uptake is not inhibited by extremely high concentrations of Cl^-.

Estimates of electrochemical activity of phosphate support the view that there is need for active transport, since the potential difference ψ_{vo} is usually large and negative, and phosphate concentration in the cell higher than outside. Jackson and Hagen (1960) found at least 95% of PO_4 entering the root from $10\,\mu M$ solution was incorporated into organic compounds. Their data and that of Loughman and Russell (1957) show that inorganic phosphate ranged from 10 to 60% of total P as external concentration (and internal content) increased. A more extensive treatment of P content and uptake is given by Bieleski (1973).

6.2 Sulfate

Sulfate uptake to plant roots has been discussed in relation to kinetic analysis of uptake in *3.2.3.1*, but the general point may be made that the mechanism seems to have a very high affinity for SO_4^{2-} ($K_m \simeq 5\,\mu M$) and thus like phosphate rather than Cl^-. However, there seems some uncertainty about K_m for SO_4^{2-} (as for phosphate). Nissen (1971) identifies components (phases) at extremely low external concentration (10^{-9} M) and Reisenauer (1968) suggests $4\,\mu M$ is an adequate concentration for growth of wheat. Epstein (1972) quotes a value for K_m of $6\,\mu M$ (from leaf tissue, Jeschke and Simonis, 1965).

Uptake of SO_4^{2-} can be inhibited 60% to 70% by DNP, KCN, and sodium azide (Table *3.14*) and on these and electro-chemical grounds appears to be an active process. Sulfate is also inhibited competitively by SeO_4^{2-}, and uptake of SeO_4^{2-} is reduced too by inhibitors of energy metabolism.

Persson (1969) analyzed distribution of SO_4^{2-} taken up to roots and showed that over and above the free space content there was another absorbed component dependent on metabolism. This "labile-bound" component was inhibited by DNP and proportional to active SO_4^{2-} uptake by the root at different SO_4^{2-} concentrations in solution and at different temperatures. The amount of SO_4^{2-} bound in this way (20% of uptake) seems too large for it to be a carrier, but its location will be of considerable interest. (A similar component in Rb^+ uptake has been described by Ighe and Pettersson, 1974.)

6.3 Nitrate

Nitrate uptake in some ways resembles uptake of Cl^-, since NO_3^- can be taken up to the same extent as Cl^- by low-salt roots from KNO_3. However, NO_3^- is taken up preferentially to Cl^- from mixtures and has been suggested to be transported by a separate system (CRAM, 1973b, SMITH, 1973).

A particular difference between NO_3^- and Cl^- uptakes is that NO_3^- can be reduced, involving nitrate reductase. There seems to be regulation of the level of reductase by the level of NO_3^- present and by interaction with uptake of reduced nitrogen compounds (amino acids, NH_4^+) that can replace NO_3^- as the plants N-source (FILNER, 1969). Nitrate in the cell has been shown by efflux analysis to be in a small (cytoplasmic) pool that can activate NO_3^--reductase and a larger inactive (vacuolar) pool (FERRARI et al., 1973). The NO_3^--transport system has also been suggested to be inducible (JACKSON et al., 1973). Interaction of uptake of NO_3^- and its efflux is discussed in relation to the cytoplasmic pool by MORGAN et al. (1973). There is evidence too for an enzyme that de-activates nitrate reductase (WALLACE, 1973). (See too the discussion in Part A, *11*.3.2.2.)

6.4 Borate and Silica

These substances deserve mention because their uptake appears to be passive. In intact plants the total uptake of silica seems to be proportional to transpiration (e.g. oats; JONES and HANDRECK, 1965). Other plants show some restriction of silica uptake (SiO_2) by the roots in ways analogous to nutrient ions (BARBER and SHONE, 1966; HANDRECK and JONES, 1967). Borate too has been claimed to be taken up passively (BINGHAM et al., 1970). Though the root may show some level of control over uptake there is clear dependence of much of the uptake on transpiration, and this should be a factor affecting uptake of uncharged organic molecules too, providing they can pass across the stele, though this is more properly a topic for the next Chapter (*3.4*).

7. Measurement of Absorption by the Root

Many of the measurements of rates of uptake discussed so far were made using tracers, involving certain difficulties of interpretation (Part A, 5.5.3.2). Rates of uptake measured in this way, or as changes in vacuolar content of the root tend to ignore the large amounts of ions transported across the root to the xylem. For example, in the experiments of LONERAGAN and SNOWBALL (1969b) (Fig. *3.22*) only 5% of the Ca^{2+} *absorbed* by the root was retained in the roots and 95% passed to the shoot. Comparable data (ASHER and OZANNE, 1967; ASHER and LONERAGAN, 1967) show that only 16% of K^+ and 25% of phosphate absorbed are retained in the roots. Absorption of ions has to meet the large demand of the shoot as well as the local requirements of the roots.

Net rates of absorption of nutrients by plant roots have been estimated in many cases from changes in content of root and shoot. The values are necessarily averages over several days but are nonetheless useful indications of the scale of the absorption process. Equations relating uptake to content and weight of roots are given by WILLIAMS (1948) and discussed, for example, by LONERAGAN (1968); equivalent equations are given below (Eqs. (*3.2* to 5)).

The net rate of uptake relative to root weight (J_R) is equal to $\dfrac{1}{W_R} \cdot \dfrac{dM}{dt}$, where W_R is root weight and M is the content of the plant. An average rate of uptake is therefore

$$J_R = \frac{1}{\overline{W}_R} \cdot \frac{M_2 - M_1}{t_2 - t_1} \tag{3.2}$$

where $\overline{W}_R$ is average root weight. For young plants growing exponentially $\overline{W}_R = (W_{R2} - W_{R1})/\ln(W_{R2}/W_{R1})$. Hence in this case

$$J_R = \frac{(M_2 - M_1)}{(t_2 - t_1)(W_{R2} - W_{R1})} \cdot \ln(W_{R2}/W_{R1}) \tag{3.3}$$

Note that if the relative content, M/W, is constant (X) then

$$J_R = X \cdot \frac{W}{W_R} \cdot \frac{\ln(W_{R2}/W_{R1})}{(t_2 - t_1)} = X \cdot \frac{W}{W_R} \cdot R \tag{3.4}$$

where R is relative growth rate, provided W/W_R is constant.

Note that J_R is an average made up of uptake to each part of the plant, e.g.

$$J_R = J_{Ra} + J_{Rb} + \cdots J_{Rn}$$

where a ...n refer to different parts of the plant and if X_n is constant then

$$J_n = \frac{W_n}{W_R} \cdot X_n \cdot R_n \tag{3.5}$$

Alternatively if X_n is not constant individual J_{Rn} can be evaluated as for Eq. (3.2) and summed to give total uptake to the shoot.

Fig. *3.23* gives rates of absorption plotted on a logarithmic scale in order to include a large range of rates and concentrations, without over-emphasizing higher values.

When using concentrations as low as 10 or 100 µM a major experimental problem in measuring uptake is supplying adequate salt to the roots at the required concentration, since uptake depletes the solution. ASHER et al. (1965) overcame this problem by growing plants in large volumes of solution that could be circulated past the roots, and have used this technique since for various plants and nutrients. CLEMENT et al. (1974) describe an adaptation to flowing culture systems using ion-selective electrodes to monitor concentrations, and show that there was little difference in uptake or growth of *Lolium perenne* plants between 7 µM and 710 µM NO_3^- (see also Fig. *3.23*). At concentrations above 5 mM of K^+ and NO_3^- or 1 mM of other major ions, solutions changed at frequent intervals seem adequate to maintain the nutrient supply; the volumes used are, of course, a matter of individual calculation.

In this type of study, plant growth can change as well as rate of uptake, and there is a tendency for plant size to match the availability of nutrient. Thus over the dashed parts of the curves, growth (and so uptake *per plant*) was impaired, but there are ranges in which the rates of uptake *relative to root weight* were unaf-

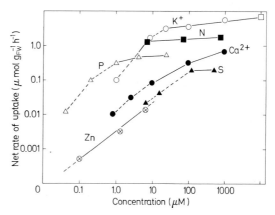

Fig. *3.23*. Rates of net uptake of various elements calculated from change in external concentration or internal content. Dashed lines show where growth limitation was observed. Note this is a log/log plot to increase the range of data given. (○) K$^+$, data from ASHER and OZANNE, 1967, assuming $R_w = 0.13$ d^{-1}, mean of 14 spp; (□) from PITMAN, 1972b for same R_w for barley; (△) P from ASHER and LONERAGAN, 1967, LONERAGAN and ASHER, 1967, mean of 8 spp; (■) N from CLEMENT et al., 1974 for *Lolium perenne*; (●) Ca^{2+} from LONERAGAN and SNOWBALL, 1969a, b, mean of 30 spp; (▲) S from BOUMA, 1967a, b, c, *Trifolium subterraneum*; (⊗) Zn^{2+} from CARROLL and LONERAGAN, 1968, 1969, mean of 8 spp

fected. At extremely low concentrations, absorption will be low due to some indeterminate combination of reduced transport efficiency, reduced demand due to reduced growth, and diffusion through unstirred layers of solution to the root surface.

Net rates of absorption measured in this way determine the *demand* of plants growing under conditions of steady nutrient supply in solutions. There has been particular interest in comparing the relationship of absorption to external concentration with the kinetics of uptake described in Chap. *3.2*. The latter data should represent the potential ability of the roots to supply ions.

In the low range of concentration values for K_m are: K$^+$, 10–20 μM (Table *3.5*); Cl$^-$, 13 μM (Table *3.5*); phosphate, 6 μM (CARTER and LATHWELL, 1967a). As discussed in *3.3.6.2*, for SO$_4^{2-}$ K_m is likely to be like that of phosphate; for nitrate K_m is likely to be about 10 μM (BARLEY, 1970).

There is generally good agreement between those values and the data in Fig. *3.23* for solution studies. The exceptions are phosphate, for which K_m seems lower than 6 μM and SO$_4^{2-}$ for which K_m seems much higher. However the SO$_4^{2-}$ data were obtained using soils where diffusion may have been limiting uptake. Absorption of K$^+$ reached its maximum rate at about 25 μM. Nitrate uptake was 70% of maximum at 7 μM (CLEMENT et al., 1974) though LYCKLAMA (1963) suggested that K_m for NO$_3^-$ uptake was about 40 μM. The value of K_m for NO$_3^-$ is likely to vary between species and be affected too by availability of N as NH$_4^+$ (cf. Table *3.5*).

At lower concentrations (Mechanism I) there seems to be reasonable agreement between rates of absorption measured using whole plants in solutions and rates of absorption measured with low-salt roots.

At higher concentrations the potential rates of uptake predicted by Mechanism II (*3.2.3.3*) are not realized. The rate of K$^+$ absorption between a 100 μM and

10 mM increased by 35% (Fig. *3.*23) whereas uptake to low salt roots increases by about 300% over this range (Fig. *3.*19). The implication is that the uptake is in some way restricted to meet the demand of the shoot (not that Mechanism II should be discounted!). Possible means of regulation of absorption is a separate problem.

The uptake of nutrients from the soil involves diffusion to the root, which may be a rate-limiting process. In the plant-soil system different units are commonly used with uptake being expressed relative to root length and averaged over the whole root system. The concept of "plant demand" has been introduced to express the requirement of the plant and its relation to growth (Nye and Tinker, 1969). Uptake of nutrients from soils is a topic that warrants more space than available here to do it justice. The approach is discussed by Barley (1970) and Milthorpe and Moorby (1974).

8. Control of Absorption

In *3.*3.3.3 and *3.*3.4.1 it was shown that the levels of major cations in the root cells can be largely independent of external concentrations, due to the relation between ϕ_{cv} and c_o (Fig. *3.*18). As discussed (Part A, Chap. *11*) an explanation could be feedback control between vacuolar concentrations and a process affecting ϕ_{cv}. One obvious process is the supply of energy to ϕ_{cv} and other fluxes in the cell. Bowling (1968) showed that cooling the stem of *Helianthus* plants reduced K^+ uptake by the roots by 60%, due to reduction in translocation of sugars (transpiration was unaffected). Hatrick and Bowling (1973) measured the effect of light and darkness on uptake of Rb^+ to barley seedlings and showed there was a correlation between translocation of sugar to the root and the rate of respiration. A direct effect of illumination on influx is shown in Fig. *3.*24, which gives measurements of $\Phi_{ov,K}$ for roots of barley seedlings growing in (2 h light/22 h dark) light regime. There was a lag of about an hour after illumination before $\Phi_{ov,K}$ increased. The same delay was found for appearance of ^{14}C-labeled sugars in the root following a pulse of $^{14}CO_2$ supplied to the shoot at the start of the light period.

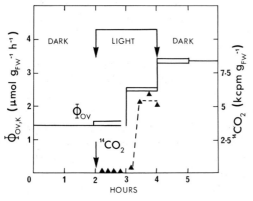

Fig. *3.*24. Barley seedlings growing in a 2/22 h light regime (light/dark) were given a 15 min pulse of $^{14}CO_2$ around the shoots as the light was switched on. Measurements were then made of the radioactivity in the roots (▲) and separate measurements showed this was due to sugars (>95% of radioactivity). On similar plants, measurements were made of the rates of ^{86}Rb accumulation (Φ_{ov}) over periods of 1 h. The plants were grown on full nutrient solution containing 10 mM K^+. (From Pitman and Cram, 1973)

Dependence of fluxes on "metabolic status" of the plant is also shown in Table 3.15. Comparisons were made of various fluxes in barley seedlings growing at a very low and a high relative growth rate produced by varying the photoperiod. The active fluxes into the cell (ϕ_{oc}, ϕ_{cv}) and transport from root to shoot (ϕ_{cx}) were much reduced at the lower relative growth rate. The passive flux out of the cell (ϕ_{co}) was not affected. It is interesting that relative growth rate affected root fluxes but not those into cells of the leaf.

More generally, it was found that the concentration of K^+ in the shoot was independent of growth rate so that absorption was proportional to relative growth rate (Eq. 3.4). Variation in size of the root system did not affect the concentration of K^+ in the shoot supporting the view that this correlation was due to regulation of input to the shoot (ϕ_{cx}) (which is the topic of Chap. 3.4), but Table 3.15 shows that parallel changes could be expected in ϕ_{oc}.

Table 3.15. Effect of relative growth rate on fluxes in root and leaf cells of barley seedlings[a]

Light period	2 h	16 h
Relative growth rate (d^{-1})	0 ± 0.05	0.19 ± 0.05
root fluxes (μmol g_{FW}^{-1} h^{-1})		
ϕ_{oc}	3.5	9.3
ϕ_{co}	2.2	2.4
ϕ_{cv}	0.9	7.5
ϕ_{vc}	0.6	6.5
ϕ_{cx}	1.0	5.9
$\Phi_{ov, Cl}$ (leaf slices)	1.8	2.2

[a] Seedlings grown on 2/22 and 16/8 h light/dark (PITMAN, 1972b; PITMAN et al., 1974).

Though rates of absorption correlate with relative growth rate it seems some other factor is involved than metabolism at higher relative growth rates. Thus rates of absorption were proportional to sugar content up to 5 μmol g_{FW}^{-1} but further increase in sugar content led to no further increase in absorption (PITMAN and CRAM, 1973). There is, too, the difficulty of explaining how rates of absorption at high concentrations can be less than the potential rates of uptake, whether estimated from kinetic studies or from determination of P_K and ψ_{vo} (Fig. 3.19).

Regulation of ϕ_{oc} may be due to feedback *via* the symplasm from ϕ_{cx}. This only transfers the problem from one site to another, but as discussed in Chaps. 3.4 and 7, ϕ_{cx} is affected by cytokinins and ABA providing a possible means of control *via* levels of plant hormones.

Many plants show abnormally high rates of nutrient uptake when transferred from deficient to adequate levels of the nutrient (S, *Trifolium subterraneum*, BOUMA, 1967a; P, pine seedlings, BOWEN, 1970; Fe, soybean and tomato, BROWN and CHANEY, 1971). It is suggested that this behavior is due to an increased number of sites for transport in the deficient plants, though this is an area in which there seem to have been few studies.

9. Conclusions

The purpose of this Chapter has been to bring together data from kinetic analysis (*3.2*) and other work using plant roots so that processes involved in transport in roots can be compared with transport in other kinds of cells.

At low concentrations there is excellent agreement between kinetic measurements (with low-salt roots) and the rates of absorption to whole plants (Fig. *3.23*). At higher concentrations (Mechanism II; *3.2.3.3*) there is a difference between rates of uptake measured in kinetic analysis and as net absorption. The reason seems to be that kinetic analysis determines some combination of ϕ_{oc} and ϕ_{cv} (*3.3.3.3*) under conditions of net accumulation in the vacuoles ($\phi_{cv} >> \phi_{vc}$), whereas net absorption measures the demand by the whole plant, which is dominated by uptake to the shoot and so by ϕ_{cx}. In the higher concentration range ϕ_{oc} can be very much greater than ϕ_{cx} (and ϕ_{cv}), and absorption by the plant from solutions does not reach the potential rates of absorption set by ϕ_{oc}. This distinction should not be taken as a contradiction or "refutation" of kinetic studies. It emphasizes that different experiments can measure particular expressions of the combination of fluxes in the root and that there is a need for integration of different approaches to understand the flexibility and properties of the whole system.

In comparison with algal and fungal cells, one striking feature has been the involvement of H^+ fluxes (or external pH) in many of the transport processes, including those of ions and organic molecules. Fluxes of H^+ appear to act as a driving force for transport. Local concentrations of H^+ can modify membrane permeability (*3.3.1*) or the activity of carrier molecules (Fig. *3.16*). Under certain conditions root cells show K^+ selectivity relative to Na^+ (as in algal cells) and Na^+ efflux seems to be complementary to H^+ efflux (as in *Neurospora,* Part A, *7.2*). Like fungal cells, the root surface has some reducing ability (Part A, *7.2.2.3*).

A major difference of root cells from algal and fungal cells is the interdependence of K^+ uptake and organic acid synthesis in root cells. Again, involvement of $[H^+]_c$ is implied, as discussed in more detail in Part A, Chaps. *12* and *13*.

Transport of univalent cations has been more intensively studied than that of divalent cations. Much of the most useful work on divalent cations has come from experiments with whole plants, stimulated by the relevance to agricultural aspects of plant nutrition. Though the results seem very complex, comparison of different divalent cations has the promise of finding some useful generalizations about the mechanism of absorption.

The next Chapter (*3.4*) elaborates on transport of ions across the root, a further difference between algal cells and root cells. The behavior of the root as an absorptive organ means that the demands of the shoot (ϕ_{cx}) and the flow of water across the root interact with fluxes of ions and other materials across the cell surfaces. This absorptive role of the root has been the reason for much of the studies of uptake by roots, and it cannot be stressed too much that the physiological studies discussed in this Chapter take a limited approach to the problem. Plant nutrition involves the interaction of the plant and the soil ecosystem to which the root has become adapted, as well as the physiology of transport processes.

References

AMBLER, J.E., BROWN, J.C., GAUCH, H.G.: Effect of zinc on translocation of iron in soybean plants. Plant Physiol. **46**, 320–323 (1970).

ANDERSON, W.P., HENDRIX, D.L., HIGINBOTHAM, N.: Higher plant cell membrane resistance by a single intracellular electrode method. Plant Physiol. **53**, 122–124 (1974).

ASHER, C.J., LONERAGAN, J.F.: Response of plants to phosphate concentration in solution culture: I. Growth and phosphorus content. Soil Sci. **103**, 225–233 (1967).

ASHER, C.J., OZANNE, P.G.: Growth and potassium content of plants in solution cultures maintained at constant potassium concentrations. Soil Sci. **103**, 155–161 (1967).

ASHER, C.J., OZANNE, P.G., LONERAGAN, J.F.: A method for controlling the ionic environment of plant roots. Soil Sci. **93**, 39–49 (1965).

BAKER, D.A.: The effect of CCCP on ion fluxes in the stele and cortex of maize roots. Planta **122**, 293–299 (1973).

BAKER, D.A., HALL, J.L.: Ion transport in plant cells and tissues. Amsterdam: Elsevier 1975.

BARBER, D.A., SHONE, M.G.T.: The absorption of silica from aqueous solutions by plants. J. Exptl. Bot. **17**, 569 (1966).

BARLEY, K.P.: The configuration of the root system in relation to nutrient uptake. Advan. Agron. **22**, (1970).

BIELESKI, R.L.: Phosphate pools, phosphate transport, and phosphate availability. Ann. Rev. Plant Physiol. **24**, 225–252 (1973).

BINGHAM, F.T., ELSEEWI, A., OERTLI, J.J.: Characteristics of boron absorption by excised barley roots. Soil Sci. Soc. Am. Proc. **34**, 613–617 (1970).

BOUMA, D.: Growth changes of subterranean clover during recovery from phosphorus and sulphur stresses. Australian J. Biol. Sci. **20**, 51–66 (1967a).

BOUMA, D.: Nutrient uptake and distribution in subterranean clover during recovery from nutritional stresses. I. Experiments with phosphorus. Australian J. Biol. Sci. **20**, 601–612 (1967b).

BOUMA, D.: Nutrient uptake and distribution in subterranean clover during recovery from nutritional stresses. II. Experiments with sulphur. Australian J. Biol. Sci. **20**, 613–622 (1967c).

BOWEN, G.D.: Early detection of phosphate deficiency in plants. Soil Sci. Plant Analysis **1**, 293–298 (1970).

BOWLING, D.J.F.: Translocation at 0° C in *Helianthus annuus*. J. Exptl. Bot. **19**, 381–388 (1968).

BOWLING, D.J.F., ANSARI, A.Q.: Evidence for a sodium influx pump in sunflower roots. Planta **98**, 323–329 (1971).

BOWLING, D.J.F., ANSARI, A.Q.: Control of sodium transport in sunflower roots. J. Exptl. Bot. **23**, 241 (1972).

BROWN, J.C., CHANEY, R.L.: Effect of iron on transport of citrate into the xylem of soybeans and tomatoes. Plant Physiol. **47**, 836–840 (1971).

CARROLL, M.D., LONERAGAN, J.F.: Response of plant species to concentrations of zinc in solution 1. Growth and zinc content of plants. Australian J. Agr. Res. **19**, 859–868 (1968).

CARROLL, M.D., LONERAGAN, J.F.: Response of plant species to concentrations of zinc in solution. II. Rates of zinc absorption and their relation to growth. Australian J. Agr. Res. **20**, 457–463 (1969).

CARTER, O.G., LATHWELL, D.J.: Effects of temperature on orthophosphate absorption by excised corn roots. Plant Physiol. **42**, 1407–1412 (1967a).

CARTER, O.G., LATHWELL, D.J.: Effect of chloride on phosphorus uptake by corn roots. Agron. J. **59**, 250–253 (1967b).

CHANEY, R.L., BROWN, J.C., TIFFIN, L.O.: Obligatory reduction of Fe chelates in Fe uptake by soybeans. Plant Physiol. **50**, 208–213 (1972).

CHAUDHRY, F.M., LONERAGAN, J.F.: Zinc absorption by wheat seedlings: II. Inhibition by hydrogen ions and by micronutrient cations. Soil Sci. Soc. Am. Proc. **36**, 327–331 (1972).

CLARKSON, D.T.: Ion transport and cell structure in plants. London: McGraw-Hill 1974.

CLEMENT, C.R., HOPPER, M.J., CANAWAY, R.J., JONES, L.H.P.: A system for measuring the uptake of ions by plants from flowing solutions of controlled composition. J. Exptl. Bot. **25**, 81–99 (1974).

COLLANDER, R.: Selective absorption of cations by higher plants. Plant Physiol. **16**, 691–720 (1941).

CRAM, W.J.: Chloride fluxes in cells of the isolated root cortex of *Zea mays*. Australian J. Biol. Sci. **26**, 757–779 (1973a).

CRAM, W.J.: Internal factors regulating nitrate and chloride influx in plant cells. J. Exptl. Bot. **24**, 328–341 (1973b).

CRAM, W.J., LATIES, G.G.: The use of short-term and quasisteady influx in estimating plasma-lemma and tonoplast influx in barley root cells at various external and internal chloride concentrations. Australian J. Biol. Sci. **24**, 633–646 (1971).

DENNY, P., WEEKS, D.C.: Electrochemical potential gradients of ions in an aquatic angiosperm, *Potamogeton schweinfurthii* (Benn.). New Phytologist **67**, 875–882 (1968).

DIEST, A. VAN, SCHUFFELEN, A.C.: Uptake of manganese by oats and sunflower. In: Trans. int. Soc. Soil Sci., Comm. II and IV (JACKS, G.V., ed.). Aberdeen: 1960.

DREW, M.C., BIDDULPH, O.: Effect of metabolic inhibitors and temperature on uptake and translocation of ^{45}Ca and ^{42}K by intact bean plants. Plant Physiol. **48**, 426–432 (1971).

DUNLOP, J., BOWLING, D.J.F.: The movement of ions to the xylem exudate of maize roots I. Profiles of membrane potential and vacuolar potassium activity across the root. J. Exptl. Bot. **22**, 434–444 (1971).

ELMSTROM, G.W., HOWARD, F.D.: Promotion and inhibition of iron accumulation in soybean plants. Plant Physiol. **45**, 327–330 (1970).

EPSTEIN, E.: The essential role of calcium in selective cation transport by plant cells. Plant Physiol. **36**, 437–444 (1961).

EPSTEIN, E.: Dual pattern of ion absorption by plant cells and by plants. Nature **212**, 1324–1327 (1966).

EPSTEIN, E.: Mineral nutrition of plants: principles and perspectives. New York-London-Sydney-Toronto: John Wiley and Sons, Inc. 1972.

ETHERTON, B.: Steady-state sodium and rubidium effluxes in *Pisum sativum* roots. Plant Physiol. **42**, 685–690 (1967).

ETHERTON, B.: Vacuolar and cytoplasmic potassium concentrations in pea roots in relation to cell-to-medium electrical potentials. Plant Physiol. **43**, 838–840 (1968).

ETHERTON, B.: Effect of indole-3-acetic acid on membrane potentials of oat coleoptile cells. Plant Physiol. **45**, 527–528 (1970).

ETHERTON, B., HIGINBOTHAM, N.: Transmembrane potential measurements of cells of higher plants as related to salt uptake. Science **131**, 409–410 (1960).

EVANS, H.J., SORGER, G.: Role of mineral elements with emphasis on the univalent cations. Ann. Rev. Plant Physiol. **17**, 47–76 (1966).

FERRARI, T.E., YODER, O.C., FILNER, P.: Anaerobic nitrite production by plant cells and tissues: evidence for two nitrate pools. Plant Physiol. **51**, 423–431 (1973).

FILNER, P.: Control of nutrient assimilation, a growthregulating mechanism in cultured plant cells. Develop. Biol., Suppl. **3**, 206–226 (1969).

GREENWAY, H.: Plant response to saline substrates. VII. Growth and ion uptake throughout plant development in two varieties of *Hordeum vulgare*. Australian J. Biol. Sci. **18**, 763 (1965).

GREENWAY, H., GUNN, A., THOMAS, D.A.: Plant response to saline substrates. VIII. Regulation of ion concentrations in salt-sensitive and halophytic species. Australian J. Biol. Sci. **19**, 741–756 (1966).

HAGEN, C.E., HOPKINS, H.T.: Ionic species in orthophosphate absorption by barley roots. Plant Physiol. **30**, 193–199 (1955).

HAGEN, C.E., LEGGETT, J.C., JACKSON, P.C.: The sites of orthophosphate uptake by barley roots. Proc. Natl. Acad. Sci. (U.S.) **43**, 496–506 (1957).

HANDRECK, K.A., JONES, L.H.P.: Uptake of monosilicic acid by *Trifolium incarnatum*. Australian J. Biol. Sci. **20**, 483–485 (1967).

HATRICK, A.A., BOWLING, D.J.F.: A study of the relationship between root and shoot metabolism. J. Exptl. Bot. **24**, 607–613 (1973).

HIATT, A.J.: Relationship of cell sap pH to organic acid change during ion uptake. Plant Physiol. **42**, 294–298 (1967).

HIATT, A.J., HENDRICKS, S.B.: The role of CO_2 fixation in accumulation of ions by barley roots. Z. Pflanzenphysiol. **56**, 220–232 (1967).

HIGINBOTHAM, N.: The mineral absorption process in plants. Botan. Rev. **39**, 15–69 (1973).

HIGINBOTHAM, N., ANDERSON, W.P.: Electrogenic pumps in higher plant cells. Canad. J. Bot. **52**, 1011–1021 (1974).

HIGINBOTHAM, N., ETHERTON, B., FOSTER, R.J.: Mineral ion contents and cell trans-membrane electropotentials of pea and oat seedling tissue. Plant Physiol. **42**, 37–46 (1967).

HIGINBOTHAM, N., GRAVES, J.S., DAVIS, R.F.: Evidence for an electrogenic ion transport pump in cells of higher plants. J. Membrane Biol. **3**, 210–222 (1970).

HODGES, T.K., DARDING, R.L., WEIDNER, T.: Gramicidin-D-stimulated influx of monovalent cations into plant roots. Planta **97**, 245–256 (1971).

HOLFORD, I.C.R.: Comparative requirements of sulphur by cereals and legumes. Australian. J. Agr. Res. **22**, 879–884 (1971).

HUTCHIN, M.E., VAUGHAN, B.E.: Relation between calcium and strontium transport rates as determined simultaneously in the primary root of *Zea mays*. Plant Physiol. **42**, 644–650 (1967).

IGHE, U., PETTERSSON, S.: Metabolism-linked binding of rubidium in the free space of wheat roots and its relation to active uptake. Physiol. Plantarum **30**, 24–29 (1974).

JACKSON, P.C., ADAMS, H.R.: Cation — anion balance during potassium and sodium absorption by barley roots. J. Gen. Physiol. **46**, 369–386 (1963).

JACKSON, P.C., HAGEN, C.E.: Products of orthophosphate absorption by barley roots. Plant Physiol. **35**, 326 (1960).

JACKSON, W.A., FLESHER, D., HAGEMAN, R.H.: Nitrate uptake by dark-grown corn seedlings. Some characteristics of apparent induction. Plant Physiol. **51**, 120–127 (1973).

JACOBY, B., PLESSNER, O.E.: Oligomycin effect on ion absorption by excised barley roots and on their ATP content. Planta **90**, 215–221 (1970).

JEFFERIES, R.L.: The ionic relations of seedlings of the halophyte *Triglochin maritima* L. In: Ion transport in plants (W.P. ANDERSON, ed.), p. 297–321. London-New York: Academic Press 1973.

JEFFREY, D.W.: Phosphate nutrition of Australian heath plants II. The formation of polyphosphate by five heath species. Australian J. Bot. **16**, 603–613 (1968).

JESCHKE, W.D.: Evidence for a K^+-stimulated Na^+ efflux at the plasmalemma of barley root cells. Planta **94**, 240–245 (1970).

JESCHKE, W.D.: Evidence for a K^+-stimulated Na^+ efflux at the plasmalemma of barley root cells. Planta **94**, 240–245 (1970).

JESCHKE, W.D.: Wirkung von K^+ auf die Fluxe und den Transport von Na^+ in Gerstenwurzeln, K^+-stimulierter Na^+-Efflux in der Wurzelrinde. Planta **106**, 73–90 (1972).

JESCHKE, W.D.: K^+-stimulated Na^+ efflux and selective transport in barley roots. In: Ion transport in plants (W.P. ANDERSON, ed.), p. 285–296. London-New York: Academic Press 1973.

JESCHKE, W.K., SIMONIS, W.: Über die Aufnahme von Phosphat- und Sulfationen durch Blätter von *Elodea densa* und ihre Beeinflussung durch Licht, Temperatur und Außenkonzentration. Planta **67**, 6–32 (1965).

JOHANSEN, C., EDWARDS, D.G., LONERAGAN, J.F.: Interactions between potassium and calcium in their absorption by intact barley plants. I. Effects of potassium on calcium absorption. Plant Physiol. **43**, 1717–1721 (1968a).

JOHANSEN, C., EDWARDS, D.G., LONERAGAN, J.F.: Interactions between potassium and calcium in their absorption by intact barley plants. II. Effects of calcium and potassium concentration on potassium absorption. Plant Physiol. **43**, 1722–1726 (1968b).

JONES, L.H.P., HANDRECK, K.A.: Studies of silica in the oat plant. III. Uptake of silica from soils by the plant. Plant Soil **23**, 79–96 (1965).

KANNAN, S.: Kinetics of iron absorption by excised rice roots. Planta **96**, 262–270 (1971).

KIRKBY, E.A., MENGEL, K.: Ionic balance in different tissues of the tomato plant in relation to nitrate, urea, or ammonium nitrate. Plant Physiol. **42**, 6–14 (1967).

KOMOR, E., TANNER, W.J.: The hexose-proton cotransport system of *Chlorella*. pH-dependent change in K_m values and translocation constants of the uptake system. J. Gen. Physiol. **64**, 568–581 (1974).

KOMOR, E., TANNER, W.J.: Simulation of a high- and low-affinity sugar-uptake system in *Chlorella* by a pH-dependent change in the K_m of the uptake system. Planta **123**, 195–198 (1975).

LÄUCHLI, A., EPSTEIN, E.: Lateral transport of ions into the xylem of corn roots I. Kinetics and energetics. Plant Physiol. **48**, 111–117 (1971).

LÄUCHLI, A., LÜTTGE, U., PITMAN, M.G.: Ion uptake and transport through barley seedlings: differential effect of cycloheximide. Z. Naturforsch. **28c**, 431–434 (1973).

Lazaroff, N., Pitman, M.G.: Calcium and magnesium uptake by barley seedlings. Australian J. Biol. Sci. **19**, 991–1005 (1966).

Leggett, J.E., Frere, M.H.: Growth and nutrient uptake by soybean plants in nutrient salts of graded concentration. Plant Physiol. **48**, 457–460 (1971).

Leggett, J.E., Gilbert, W.A.: Localization of the Ca-mediated apparent ion selectivity in the cross-sectional volume of soybean roots. Plant Physiol. **42**, 1658–1664 (1967).

Leggett, J.E., Gilbert, W.A.: Magnesium uptake by soybeans. Plant Physiol. **44**, 1182–1186 (1969).

Leigh, R.A., Wyn Jones, R.G.: The effect of increased internal ion concentration upon the ion uptake isotherms of excised maize root segments. J. Exptl. Botany **24**, 787–795 (1973).

Lin, W., Hanson, J.B.: Increase in electrogenic membrane potential with washing of corn root tissue. Plant Physiol. **54**, 799–801 (1974).

Lonergan, J.F.: Nutrient concentration, nutrient flux, and plant growth. Trans. 9th Int. Cong. Soil Sci., p. 173–182 (1968).

Lonergan, J.F., Asher, C.J.: Response of plants to phosphate concentration in solution culture. II. Rate of phosphate absorption and its relation to growth. Soil Sci. **103**, 311–318 (1967).

Lonergan, J.F., Snowball, K.: Calcium requirements of plants. Australian J. Agr. Res. **20**, 465–467 (1969a).

Lonergan, J.F., Snowball, K.: Rate of calcium absorption of plant roots and its relation to growth. Australian J. Agr. Res. **20**, 479–490 (1969b).

Loughman, B.C., Russell, R.S.: The absorption and utilization of phosphate by young barley plants. IV. The initial stages of phosphate metabolism in roots. J. Exptl. Bot. **8**, 280–293 (1957).

Lüttge, U.: Stofftransport der Pflanzen. Berlin-Heidelberg-New York: Springer 1973.

Lüttge, U., Läuchli, A., Ball, E., Pitman, M.G.: Cycloheximide: A specific inhibitor of protein synthesis and intercellular ion transport in plant roots. Experientia **30**, 470–471 (1974).

Lycklama, J.C.: The absorption of ammonium and nitrate by perennial rye-grass. Acta Botan. Neerl. **12**, 361–423 (1963).

Maas, E.V.: Calcium uptake by excised maize roots and interactions with alkali cations. Plant Physiol. **44**, 985–989 (1969).

Maas, E.V., Moore, D.P., Mason, B.J.: Manganese absorption by excised barley roots. Plant Physiol. **43**, 527–530 (1968).

Maas, E.V., Moore, D.P., Mason, B.J.: Influence of Ca and Mg on Mn absorption. Plant Physiol. **44**, 796–800 (1969).

Maas, E.V., Ogata, G.: Absorption of magnesium and chloride by excised corn roots. Plant Physiol. **47**, 357–360 (1971).

Marrè, E., Lado, P., Ferroni, A., Ballarin Denti, A.: Transmembrane potential increase induced by auxin, benzyladenine and fusicoccin. Correlation with proton extrusion and cell enlargement. Plant Sci. Letters **2**, 257–265 (1974).

Marschner, H.: Einfluß von Calcium auf die Natrium-Aufnahme und die Kalium-Abgabe isolierter Gerstenwurzeln. Z. Pflanzenernähr. Düng. Bodenk. **107**, 20–32 (1964).

Marschner, H., Mengel, K.: Der Einfluß von Ca- und H-Ionen bei unterschiedlichen Stoffwechselbedingungen auf die Membranpermeabilität junger Gerstenwurzeln. Z. Pflanzenernähr. Düng. Bodenk. **112**, 39–49 (1966).

Milthorpe, F.L., Moorby, J.: An introduction to crop physiology. Cambridge: Cambridge University Press 1974.

Moore, D.P., Mason, B.J., Maas, E.V.: Accumulation of calcium in exudate of individual barley roots. Plant Physiol. **40**, 641–644 (1965).

Moore, D.P., Overstreet, R., Jacobson, L.: Uptake of magnesium and its interaction with calcium in excised barley roots. Plant Physiol. **36**, 290–295 (1961).

Morgan, M.A., Volk, R.J., Jackson, W.A.: Simultaneous influx and efflux of nitrate during uptake by perennial ryegrass. Plant Physiol. **51**, 267–272 (1973).

Munns, D.N., Johnson, C.M., Jacobson, L.: Uptake and distribution of manganese in oat plants. I. Varietal variation. Plant Soil **19**, 115–126 (1963).

Nassery, H., Baker, D.A.: Extrusion of sodium ions by barley roots. I. Characteristics of the extrusion mechanism. Ann. Bot. (London), N.S. **36**, 881–887 (1972a).

NASSERY, H., BAKER, D.A.: Extrusion of sodium ions by barley roots II. Localisation of the extrusion mechanism and its relation to long-distance sodium ion transport. Ann. Bot. (London), N.S. **36**, 889–895 (1972b).

NISSEN, P.: Uptake of sulphate by roots and leaf slices of barley: mediated by single, multiphasic mechanisms. Physiol. Plantarum **24**, 315–324 (1971).

NYE, P.H., TINKER, P.B.: The concept of a root demand coefficient. J. Appl. Ecol. **6**, 293–300 (1969).

PALLAGHY, C.K., SCOTT, B.I.H.: The electrochemical state of cells of broad bean roots. Australian J. Biol. Sci. **22**, 585–600 (1969).

PERSSON, L.: Labile bound sulphate in wheat roots: localisation, nature and possible connection to the active transport mechanism. Physiol. Plantarum **22**, 959–976 (1969).

PETTERSON, S.: Active and passive components in sulphate uptake in sunflower plants. Physiol. Plantarum **19**, 459–492 (1966).

PITMAN, M.G.: Sodium and potassium uptake by seedlings of *Hordeum vulgare*. Australian J. Biol. Sci. **18**, 10–24 (1965a).

PITMAN, M.G.: Transpiration and the selective uptake of potassium by barley seedlings (*Hordeum vulgare* cv. Bolivia). Australian J. Biol. Sci. **18**, 987–998 (1965b).

PITMAN, M.G.: Uptake of potassium and sodium by seedlings of *Sinapis* alba. Australian J. Biol. Sci. **19**, 257–269 (1966).

PITMAN, M.G.: Adaptation of barley roots to low oxygen supply and its relation to potassium and sodium uptake. Plant Physiol. **44**, 1233–1240 (1969a).

PITMAN, M.G.: Simulation of Cl^- uptake by low-salt barley roots as a test of models of salt uptake. Plant Physiol. **44**, 1417–1427 (1969b).

PITMAN, M.G.: Active H^+ efflux from cells of low-salt barley roots during salt accumulation. Plant Physiol. **45**, 787–790 (1970).

PITMAN, M.G.: Uptake and transport of ions in barley seedlings. I. Estimation of chloride fluxes in cells of excised roots. Australian J. Biol. Sci. **24**, 407–421 (1971).

PITMAN, M.G.: Uptake and transport of ions in barley seedlings. II. Evidence for two active stages in transport to the shoot. Australian J. Biol. Sci. **25**, 243–257 (1972a).

PITMAN, M.G.: Uptake and transport of ions in barley seedlings. III. Correlation of potassium transport to the shoot with plant growth. Australian J. Biol. Sci. **25**, 905–919 (1972b).

PITMAN, M.G., CRAM, W.J.: Regulation of inorganic ion transport in plants. In: Ion transport in plants (W.P. ANDERSON, ed.), p. 465–481. London-New York: Academic Press 1973.

PITMAN, M.G., COURTICE, A.C., LEE, B.: Comparison of potassium and sodium uptake by barley roots at high and low salt status. Australian J. Biol. Sci. **21**, 871–881 (1968).

PITMAN, M.G., LÜTTGE, U., LÄUCHLI, A., BALL, E.: Ion uptake to slices of barley leaves, and regulation of K content in cells of the leaves. Z. Pflanzenphysiol. **72**, 75–88 (1974).

PITMAN, M.G., MERTZ, S.M. JR., GRAVES, J.S., PIERCE, W.S., HIGINBOTHAM, N.: Electrical potential differences in cells of barley roots and their relation to ion uptake. Plant Physiol. **47**, 76–80 (1971a).

PITMAN, M.G., MOWAT, J., NAIR, H.: Interaction of processes for accumulation of salt and sugar in barley plants. Australian J. Biol. Sci. **24**, 619–631 (1971b).

PITMAN, M.G., SADDLER, H.D.W.: Active sodium and potassium transport in cells of barley roots. Proc. Natl. Acad. Sci. (U.S.) **57**, 44–49 (1967).

PITMAN, M.G., SCHAEFER, N., WILDES, R.A.: Stimulation of H^+ efflux and cation uptake by fusicoccin in barley roots. Plant Sci. Letters **4**, 323–329 (1975a).

PITMAN, M.G., SCHAEFER, N., WILDES, R.A.: Relation between permeability to potassium and sodium ions and fusicoccin-stimulated hydrogen-ion efflux in barley roots. Planta **126**, 61–73 (1975b).

RAINS, D.W., EPSTEIN, E.: Sodium absorption by barley roots: role of the dual mechanisms of alkali cation transport. Plant Physiol. **42**, 314–318 (1967a).

RAJU, K.V. VON, MARSCHNER, H., RÖMHELD, V.: Effect of iron nutritional status on ion uptake, substrate pH and production and release of organic acids and riboflavin by sunflower plants. Z. Pflanzenernähr. Düng. Bodenk. **132**, 177–190 (1972).

REISENAUER, H.M.: Growth and nutrient uptake by wheat from dilute solution cultures. Agron. Abst., p. 108 (1968).

ROBSON, A.D., LONERAGAN, J.F.: Sensitivity of annual *Medicago* species to manganese toxicity as affected by calcium and pH. Australian J. Agr. Res. **21**, 223–232 (1970).

RORISON, I.H.: Ecological inferences from laboratory experiments on mineral nutrition in ecological aspects of the mineral nutrition of plants, p. 155–177. Oxford: Blackwell 1969.

SCHAEFER, N., WILDES, R.A., PITMAN, M.G.: Inhibition by p-fluorophenylalanine of protein synthesis and of ion transport across the roots in barley seedlings. Australian J. Plant Physiol. **2**, 61–74 (1975).

SCOTT, B.I.H., GULLINE, H., PALLAGHY, C.K.: The electrochemical state of cells of broad bean roots. I. Investigations of elongating roots of young seedlings. Australian J. Biol. Sci. **21**, 185–200 (1968).

SHEPHERD, U.H., BOWLING, D.J.F.: Active accumulation of sodium by roots of five aquatic species. New Phytologist **72**, 1075–1080 (1973).

SLAYMAN, C.W.: Net potassium transport in *Neurospora:* properties of a transport mutant. Biochim. Biophys. Acta **211**, 502–512 (1970).

SMITH, F.A.: The internal control of nitrate uptake into excised barley roots with differing salt contents. New Phytologist **72**, 769–782 (1973).

SMITH, F.W.: The effect of sodium on potassium nutrition and ionic relations in rhodes grass. Australian J. Agr. Res. **25**, 407–414 (1974).

SPECHT, R.L., GROVES, R.H.: A comparison of the phosphorus nutrition of Australian heath plants and introduced economic plants. Australian J. Bot. **14**, 201 (1966).

SUTCLIFFE, J.F.: The role of protein synthesis in ion transport. In: Ion transport in plants (W.P. ANDERSON, ed.), p. 399–406. London-New York: Academic Press 1973.

UHLER, R.L., RUSSELL, R.S.: Chloramphenicol inhibition of salt absorption by intact plants. J. Exptl. Botany **14**, 431–437 (1963).

ULRICH, A.: Metabolism of non-volatile organic acids in excised barley roots as related to cation-anion balance during salt accumulation. Amer. J. Bot. **28**, 526–537 (1941).

ULRICH, A.: Metabolism of organic acids in excised barley roots as influenced by temperature, oxygen tension and salt concentration. Amer. J. Bot. **29**, 220–227 (1942).

WALLACE, W.: The distribution and characteristics of nitrate reductase and glutamate dehydrogenase in the maize seedling. Plant Physiol. **52**, 191–196 (1973).

WEIGL, J.: Wirkung von CCCP und UO_2^{2+} auf die Ionenfluxe in Wurzeln. Planta **91**, 270–273 (1970).

WELCH, R.M.: Vanadium uptake by plants. Absorption kinetics and the effects of pH, metabolic inhibitors, and other anions and cations. Plant Physiol. **51**, 828–832 (1973).

WILDES, R.A., PITMAN, M.G.: Pyruvic kinase activity in roots of barley seedlings in relation to salt status. Z. Pflanzenphysiol. **76**, 69–75 (1975).

WILLIAMS, R.F.: The effects of phosphorus supply on the rates of intake of phosphorus and nitrogen and upon certain aspects of phosphorus metabolism in gramineous plants. Australian J. Sci. Res. **1**, 333–361 (1948).

3.4 Transport through Roots

W.P. ANDERSON

1. Definition of Scope

Chapter *3.4* is chiefly concerned with collecting and reviewing the information available to us from studies of xylem exudation from excised root systems. The emphasis will be on the mechanisms by which the plant root transports material (principally salts and water) from the external solution to the xylem stream. Ion absorption into the root cells is discussed elsewhere in this volume (see *3.2* and *3.3*).

Although it is true that the excised, exuding root is the best experimental system for definitive measurement, the question may be raised as to how closely such results mirror the through-put of salts and water from root to shoot in an intact plant. Therefore data on intact and transpiring plants will also be included.

2. Ion Transport

2.1 Exudation Experiments

The experimental techniques of measuring exudation parameters from excised roots are extremely simple; the cut end of the root or root system is inserted and sealed into a glass tube of appropriate diameter which collects the xylem exudate. The rate of production of exudate, with the root bathed in (usually aerated) solution of known chemical composition, can then be monitored at intervals by measuring the height of the column within the collecting tube. At various times the collected exudate can be removed for chemical analysis. The metal cations are normally assayed by flame photometry (emission or absorption) and the anions by various methods which can be found in the appropriate citations given in Tables *3.16* and *3.17*. Total osmotic content of the exudate can be measured on a commercial osmometer; there is usually no trouble in obtaining the necessary sample volume. Total exudate ion content can often be conveniently determined by conductivity measurements.

In such preparations ion transport through the root from the external solution to the xylem stream is given by the exudation ion flux which is simply derived from the primary parameters of exudate concentration and volume exudation rate as

$$J_s = J_v \cdot c_s \qquad (3.6)$$

where J_s (mol m^{-2} s^{-1}) is the exudation flux of ion "s" from the xylem, c_s (mM = mol m^{-3}) is the exudate concentration of ion "s" and J_v (m^3 m^{-2} s^{-1})

Table 3.16. Exudation data from excised root preparations

Roots and solution	Water flux (mm³ m⁻² s⁻¹)	Exudate concentration mM (=mol m⁻³)			Exudate ion flux (nmol m⁻² s⁻¹)			Ref.
		K⁺	Cl⁻	SO₄²⁻	K⁺	Cl⁻	SO₄²⁻	
Zea mays								
1 mM KCl	7.22	20.4	25.1		147	181		House and Findlay (1966a)
10 mM KCl	6.92	28.4	30.7		196	212		House and Findlay (1966a)
50 mM KCl	3.60	54.3	57.6		196	207		House and Findlay (1966a)
0.1 mM K₂SO₄	1.94	13.8		2.9	27		6	Anderson and Collins (1969)
10 mM K₂SO₄	1.66	25.2		6.2	42		10	Anderson and Collins (1969)
Avena fatua								
1 mM KCl	1.66	5.9			9.8			J.C. Collins (personal communication)
10 mM KCl	1.11	10.4			11.5			
50 mM KCl	0.55	63.7			35.3			
Avena sativa								
1 mM KCl	3.32	5.0			16.6			J.C. Collins (personal communication)
10 mM KCl	1.94	8.4			16.3			
50 mM KCl	1.11	45.3			50.2			
Allium cepa		K⁺	Na⁺		K⁺			
1 mM KCl	3.6	12.5	5.6		45.0	20.2		Hay and Anderson (1972)
10 mM KCl	3.3	12.3	9.1		40.6	30.0		
Hordeum vulgare (K⁺+Na⁺) (mM)		K⁺	Na⁺		K⁺	Na⁺		
0.5+9.5	4.3	25	12		107	51		Pitman (1965a)
1.0+9.0	4.3	24	8		105	34		
2.5+7.5	3.8	30	4		115	15		
6.0+4.0	4.2	35	1.6		147	7		
8.0+2.0	4.0	36	1.0		145	4		
Ricinus communis	(mm³ g$_{FW}^{-1}$ s⁻¹)				K⁺ (nmol g$_{FW}^{-1}$ s⁻¹)			
1 mM KNO₃+0.1 mM CaCl₂	6.76	12.2			82			Minchin and Baker (1973)
1 mM KNO₃+2.5 mM CaCl₂	4.72	7.5			35			
5 mM KNO₃+0.1 mM CaCl₂	4.33	15.6			68			
5 mM KNO₃+2.5 mM CaCl₂	3.39	16.4			56			

Table 3.17. Exudation data from excised root preparations in nutrient solutions

Plant	Concentrations								Ref.
	K^+	Na^+	Ca^{2+}	Mg^{2+}	Cl^-	NO_3^-	SO_4^{2-}	PO_4	
Zea mays									
in solution ($0.1\times$)	0.1	0.1	0.1	–	–	0.2	–	–	DAVIS and
in exudate	14.7	0.8	1.4	–	–	5.6	–	–	HIGINBOTHAM
									(1969)
in solution ($1\times$)	1.0	1.0	1.0	0.25	1.0	1.0	–	–	
in exudate	19.4	1.2	3.3	1.4	3.1	9.7	–	–	
Sinapis alba									
in solution	1.0	–	0.1	–	1.2	–	–	–	COLLINS and
in exudate	4.4	–	0.6	–	5.1	–	–	–	KERRIGAN
									(1973)
in solution	10.0	–	0.1	–	10.2	–	–	–	
in exudate	12.4	–	1.3	–	13.2	–	–	–	
Ricinus communis									
in solution	0.57	0.33	0.40	0.22	0.13	1.42	0.37	0.2	BOWLING et al.
in exudate	5.13	0.83	4.50	1.70	1.1	11.8	0.77	1.82	(1966)

is the flux of exudate volume from the xylem. Normalization of these fluxes is usually made over unit area of root surface, but in some cases with extensively branched root system, fluxes are normalized on a dry or fresh weight basis (see Table 3.2).

Tables 3.16 and 3.17 show exudation data on a number of different species. In general there is much inter-species variation in initial exudation behavior, i.e. in the first few hours after excision. Most species show an initial lag, from 1 h in the case of *Zea* roots grown and allowed to exude from 1 mM KCl/0.1 mM $CaCl_2$, to as long as 8 h in the case on *Allium* roots under similar conditions. In all species the initial lag phase, during which exudation starts and builds up to a steady rate, is increased by switching, after excision, roots grown on dilute solution to a more concentrated solution. Note finally that transport through the root to the xylem (J_s) does not continously increase as the external concentration increases. Fig. 3.25 shows the Cl^- flux from the xylem of excised *Zea* roots in a range of KCl external

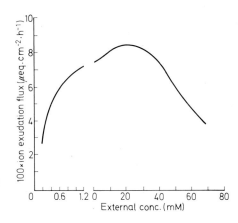

Fig. 3.25. Exudation Cl^- flux from excised maize roots at varying external KCl concentrations. Reproduced from HOUSE and FINDLAY (1966a)

concentrations. Exudate concentration continuously increases but as the rate of volume production falls at high external concentration, the Cl^- exudation flux goes through a broad maximum. The situation is more complex with other ions which are less osmotically important. Compare these data for J_s in excised roots with the net uptake rates to whole plants in Fig. *3.23* (J_s is equivalent to Φ_{ox}).

2.2 Radiotracer Experiments and Transpirational Effects

2.2.1 Tracer Determinations of Ion Export from Excised Roots

The excised root preparation described in *3.4.2.1* can also be used for radiotracer experiments; isotope is added to the external solution and the time course of radioactivity build-up in both exudate and tissue may be followed in an obvious manner. Assay techniques for the various isotopes can be found in the literature citations. The major caution is that for most ions the exudate and tissue-specific activities cannot be assumed to equal those of the external solution for at least 24 h after addition of the isotope.

A modification of the excised root preparation has been developed for tracer flux studies (WEIGL, 1969; PITMAN, 1971), where the excised root is sealed into a two- or three-compartment chamber, so that the only communication between compartments is transport through the root. Using this technique, the most satisfactory tracer-flux analyses have been made.

Radiotracer studies have also clearly shown that ion uptake to the xylem varies at different zones of the root from the tip. BOWEN and ROVIRA (1967) have shown that PO_4 uptake in different zones of wheat roots is significantly different, irrespective of transpirational effects; one major absorption peak occurs at 3 cm from the root tip and another at a higher region corresponding to the development of lateral root primordia. Further information on the absorptive ability of barley root systems is available in RUSSEL and SANDERSON (1967) where it was found that there was little translocation from the apical 3 mm of roots, while in the older regions of barley root systems, there is greater PO_4 uptake by the nodal axes than from the seminal axes or from the lateral primordia. This variation in absorptive and translocating capacity has been known since the days of the pioneering work of PREVOT and STEWARD (1936). (See also the discussion of ion transport and transpiration in Chap. *10.2*.)

2.2.2 Ion Transport in Relation to Translocation

Estimates of ion translocation from the root to the shoot of an intact, transpiring plant are generally made by either of two techniques. Radio isotope may be applied to the root solution and its passage through the stem and into the leaves followed by placing an end window counter against the plant. There are many uncertainties, e.g. counting efficiency, tissue-specific activity, but for comparative work on similar plants at differing transpiration rates, this method can provide useful data. The second common technique is to harvest the leaves after the plant has been exposed to a known regime of transpirational stress, digest the tissue and assay for the ions, either by chemical concentration analysis or again by radioactive counting.

Alternatively, experiments have been designed where transpirational effects on ion translocation are simulated by applying either a hydrostatic or osmotic pressure water stress to de-topped root systems; in Table 3.18 there is a selection of data from three species. The general rule is that the xylem sap becomes diluted at increased rates of water flow, but nevertheless the net effect of higher transpiration rates is to increase ion translocation to the shoot. The flux of an ion through the root and into the xylem under these conditions may be written as

$$J_s = J_v(1-\sigma)\bar{c}_s + J_{pass} + J_{act} \tag{3.7}$$

in the usual formalism of irreversible thermodynamics. The first term gives the component of the ion flux coupled to the volume flux, J_v, which is essentially the water flow rate, or transpiration rate in this case; σ is the effective reflexion coefficient across the root from the external solution to the xylem and $\bar{c}_s$ is the mean solute concentration associated with the effective membrane having the reflexion coefficient σ. (The problem of there being a CURRAN and MCINTOSH (1962) two-membrane system in the root will not be taken up here, but see later in 3.4.3.2.) The second term, J_{pass}, is simply the passive flux of the ion under the physical driving force of its electrochemical potential gradient. At this point the complication of streaming potential will not be taken up explicitly. The third term measures the active flux of the ion being delivered to the xylem by carrier mechanisms coupled to the metabolic reactions of the cells. It is possible that this active flux may be affected by the water flow through some coupling mechanism which is as yet not at all understood.

The experimental problems in transpiration simulation experiments are considerable; firstly it is by no means certain that the water potential gradient applied to an excised root system by hydrostatic pressure differences between the external solution and the xylem sap is effectively the same as the natural stress in an intact transpiring plant. In the latter case the pressure gradient is purely between the xylem and the external solution, with no tendency for the intercellular air spaces of the root cortex to become water-filled. The second point of divergence is that in excised root systems there can be no circulation of material from the leaves down again to the roots, which movement in the intact plant may have important effects, both for recirculation of ions and for distributing hormones (e.g. ABA) to control ion uptake and translocation.

Ion selectivity is also known to be affected by transpiration rate. PITMAN (1965b) has shown that higher transpiration rates increase the rate of Na^+ translocation to the shoot and of Na^+ uptake to the root system, at the expense of K^+. The total $(K^+ + Na^+)$ uptake is apparently unaffected by transpiration rate in barley roots. The overall picture of $K^+ - Na^+$ selectivity is however complicated by the salt status of the seedlings (PITMAN et al., 1968) and the interested reader should also consult Chap. 3.3.

2.3 The Role of the Cortex

The concern in 3.4.2.3 to 3.4.2.5 is with the ion transport function of cortex, endodermis and stele; water transport and exudation will be discussed later. The emphasis

Table 3.18. Ion translocation from the root at different J_v. Note different units used for ion translocation and xylem sap content

Plant preparation	Solution	Water flow (J_v) $(mm^3 m^{-2} s^{-1})$	Ion transport (pmol $g_{FW}^{-1} s^{-1}$; K^+)	Xylem sap concentration	Ref.
Ricinus communis	0.5 mM KNO_3		20	—	Baker and Weatherley (1969)
	" + 10 mM mannitol		11	—	
	5.0 mM KNO_3		32	—	
	" + 10 mM mannitol		27	—	
Lyco-persicum esculentum	Hoagland + 6 ppm ^{32}P		(g ^{32}P in shoot after 4 h exposure)	(ppm ^{32}P in xylem sap)	Klepper and Greenway (1968)
		0.27		38	
		1.35		15	
		1.76		10	
	^{32}P pulse; −0.4 bar manni-tol solution	—	7.7	—	Greenway and Klepper (1968)
	^{32}P pulse; −5.4 bar manni-tol solution	—	3.5	—	
			(pmol $m^{-2} s^{-1}$)	(mM; SO_4^{2-})	
Helianthus annuus	0.05 mM SO_4^{2-}	$13.9 \cdot 10^{-3}$	1.25	0.09	Pettersson (1966)
		27–$162 \cdot 10^{-3}$	—	0.04	
	0.5 mM SO_4^{2-}	$13.9 \cdot 10^{-3}$	4.17	0.3	
		27–$162 \cdot 10^{-3}$	—	0.15	
	2.5 mM SO_4^{2-}	$13.9 \cdot 10^{-3}$	16.6	1.2	
		42–$162 \cdot 10^{-3}$	—	0.4	
	25 mM SO_4^{2-}	$13.9 \cdot 10^{-3}$	209	15.1	
		$27 \cdot 10^{-3}$	216	8.0	
		$56 \cdot 10^{-3}$	235	4.2	
			(nmol $m^{-2} s^{-1}$)	(mM; Cl^-)	
Hordeum vulgare	50 mM NaCl + nutrient solution	2.2	34	15.5	Greenway (1965) (assuming 1 g_{DW} of shoot = 0.03 m^2 root surface)
		2.2	31	14	
		4.1	41	10	
		5.0	40	8	
		8.7	70	8	
		13.0	65	5	
		14.5	43	3	
		15.0	60	4	
		18.0	72	4	
		19.0	76	4	
		35.0	70	2	

will be on through-transport from the external solution to the xylem and hence to the shoot. Ion uptake *per se* is dealt with in Chaps. *3.2* and *3.3*.

There are two parallel pathways for radial movement across the cortex, in the extra-cellular space or apoplasm, and in the symplasm. The root apoplasm is a free diffusional continuity of the external solution, but with its properties modified by the fixed negative charges in the cell walls, chiefly the ionized carboxyl end groups of pectic material (see *1.2.1.2* and Part A, *5.3*).

The parameters of transport in the free space or apoplasm have been reviewed in detail elsewhere (Chaps. *1* and *3.1*; Part A, *5.3*) and will not be taken up again here. In general it can be safely concluded that diffusional constraints in the apoplasm will not rate-control ion uptake, except perhaps for multi-valent ions in low ionic strength external solutions. Thus it seems that the apoplasm is a relatively low resistance pathway across the root cortex (*1.4.2.2.1*), and as such *could* provide a pathway for the ion fluxes to the xylem.

BANGE (1973) has suggested that there may be a resistance to diffusion at the outer boundary of the cortex (see also VAKHMISTROV, 1967). However, further work is needed to establish the experimental conditions under which such a barrier becomes rate-limiting, since there are many results in the literature that show no diffusional resistance at the root surface.

The effectiveness of the symplasm as a transit pathway across the cortex is the second quantity to be considered. There is information from physiological experiments (ARISZ, 1956), electrical measurements (SPANSWICK, 1972) and theoretical assessments (TYREE, 1970) to suggest that transport of ions through the symplasm can maintain the fluxes to the xylem. Symplasmic transport has been reviewed earlier in this volume (Chap. *2*) in much more detail than is given here.

Briefly, there are two distinct areas for consideration. First, the mechanistic consideration of how the ions move through the peripheral cytoplasm layer of a cell and how they pass through a plasmodesma from one cell to the next, and second the anatomical consideration of what are the dimensions of a plasmodesma, how many plasmodesmata there are per cell, and how they are distributed on the different walls of the cell. Further information in this area is urgently required; the recent work by GUNNING et al. (1974) on plasmodesmatal frequency in transfer cells is a model of what is needed. However a certain amount of information is available (see Table *2.1*).

The mechanism of symplasmatic ion transport is chiefly diffusional, although there will be a small convective component within a single cell as a result of cyclosis. Nevertheless, it is quite clear (TYREE, 1970) that the rate-controlling process of symplasmatic ion transport is plasmodesmatal transit from cell to cell, and therefore the cessation of cyclosis, *by itself*, would not reduce ion fluxes through the symplasm.

Direct experimental evidence of symplasmatic coupling in the maize root cortex has been observed by SPANSWICK (1972; *2.5.1*). Microelectrodes were inserted into two neighboring cortical cells and current pulses were injected. By observing the resultant voltage signals the electrical resistance of the symplasm could be deduced. The voltage coupling ratio of 0.24 in *Zea* root cortex certainly demonstrates a relatively low resistance symplasm. Further, SPANSWICK (1972) measured almost zero attenuation through the cytoplasm of a single cell, confirming TYREE's (1970) theoretical prediction that plasmodesmatal transit constitutes the rate-control on symplasmic transport. But it must be realized that overall radial ion transport across the root may be rate-controlled by the membrane penetration processes of loading and unloading the symplasm; it is only within the symplasm that transport rates are controlled by plasmodesmatal transit.

Finally, what is the role of the cortical cell vacuoles which occupy some 90% of the cortex volume in most cereal species? The implication of much work has been that symplasmic transport essentially by-passes the vacuoles, which are sequestered from the main symplasm by the low permeability of the tonoplast. As will be seen below (*3.4.2.7*) when the effects of pre-treatment are discussed, this may not be generally true, and under certain conditions where ion uptake at the plasmalemma is rate-controlling symplasmic transport, this rate-control may be so severe that ion supply from the vacuoles to the symplasm takes over as the dominant loading mechanism to the symplasm.

2.4 The Role of the Endodermis

The traditional role of the endodermis (CRAFTS and BROYER, 1938; VAN FLEET, 1961) assumes that the Casparian band acts as an effective barrier to apoplasmic movement of both ions and water from the root cortex to the stele (*1.4.1.2, 1.4.2.2.1*). Transport to any space within the stele must therefore be through the endodermal cell membranes. Fig. *3.26* shows grain counts obtained from microautoradiographs of $^{35}SO_4$ transport through maize roots. In the presence of the metabolic inhibitor NaN_3, $^{35}SO_4$-transport is limited by the endodermis (see also Fig. *1.13*).

Apoplasmic ion fluxes across the cortex would have to be taken up at the endodermis to continue their passage through the endodermal cytoplasm to the parenchyma of the stele. To maintain the total xylem ion flux by this mechanism would require, for K^+ in young excised maize roots, endodermal membrane fluxes of about 400 nmol m^{-2} s^{-1}, somewhat larger than normally measured in root cells (see *3.3.3*). Later in this article it will be argued that a purely apoplasmic

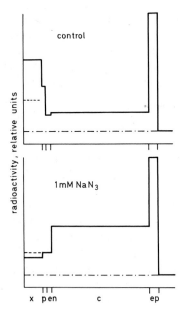

Fig. *3.26*. Grain counts obtained from microautoradiographs showing the distribution of radioactivity in maize roots after the roots were transporting ^{35}S labeled SO_4^{2-} for 1–3 h in the absence and in the presence of 1 mM NaN_3 (azide) respectively. In the presence of the inhibitor the transport to the stele across the endodermis is highly reduced. (From WEIGL and LÜTTGE, 1962.) *x* xylem, *p* pericycle, *en* endodermis, *c* cortex, *ep* epidermis; the solid line represents radioactivity in the above-mentioned tissues, line of short dashes: radioactivity in the phloem, line of dots and dashes: average background

ion flux across the cortex, with membrane transport of material solely at the endodermis, is quite unrealistic.

Ion fluxes which cross the cortex in the symplasm after uptake at the epidermal and outer cortical cells, do not have to cross the endodermal membranes and simply continue radially inward in the symplasm which is continuous through the endodermis and pericycle to the parenchyma of the stele. HELDER and BOERMA (1969) measured an average of $2 \cdot 10^4$ plasmodesmata per cell in the endodermis of young barley roots. CLARKSON et al. (1971) have conducted a similar study of the tertiary endodermis of older regions of barley roots. They estimate an average of $1.25 \cdot 10^{10}$ pits per m^2 of inner tangential wall, each pit containing on average 54 plasmodesmata, so that the estimated plasmodesmatal frequency is $6.7 \cdot 10^{11} \, m^{-2}$ of the total endodermal surface area. Comparison of this mature endodermal frequency with the measurements of HELDER and BOERMA (1969) on young endodermal cells of the same species, reveals a reduction in frequency with age of about $4 \cdot 10^{12} \, m^{-2}$ of the total endodermal surface area. CLARKSON et al. (1971) suggest explanations for it and calculate from their measurements on mature cells that the xylem exudation fluxes of water and of an anion (phosphate) can reasonably be expected to pass through these plasmodesmata. They further conclude that the passage cells in the mature endodermis, considered by LUNDEGÅRDH (1950) among others to facilitate endodermal transit, occur so infrequently that only small fractions of the total fluxes will pass through them. The symplasmic pathway, even through mature endodermal cells will carry the bulk of the material in transit to the xylem. Further to the function of passage cells, there is autoradiographic evidence (ZIEGLER et al., 1963) that they may serve to release ions from the stele to the cortex.

2.5 The Role of the Stele

The traditional interpretation of the stele function is found in the classic exposition by CRAFTS and BROYER (1938). Dead, empty xylem vessels provide a low-resistance conduit for the xylem translocation stream, while the parenchyma cells are under anaerobic stress and consequently are physiologically inactive and "leaky" to solutes. According to this view, the ions accumulated in the epidermis and cortex, and transported radially through the symplasm, generally perfuse the volume of the stele and are prevented from diffusing back into the cortex by the Casparian band of the endodermis.

Root-pressure exudation can be reasonably understood on this proposal; the excess solute concentration in the stele, compared with the external solution, results in an osmotic driving force for water entry across the endodermal membranes. The water then flows away along the dead xylem vessels, and carries with it the "average" ionic concentrations in the stele. In this manner, the observed exudation of hypertonic fluid from the xylem of excised roots can be explained.

The "leaky stele" model just described has received experimental support from several sources. It is clear that freshly isolated steles have only limited ion-uptake capability, which becomes enhanced in steles kept for several hours after isolation in aerated solution (LATIES and BUDD, 1964; LÜTTGE and LATIES, 1967; HALL et al., 1971). The inference from these observations is that steles *in situ* have only poorly developed ion-carrier mechanisms at the plasmalemma, and therefore are leaky to ions.

Whatever the reason for this presumed leakiness, it seems unlikely to be *in situ* oxygen deficit; although Hall et al. (1971) measured a lower respiration rate in freshly isolated maize stele than in cortex, the rate is not negligible and seems sufficient to supply ion transport requirement (Anderson, 1972). Moreover, there is the conflicting measurement by Yu and Kramer (1969), that freshly isolated stele respires more rapidly than cortex. Fiscus and Kramer (1970) have evidence that oxygen tension in the maize stele *in situ* is sub-optimal, but suggest the respiratory reactions are adjusted to this lower tension. Bowling (1973b) has measured the O_2 partial pressure in the cells of sunflower roots by an electrode method, to find a gradient of 0.20 bar at the epidermis to 0.17 bar at the pericycle, to 0.165 bar at the protoxylem, not a very large gradient. One cannot but agree with him that this evidence causes grave doubts about the validity of the original Crafts and Broyer (1938) explanation for leakiness in the stele.

The alternative proposal is that the parenchyma of the stele actively secretes ions across the plasmalemma, rather than there being simply passive leakage. This point of view has been argued by Läuchli and Epstein (1971), and by Läuchli et al. (1971) on the results of several experimental techniques, radioactive pulse labeling, electron probe assays of ion concentration profiles across the root, and inhibitor studies. These authors conclude that ion transport to the xylem requires two carrier-mediated membrane events, uptake at the plasmalemma of the outer cortical cells and secretion at the plasmalemma of the xylem parenchyma.

A similar conclusion has been reached independently by Pitman (1972) where the evidence is a comprehensive flux analysis for $^{36}Cl^-$ in excised barley roots after a variety of pre-treatments and with or without the addition of the uncoupler CCCP. Xylem exudation is inhibited by CCCP but the plasmalemma and tonoplast Cl^- effluxes in the cortex are not affected. Since the cortical vacuoles could therefore supply the symplasm in pre-loaded tissue even after the application of CCCP, Pitman (1972) argues that the observed rapid cessation of exudation must be due to CCCP inhibition of active secretion by the xylem parenchyma (Fig. *3.27*). In comparison, exudation of Cl^- (and K^+) from the stele continued for at least 2.5 h when the roots were transferred to a solution containing no KCl (0.5 mM $CaSO_4$), thus inhibiting influx of K^+ and Cl^- at the plasmalemma. (See also Fig. *3.30* and *3.4.2.7*

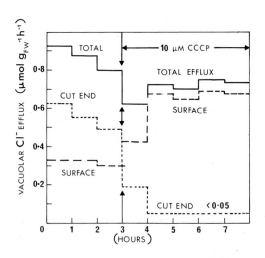

Fig. *3.27*. Effect of CCCP on the efflux of $^{36}Cl^-$ (Φ_{vo}) from barley roots. Tracer was collected separately from the surface of the root and from the cut end. CCCP was added after 3 h. Note the strong inhibition of $^{36}Cl^-$ efflux from the cut end, but that there was little effect of CCCP on the total efflux (Pitman, 1972)

for further examples of supply of ions to the exudate from the vacuoles of root cells.)

Experiments with isolated steles do not support the claims for active secretion from the stele. BAKER (1973a, b) using isolated maize steles has shown that ^{86}Rb efflux is enhanced by low temperature and by CCCP, both treatments which would be expected to block secretion from the stele.

The main problem in use of CCCP is that it uncouples respiration and so can affect all ATP-dependent processes. Evidence of a different kind for a separate process of ion transport in the stele and at the plasmalemma comes from studies with inhibitors of protein synthesis, and with certain phytohormones. Thus the inhibitor of protein synthesis CHM and the amino acid analog p-fluorophenyl alanine (FPA) both inhibit transport across the root to the stele, but have no effect on uptake to the cortical cells in short-term experiments (LÄUCHLI et al., 1973; LÜTTGE et al., 1974; SCHAEFER et al., 1975; 8.3.1.4). The phytohormone ABA under certain conditions also inhibits transport across the root without inhibiting the influx at the plasmalemma of the cortical cells (CRAM and PITMAN, 1972; PITMAN et al., 1974a; see also Table 3.19). Irrespective of how these substances affect transport across the root there is clear distinction between influx at the outer surface of the root and transport to the xylem.

In considering differential inhibitor effects on uptake and exudation it must be borne in mind that the inhibitor may possibly affect symplasmic transport rather than act on a site within the stele. Although it seems true that symplasmic transport is chiefly diffusive, it is still reasonable to suppose that continued functioning of the plasmodesmata as effective transit channels may require continuing membrane turnover and energy supply. (Though it is interesting to note that while CCCP inhibits Cl$^-$ transport across the root to about 97% in excised roots, it is only inhibited to about 70% in intact transpiring plants (LÜTTGE and LATIES, 1967; PITMAN, 1972.)

Separation of stele and cortex of roots in which transport (but not uptake) was inhibited by ABA or FPA shows that these compounds do not block movement across the endodermis (PITMAN et al., 1974b; SCHAEFER et al., 1975). These authors therefore argued that transport in the symplasm was not the limiting step, but that ABA and FPA were acting on sites within the stele. They suggested that release to the xylem was controlled by a protein which turns over relatively rapidly at the plasmalemma of the xylem parenchyma. This view on transport of K$^+$ and Cl$^-$ can be compared with transport of Fe in certain maize mutants. Thus CLARK et al. (1973) showed that a maize mutant lacked Fe in the shoot, due to inability to transport Fe from the root to the xylem; uptake of Fe to the root was the same in the mutant and a normal plant.

Under certain conditions and in certain species, xylem parenchyma may also be important in selective re-absorption of ions from the xylem stream, usually in those regions of the root above the chief absorbing zones. Na$^+$ re-absorption may be important for translocation selectivity in certain species. However, in some reported experiments, insufficient care to distinguish between tracer exchange and net re-absorption may have led to misconceptions about the magnitude of this effect.

These observations lead to consideration of the function of xylem vessels. In all the prior arguments it has been assumed that the vessels are inert conduits,

Table *3.19.* Hormone effects on ion and water transport through excised roots. Note different units used

Material and treatment	Water flux (mm^3 m^{-2} s^{-1})	Exudate or transport			Uptake to tissue			Ref.
		K^+	Ca^{2+}	Cl^-	K^+	Ca^{2+}	Cl^-	
		(mM)			(mol m^{-3})			
Zea mays								
Control (0–24 h)	4.64	22.1	1.8	23.1	69.4	2.9	41.5	COLLINS and
Kinetin (1 µM)	1.02	16.1	2.9	15.4	94.2	2.5	69.3	KERRIGAN
ABA (1 µM)	7.34	20.5	1.1	17.2	91.5	3.7	84.0	(1973)
Control (0–3 h)	3.60	13.8	—	—	—	—	—	CRAM and
ABA (80 µM)	0.60	10.7	—	—	—	—	—	PITMAN (1972)
Hordeum vulgare		(nmol g_{FW}^{-1} s^{-1})			(nmol g_{DW}^{-1} s^{-1})			
Control (0–4 h)	—	0.53	—	0.53	—	—	—	CRAM and
ABA (80 µM)	—	0.11	—	0.11	—	—	—	PITMAN (1972)
Control (0–4 h)	—	1.2	—	—	1.3	—	—	PITMAN et al.
Benzyladenine (41 µM)	—	0.6	—	—	1.2	—	—	(1974 a)
Control (0–4 h)	—	0.8	—	—	0.50	—	—	
GA (10 nM)	—	0.7	—	—	0.47	—	—	
GA (10 µM)	—	0.75	—	—	0.55	—	—	
Control (0–4 h)	—	0.9	—	—	0.59	—	—	
IAA (1 µM)	—	0.85	—	—	0.55	—	—	
IAA (10 µM)	—	0.85	—	—	0.55	—	—	

	Water flux (mm^3 g_{DW}^{-1} s^{-1})	Osmotic pressure of exudate (bars)	Ref.
Lycopersicum esculentum			TAL and IMBER (1971)
Control	$0.58 \cdot 10^{-3}$	0.73	
Kinetin (1 g m^{-3})	$0.21 \cdot 10^{-3}$	1.14	
ABA (10 g m^{-3})	$1.18 \cdot 10^{-3}$	0.74	
IAA (10 g m^{-3})	$0.76 \cdot 10^{-3}$	0.62	

	(nmol g_{FW}^{-1} s^{-1})	Ref.
Pisum sativum		LÜTTGE et al. (1968)
Control		
1 h	1.61	
4 h	1.65	
5.5 h	1.71	
+GA		
1 hr	1.54	
4 h	1.59	
5.5 h	2.38	

but it has been proposed from time to time that the vessels are more actively involved. HYLMÖ (1953) suggested in his test-tube hypothesis, that root pressure is developed across the vessel membranes. ANDERSON and HOUSE (1967) observed a correlation between xylem K^+ transport in segments of maize root and the number of vessels still containing membrane-bound cytoplasm, and suggested that the final step in radial ion transport was across the vessel membranes. SCOTT (1965) reached a similar conclusion in *Ricinus* roots and stated that all solute movement in the root hair zone is into living, nucleated xylem vessels. HIGINBOTHAM et al. (1973) offer further support for this point of view in young maize roots. They estimate that most of the outer-ring metaxylem vessels are not yet open at 10 cm from the tip, using an ink perfusion technique. However, as pointed out earlier (*3.1.2.1*) there seems to be wide variation between species; LÄUCHLI et al. (1974) found no cytoplasm in xylem beyond a few millimetres of the tip in barley roots, though transport through the xylem continued when the terminal 1 cm was cut from the roots.

Perhaps the most serious criticism of all such proposals has been pointed out by ANDERSON et al. (1970)[1]. In maize, if exudation is from the outer ring metaxylem, and if the exudation ion fluxes must cross living xylem vessel membranes, then these membranes must be able to conduct K^+ fluxes of several thousands nmol $m^{-2} s^{-1}$, many times larger than "typical" values for other root cells. The discrepancy will be even greater for anion fluxes. Although Cl^- fluxes of up to 8 μmol $m^{-2} s^{-1}$ have been measured in *Acetabularia mediterranea* (SADDLER, 1970), such values have never been observed in non-marine species. CRAM (1973) has measured Cl^- fluxes in root cortex, where the largest value he recorded was 100 nmol $m^{-2} s^{-1}$.

2.6 Electrophysiology of Roots

It is clearly established that there is a potential difference between an excised root xylem exudate and the bathing solution, having values of a few tens of millivolts, exudate negative. A comprehensive study on a single species was made by DAVIS and HIGINBOTHAM (1969) from which Fig. *3.28* is taken. These authors, together with SHONE (1969) are responsible for the current interpretation of the exudate potential having two additive components, a diffusion potential and an electrogenic potential. See Part A, *4.2.6* and *6.6.6* for an explanation of these terms.

The diffusion potential can be fitted by the Goldman equation with K^+ making the major contribution, just as has been found with single-cell potentials. The electrogenically pumped ions have not yet been identified, but the observed hyperpolarization makes it certain that anions are pumped inward electrogenically, or cations are pumped out. The only cation likely to be pumped out is H^+, but the whole situation requires further resolution.

Data on a variety of species is now available (see Table *3.20*). The information on potential profiles across the root can be suitably summarized in Fig. *3.29*. Although the values of the cell electropotentials seem rather low, these results seem to indicate that the xylem exudate potential is generated at the epidermal cells, the symplasm being an essentially equipotential network. This result is understandable in the light of SPANSWICK's (1972) resistance measurements in the symplasm.

---

[1] There are arithmetical errors at this point in the original paper; details from the present author who admits sole responsibility. The corrections (p 455 at bottom) are: $4.36 \cdot 10^{-11}$ mol $cm^{-2} s^{-1}$ for the stele; vessel surface area per cm of mature root $= 0.014$ cm^2.

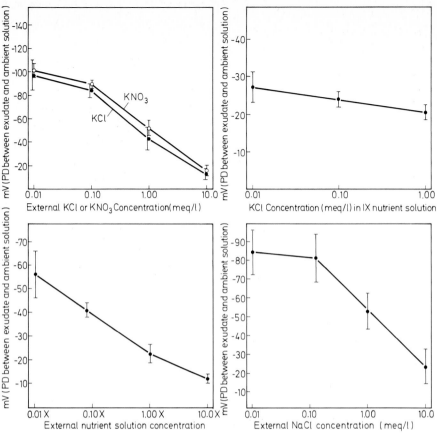

Fig. 3.28. A series of four graphs showing the variation in xylem exudate electropotential with external solution. The medium designated $1\times$ has the following composition: 1 mM KCl, 1 mM $Ca(NO_3)_2$, 0.25 mM $MgSO_4$, NaH_2PO_4 and Na_2HPO_4 to give 1 mM Na^+ and pH 5.5–5.7. Reproduced from Davis and Higinbotham (1969)

Table 3.20. Xylem exudate electropotentials of roots

Material and conditions		Exudate potential (mV) (Exudate negative)
Zea mays: bathed in	0.1 mM KCl[a]	50
	0.3 mM KCl[a]	48
	1.0 mM KCl[a]	30
	3.0 mM KCl[a]	25
	10.0 mM KCl[a]	18
Zea mays: bathed in	0.2 mM KCl[b]	58
	0.2 mM KCl, with 10^{-5} DNP added[b]	32
Zea mays: bathed in	$1\times$ solution[c]	49
	$1\times$ solution with 1 mM CN added[c]	32
Ricinus communis:	bathed in $^1/_{10}$ Stout and Arnon's solution[d]	58
Helianthus annuus:	bathed in $^1/_{10}$ culture solution[e]	48
	bathed in full strength culture solution	34

The source references are: [a] Dunlop and Bowling (1971); [b] Shone (1969); [c] Davis and Higinbotham (1969); [d] Bowling and Spanswick (1964); [e] Bowling (1966). The solution described as $1\times$ is the same as referred to in Fig. 3.28.

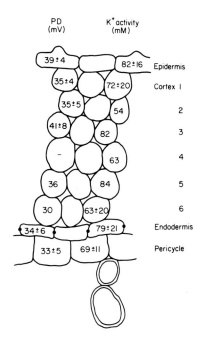

Fig. 3.29. Sketch of a section through a maize root, detailing the cell electropotentials and the K^+ activities as obtained with a K^+-sensitive microelectrode. Reproduced from BOWLING (1973a)

2.7 Effects of Pretreatment and Salt Status

The xylem exudate composition from excised roots depends on the previous growth conditions of the seedling or plant, as well as on the composition of the bathing solution during exudation. The implications of this observation may be summarized as: (i) ion transport to the xylem may be supplied at least in part, by ion reservoirs within the root, (ii) membrane-located carriers for certain ions may not be present, or may have much reduced activity, if those specific ions have not been present in the growth solution, and (iii) membrane-located carriers of ions present at high concentration during growth may have reduced activity for allosteric reasons, or simply because of insufficient energy supply (see later).

A dramatic demonstration of possibility (i) can be found in JARVIS and HOUSE (1970). Excised maize roots, grown as seedlings in 1 mM $KCl/0.1$ mM $CaCl_2$, were suspended in moist air and were observed to exude; the volume rate was 28% of controls and the K^+ exudation flux was 33%, over a 24 h period (see also 3.4.2.5). The K^+ flux can only have been supplied from the reserve in the root. A similar effect can be noticed in tracer exudation experiments; HODGES and VAADIA (1964) showed that exudate specific activity of ^{36}Cl in onion roots increases relatively slowly. In-coming ^{36}Cl in the symplasm must be exchanging with the vacuoles during transit. Conversely, pre-loaded roots exude ^{36}Cl when in external solution containing no isotope (Fig. 3.30).

There are many reports of the effects of salt status on exudation. The most recent, and very comprehensive study is by MEIRI (1973) on maize. Table 3.21 contains a synopsis of these data. In general, there are adjustments in tissue salt status following the external solution change, with resultant long term transients in exudate

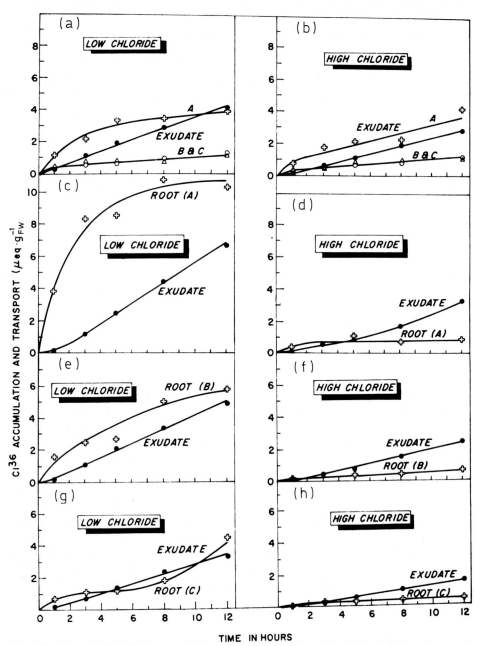

Fig. *3.30.* (a) and (b): ^{36}Cl accumulation by various zones of roots of *Allium cepa* and transport to the exudate with time for low-Cl$^-$ roots (grown on 0.3 mM CaSO$_4$) and high-Cl$^-$ roots (18 h in 0.3 mM CaSO$_4$ +2.0 mM KCl). All parts of the root were exposed to 2.0 mM KCl plus 0.3 mM CaSO$_4$ labeled with ^{36}Cl. *A*, *B* and *C* refer to root-tip zone (0–3 cm length) "middle zone" (3–6 cm) and basal zone (6–8 cm). Open symbols show accumulation and ($\bullet$) is exudation. (c) and (d): ^{36}Cl$^-$ accumulation by root zone *A* and transport to the exudate when only zone *A* was exposed to labeled solution. Other zones were in 0.3 mM CaSO$_4$. (e) and (f): As (c) and (d) but root zone *B* was exposed to labeled solution. (g) and (h) as (c) and (d) but root zone *C* was exposed to tracer (HODGES and VAADIA, 1964)

Table 3.21

Exudate ion concentrations (mM) as affected by growth and exudation media of excised maize roots

Exudation medium (mM KCl)	Growth media (mM KCl)									
	0.1		1.0		10.0		25.0		50.0	
	K	Cl	K	Cl	K	Cl	K	Cl	K	Cl
0.1	18	15	17	14	13	11	17	15	19	18
1.0	22	16	20	16	18	13	22	15	21	15
10.0	23	16	28	23	27	21	26	20	25	21
25.0	38	29	41	35	38	33	39	32	35	26
50.0	54	44	52	45	50	45	52	45	52	44

Maize root tissue ion concentrations (mM) as affected by growth and exudation after 24 h exudation

0.1	65	35	54	39	88	47	88	60	102	63
1.0	77	33	85	46	100	44	95	52	105	66
10.0	78	48	100	63	115	68	108	65	113	71
25.0	107	54	112	68	130	70	120	68	130	82
50.0	115	75	110	90	138	105	132	102	143	98

Both the growth and exudation media also contained 0.1 mM $CaCl_2$. These data are estimated by the author from MEIRI (1973).

concentrations. The effects of salt status on ion uptake in roots has been discussed in 3.3.1 and in PITMAN and CRAM (1973).

Finally, there is evidence from ANDERSON et al. (1974) that the xylem exudate of maize roots is affected by the chemical composition of the growth medium. Maize roots grown in KNO_3 and then switched to KCl solution after excision, are found to exude significant amounts of the growth-supplied anion for up to 24 h after excision.

The only reservoir in the root of sufficient capacity to account for the effects detailed above is the cortical cell vacuoles. It is implied further that ion exchange between the vacuoles of the cortex and the symplasm is, in general, rapid enough to allow significant modification of transport of ions to the xylem by the vacuolar ion contents. The overall flux situation is as given by PITMAN (1971); see also Table 3.15.

2.8 Transport of Trace Metals

There are problems in all trace metal transport studies, since all are toxic, or at least deleterious to the root, at concentrations which are easily assayed. Iron has been by far the most studied and is in one sense untypical, because so many ferric salts with anions commonly found in plants are highly insoluble.

Iron must therefore be supplied as a high stability chelate; for a brief review of the chemistry of iron chelated by ethylene diamine tetra-acetic acid (EDTA) see Beckett and Anderson (1973). There has been a certain amount of controversy as to whether the iron must be displaced from the EDTA before uptake by the root, or whether the Fe-EDTA complex is taken up as a single entity. One school of thought (e.g. Tiffin, 1970) believes that the iron is displaced by competition with other chelating agents in the membranes, presumably the carrier molecules themselves. The second opinion (e.g. Hill-Cottingham and Lloyd-Jones, 1965) is that the Fe-EDTA complex is transported across the membrane as a single entity. Beckett and Anderson (1973) show in young maize roots that iron is apparently displaced from EDTA and taken up alone in low external Fe-EDTA concentrations while at higher concentrations the Fe-EDTA complex is transported. However, it is known that EDTA can be metabolized to the eventual production of CO_2 in maize roots, which fact in itself may account for these results.

Tiffin (1967, 1970) has shown that in mature root systems of tomato and soybean, iron is translocated in the xylem as citrate, with which it forms a sufficiently stable chelate to prevent precipitation with other anions. It should be noted that iron must always remain complexed with something to keep it soluble, all the way from EDTA in the external solution to its chief eventual fate in the porphyrin ring of the haem proteins.

Tiffin (1967) has also given data for the xylem translocation in mature tomato root systems, of manganese, cobalt and zinc. All are found to be translocated as simple hydrated cations. Calcium and magnesium are likewise found as simple cations in the translocation stream, and it is assumed, although very little is known in detail, that all are taken up at the cell membranes as simple cations by carrier mechanisms similar to those which transport the major nutrient ions.

2.9 Effects of Hormones on Ion Transport

This is an area of increasing interest with a heartening rate of progress in phenomenological characterization, although the interpretation of hormonal action is still in its infancy. As might be expected there is at present a degree of dissent concerning the effect of hormones on exudation behavior but time will doubtlessly resolve the difficulties. For example, Tal and Imber (1971), Glinka (1973) and Collins and Kerrigan (1973) found that abscisic acid (ABA) in the concentration range around 10^{-6} M produced increased exudation rates in tomato, sunflower and maize roots respectively; Cram and Pitman (1972) found that similar concentrations of ABA caused the exudation rate to decrease in barley and maize roots. More recently Pitman et al. (1974a) have produced evidence that the ABA effect on exudation depends dramatically on temperature and on the culture solution in which the roots were grown. Table *3*.19 contains the information gathered to date on the effects of several hormones on ion transport. Note that there is general agreement that ABA causes tissue concentrations to rise, chiefly a reflection of increased concentrations in the cortical cell vacuoles. Kinetin has a similar effect on tissue concentration, but causes exudation ion fluxes to decrease. Indole acetic acid (IAA) has a somewhat less clear-cut effect on ion transport, but a rather more definite effect on water transport. The effect of all these hormones on water movement will be taken up later. If any general comment can be made it might be that symplasmic ion transport seems to be sensitive to hormone levels; if this transport is essentially diffusive, the implication is then that plasmodesmatal structure is in some way altered (see also *3.4.2.5*).

Finally, the root is an important natural source of certain hormones in the intact plant. Exudation fluid has been shown to contain plant hormones, particularly cytokinins and giberellins, and the production of these compounds has a sharp temperature maximum (ATKIN et al., 1973). The xylem stream seems to serve for the translocation of these hormones from the root to the shoot in the intact plant.

3. Water Transport

3.1 Root Pressure Exudation

The experimental technique has been described in *3.4.2.1*. It is now universally accepted that the fluid exuded from the cut end of a root or root system is chiefly the result of osmotic water flow from the external solution to the xylem. The most commonly used phenomenological equation is due to ARISZ et al. (1951) and in its more modern form to HOUSE and FINDLAY (1966a).

$$J_v = -L_p \sigma RT \Delta c_s + J_o \qquad (3.8)$$

where J_v is the observed volume exudation flux, L_p is the overall hydraulic conductivity of the root from external solution to the xylem, σ is the reflexion coefficient of the solute, Δc_s is the solute concentration difference between external solution and exudate, R is the gas constant and T is the temperature. The additional term J_o, the so-called non-osmotic term, has been found necessary to complete the description by many groups of workers in a large number of species. It can be shown to be necessary by direct observation of continued exudation from a root bathed in a solution made iso-osmotic with the xylem fluid by the addition of an impermeant solute such as mannitol. Alternatively, extrapolation of normal hypertonic exudation data to zero concentration difference yields a non-zero volume flux value.

There are only two matters of continuing controversy in this area of root physiology, the first concerning the non-osmotic flux J_o, and the second concerning the location of the barrier across which the osmotic pressure is generated. Exudation data from roots of many different species, bathed in media containing all the usual inorganic solutes, yield values for hydraulic conductivity in the range 10^{-8} to 10^{-9} m s^{-1} bar and values of σ for both normal ionic solutes which appear in the exudate and "impermeant" solutes like mannitol, in the range 0.8–1.0. There is an impressive consensus of experimental verification in respect of these values.

3.2 The Non-Osmotic Water Flux

Interpretation of the non-osmotic term J_o has occupied several workers, but first it should be realized that its magnitude is in doubt and can indeed be set equal to zero. ANDERSON et al. (1970) were able to account for the total water exudation flux on a standing gradient osmotic model, provided that the hydraulic conductivity and the xylem salt flux may vary with distance from the root tip. There is direct experimental evidence that both do vary in this manner. On this model J_o may be zero, but of course it is not necessarily zero. Isotonic exudation can be explained in analogous fashion to isotonic water flow across the toad bladder, the original standing gradient osmotic model (DIAMOND and BOSSERT, 1967).

Others, assuming a non-zero value for J_o have proposed explanations to account for it. HYLMÖ (1953) suggested electroosmosis as a driving mechanism, a suggestion which has recently been taken up by TYREE (1973). He demonstrates that the magnitude of J_o in maize roots could be accounted for by an electro-osmotic flow, although his value assumed for electro-osmotic efficiency in the cell wall seems rather high. However, the details are unimportant because the actual value of J_o is essentially unknown; the apparent value will be modified to some extent by effects of a realistic standing osmotic gradient system along the root.

GINSBURG and GINZBURG (1971) have produced evidence that maize cortical sleeves (primary roots with the stele removed to leave a hollow cylinder of cortex) exude an isotonic or even hypotonic fluid which is inhibited by cyanide. GINSBURG (1971) has interpreted this effect and also the J_o term of normal root exudation in terms of the CURRAN and MCINTOSH (1962) model of two membranes in series with differing reflexion coefficients.

3.3 Location of the Osmotic Barrier

The traditional barrier is the endodermis where the Casparian band is thought to force the water flow through the cell membranes, and indeed the measured values of L_p from exudation experiments correlate well with what is expected if a single-cell layer is the effective barrier. Attempts at localization by observing initial transients in water flow after step function alterations to the external concentration (ARISZ et al., 1951; HOUSE and FINDLAY, 1966b) have placed the chief osmotic barrier just within the endodermis in tomato and maize roots. The estimated volume of the osmotically sequestered compartment in maize roots (HOUSE and FINDLAY, 1966b) correlates well with the total xylem volume of the root. However it is possible that both the above estimates must be somewhat revised to take account of possible expansion or contraction in the root-tissue volume, which in itself will affect the exudation, because of turgor adjustments in the cortical cells when the external concentration is changed.

There are two parallel pathways for water movement across the root cortex, in the apoplasm or in the symplasm. If the major osmotic event is at the endodermis, then the apoplasm flux must be dominant. The alternative of water transport being chiefly in the symplasm requires the major osmotic event to be at the plasmalemma of the epidermal and outer cortical cells; this seems to be ruled out by the osmotic transient experiments (e.g. HOUSE and FINDLAY, 1966b) which are highly unlikely to have so misplaced the barrier.

3.4 Transpirational Effects

Transpiration exerts its primary effect on water transport by the simple addition of a hydrostatic component to the total driving force on water. Under most environmental conditions this "transpiration tension" is by far the largest force on water through the root of the intact plant. A good exposition of this whole subject can be found in SLATYER (1967). But briefly the Equation given earlier for root pressure exudation is simply modified by the addition of the pressure term to read

$$J_v = L_p(\Delta P - \sigma RT\Delta c_s) + J_o \qquad (3.9)$$

where ΔP is the hydrostatic pressure difference between the root external solution and the xylem stream. The additional complexity of including a matric pressure term will not be discussed, but see Part A, Chap. 2 for a more detailed treatment.

There is however a complicating factor in that several reports have shown that the hydraulic conductivity of the root, L_p, varies as P varies (BROUWER, 1954; MEES and WEATHERLEY, 1957). Root hydraulic conductivity apparently increases by five times as ΔP increases by about 2 bar in tomato-root systems. Two explanations have been offered for this observation, firstly that the actual membrane permeability of the cells increases and secondly that there is an increase in the effective absorbing zone of the root. MEIRI and ANDERSON (1970) observed an apparently similar effect in excised primary maize roots, but were able to correlate it with the relatively trivial artifact of infiltration of the inter-cellular spaces of the root, which are normally air-filled; application of suction pressure to the cut end of the root caused a progressive increase in root density as a result of filling the cortical air spaces with water. This additional conduit for water movement may explain the observed increase in L_p.

A secondary effect of increased water flow through the root as a result of transpiration tension, is the modification produced in the xylem ion fluxes. Table 3.18 contains data on several species. In many experiments there is evidence that increased transpiration results in increased xylem translocation of ions (e.g. PETTERSSON, 1966; LAZAROFF and PITMAN, 1966). Further, in barley roots PITMAN (1965b) has shown that the selective uptake of $K^+ : Na^+$ is decreased by increasing transpiration rate; total $K^+ + Na^+$ uptake is unaffected, but Na^+ uptake is increased and K^+ decreased. See also 3.4.2.2.

The simplest explanation of the enhancement of ion translocation by transpiration is that the xylem compartment of the root is flushed through more effectively by the increased volume flow, and therefore is a more effective sink, for both passive and active ion transport. However, this cannot be the whole explanation and more work is required. (But see also 3.4.2.2.2.)

3.5 Effects of Hormones on Water Transport

As mentioned earlier in connexion with hormone effects on ion transport, there is some divergence in the published experimental results on the effect of ABA, but apparently the effect is highly sensitive to both temperature and culture conditions. CRAM and PITMAN (1972) find a decrease in volume exudation rate from maize roots, which is attributed to a decrease in salt transport to the xylem, with no effect on root hydraulic conductivity. GLINKA (1973) finds an increase in exudation rate after ABA application, and ascribes it to an increase in root L_p, but without justification because no values of exudate concentration are given. TAL and IMBER

(1971) and Collins and Kerrigan (1973) have measured both volume exudation rates and exudate concentrations and both give clear evidence that indeed L_p has increased upon addition of ABA. Interestingly, Tal and Imber (1971) applied ABA as a spray to the leaves of the plant prior to decapitation, while Collins and Kerrigan (1973) added ABA to the solution bathing the excised roots. ABA is known to be translocated from the leaves to the roots of cotton seedlings (Shindy et al., 1973).

In both the above-mentioned publications (Tal and Imber, 1971; Collins and Kerrigan, 1973) kinetin is found, on the same criteria of water flux and exudate concentration, to cause L_p to decrease. The effect of kinetin apparently takes longer to become established, 20 h as opposed to an hour or two for the ABA effect (Collins and Kerrigan, 1973), IAA has also been found to cause an increased hydraulic conductivity in the excised roots of pea and sunflower (Skoog et al., 1938), in tobacco (Joven, 1964) and in tomato (Tal and Imber, 1971).

4. Transport of Organic Solutes

Various organic compounds are found in the exuding sap of plant roots, such as carboxylic acids, amino acids and amides (together with trace amounts of phytohormones). Carboxylic acid content depends on cation and anion balance, but these compounds have also been implicated in transport of Fe and other cations (Brown and Tiffin, 1965; Tiffin, 1967). Under certain conditions of abnormal balance in the external solution (e.g. with SO_4^{2-} as the only anion) significant amounts of Krebs cycle acids have been observed in the exudation from maize roots (Anderson and Collins, 1969). These are undoubtedly produced endogenously probably as a response to increased cation uptake by K^+-H^+ exchange in the presence of the sulfate anion which is only slowly transported by maize. The effect is likely to be an acceleration of the naturally operative mechanism which provides citrate to keep soluble translocated iron.

Nitrogenous compounds are normally present in significant amounts in the xylem sap of leguminous plants where they can reach concentrations of up to 200 mM (Gunning et al., 1974).

There is also information on translocation through roots of various synthetic organic molecules, herbicides and the like, which are added to the root medium (Sheets, 1961; Wax and Behrens, 1965; Van Oorschot, 1970; Shone and Wood, 1972; Shone et al., 1973). All these reports indicate a broad correlation between transpiration rate and the rate of translocation of the herbicide. Transport through the root to the xylem stream is thought to be passive; many of the triazines for example are lipophilic and hence are likely to be membrane-permeable. The correlation with transpiration rate is likely to be due to a more efficient flushing of the xylem when transpiration is rapid. (See also *3.3.6.4*.)

5. Summary of Transport Mechanisms

Any mechanism, to account properly for transport through the plant root, must consider the three dimensionality of the root. In practice, this means that we require a description of transport in a radial plane normal to the longitudinal axis of the root, and then we must superimpose on this radial model the possibility that variation may occur in the radial situation as we examine the root at different distance from the root tip. The main features of the current radial model for young, absorbing regions of the root are set out in Fig. 3.31. These proposals are essentially similar to those given by many other workers and are, at least tacitly, assumed to apply to the general functioning of all the higher plants.

With regard to radial transport across the cortex, we must distinguish between ions and water; cortical transit of ions is mainly symplasmic, while water transit may be chiefly apoplasmic. Consider that TYREE's (1970) calculation of symplasmic transport parameters and the measured values for apoplasmic ion diffusion (e.g. PITMAN, 1965a) indicate that these alternative pathways are approximately equally conductive to ions, diffusion being the chief movement in both. Therefore, *membrane penetration is the rate-controlling event for ion transport to the xylem,* because the endodermal Casparian band effectively blocks apoplasmic movement. It is true that a transpiration-dependent component of ion transport to the xylem remains after the root metabolism is inhibited (about 20%, LÜTTGE and LATIES, 1967), but even so this apparently passive transport will be membrane rate-controlled.

If the ion transport capability of the endodermis is not different from that of the epidermal and cortical cells, then the net xylem ion fluxes will be carried in the symplasm pathway *versus* the apoplasm-endodermis pathway, in direct proportion to the ratio of the total symplasm membrane surface and the endodermal membrane surface. On this criterion it is clear that somewhat more than 90% of the xylem ion fluxes is carried radially in the symplasm.

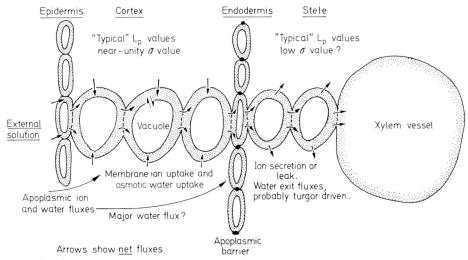

Fig. 3.31. Diagrammatic sketch of root cross section showing the usually accepted mechanisms of ion and water through-put to the xylem vessels

For water, on the other hand, it is likely that the apoplasm has a significantly higher hydraulic conductivity than the symplasm, so that the chief water flow across the root cortex may well be apoplasmic, driven by the osmotic gradient of root pressure which develops across the endodermis and by whatever hydrostatic gradient there is between the xylem and the external medium as a result of transpiration. Reported correlations between solute translocation and transpiration rate can be explained as being due to (i) more rapid flushing of the root xylem causing more efficient sinks for radial solute transport, and (ii) enhanced endodermal uptake of certain solutes (passively transported) because of the increased solute flux in the apoplasm, carried by convection in the water flux.

In the stele, the most common proposal is that the parenchyma either leaks or secretes solutes which then diffuse or are swept along into the xylem vessels. Long-distance translocation along the dead, fully-perforated xylem then occurs. Note that another way of saying the membranes are leaky is to say that σ for all solutes is small. We then may predict a coupling in the stele between water flow and solute flow as

$$J_s = J_v(1-\sigma)\, c_{s,\,m} \qquad\qquad (3.10)$$

where J_s is the solute flux and $c_{s,\,m}$ is the mean solute concentration in the membrane. If σ is indeed close to zero, it seems therefore that the root symplasm would rapidly be emptied of small solutes under prolonged conditions of high transpiration, unless it could in some manner regulate plasmodesmatal transport. The alternative supposition is that the parenchyma cells of the stele do not simply leak solutes, but secrete them under metabolic control through normal, non-leaky membranes. This difficulty would also arise if the symplasm extended through to the xylem vessels, which were lined with cytoplasm, or perhaps with parietal cytoplasm similar to that in mature sieve elements.

The parameter values (for permeability, ion-transport rates, reflexion coefficients and the like) of this radial model will in general vary as one progresses along the root from the tip. Gradients will be set up within the xylem as a result. At a steady state, this situation will constitute a standing gradient osmotic flow and can be described by the model introduced by Anderson et al. (1970).

References

Anderson, W.P.: Ion transport in the cells of higher plant tissues. Ann. Rev. Plant Physiol. **23**, 51–72 (1972).

Anderson, W.P., Aikman, D.P., Meiri, A.: Excised root exudation: a standing gradient osmotic flow. Proc. Roy. Soc. (London), Ser. B **174**, 445–458 (1970).

Anderson, W.P., Collins, J.C.: The exudation from excised maize roots bathed in sulphate media. J. Exptl. Bot. **20**, 72–80 (1969).

Anderson, W.P., Goodwin, L., Hay, R.K.M.: Evidence for vacuole involvement in xylem ion supply in the excised primary roots of two species, *Zea mays* and *Allium cepa*. In: Structure and function of primary root tissues (J. Kolek, ed.), Symp. Proc. Czechoslovakia, p. 379–388. Bratislava: Veda 1974.

Anderson, W.P., House, C.R.: A correlation between structure and function in the root of *Zea mays*. J. Exptl. Bot. **18**, 544–555 (1967).

ARISZ, W.H.: Significance of the symplasm theory for transport in the root. Protoplasma **46**, 5–62 (1956).

ARISZ, W.H., HELDER, R.J., VAN NIE, R.: Analysis of the exudation process in tomato plants. J. Exptl. Bot. **2**, 257–297 (1951).

ATKIN, R.K., BARTON, G.E., ROBINSON, D.K.: Effect of root growing temperature on growth substances in xylem exudate of Zea mays. J. Exptl. Bot. **24**, 475–487 (1973).

BAKER, D.A.: The radial transport of ions in maize roots. In: Ion transport in plants (W.P. ANDERSON, ed.), p. 511–517. London-New York: Academic Press 1973a.

BAKER, D.A.: The effect of CCCP on ion fluxes in the stele and cortex of maize roots. Planta **112**, 293–299 (1973b).

BAKER, D.A., WEATHERLEY, P.E.: Water and solute transport by exuding root systems of Ricinus communis. J. Exptl. Bot. **20**, 485–496 (1969).

BANGE, G.G.J.: Diffusion and absorption of ions in plant tissue. III. The role of the root cortex cells in ion absorption. Acta Botan. Neerl. **22**, 529–542 (1973).

BECKETT, J.T., ANDERSON, W.P.: Ferric EDTA absorption by maize roots. In: Ion transport in plants (W.P. ANDERSON, ed.), p. 595–607. London-New York: Academic Press 1973.

BOWEN, G.D., ROVIRA, A.D.: Phosphate uptake along attached and excised wheat roots measured by an automatic scanning method. Australian J. Biol. Sci. **20**, 369–378 (1967).

BOWLING, D.J.F.: Active transport of ions across sunflower roots. Planta (Berl.) **69**, 377–382 (1966).

BOWLING, D.J.F.: The origin of the trans-root potential and the transfer of ions to the xylem of sunflower roots. In: Ion transport in plants (W.P. ANDERSON, ed.), p. 483–491. London-New York: Academic Press 1973a.

BOWLING, D.J.F.: Measurement of a gradient of oxygen partial pressure across the intact root. Planta **111**, 323–328 (1973b).

BOWLING, D.J.F., MACKLON, A.E.S., SPANSWICK, R.M.: Active and passive transport of the major nutrient ions across the root of Ricinus communis. J. Exptl. Bot. **17**, 410–416 (1966).

BOWLING, D.J.F., SPANSWICK, R.M.: Active transport of ions across the root of Ricinus communis. J. Exptl. Bot. **15**, 422–427 (1964).

BROUWER, R.: The regulation influence of transpiration and suction tension on the water and salt uptake by the roots of intact Vicia faba plants. Acta Botan. Neerl. **3**, 264–312 (1954).

BROWN, J.C., TIFFIN, L.O.: Iron stress as related to the iron and citrate occuring in stem exudate. Plant Physiol. **40**, 395–400 (1965).

CLARK, R.B., TIFFIN, L.O., BROWN, J.C.: Organic acids and iron translocation in maize genotypes. Plant Physiol. **52**, 147–150 (1973).

CLARKSON, D.T., ROBARDS, A.W., SANDERSON, J.: The tertiary endodermis in barley roots: fine structure in relation to transport of ions and water. Planta **96**, 292–305 (1971).

COLLINS, J.C., KERRIGAN, A.P.: Hormonal control of ion movements in the plant root. In: Ion transport in plants (W.P. ANDERSON, ed.), p. 589–593. London-New York: Academic Press 1973.

CRAFTS, A.S., BROYER, T.C.: Migration of salts and water into xylem of the roots of higher plants. Amer. J. Bot. **24**, 415–431 (1938).

CRAM, W.J.: Chloride fluxes in cells of the isolated root cortex of Zea mays. Australian J. Biol. Sci. **26**, 757–779 (1973).

CRAM, W.J., PITMAN, M.G.: The action of abscisic acid on ion uptake and water flow in plant roots. Australian J. Biol. Sci. **25**, 1125–1132 (1972).

CURRAN, P.F., McINTOSH, J.R.: A model system for biological water transport. Nature **193**, 347–348 (1962).

DAVIS, R.F., HIGINBOTHAM, N.: Effects of external cations and respiratory inhibitors on electrical potential of the xylem exudate of excised corn roots. Plant Physiol. **44**, 1383–1392 (1969).

DIAMOND, J.M., BOSSERT, W.H.: Standing gradient osmotic flow: A mechanism for coupling of water and solute transport in epithelia. J. Gen. Physiol. **50**, 2061–2083 (1967).

DUNLOP, J., BOWLING, D.J.F.: The movement of ions to the xylem exudate of maize roots 1. Profiles of membrane potential and vacuolar potassium activity across the root. J. Exptl. Bot. **22**, 434–444 (1971).

FISCUS, E.L., KRAMER, P.J.: Radial movement of oxygen in plant roots. Plant Physiol. **45**, 667–669 (1970).

Fleet, D.S. van: Histochemistry and function of the endodermis. Botan. Rev. **27**, 165–220 (1961).

Ginsburg, H.: Model for iso-osmotic water flow in plant roots. J. Theoret. Biol. **32**, 147–158 (1971).

Ginsburg, H., Ginzburg, B.Z.: Evidence for an active water transport in a corn root preparation. J. Membrane Biol. **4**, 29–41 (1971).

Glinka, Z.: ABA effect on root exudation related to increased permeability to water. Plant Physiol. **51**, 217–219 (1973).

Greenway, H.: Plant responses to saline substrates. IV. Chloride uptake by *Hordeum vulgare* as affected by inhibitors, transpiration, and nutrients in the medium. Australian J. Biol. Sci. **18**, 249–268 (1965).

Greenway, H., Klepper, B.: Phosphorus transport to the xylem and its regulation by water flow. Planta **83**, 119–136 (1968).

Gunning, B.E.S., Pate, J.S., Minchin, F.R., Marks, I.: Quantitative aspects of transfer cell structure in relation to vein loading in leaves and solute transport in legume nodules. Soc. Exptl. Biol. Symp. **28**, 87–126 (1974).

Hall, J.H., Sexton, R., Baker, D.A.: Metabolic change in washed isolated steles. Planta **96**, 54–61 (1971).

Hay, R.K.M., Anderson, W.P.: Characterisation of exidation from excised roots of onion, *Allium cepa*. Ion fluxes. J. Exptl. Bot. **23**, 577–584 (1972).

Helder, R.J., Boerma, J.: An electron microscopical study of the plasmodesmata in the roots of young barley seedlings. Acta Botan. Neerl. **18**, 99–107 (1969).

Higinbotham, N., Davis, R.F., Mertz, S.M., Shumway, L.K.: Some evidence that radial transport in maize roots is into living vessels. In: Ion transport in plants (W.P. Anderson, ed.), p. 493–506. London-New York: Academic Press 1973.

Hill-Cottingham, D.G., Lloyd-Jones, C.P.: The behaviour of iron chelating agents with plants. J. Exptl. Bot. **16**, 233–242 (1965).

Hodges, T.K., Vaadia, Y.: Uptake and transport of radiochloride and tritiated water by various zones of onion roots of different chloride status. Plant Physiol. **39**, 104–108 (1964).

House, C.R., Findlay, N.: Water transport in isolated maize roots. J. Exptl. Bot. **17**, 344–354 (1966a).

House, C.R., Findlay, N.: Analysis of transient changes in the fluid exudation rate from isolated maize roots. J. Exptl. Bot. **17**, 627–640 (1966b).

Hylmö, B.: Transpiration and ion absorption. Physiol. Plantarum **6**, 333–405 (1953).

Jarvis, P., House, C.R.: Evidence for symplasmic ion transport in maize roots. J. Exptl. Bot. **21**, 83–90 (1970).

Joven, C.B.: Certain aspects of the exudation process in tobacco plants. Plant Physiol., Diss. Abstr. **25**, 5519–5520 (1964).

Klepper, B., Greenway, H.: Effects of water stress on phosphorus transport to the xylem. Planta **80**, 142–146 (1968).

Läuchli, A., Epstein, E.: Lateral transport of ions into the xylem of corn roots. I. Kinetics and energetics. Plant Physiol. **48**, 111–117 (1971).

Läuchli, A., Kramer, D., Pitman, M.G., Lüttge, U.: Ultrastructure of xylem parenchyma cells of barley roots in relation to ion transport to the xylem. Planta **119**, 85–99 (1974).

Läuchli, A., Lüttge, U., Pitman, M.G.: Ion uptake and transport through barley seedlings: differential effect of cycloheximide. Z. Naturforsch. **28c**, 431–434 (1973).

Läuchli, A., Spurr, A.R., Epstein, E.: Lateral transport of ions into the xylem of corn roots. II. Evaluation of a stelar pump. Plant Physiol. **48**, 118–124 (1971).

Laties, G.G., Budd, K.: The development of differential permeability in isolated steles of corn roots. Proc. Natl. Acad. Sci. U.S. **52**, 462–469 (1964).

Lazaroff, N., Pitman, M.G.: Calcium and magnesium uptake by barley seedlings. Australian J. Biol. Sci. **19**, 991–1005 (1966).

Lüttge, U., Bauer, K., Köhler, D.: Frühwirkungen von Gibberellinsäure auf Membrantransporte in jungen Erbsenpflanzen. Biochim. Biophys. Acta **150**, 452–478 (1968).

Lüttge, U., Laties, G.G.: Absorption and long-distance transport by isolated stele of maize roots in relation to the dual mechanism of ion absorption. Planta **74**, 172–187 (1967).

Lüttge, U., Läuchli, A., Ball, E., Pitman, M.G.: Cycloheximide: A specific inhibitor of protein synthesis and intercellular ion transport in plant roots. Experientia **30**, 470–471 (1974).

LUNDEGÅRDH, H.: Translocation of salts and water through wheat roots. Physiol. Plantarum **3**, 103–151 (1950).

MEES, G.C., WEATHERLEY, P.E.: The mechanism of water absorption by roots. II. The role of hydrostatic pressure gradients across the cortex. Proc. Roy. Soc. (London), Ser. B **147**, 381–391 (1957).

MEIRI, A.: Potassium and chloride accumulation and transport by excised maize roots of different salt status. In: Ion transport in plants (W.P. ANDERSON, ed.), p. 519–530. London-New York: Academic Press 1973.

MEIRI, A., ANDERSON, W.P.: Observations on the effects of pressure differences between the bathing medium and the exudates of excised maize roots. J. Exptl. Bot. **21**, 899–907 (1970).

MINCHIN, F.R., BAKER, D.A.: The influence of calcium on potassium fluxes across the root of *Ricinus communis*. Planta **113**, 97–104 (1973).

OORSCHOT, J.L.P. VAN: Effect of transpiration rate of bean plants on inhibition of photosynthesis by some root applied herbicides. Weed Res. **10**, 230–242 (1970).

PETTERSSON, S.: Artificially induced water and sulphate transport through sunflower roots. Physiol. Plantarum **19**, 581–601 (1966).

PITMAN, M.G.: Ion exchange and diffusion in roots of *Hordeum vulgare*. Australian J. Biol. Sci. **18**, 541–546 (1965a).

PITMAN, M.G.: Transpiration and the selective uptake of potassium by barley seedlings (*Hordeum vulgare cv.* Bolivia). Australian J. Biol. Sci. **18**, 987–998 (1965b).

PITMAN, M.G.: Uptake and transport of ions in barley seedlings. I. Estimation of chloride fluxes in cells of excised roots. Australian J. Biol. Sci. **24**, 407–421 (1971).

PITMAN, M.G.: Uptake and transport of ions in barley seedlings. II. Evidence for two active stages in transport to the shoot. Australian J. Biol. Sci. **25**, 243–257 (1972).

PITMAN, M.G., COURTICE, A.C., LEE, B.: Comparison of potassium and sodium uptake by barley roots at high and low salt status. Australian J. Biol. Sci. **21**, 871–881 (1968).

PITMAN, M.G., CRAM, W.J.: Regulation of inorganic ion transport in plants. In: Ion transport in plants (W.P. ANDERSON, ed.), p. 465–481. London-New York: Academic Press 1973.

PITMAN, M.G., LÜTTGE, U., LÄUCHLI, A., BALL, E.: Action of abscisic acid on ion transport as affected by root temperature and nutrient status. J. Exptl. Bot. **25**, 147–155 (1974a).

PITMAN, M.G., SCHAEFER, N., WILDES, R.A.: Effect of abscisic acid on fluxes of ions in barley roots. In: Membrane transport in plants (U. ZIMMERMANN, J. DAINTY, eds.), p. 391–396. Berlin-Heidelberg-New York: Springer 1974b.

PREVOT, P.M., STEWARD, F.C.: Salient features of the root system relative to the problem of salt absorption. Plant Physiol. **11**, 509–534 (1936).

RUSSELL, R.S., SANDERSON, J.: Nutrient uptake by different parts of the intact root of plants. J. Exptl. Bot. **18**, 491–508 (1967).

SADDLER, H.D.W.: The membrane potential of *Acetabularia mediterranea*. J. Gen. Physiol. **55**, 802–821 (1970).

SCHAEFER, N., WILDES, R.A., PITMAN, M.G.: Inhibition by p-fluorophenylalanine of protein synthesis and of ion transport across the roots in barley seedlings. Australian J. Plant Physiol. **2**, 61–73 (1975).

SCOTT, F.M.: The fine structure of xylem vessels. In: Ecology of soilborne pathogens (K.F. BAKER, W.C. SYNDER, eds.), p. 145–153. Berkeley: University of California Press. 1965.

SHEETS, T.J.: Uptake and distribution of simazine by oat and cotton seedlings. Weeds **9**, 1–13 (1961).

SHINDY, W.W., ASMUNDSON, C.M., SMITH, O.E., KUMAMOTO, J.: Absorption and distribution of high specific radioactivity 2-^{14}C-ABA in cotton seedlings. Plant Physiol. **52**, 443–447 (1973).

SHONE, M.G.T.: Origins of the electrical potential difference between the xylem sap of maize roots and the external solution. J. Exptl. Bot. **20**, 698–716 (1969).

SHONE, M.G.T., CLARKSON, D.T., SANDERSON, J., WOOD, A.V.: A comparison of the uptake and translocation of some organic molecules and ions in higher plants. In: Ion transport in plants (W.P. ANDERSON, ed.), p. 571–582. London-New York: Academic Press 1973.

SHONE, M.G.T., WOOD, A.V.: Factors affecting absorption and translocation of simazine by barley. J. Exptl. Bot. **23**, 141–151 (1972).

SKOOG, F., BROYER, T.C., GROSSENBACHER, K.A.: Effect of auxin on rates, periodicity and osmotic relations in exudation. Amer. J. Bot. **26**, 749–757 (1938).

SLATYER, R.O.: Plant-water relations. London-New York: Academic Press 1967.

SPANSWICK, R.M.: Electrical coupling between cells of higher plants: a direct demonstration of intercellular communication. Planta **102**, 215–227 (1972).

TAL, M., IMBER, D.: Abnormal stomatal behaviour and hormonal imbalance in *flacca,* a wilty mutant of tomato. III. Hormonal effects on the water status in the plant. Plant Physiol. **47**, 849–850 (1971).

TIFFIN, L.O.: Translocation of manganese, iron, cobalt and zinc in tomato. Plant Physiol. **42**, 1427–1432 (1967).

TIFFIN, L.O.: Translocation of iron citrate and phosphorus in the xylem exudate of soybean. Plant Physiol. **45**, 280–283 (1970).

TYREE, M.T.: The symplast concept: a general theory of symplastic transport according to the thermodynamics of irreversible processes. J. Theoret. Biol. **26**, 181–214 (1970).

TYREE, M.T.: An alternative explanation for the apparently active water exudation in excised roots. J. Exptl. Bot. **24**, 111–115 (1973).

VAKHMISTROV, D.B.: On the function of apparent free space in plant roots. A study of the absorbing power of epidermal and cortical cells in barley roots. Soviet Plant Physiol. **14**, 103–107 (1967).

WAX, L.M., BEHRENS, R.: Absorption and translocation of atrazine in quackgrass. Weeds **13**, 107–109 (1965).

WEIGL, J.: Efflux und Transport von Cl^- und Rb^+ in Maiswurzeln. Wirkung von Außenkonzentration, Ca^{++}, EDTA und IES. Planta **84**, 311–323 (1969).

WEIGL, J., LÜTTGE, U.: Mikroautoradiographische Untersuchungen über die Aufnahme von $^{35}SO_4^{--}$ durch Wurzeln von *Zea mays* L. Die Funktion der primären Endodermis. Planta **59**, 15–28 (1962).

YU, G.H., KRAMER, P.J.: Radial transport of ions in roots. Plant Physiol. **44**, 1095–1100 (1969).

ZIEGLER, H., WEIGL, J., LÜTTGE, U.: Mikroautoradiographischer Nachweis der Wanderung von $^{35}SO_4$ durch die Tertiärendodermis der *Iris*-Wurzel. Protoplasma **56**, 362–370 (1963).

4. Transport Processes in Leaves

4.1 General Introduction

U. LÜTTGE and M.G. PITMAN

Transport is an important feature of the contact between plants and their environment. In many lower plants, the plant body, or the population of cells is predominantly of a single type and this cell type mediates all the exchanges between the plant and its environment. This is partly the reason why certain algal cells (Part A, Chap. *6*) and fungal hyphae (Part A, Chap. *7*) have become standard systems for study of transport processes (e.g. *Chara, Chlorella, Neurospora*).

In higher plants transport of the plants requirements takes place selectively in the different parts of the plant. Evolution has led to the functional specialization of roots in the *soil* environment and leaves in the *aerial* environment. Supply to different parts of the plant is maintained by transport systems, the xylem and phloem (that are generally the topic of Volume 1 of this series).

Because they are adapted to absorption of nutrients and water from the soil, roots have been the classical material for study of ion-transport mechanisms. Results obtained from study of intact roots have been generalized to explain the ionic relations of individual cells (e.g. Chap. *3.2*). In contrast, leaves have been used primarily to study gas exchange, the absorption of solar energy, the partial reactions of photosynthesis and the synthesis of carbohydrates. Though gas exchange and exchange of energy are, strictly speaking, transport processes between the plant and its environment, they are not within the scope of this Volume.

For a long time, plant physiology has been dominated by the different experimental goals of studies using roots on the one hand and leaves on the other. Indeed, there seems to have been an attitude that the processes of solute transport between cells and the medium that are characteristic of roots (and algae and fungi) are less important in leaves. Early attempts to study transport in leaf cells was also discouraged by a major experimental difficulty. The cuticle is an effective barrier to diffusion into the leaf. Physiologists were suspicious that any way of mutilating the leaf to expose the cells was likely to introduce an artificial and physiologically irrelevant situation due to damage, wound responses or induced developmental changes.

A more natural problem appeared to be that of the ion relations of leaves of aquatic angiosperms, which have secondarily evolved adaptation to absorption of nutrients by the leaves. Deflected from the study of aerial leaves, many physiologists therefore turned their attention to the green leaf tissues of these plants. Early investigations showed that green cells of higher plants possess transport processes mediating exchange with the environment much as in algae cells (*4.2.2.1*; *4.2.3.2.1*; *4.2.3.4.1*). This work has also shown the need to consider transport *between* cells of the leaf, and possible differences between transport at the plasmalemma and tonoplast.

Work with aerial leaves of higher plants (Fig. *4.*1) received a boost about ten years ago when SMITH and EPSTEIN (1964) showed that meaningful results could

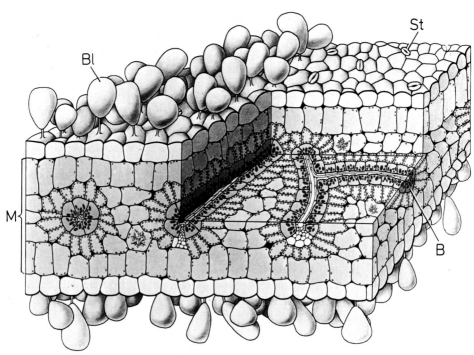

Fig. *4.*1. Leaf anatomy (*Atriplex spongiosa*): Stomata (*St*) and salt accumulating bladders (*Bl*; "salt hairs" see *5.2.*3) in the epidermis; various cell types in the mesophyll (*M*) and bundles with phloem and xylem strands and green bundle sheath cells (*B*). (From Lüttge, 1974)

be obtained using leaf *slices* (*4.2.*2.2.1). Since that time, interest in ionic relations of cells of aerial leaves has increased and it has become clear that like roots, algae and hyphae, leaf cells must have mechanisms for ion uptake and their ionic relations are determined by influxes and effluxes across the cell membranes. Symplasmic transport has been shown to be as much a part of ion movement through the leaf as through the root. We can think of the symplast of the leaf as having an *internal* supply of ions and water *via* the xylem and apoplasmic space corresponding to the *outer medium-free space* system to which the symplast of roots is exposed. A major difference, of course, is that ion transport in green leaf cells can potentially draw on photosynthetic reactions as a source of energy. Chap. *4.2* reports results of increasing amounts of experimental work that draws its justification from these ideas.

However, this progress in elucidating similarities in ionic relations of roots, algae, fungal hyphae and water-plant leaves on the one hand and aerial leaves on the other should not be allowed to obscure the differences. An important distinction is that the leaf has little contact with the external ionic environment, but only "sees" ions that have passed control sites in the roots. Hence it is critical to investigate the extent to which ionic relations of the leaves are controlled either by the activity of leaf cells or by the supply of ions from the roots. Regulation of the ion content of the leaf by *leaf* activities may involve redistribution in the phloem, and also control over the activity of the roots.

There is some evidence from experiments with isolated leaf slices, that the content of shoots of some plants is in fact mainly determined by control of the import to the shoot. For example, reduction in the relative growth rate of barley seedlings also reduced uptake by roots and transport to the shoot, but had no effect on fluxes into leaf cells (PITMAN, 1972; PITMAN et al., 1974a). The control over import to the shoot appears to be due partly to supply of photosynthetic products (HATRICK and BOWLING, 1973) but also and more specifically to hormonal messengers (e.g. PITMAN et al., 1974b).

One factor which seems to have considerable effect on the ionic concentration of leaf cells of many species is the external water potential. Usually, the concentration of the major osmotic ions (K^+ and Na^+) is found to increase in the leaves as external water stress becomes more severe. In an extreme case, the growth of salt bush (*Atriplex halimus*) under even moderate water stress has been shown to depend on the availability of ions (e.g. NaCl) which maintain a high osmotic pressure in cells of the leaves (GALE and POLJAKOFF-MAYBER, 1970). At the other extreme, many cereals show little response of concentration in the leaves to external water potential.

In some specific cases there are highly independent and intrinsically regulated ion-transport mechanisms in leaves. One example is the salt gland, which has evolved in certain halophytes. The secretion of salt from these glands reduces the level of NaCl stress. Since the plants need to absorb nutrient ions from saline media including sea water, but also saline soils in the arid regions, the secretion of NaCl reduces the degree of discrimination needed between Na^+ and other ions taken up by the roots.

A second example is the guard cells and, in some cases, the subsidiary cells of the stomatal complex. Highly specific ion fluxes are the basis of the regulation of stomatal opening. These fluxes respond to a number of internal and environmental factors and are thus indispensible features of the function of aerial leaves of higher plants and the regulation of plant growth (Chap. *4.3*).

In Chap. *4.2* we are concerned with the ionic relations of cells of the leaf and in particular with the coupling between transport and photosynthesis. In Chap. *4.3* emphasis is put on the specific transport processes in the stomata. Transport processes in salt glands and the relation between secretion from the glands and uptake to the whole plant are discussed later in Chap. *5*. Integration of transport in the root and uptake in the leaves is considered in Chap. *10*.

References

GALE, J., POLJAKOFF-MAYBER, A.: Interrelations between growth and photosynthesis of salt bush (*Atriplex halimus* L.) grown in saline media. Australian J. Biol. Sci. **23**, 937–945 (1970).
HATRICK, A.A., BOWLING, D.J.F.: A study of the relationship between root and shoot metabolism. J. Exptl. Bot. **24**, 607–613 (1973).
LÜTTGE, U.: Co-operation of organs in intact higher plants: A review. In: Membrane transport in plants (U. ZIMMERMANN, J. DAINTY, eds.). Berlin-Heidelberg-New York: Springer 1974.
PITMAN, M.G.: Uptake and transport of ions in barley seedlings. III. Correlation between transport to the shoot and relative growth rate. Australian J. Biol. Sci. **25**, 905–919 (1972).
PITMAN, M.G., LÜTTGE, U., LÄUCHLI, A., BALL, E.: Ion uptake to slices of barley leaves, and regulation of K content in cells of the leaves. Z. Pflanzenphysiol. **72**, 75–88 (1974a).
PITMAN, M.G., LÜTTGE, U., LÄUCHLI, A., BALL, E.: Action of abscisic acid on ion transport as affected by root temperature and nutrient status. J. Exptl. Bot. **25**, 147–155 (1974b).
SMITH, R.C., EPSTEIN, E.: Ion absorption by shoot tissue: technique and first findings with excised leaf tissue of corn. Plant Physiol. **39**, 338–341 (1964).

4.2 Ionic Relations of Leaf Cells

W.D. Jeschke

1. Introduction

Although leaves are not usually thought of as organs specialized for ion transport, this process is essential for all leaves. Apart from provision of nutrients to cells of the leaf, transport of ions into the cells leads to accumulation of high concentrations of ions in the vacuoles, commonly about 200 mM but often nearer 500 mM in leaves of plants from deserts or sea-shore. This concentration of salt makes up most of the osmotic water potential of the cell sap and so determines the extent to which water potential of the leaf can be generated, and the ability of the plant to absorb water from the soil.

In certain plants, such as submerged hydrophytes or mosses, uptake of ions to the leaf may take place at the leaf surface from the surrounding or adhering medium (SCULTHORPE, 1967) although also in hydrophytes, roots contribute to the mineral nutrition (BRISTOW and WHITCOMBE, 1971; WAISEL and SHAPIRA, 1971). Even in aerial leaves of angiosperms, uptake of applied ions or salt may occur from dust or rain water (FRANKE, 1967; WITTWER and BUKOVAC, 1969). Indeed foliar absorption is normal for certain epiphytic plants (e.g. Bromeliaceae) where ions and water are absorbed by specialized absorptive scales on the leaves. In other plants, ion transport is important for excretion of salt from specialized glands on the leaves (Chap. 5), but the common site of transport is at the plasmalemma of cells of the leaf where ions supplied by the xylem or phloem are accumulated.

One difference between leaves and other parts of the higher plant is the possibility for photosynthesis to provide a source of energy for ion transport. Leaves, like algae (Part A, Chap. 6) thus provide a material in which the energetic linkages of ion transport can be conveniently investigated because the energy supply may be deliberately changed within a physiological range by controlling light quality and intensity and the supply of CO_2.

Apart from their photosynthetic energy supply, leaves differ from roots in being closed systems not in immediate contact with nutrients in the environment. Ions are supplied to the leaf in the xylem and phloem and are exported from the leaf in the phloem. Some small exchange may also take place across the cuticle. The rates of these various processes will depend on the stage of development of the leaf (STEWARD and SUTCLIFFE, 1959; STEWARD and MOTT, 1970; JACOBY and DAGAN, 1969; ILAN et al., 1971).

The early work has been reviewed by BRAUNER (1956), STEEMAN NIELSEN (1960), SCULTHORPE (1967), HELDER (1967), later work also by LÜTTGE (1969, 1973b), MAC-ROBBIE (1970) and HIGINBOTHAM (1973a).

2. Experimental Approaches

2.1 Hydrophyte and Moss Leaves

An advantage of using these plants is that the leaves consist only of one (numerous mosses), two (*Elodea*) or a few layers of cells (*Limnophila, Vallisneria*), and a cuticle is absent or very thin (SCULTHORPE, 1967). In one- or two-layered leaves every cell is in direct contact with the external solution and the unthickened cell walls offer little resistance. Diffusion in the unstirred layer and/or the cell wall limited the $H_2PO_4^-$ or SO_4^{2-} uptake by *Elodea* leaves only below 0.1 µM external concentration (JESCHKE and SIMONIS, 1965) (cf. roots, *3.1.2.2*).

In the pioneering work by ARISZ and colleagues, net Cl^- uptake was measured to *Vallisneria* pieces (ARISZ, 1947a) and shown to be to the vacuole by the increase in osmotic pressure. More recently, tracers have been used to estimate fluxes. Measurements of individual fluxes by efflux analysis are few (LÜTTGE and BAUER, 1968; JESCHKE, 1971; HELLER et al., 1973; see *4.2.3.4.6*). Instead, short term tracer uptake has been used to estimate ϕ_{oc} (see Part A, *5.4.4*) where free space exchange was rapid (*Elodea*, JESCHKE, 1970a; *Lemna*, YOUNG and SIMS, 1972). Alternatively various studies have measured long term uptake with long rinses that tend to estimate Φ_{ov} the net tracer flux to the vacuole (Part A, *5.4.4*), particularly where free space exchange is slow (e.g. *Vallisneria gigantea*, KYLIN, 1957, 1960d; see Table *4.4*).

2.2 Aerial Leaves of Angiosperms

2.2.1 Free Space, Ion Uptake and Flux Measurements

Uptake of ions through the epidermis, and loss by leaching from leaves is important in some biological situations (reviews by WITTWER and BUKOVAC, 1969; FRANKE, 1967; TUKEY, 1970). The cuticle appears to act as a charged "membrane" that is more permeable to univalent than multivalent ions (MCFARLANE and BERRY, 1974). However, rates of movement across the cuticle or through stomata are too slow to allow study of ion transport processes in cells within the leaf from measurement of uptake to intact leaves. Instead, different approaches have been needed, and perhaps the most convenient method has been the use of narrow leaf slices.

Ions diffuse through the cut edges of the leaf slice to cells within the tissue. Experimentally, a problem is to decide on the thickness of the slice; narrower slices allow more ready access by diffusion to the cells but contain a higher proportion of damaged cells. SMITH and EPSTEIN (1964a) using *Zea mays* found maximum rates of uptake on a fresh-weight basis were obtained with slices 0.3 mm wide (see also OSMOND, 1968; KANNAN, 1970). Damage was estimated by MacDONALD and MACKLON (1972) to extend 0.12 mm from a cut edge of wheat leaves, and a convenient width to be 1.5 mm. Similarly PITMAN et al. (1974a) found 30% of cells damaged in slices of barley leaves 0.9 mm wide. Nevertheless, rates of ion uptake to leaf slices are comparable with those from roots (SMITH and EPSTEIN, 1964a), and cells are able to retain previously accumulated ions (RAINS, 1968; PITMAN et al., 1974b). MORROD (1974) has described experiments in which the epidermis was removed to reduce the diffusion resistance.

The rate of free-space exchange has been estimated for Na^+ in *Atriplex* leaf slices (OSMOND 1968) and for K^+ and Cl^- in barley leaf slices (PITMAN et al., 1974a). Diffusion coefficients were estimated to be 1 to $3 \cdot 10^{-11}$ m^2 s^{-1} for Na^+ in *Atriplex*; $5.5 \cdot 10^{-10}$ m^2 s^{-1} for Cl^- and $4 \cdot 10^{-10}$ m^2 s^{-1} for K^+ in barley. PITMAN et al. (1974a) found the Donnan free space (DFS) was in a volume of 0.013 cm^3 g$_{FW}^{-1}$, which was close to the estimated volume of the cell walls: larger volumes of 3 to 5% were given by KYLIN (1960d) and by CROWDY and TANTON (1970). PITMAN et al. (1974a) showed the DFS of barley leaves to be a cation exchange system containing 3 µeq g_{FW}^{-1} fixed anions at a concentration of 300 eq m^{-3}. Van Steveninck and Chenoweth (1972) studying localization of Cl$^-$ in barley leaves found it to be excluded from the cell walls, supporting the view that they contain a cation exchange system. In leaf slices, water in the intercellular spaces seems to be the main pathway to supply of ions to the cell: in the intact leaf the cell wall may be more important and hence transport of univalent ions strongly affected by Ca^{2+} level in the exchange phase (Pitman et al., 1974b; Lüttge et al., 1974).

Pitman et al. (1974a) measured the efficiency of short rinses and found 2 min at 0° C to be sufficient to remove about 50% of the radioactivity (^{36}Cl and ^{86}Rb) from the free space of 1 mm barley leaf slices, cf. also MacDonald and Macklon (1972). These short rinsing periods are important as they permitted an estimation of ϕ_{oc} in addition to Φ_{ov} if influxes were measured for short and long uptake periods.

A difficulty with using leaf slices is that the rate of photosynthesis may be limited by diffusion of CO_2 inwards (or of OH$^-$ outwards in HCO_3^- buffer solutions). Rates of CO_2 fixation in intact leaves and in leaf slices floating on solution can be as high as 300 µmol g_{FW}^{-1} h^{-1}, whereas measured rates of photosynthesis of submerged leaf slices are commonly only 30–50 µmol g_{FW}^{-1} h^{-1} (but see also Jones and Osmond, 1973). It can be calculated from the diffusion coefficient for CO_2 in water that rates of uptake of 300 µmol g_{FW}^{-1} h^{-1} could not be met by diffusion of CO_2 in water filled air spaces of the leaf except at abnormally high concentrations (see also p. 3 bottom). This point has been considered by MacDonald (1975) and MacDonald and Macklon (1975); low rates of O_2 evolution achieved using leaf slices may mean that comparison of ion uptake in light and dark with this material may not be satisfactory as a test of dependence on photosynthetic energy supply.

2.2.2 Extrapolation to the Intact Leaf

Pitman et al. (1974b) showed there was no change in rate of tracer uptake within the first few minutes of cutting leaf slices, and claimed *freshly cut* leaf slices had the same rates of absorption as the fresh leaf. It is clear, though, that over periods of many hours, developmental changes may set in and fluxes may change (*4.2.3.2.1*).

A most important question in evaluating the ionic relations of aerial leaves is the question of the ion content in the free space of the intact leaf, i.e. of the natural environment of the leaf cells. As has been extensively dicussed by Robinson (1971) the content of the extracellular compartment will be determined by the import through the vascular bundles and the export of ions, i.e. to the cells by uptake, to the outside by leaching (Tukey, 1970) or by glands (see Chap. *5.2*) or export by retranslocation *via* the phloem (Stenlid, 1958; Greenway and Pitman, 1965).

According to Robinson (1971), a minimum value for the concentration in the leaf free space will be given by the concentration in the xylem sap (Atkinson et al., 1967)—for examples see Chap. *3.4*—provided that ions enter the leaf *via* the xylem vessels as fast or faster than they are taken up or removed. However, if ion uptake exceeds the supply by the xylem, the free space content will be lower than a Donnan system in equilibrium with the xylem sap. This certainly may be the situation in young expanding leaves. Good indications for this may be derived from the ion concentration in the guttation fluid, which may be very low (Kramer, 1969; Stenlid, 1958, Perrin, 1972) indicating a high ion-transport capacity of the leaf cells or the epithem cells in the hydathodes.

Scholander et al. (1966) attempted to determine the extracellular ion content by forcing the sap backwards out of twigs by application of pressure and obtained relatively low ion concentrations even in halophytes and mangroves. By a perfusion technique Bernstein (1971) found an extracellular ion concentration of 2 eq m^{-3} in *Ricinus* and *Helianthus* and 10 eq m^{-3} in *Brassica* leaves.

Pitman et al. (1974b) arrived at a free space content corresponding to a Donnan phase in equilibrium with 5 mM KCl+0.5 mM CaSO$_4$ in barley leaves, which compares favorably with the K$^+$ concentration in the xylem sap of glycophytes, which ranges from about 0.5 to 5 mM depending on the rate of transpiration (Bollard, 1953; Stewart and Sutcliffe, 1959; Sutcliffe, 1962; Greenway, 1965).

3. Results

3.1 Ion Concentrations and Electropotentials

Only the average of the vacuolar and cytoplasmic ion concentration can be determined by direct analysis as in other higher plant cells. Table 4.1 gives these concentrations and electrical potentials between the vacuole and the external solution ψ_{vo}. In all cases save the halophyte *Atriplex* (OSMOND et al., 1969) the K^+ concentration exceeds that of Na^+. However, this is not a valid estimate of the ability of terrestial *leaf cells* to discriminate between K^+ and Na^+ as the ratio found in the leaf is determined by the supply of K^+ and Na^+ to the leaf in the xylem and by retranslocations of K^+ from old to young leaves; compare Table 4.2. A better estimate of the inherent selectivity comes from use of leaf slices, which also show selective K^+ uptake (SMITH and EPSTEIN, 1964b; RAINS and EPSTEIN, 1967 and see Fig. 4.2 and 4.2.3.2.1).

Hydrophyte and moss leaves, which may be brought to or close to an equilibrium with the external solution also show preferential uptake of K^+ relative to Na^+ (Table 4.2) and the ion distribution can be related to the membrane potential. However, in several species the resting potential is considerably hyperpolarized in the light (4.2.3.5) which would alter the driving forces acting on ions. In Table 4.1 the driving force $\Delta\psi_j$ was calculated as the difference of the resting potential in the light $_L\psi_{vo}$ and the Nernst potential of the ion ψ_j. As with coleoptile or epicotyl cells (HIGINBOTHAM et al., 1967) K^+ appears to be close to equilibrium,

Table 4.1. Ionic concentrations (mM) in leaf cells (i) and the external solution (o) and the electrical potential between the vacuole and external solution in light (L) ($_L\psi_{vo}$) and dark (D) ($_D\psi_{vo}$). The driving forces in the light $\Delta\psi_j$ have been calculated as the difference between ψ_{vo} and the Nernst potential of each ion ψ_j

Species		K^+	Na^+	Ca^{2+}	Cl^-	$H_2PO_4^-$	SO_4^{2-}	ψ_{vo} (mV)
Atriplex	i	90	80	—	ca. 0	—	—	L −105
spongiosa[a]	o	6	5	—	5	—	—	D −83
Hookeria	i	210	3.9	22	27	—	—	−145
lucens[b]	o	0.1	1.0	0.3	1.7	—	—	(−200)
	$\Delta\psi_j$	+40	−117	−84	−215	—	—	
Elodea	i	102	31	1.6	24	5.6	11	L −148
densa[c]	o	0.3	1.3	1.3	1.1	0.01	0.2	D −94
	$\Delta\psi_j$	+3	−43	−146	−227	−310	−198	
Vallisneria	i	80	30					L −200
spiralis[d]	o	0.1	35	0.1	10.3	—	—	D −90
	$\Delta\psi_j$	−28	−204					

[a] OSMOND et al. (1969), the external solution is the growth medium to which the roots are exposed in this case, ψ_{vo} are values for bladder cells;
[b] SINCLAIR (1967) ions determined in extracted leaves;
[c] JESCHKE and SIMONIS (1965), JESCHKE (1970a). Cations and Cl^- determined in cell sap prepared by ultracentrifugation, $H_2PO_4^-$ and SO_4^{2-} measured in extracts.
[d] BENTRUP et al. (1973), K^+ and Na^+ determined in sap prepared in ultracentrifuge.

Table *4.2*. K-Na-selectivity in leaves. The selectivity is calculated as K/Na in the cells divided by K/Na in the medium (i.e. as $S_{K, Na}$ in Table *3*.7)

Species	K_o (mM)	Na_o (mM)	Selectivity $\dfrac{K_i \cdot Na_o}{K_o \cdot Na_i}$	Ref.
Ceratophyllum demersum	0.15	3.2	140	Collander (1941)
Lemna trisulca, fronds	0.19	6.5	95	Collander (1941)
Hookeria lucens	0.1	1.0	570	Sinclair (1967)
Elodea densa	0.3	1.3	14	Jeschke (1970a)
Vallisneria spiralis	0.1	39	930	Bentrup et al. (1973)
Hordeum vulgare	0.5[a]	9.5[a]	65[b]	Pitman (1965)
Avena sativa, coleoptiles	1	1	10	Higinbotham et al. (1967)

[a] Concentration in the culture solution.
[b] Selectivity in the shoot compared to the culture solution. The selectivity of the root relative to the solution was 15 and in the xylem exudate from the root it was 40. Retranslocation of K^+ from old to younger leaves increases the selectivity for the younger leaves compared with that for the shoot as a whole.

while the driving force on other cations is inwardly and that on anions outwardly directed. This would demand anion and possibly K^+ influx pumps and at least a Na^+ efflux pump at the plasmalemma, if ψ_{cv} is small, as is probable (Higinbotham, 1973b).

In contrast to algae, direct information on ion distribution in the major compartment of leaf cells is not available. Measurements with the electron microprobe (Läuchli and Lüttge, 1968; Waisel and Eshel, 1971; Pallaghy, 1973; Ramati et al., 1973) have not as yet yielded concentrations due to uncertainties in calibration and to sample thickness. Estimates of cytoplasmic and vacuolar ion contents may be derived from flux measurements using the compartmental analysis (Part A, Chap. 5). Despite the draw-backs discussed in that Chapter, this method offers at present the best and reasonably consistent estimates of fluxes and of cytoplasmic ion concentrations.

Using this method the ion distribution has been thoroughly studied in a non-green leaf tissue *viz.* oat coleoptiles (Pierce and Higinbotham, 1970) indicating that the K^+/Na^+ selectivity is set up at the plasmalemma with active K^+ and Cl^- influxes and an active Na^+ efflux while the influxes at the tonoplast were active for the 3 ions. Similar measurements using *Elodea* leaves (Jeschke, 1971) showed differences in the behavior of the cytoplasmic content of K^+ or Cl^- and that of Na^+. After equilibration with 1 mM experimental solutions for 18–24 h in the dark or light, the cytoplasmic K^+ and Cl^- contents were higher in the light compared to dark, while the cytoplasmic Na^+ content appeared to be lower in the dark. This points to similar active plasmalemma fluxes in *Elodea* as suggested for oat coleoptiles by Pierce and Higinbotham (1970). The vacuolar concentrations were less affected by light but were increased for all three ions (for the K^+ contents see Table *4.5*).

Certainly the cytoplasm of leaf cells is a heterogeneous phase containing the chloroplasts as a large intracellular compartment. In Table *4.3* estimated ion con-

Table 4.3. Estimated ion concentrations in chloroplasts and in the vacuole (in brackets)

Species	K^+	Na^+	Ca^{2+}	Mg^{2+}	Cl^-	$H_2PO_4^-$
Lagarosiphon major[a] = *Elodea crispa*	350	150	14	35	85	
Elodea densa[g] Dark	240	310	150		530	
Light	200	170	110		320	19[c]
	(105)	(66)			(28)	(5.4)
Limonium vulgare[b]	275	100	84[a]	100[a]	270	
Cochlearia officinalis[d]					320	
					(150)	
Beta vulgaris[a]	250	195	13	60	65	
Pisum sativum[e]	99	10	15	16	92	
Hordeum vulgare[f]					170	
					(8)	

From data of [a] LARKUM (1968), [b] LARKUM and HILL (1970), [c] ULLRICH et al. (1965), [d] KAPPEN and ULLRICH (1970), [e] NOBEL (1969b), [f] van STEVENINCK and CHENOWETH (1972), [g] DE PHILIPPIS and PALLAGHY (1973). The concentrations [a] and [b] are estimated from the values in [a] and [b] using a water content of 2 μl/mg dry weight [a]. The data for *Limonium*[b] are for low salt status. [a, b, c, d] were determined in nonaqueously isolated chloroplasts, [e] from chloroplasts isolated rapidly in an aqueous medium, [f] determined by histochemical and [g] by electron microprobe analysis.

centrations in higher plant chloroplasts are given. Most of these results were derived from measurements on nonaqueously isolated chloroplasts (STOCKING and ONGUN, 1962; ULLRICH et al., 1965; LARKUM, 1968; KAPPEN and ULLRICH, 1970), a method which does not allow a comparison with the concentration in the bulk cytoplasm. But the data clearly indicate high ionic concentrations in the chloroplasts. These results were confirmed by histochemical and electron probe analyses (VAN STEVENINCK and CHENOWETH, 1972; DE PHILIPPIS and PALLAGHY, 1973, Table 4.3). Although the ion concentrations given by DE PHILIPPIS and PALLAGHY appear rather high for the fresh water species *Elodea,* their method is very promising and it was possible to show lower chloroplastic ion contents in the light (Table 4.3) in agreement to analyses in rapidly isolated chloroplasts (NOBEL, 1969b). So the chloroplasts not only are a major cytoplasmic compartment, but they regulate their ionic content in light and darkness (PACKER et al., 1970).

3.2 Effect of Light on Ion Fluxes

It was shown above that anion and K^+ influxes are active, and hence it might be expected that there would be an increase in flux when photosynthesis is stimulated. This has been shown already by the early investigators (ARENS, 1930, 1933, 1936; INGOLD, 1936; GESSNER, 1943; ARISZ, 1947; STEEMANN NIELSEN, 1951; LOWENHAUPT, 1958a, b; GRUBE, 1953; SIMONIS and GRUBE, 1952). However, light does not always increase ion fluxes and in many cases, fluxes are the same in light and in dark. Presumably the different behavior represents different degrees of coupling between photosynthesis and respiration as a source of metabolic energy (Table 4.4). It should be pointed out that in most cases Φ_{ov} was measured (Table 4.4). As $\Phi_{ov} = \phi_{cv} \cdot \phi_{oc}$ ($\phi_{co} + \phi_{cv}$) a change in Φ_{ov} (or lack of it) does not necessarily imply a corresponding change in ϕ_{oc}.

3.2.1 Comparison of Fluxes in Light and Dark

Column 3 of Table 4.4 shows that whether light stimulates the influx depends on the ion investigated and on the plant species used.

Some of the differences in response of hydrophytes could be due to different pretreatment. Leaves of *Vallisneria* and *Elodea* were cultivated in solutions low in Cl^- and the Cl^- influx was light-stimulated at low Cl^- concentrations (Table 4.4), whereas in *Limnophila* they were grown in higher Cl^- concentrations than used in the experiments and light failed to stimulate Cl^- influx (PENTH and WEIGL, 1969a). When Φ_{ov} was measured at higher concentrations than in the culture solution, the Cl^- (and SO_4^{2-}) influxes in *Limnophila* also were stimulated by light (PENTH and WEIGL, 1969a). An effect of the pretreatment on the Cl^- uptake by *Vallisneria* was observed by ARISZ (1947a) and KYLIN (1960a) and extensively studied by SOL (1958). Pretreatments of leaves for 24 h in water in the light increased the subsequent Cl^- uptake in the light or in the dark, pretreatments in Cl^- solutions were inhibitory (SOL, 1958). In *Elodea* the $H_2PO_4^-$ uptake was strongly light-stimulated at 1 mM when leaves grown in 0.04 mM phosphate and pretreated 2 h in phosphate-free solutions were used (GRÜNSFELDER and SIMONIS, 1973). But it was not affected by light when the leaves were pretreated in 1 mM phosphate for 14–20 h to achieve almost steady-state conditions (JESCHKE and SIMONIS, 1965).

Table 4.4. Ion fluxes or uptake in hydrophyte and aerial leaves in the light (L) and dark (D) (Column and the effect of inhibitors on metabolic reactions and the ion fluxes (Columns 4 and 5); wh available, the inhibition is described by the concentration $[I_{50}]$ in µM of the inhibitor producing 50% inhibition, otherwise the inhibition in % of the control at a concentration in µM is indicat Col. 1: ϕ_{oc} = plasmalemma influx, Φ_{ov} = influx to the vacuole, $\phi_{oc} \rightarrow \Phi_{ov}$ = intermediate between and Φ_{ov}; Col. 2: Ca = pretreatment in $CaSO_4$-solution; $CaSO_4$ was present during pretreatment in other solutions, unless otherwise indicated; Col. 4 and 5: $O_2 = O_2$-evolution, $^{14}C = {}^{14}CO_2$ fixati

	Species. leaf age: y = young n.s. = not specified, C_3 = C_3-species C_4 = C_4-species	1 Ion, ext. concentration (mM). Flux determined	2 Conc. in culture (mM). Pretreatment	3 Influx $\left(\dfrac{\mu mol}{g^{-1} h^{-1}}\right)$ in L, air D, air	4 Effect of DCMU on		
					Photo-synthesis $[I_{50}]$ (µM)	Ion transport Conditions	$[I_{50}]$ (µM)
a)	*Vallisneria spiralis*	Cl^-, 1 Φ_{ov}	"soil" 16 h Ca	5.7 1.4	O_2 0.11	L, air	0.5
		Rb^+, 0.2 Φ_{ov}	"soil" 16 h Ca	4.4 0.75		L, air	0.3

Hence, a strong stimulation of influxes by light may be restricted to conditions under which a net ion uptake into leaf cells occurs. This is evident also when the effect of light on the initial K^+ influx ϕ_{oc} in *Elodea* leaves pretreated in K^+-free solutions (Table 4.4b) is compared to the small effect on K^+ influx ϕ_{oc} measured under quasisteady state conditions (Table 4.5). When related to the surface of the *Elodea* leaves the K^+ fluxes are: ϕ_{oc} initial 140 and 2.4 and ϕ_{oc} steady 7.5 and 3.8 nmol m^{-2} s^{-1} in the light and dark respectively. The quasisteady fluxes compare well with ϕ_{oc} in fresh water algal cells (Part A, Chap. 6). It is mainly the influx in the light that appears to be enhanced by the pretreatment.

Similarly, stimulation of ion influx in aerial leaves was found using young leaves of plants grown in solutions low in the respective ion (but see SCHEIDECKER et al., 1971), and again response to light may be correlated with *net ion uptake*. This observation would explain the lack of a stimulation in maize leaves grown on full nutrient solution (SCHÖCH and LÜTTGE, 1974, Table 4.4d) as opposed to the results of RAINS (1968, Table 4.4d), and could be the reason for the small stimulation generally found in aerial compared to submerged leaves, as the former are exposed to the xylem sap which contains the ions at relatively higher concentrations than in the growth medium.

Leaf age appears to affect the response to light (ROBINSON and SMITH, 1970; HORTON and BRUCE, 1972). In *Phaseolus* the K^+ influx decreased while the Na^+ influx increased during leaf maturation (Fig. 4.2). Due to this divergent behavior, the K/Na selectivity appeared to be a function of leaf age. Although a dependence of the light response on leaf age has not been studied explicitly, light stimulated the K^+ and Rb^+ but not the Na^+ influx in young growing leaves (JACOBY et al., 1973, Table 4.4f).

= respiration. Col. 5: Results denoted by an asterisk have been obtained with FCCP, the other data [wi]th CCCP. The light intensity is indicated because the effect of CCCP and FCCP strongly depends [on] intensity. Hence [I_{50}] of FCCP in lines c and d are comparable to [I_{50}] of CCCP, line b, [alt]hough FCCP is the more effective uncoupler. The intensities in lines c and d are extremely high: [≈] 130000 erg cm^{-2} s^{-1}. [b] at high concentrations CCCP produced a net Cl^- efflux and an *increase* in [th]e ^{86}Rb influx

[Ef]fect of CCCP or FCCP(*) on				6 Other effective inhibitors	7 Ref.
[Ph]oto-nthesis [O₂] or [r]espi-[ra]tion(R) [I_{50}] (µM)	ATP-level [I_{50}] (µM) L D	Ion transport Condi-tions L D	[I_{50}] (µM) L D		
		L, 1000 lx, air D, air	1.2[b] 0.4[b]	KCN, UO_2^{2+}, arsenate; azide, DNP D: CO	PRINS (1970, 1974) ARISZ (1958) VAN LOOKEREN CAMPAGNE (1957b)
		L, air D, air	1.1[b] ~0.3[b]		PRINS (1970, 1974) [b] see caption

Table 4.4 (continued)

	Species. leaf age: y = young n.s. = not specified, C_3 = C_3-species C_4 = C_4-species	1 Ion, ext. concentration (mM). Flux determined	2 Conc. in culture (mM). Pretreatment	3 Influx (μmol g^{-1} h^{-1}) in L, air D, air	4 Effect of DCMU on Photosynthesis [I_{50}] (μM)	Ion transport Conditions	[I_{50}] (μM)
b)	Elodea densa	Cl^-, 0.2 ϕ_{oc}	0.01 Cl 14 h 0.01 Cl	1.9 0.4	O_2 0.11	L, N_2 L, air, 3% CO_2 D, air	1.1 0.25 100
		K^+, 0.2 ϕ_{oc}	0.4 K 14 h 0 K	15 0.25	O_2 0.11	L, N_2 L, air, 3% CO_2	1.1 0.5
		$H_2PO_4^-$, 0.01 $\phi_{oc} \rightarrow \Phi_{ov}$	0.04 14 h 0.01 $H_2PO_4^-$	0.07 0.055	O_2 0.11	L, air L, N_2	3[1] no inhib
		SO_4^{2-}, 0.01 $\phi_{oc} \rightarrow \Phi_{ov}$	0.16 14 h 0.01 SO_4^{2-}	0.036 0.028		L, N_2	no inhib
c)	Atriplex hastata n.s., C_3	Cl^-, 5 Φ_{ov}	soil 12 h Ca	0.2 0.15			
	Oenothera n.s., C_3	Cl^-, 5 Φ_{ov}	soil 12 h Ca	4.2[1] 3.5			
d)	Atriplex spongiosa n.s., C_4	Cl^-, 5 Φ_{ov}	5 Cl 12 h Ca	1.4 0.65	^{14}C 1 μM: 7%	L, air	1 μM: 31%
	Zea mays y, C_4	Cl^-, 5 Φ_{ov}	soil 12 h Ca	1.26 1.0			
		K^+, 0.2 Φ_{ov}	0.3 K 1 h Ca	3.5 2		L, air	1 μM: no inhib
		K^+, 5 Φ_{ov}	2.7 K 0.5 h 5 K	0.9 0.9			
e)	Pisum sativum y	K^+, 5 net uptake	n.s. 1 h ($-$Ca)	7.6 0.05		L, air	10 μM: 23%
		Cl^-, 5 net uptake	n.s. 1 h ($-$Ca)	1.9 0.1		L, air	10 μM: 13%
f)	Phaseolus vulgaris y (7 days)	K^+, 0.1 $\phi_{oc} \rightarrow \Phi_{ov}$	0.6 K 2 h Ca	1.4 0.69			
	y (11 days)	Na^+, 0.1 $\phi_{oc} \rightarrow \Phi_{ov}$	0 Na 2 h Ca	0.35 0.36			

		Effect of CCCP or FCCP(*) on			6 Other effective inhibitors	7 Ref.
oto-athesis) or spi-ion (R)] (µM) L D	ATP-level [I$_{50}$] (µM) L D	Ion transport Conditions L D	[I$_{50}$] (µM)			
0.2		L, N$_2$ ± CO$_2$ 3750 erg cm^{-2} s^{-1}	0.2		DNP, Dio-9, atebrin, oligomycin[1] DSPD[2],	JESCHKE (1967, 1972a, b) JESCHKE and SIMONIS (1967) [1] WEIGL (1967) (*E. canadensis*) [2] GIMMLER et al. (1968)
		L, N$_2$ D, air	0.8 ~0.1		DNP, Dio-9, atebrin	JESCHKE (1970a, 1972c)
		L, air ~25000 erg cm^{-2} s^{-1} D, air	0.8[1] 0.2		oligomycin[2]	JESCHKE and SIMONIS (1965) [1] GRÜNSFELDER and SIMONIS (1973) [2] WEIGL (1967) (*E. canadensis*)
					oligomycin[1]	JESCHKE and SIMONIS (1965) [1] WEIGL (1967) (*E. canadensis*)
1.6* 0.9*	2*	L[a], air D, air	0.9* 0.2*			LÜTTGE et al. (1971a, b) [a] 130000 erg cm^{-2} s^{-1}
0.7* 0.3*		L[a], air D, air	0.7* 0.08*			LÜTTGE et al. (1971a, b) [1] in some experiments light inhibited
1* 0.2*	1* 0.2*	L[a], air D, air	1.2* 0.25			LÜTTGE et al. (1970) LÜTTGE et al. (1971a, b)
3* 0.16*	>2* <0.5*	L[a], air D, air	1.6* 0.3*			LÜTTGE et al. (1971a, b)
		L, air 1 µM: 30% D, air 1 µM: 17%			DNP: D > L, azide, KCN: D > L	RAINS, 1968
						SCHÖCH and LÜTTGE (1974)
		L, air	10 µM: 27%			NOBEL (1969a)
		L, air	10 µM: 82%			NOBEL (1969a)
		L, air (Rb$^+$)	<5			JACOBY et al. (1973)
	~1.3 ~1.0	L, air D, air	~0.8 ~0.2		D: antimycin A, oligomycin	JACOBY et al. (1973) STEINITZ and JACOBY (1974)

Table *4.5*. Fluxes (µmol g^{-1} h^{-1}) and content (µmol g^{-1}) of K$^+$ in leaves of *Elodea densa*, measured in a solution containing 1 mM K$^+$ and Na$^+$ and 0.1 mM Ca^{2+}. The leaves were pretreated for 18–24 h in this solution in dark or light. Duration of influx 3 h, from Jeschke (1971) and unpublished results

Flux	Dark	Light	Flux or content	Dark	Light
ϕ_{oc}	0.4	0.8	ϕ_{vc}	4.0	7.6
ϕ_{co}	0.9	0.7	Q_c	2.3	10
ϕ_{cv}	3.4	7.7	Q_v	140	190

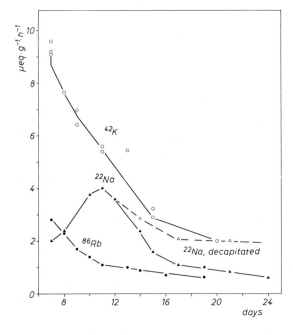

Fig. *4.2*. Absorption of ^{42}K, ^{86}Rb and ^{22}Na from 5 mM chlorides (+0.2 mM CaSO$_4$) by primary bean leaf slices as a function of leaf age. Open triangles: ^{22}Na absorption by slices from decapitated bean plants. After Jacoby et al. (1973)

A correlation between the occurrence of net ion uptake and light-stimulation is also suggested by seasonal differences (Gessner, 1943, see also Lüttge et al., 1971 b for *Spinacia*) or by a dependence on the growth conditions (Kylin, 1960a).

Other factors may influence the degree of any stimulation of influx in the light, such as intensity and wavelength of light (see *4.2.3.4*); temperature (Jeschke and Simonis, 1965; Jeschke, 1970a) and external concentration of ions (see *4.2.3.3*). The method of preparing leaf tissue may also influence the response to light: little stimulation was found using enzymatically separated leaf cells (Jyung et al., 1965), probably due to impairment of membranes by the isolation procedure (Jacoby and Dagan, 1967); strong stimulation was found with vacuum-infiltrated wheat leaves 5 cm long (MacDonald and Macklon, 1972), in which a minimal damage of the cells may be anticipated. In *Crassula* the osmotic pressure of added mannitol solution affected the degree of stimulation of SO$_4^{2-}$ uptake (Kylin, 1960b) and chopped leaves of *Pisum* suspended in isotonic sucrose solution showed almost complete light dependence of K$^+$ uptake. Lüttge et al. (1971 b) found differences in response of C$_3$- and C$_4$-plants suggesting that this may be due to biochemical adaptation.

Apart from the data given in Table *4.*4, light stimulated the influx in aerial leaves of *Hordeum* (Rb$^+$, Cl$^-$; KHOLDEBARIN and OERTLI, 1970), *Triticum* (Cl$^-$, MACDONALD and MACKLON, 1972), *Hedera* (K$^+$, SCHEIDECKER et al., 1971) and several C$_4$-plants (LÜTTGE et al., 1971b). No, or only slight light stimulation was found in *Thuidium, Hordeum, Triticum* and *Crassula* (SO$_4^{2-}$, KYLIN, 1960a), *Citrus* (K$^+$, Cl$^-$, ROBINSON and SMITH, 1970; SMITH and ROBINSON, 1971), *Hordeum* (Cl$^-$, PITMAN et al., 1974b) and *Tradescantia* (K$^+$, Cl$^-$, LÜTTGE and BALL, 1971) and several C$_3$-species (LÜTTGE et al., 1971b).

No stimulation was found also for Ca^{2+} (SCHEIDECKER et al., 1971), heavy metal ions (BOWEN, 1969; KANNAN and WITTWER, 1967) and Na$^+$ (SMITH and ROBINSON, 1971, and Table *4.*4f). This is in agreement with the direction of active Na$^+$ fluxes which would suggest a stimulation if at all of the Na$^+$ efflux, as found for *Pisum* leaves (NOBEL, 1969a). In *Atriplex* a light-independent net Na$^+$ efflux was observed (LÜTTGE et al., 1970).

Slices of leaves of higher plants often showed increased response to light when the slices had been kept in aerated solution after cutting. As with storage tissue, this change has been referred to as "ageing" (Part A, *8.*2.3). In maize leaves grown on full nutrient solution, the K$^+$ and Cl$^-$ influx was stimulated by light, when the slices were aerated ("aged") in CaSO$_4$-solution over night (LÜTTGE et al., 1971b) or at least 4 h (SCHÖCH and LÜTTGE, 1974), while it was independent of light in freshly cut slices (LÜTTGE and BALL, 1973; SCHÖCH and LÜTTGE, 1974); similar changes in the light response of the K(Rb$^+$) or Cl$^-$ influx were found in barley, even after prolonged treatment in 5 mM KCl (+CaSO$_4$) solution (PITMAN et al., 1974b); "ageing" also increased the H$_2$PO$_4^-$ (PRATT and MATTHEWS, 1971) and amino acid uptake (SHTARKSHALL et al., 1970) of leaf slices. An increase was found mainly in the uptake in the light, less so in the dark, and to some extent in the respiration (SCHÖCH and LÜTTGE, 1974). These changes are termed "adaptive ageing due to excision" (VAN STEVENINCK, Chap. *8*). Their biochemical causes are not known; an effect of bacteria (SHTARKSHALL et al., 1970), a loss of K$^+$ from the tissue (SCHÖCH and LÜTTGE, 1974; PITMAN et al., 1974b), or changes in permeability or photosynthetic capacities were excluded as possible causes; PITMAN et al. (1974b) and SCHÖCH and LÜTTGE (1974) suggested hormonal or developmental changes to be responsible instead. (For a detailed treatment see Chap. *8*).

To conclude, light-stimulation of ion uptake by aerial leaf slices is strongly dependent on the physiological state or nature of the leaves. One factor determining the light response may be the ion contents in the cells or the occurrence of net ion uptake, although this is not a sufficient determinant as in barley leaves the K$^+$ and Cl$^-$ uptake were independent of light at 5 mM external concentration where the Cl$^-$ content showed a net increase during 24 h, while the K$^+$ content was unchanged (PITMAN et al., 1974b).

3.2.2 Transport *across* Hydrophyte Leaves

Leaves of several aquatic species such as *Potamogeton* and *Elodea* are capable of a polar, light-dependent cation transport across the leaf toward the upper side (ARENS, 1933; LOWENHAUPT, 1956; STEEMANN NIELSEN, 1960; cf. Part A, *12.*5.2.1). This transport is linked to photosynthetic bicarbonate utilization and normally involves Ca^{2+} and results in CaCO$_3$ deposits on the leaves. In *Potamogeton* *lucens* Rb^+ was transported across the leaf only when present as $RbHCO_3$ at the lower leaf surface, no transport occurred when RbCl was applied (Helder and Boerma, 1972). This light-dependent transport occurred against a Rb^+ concentration gradient. But as additionally there was considerable undirected $^{86}Rb^+$-Rb^+ exchange across the leaf independent of light (Helder and Boerma, 1973), it was suggested that the cation transport may represent a passive movement in response to an active OH^- efflux at the adaxial leaf surface (Helder and Boerma, 1972, 1973; Steemann Nielsen, 1960).

3.3 Ion Fluxes, Effect of External Concentration

The effects of external concentration on ion uptake are treated in detail in Chap. *3.2*; here the aspects relevant to leaves shall be considered only. In young maize leaves which were low in K^+, the relation of the tracer K^+ and Rb^+ uptake to the external ion concentration (0.002–0.2 mM, "system I") could be described by a single Michaelis-Menten equation. This points to similar properties in leaves and roots as was confirmed by comparable rates, by a similar temperature dependence, and by similar interactions with other alkali ions (Smith and Epstein, 1964b).

In a wide range of external concentrations two kinetic systems (I and II) at low and high concentration could be distinguished ($H_2PO_4^-$, SO_4^{2-}: Jeschke and Simonis, 1965; K^+: Rains, 1968; K^+ and Rb^+: Jeschke, 1970a, c). System II may be composed of more than one saturation curve (Cl^-, $H_2PO_4^-$, SO_4^{2-}: Weigl, 1967, Penth and Weigl, 1969a; $H_2BO_3^-$: Bowen, 1968; SO_4^{2-}: Nissen, 1971). Where it was studied, a similar kinetic behavior was found under light and dark conditions.

The results indicate that the rate-limiting step in tracer ion uptake has similar properties in roots and leaves and whether the energy is supplied by dark or by light reactions. Since the plasmalemma is generally agreed to be the rate-limiting site at least in the low concentration range (Epstein, 1966; Torii and Laties, 1966), these results preclude a direct energetic participation of electron transport within the mitochondria or chloroplasts in the ion uptake by leaves (Weigl, 1967). They point to mobile molecules like ATP or NADPH as energy sources or to an energy transfer by transport metabolites (see *4.2.3.4.3*).

Although analogous kinetic systems are found in light and dark, the apparent maximum rate of influx $V_{max\,I}$ in system I often was increased by light (Rains, 1968; Jeschke and Simonis, 1965; Jeschke, 1970a, c; Figs. 2 and 3 in Weigl, 1967; Grünsfelder, 1971). The apparent Michaelis constant $K_{m,\,I}$ was either increased in the light (Rains, 1968; Jeschke, 1970a) or decreased (Penth and Weigl, 1969a; Weigl, 1967; Grünsfelder, 1971). As suggested by Weigl (1967), the effect of light could possibly be described by application of coenzyme kinetics (Netter, 1959). The increase in the apparent V_{max} and M_m of the K^+ influx in the light was predicted qualitatively by applying coenzyme kinetics and assuming that light increases the local concentration of a cofactor (e.g. ATP) acting at the transport site (Jeschke, 1970a).

In *Limnophila,* where the Cl^- uptake (Φ_{ov}) was stimulated by light only in the higher concentration range (*4.2.3.2.1*), Penth and Weigl (1969a) suggested from the isotherms that light might be stimulatory due to an effect on a passive $^{36}Cl^-$

influx. In *Elodea* the light-stimulation of the $H_2PO_4^-$ (and SO_4^{2-}) uptake was restricted to system I (JESCHKE and SIMONIS, 1965; GRÜNSFELDER, 1971), which corresponds to the natural range of $H_2PO_4^-$ concentrations (BIELESKI, 1973), while the light-stimulation and the activation energy of $H_2PO_4^-$ uptake declined in the range of system II. Similarly the inhibitor-sensitivity was smaller at higher external concentrations (NISSEN, 1971; GRÜNSFELDER and SIMONIS, 1973). These results could indicate that diffusion of tracer anions across the plasmalemma may contribute increasingly to the tracer influx in leaves at higher external concentrations (JESCHKE and SIMONIS, 1965; NISSEN, 1971; GRÜNSFELDER and SIMONIS, 1973) as in roots (TORII and LATIES, 1966).

Whereas the results given so far have suggested some similar properties for leaves and roots, JACOBY and PLESSNER (1970) found a considerably lower Cl⁻ uptake in the system I range in leaves compared to roots while comparable rates were found at higher concentrations (system II). The difference was attributed to organic acid production in the leaves (JACOBY and PLESSNER, 1970). However, the apparent K_m values (0.15 and 0.18 mM) in roots and leaves were comparable and the different rates may additionally be due to the use of low-salt roots, but leaves that were at a much higher salt status.

Indeed, the occurrence of a system I of ion uptake in aerial leaf cells (SMITH and EPSTEIN, 1964b; RAINS, 1968; JACOBY and PLESSNER, 1970) with an apparent affinity corresponding to a K_m of about 0.1 mM is quite remarkable if compared to the higher ion concentration in the apoplast as the natural environment of leaf cells (ROBINSON, 1971; PITMAN et al., 1974b). The high affinity of system I points to the capability of leaf cells to reduce the external concentration to those of the guttation liquid (KLEPPER and KAUFMANN, 1966; see also 4.2.2.2.2).

Important in this respect is a relatively low affinity of system I K^+ uptake and an unusually high K^+/Na^+-selectivity in the high concentration range found in *Avicennia* leaves, which points to an adaptation to the high Na^+ (and K^+) concentrations to which their cells are exposed under natural conditions (RAINS and EPSTEIN, 1967). An apparent preference of Na^+ before K^+ associated with a Na^+-dependent K^+ loss, which was under some metabolic control, was detected in the halophyte *Atriplex* (OSMOND, 1968), but only in the absence of Ca^{2+} (see EPSTEIN, 1961). Although this anomalous selectivity might be termed an artifact—in the presence of Ca^{2+} the tissue was K^+/Na^+-selective even in the system II-range (OSMOND, 1968)—it may be an important indication of a reversed K^+/Na^+-selective mechanism at the tonoplast by which Na^+ is transported into the vacuole (in exchange for K^+) and which was apparent only when lack of Ca^{2+} permitted an excess of Na^+ to enter the cytoplasm. Alternatively this different selectivity may reflect differences in H^+ efflux (see Chap. 3.3.1).

The general validity of the kinetic analysis and the usefulness of the shape of isotherms as indicators of transport mechanisms at a specific membrane was questioned by SMITH and ROBINSON (1971) because the "normal" dual isotherms were not observed for the K^+ and Na^+ uptake (Φ_{ov}) by mature (high salt) *Citrus* leaves and because this tissue was somewhat K^+/Na^+-selective at high concentrations. As a comparable behavior was observed in high-salt roots (PITMAN et al., 1968), the objections raised by the authors are important for the extrapolation to the processes in intact leaves. Since high ionic contents are common in fully expanded leaves, approaches with mature leaves as made by SMITH and ROBINSON

(1971) are needed in addition to those using leaves at a relatively low salt status (e.g. Rains, 1968).

A further problem of the kinetic analysis is that at low concentrations cation uptake seems to be limited by diffusion in, and Ca^{2+} content of the Donnan free space, tending to suppress the distinction usually found between mechanisms I and II (Pitman et al., 1974b). Single saturation curves were found also for the net uptake (Arisz, 1947a: Cl^-, 24 h uptake) or net uptake rates (Young and Sims, 1972: K^+ in growing *Lemna* fronds; Pitman et al., 1974b: K^+ and Cl^- in barley leaf slices) within a physiological concentration range (up to 4 mM, Arisz, 1974a).

Beyond this range there was a linear increase in the net K^+ and Cl^- uptake (Pitman et al., 1974b). This could represent osmotic adjustment as the increment in the internal KCl almost matched the external concentration (60 mM) after 20 h. At this time, the net uptake rate approached zero while the ^{86}Rb influx (Φ_{ov}) was still at its initial value, pointing to flux equilibrium.

3.4 Energy Supply for Ion Fluxes

3.4.1 Introduction

As there is a range of responses of ion flux in leaves to light, it has been hoped to gain some insight into the energetics of ion transport by studies with leaf material. Energy can be supplied from photosynthesis and light reactions in the chloroplasts or in other situations from glycolysis and respiration. As experimental strategies, light intensity and wavelength or CO_2 level can be varied or a range of inhibitors can be used.

That the chloroplast pigments mediate the transduction of light energy to the light-driven ion fluxes has been clearly demonstrated by showing that the action

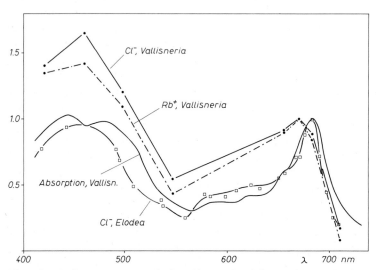

Fig. *4.3.* Action spectrum of the Cl^- and Rb^+ uptake, and absorption spectrum of *Vallisneria*, and action spectrum of Cl^- uptake by *Elodea* leaves. The ordinate is absorption or uptake (683 or 670 nm = 1) for *Vallisneria*, or relative quantum efficiency of Cl^- uptake (683 nm = 1) for *Elodea*. After Prins (1973) and Jeschke (1972c)

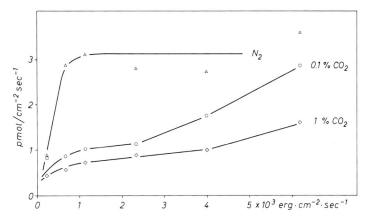

Fig. *4*.4. Dependence of the Cl^- influx by *Elodea densa* leaves on light intensity ($\lambda = 683$ nm) and the effect of CO_2 (pH 5.2). After JESCHKE and SIMONIS (1969)

spectrum of net Cl^- uptake by *Vallisneria* leaves resembles that of photosynthesis (VAN LOOKEREN CAMPAGNE, 1957a, b). This was confirmed recently for the Rb^+ and Cl^- influx in this species (PRINS, 1973) and for the Cl^- influx in *Elodea* (JESCHKE, 1972c); the former action spectrum shows a relatively high efficiency of blue light (Fig. *4*.3).

The relationship between photosynthesis and light-driven Cl^- and K^+ uptakes is also demonstrated by the close correspondence between time courses for light-induction of ion uptake and photosynthesis following a change from dark to light (VAN LOOKEREN CAMPAGNE, 1957b; JESCHKE, 1970a).

Although carbohydrates such as sucrose can promote Cl^- uptake (ARISZ and SOL, 1956; SOL, 1958), the effect of light is not due to an increased supply of carbohydrates to respiration, since light also stimulates the ion fluxes in the absence of CO_2 (ARISZ, 1947a; VAN LOOKEREN CAMPAGNE, 1957b; JESCHKE, 1967; WEIGL, 1967; RAINS, 1968) or only in N_2 (ROBINSON and SMITH, 1970; JOHANSEN and LÜTTGE, 1974). Moreover light-dependent Cl^- uptake in N_2 develops earlier than the capacity of ^{14}C-fixation in greening etiolated leaf tissue (LÜTTGE, 1973a), and ion influxes (Cl^-: VAN LOOKEREN CAMPAGNE, 1957a, b; JESCHKE and SIMONIS, 1969 and K^+: RAINS, 1968; JESCHKE, 1972c; and Rb^+: PRINS, 1974) are light-saturated at far lower intensities than photosynthesis. The saturating intensities, 1 000 lx in maize (RAINS, 1968) 1–1.5 W m^{-2} in *Elodea* (JESCHKE and SIMONIS, 1969), see Fig. *4*.4, are comparable to those in algae (RAVEN, 1968).

One can conclude therefore that ion fluxes appear to be powered by partial reactions in photosynthesis rather than by the ultimate products, but that transport may also be supplied with energy from respiration. The common factors in these processes are the *energy transfer reactions*. In these reactions, movement of protons and electrons across membranes is coupled with production of either phosphorylating (ATP) or reducing compounds (NADH, NADPH) or with the "shuttles" of energy-transferring compounds out of the organelles to the cytoplasm. The concept that ion transport may be coupled with some component partial reaction of photosynthesis at the energy transfer level has been a stimulus for much investigation.

3.4.2 Relation of Ion Transport to Energy Transfer Reactions

Working with the alga *Nitella* MACROBBIE (1965) first suggested that the energetic
linkage between the partial reactions of photosynthesis and ion fluxes may depend
on the kind of ion transported. In this alga, see Part A, *6.6.2.5*, *6.6.3.3*, *9.2*, the
Cl^- influx appears to be closely linked to noncyclic photosynthetic electron transport
or to reducing equivalents such as NADPH formed by it, while the light-driven
K^+-Na^+ exchange is powered by ATP produced by cyclic photophosphorylation.
Much effort has since been made to find whether the linkages of light-driven ion
fluxes to the energy sources in leaves are also similar to this hypothesis.

Similar alternatives for the energetic linkages—electron transport or reducing
equivalents (LUNDEGÅRDH, 1960; ROBERTSON, 1960, 1968) or else ATP—have been
envisaged also when respiration provides the energy in the dark.

This early work has developed by introduction of further criteria for separating
different components of energy flow in photosynthesis. In the commonly accepted
scheme of photosynthetic electron transport with two photosystems PS I and II,
three modes of electron flow are possible, a) noncyclic including PS I and II and
producing NADPH and ATP, b) cyclic including only PS I and producing solely
ATP, and c) pseudocyclic electron flow which involves PS I and II and O_2 as
the electron acceptor and produces ATP (pseudocyclic photophosphorylation). A
participation of PS II as in a) and c) can be detected by a strong decline in the
quantum efficiency in long wavelength light ($\lambda > 700$ nm), the "red drop", and
by an effect of DCMU, an inhibitor of PS II. A dependence on ATP is indicated
by a sensitivity to uncouplers of photophosphorylation such as CCCP or FCCP
(HEYTLER and PRICHARD, 1962), DNP (NEUMANN and JAGENDORF, 1964), atebrin
(LOSADA and ARNON, 1963) or by inhibitors of energy transfer such as Dio-9
(MCCARTHY et al., 1965) and phloridzin (IZAWA et al., 1966). A dependence on
cyclic photophosphorylation is indicated by a small decline or even a rise in the
efficiency in long wavelength light or by a light-stimulation occurring in CO_2-free
N_2 (but see ULLRICH, 1971). An inhibition by uncouplers and by DCMU, a red
drop and a dependence on O_2 would argue for a participation of pseudocyclic
phosphorylation.

3.4.3 Energy Transport and Ion Transport Mechanisms

Whether ATP or reducing equivalents carrying the energy of noncyclic electron
transport are the energy source, the site of energy transformation, i.e. the chloroplasts
or mitochondria, by all evidence is not identical to the site of energy consumption,
i.e. the plasmalemma and/or the tonoplast. This complicates models of a direct
linkage of ion fluxes to electron transport (LUNDEGÅRDH, 1960; ROBERTSON, 1960,
1968; MACROBBIE, 1965). But as the chloroplast envelope is rather impermeable
to ATP (HEBER and SANTARIUS, 1970; STOKES and WALKER, 1971) and to NAD(P)H
(HEBER and SANTARIUS, 1965), the transport of these energy sources poses complica-
tions too. As discussed in detail (see Encyclopedia of Plant Physiology, New
Series, Vol. 3), metabolic shuttles appear to mediate the transfer of energy out of the
chloroplasts.

Common to these shuttles (STOCKING and LARSON, 1969; HELDT, 1969; HEBER,
1970; HEBER and SANTARIUS, 1970; HELDT and RAPLEY, 1970) are transport metabolites such as DHAP, malate or glycolate which transfer both NAD(P)H and ATP

or only NAD(P)H into the cytoplasm. As discussed by Lüttge et al. (1971 b) and Jeschke (1972 c) the energy transfer by shuttles implies that utilization of ATP or NADH for ion transport inevitably is bound to a complementary metabolite NAD(P)H or ATP, as these are translocated too or needed in the regeneration of the transport metabolites.

ATP produced by cyclic or pseudocyclic photophosphorylation may only be carried out of the chloroplasts by a combination of two (the PGA-DHAP- and the malate-oxaloacetate-) shuttles (Heber and Krause, 1971). Operation of these shuttles at least initially, however, implies reduction of PGA, viz. noncyclic electron transport. Thus ATP formed in *cyclic* photophosphorylation would be linked to some *noncyclic* electron transport when the ATP is to be utilized in the cytoplasm.

The involvement of transport metabolites in the transfer of ATP or NAD(P)H energy out of the chloroplasts thus appears to require generally adenylate energy *and* reducing equivalents for energizing ion fluxes in the living cells. We shall see that this conclusion is in agreement with results obtained for the ion fluxes (*4.2.3.4.4*, *4.2.3.4.5*).

Since uptake of ions appears to be powered by respiratory energy transfer under certain conditions, it is worth remembering that a malate-OAA shuttle also can operate in mitochondria. This reaction allows a more sophisticated balance between chloroplasts and mitochondria than by the production of ATP, or NADH. Interactions between the processes are well known. Thus there are reports on a photoinhibition of respiration or the respiratory energy supply (Kok, 1949; Hoch et al., 1963; Kylin, 1960a; Healey and Myers, 1971; Simonis and Urbach, 1973). There are also differential effects of CCCP and other inhibitors on ion uptake in the dark and light (Kylin, 1960a; Rains, 1968; Smith and West, 1969; Kholdebarin and Oertli, 1972; Prins, 1974), which show an ability for the process to be driven by either respiratory or photosynthetic energy. *In vitro,* oxidative phosphorylation is by far more sensitive to CCCP than is photophosphorylation (Heytler and Prichard, 1962). In C_4-plants, involvement of common metabolites gives further possibilities for interaction of respiration and photosynthesis.

Asking questions about the energy supply for ion transport therefore requires knowledge of the mechanisms by which the ions cross the membranes. Cation-stimulated ATPases have been isolated from root tissues (see Part A, Chap. *10*) and with fronds of *Lemna* (Young and Sims, 1972, 1973) evidence for an association of the CCCP-sensitive K^+ uptake to an ATPase was presented. A Cl^--dependent ATPase has been found in a microsomal fraction isolated from *Limonium* leaf tissue (Hill and Hill, 1973a) and was ascribed to the electrogenic Cl^- pump of *Limonium* salt glands (see Chap. *5.2*). It was not excluded that a small fraction of the ATPase may be present in the mesophyll cells (Hill and Hill, 1973b), which would be very important if this finding could be extrapolated to less specialized species than the halophyte *Limonium*. In this case an intriguing question is posed by the affinity of the enzyme of *Limonium* which was stimulated at 100 mM NaCl, while the Cl^- transport at the plasmalemma of other species would require an affinity in the range of system I, i.e. 0.1 mM (*4.2.3.1* and *3.2.3.2*).

A way of elucidating the energy source of ion transport is the application of likely metabolites. Externally applied ATP stimulated the K^+ or Rb^+ uptake in various tissues (Jyung et al., 1965; Young and Sims, 1973; Lüttge et al., 1974) which could point to a mediation by an ATPase. However, Lüttge et al. (1974)

discussed the effects of ATP in detail and showed that ATP, as other nucleotides, acted by chelating Ca^{2+} and Mg^{2+} and thereby releasing an inhibition by the divalent ions, which may pertain also to other tissues. It appears relevant that ATP inhibited the phosphate uptake (Jyung et al., 1965) as Ca^{2+} is known to enhance phosphate uptake (Foote and Hanson, 1964). So the evidence provided by external application of ATP appears to be equivocal.

In Fig. *4.5* the main possible energetic linkages of the Cl^- and K^+ influx into leaf cells are drawn schematically, showing the interrelations between photosynthesis and respiration. Especially in cases where the influxes are independent of light as often in terrestrial leaves photosynthetic reactions and respiration can be considered as alternative energy sources (Johansen and Lüttge, 1974). As photosynthesis and respiration are interrelated not only by carbohydrate production but also by a possible photoinhibition of respiration, the cell can respond to inhibition of photosynthesis (e.g. in the dark or by DCMU) by a supply of respiratory energy to the ion fluxes. The lack of a light-stimulation additionally may be caused by a feedback of the ion transport on the energy supplies (Fig. *4.5*). As seen from Fig. *4.5*, photosynthesis and respiration are also interrelated *via* O_2 and CO_2. Hence a respiratory energy supply may not even be excluded under anaerobic conditions in the light (Johansen and Lüttge, 1974), when O_2 produced by photosynthesis could allow a residual respiration sufficient to produce ATP for ion transport.

In this respect it should be pointed out that the energy demands of ion transport in leaves are small compared to other processes (Part A, *9.4*). Even in the absence of photosynthesis, in N_2, the quantum yield of the light-driven Cl^- influx in *Elodea* was only 1 Cl^- for 13–20 absorbed quanta of red light and this ratio is much smaller under photosynthetic conditions and by far smaller in influx-saturating

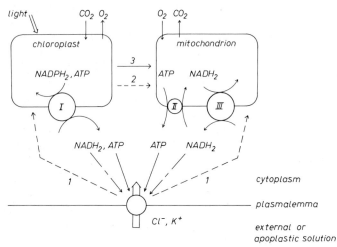

Fig. *4.5*. Possibilities of energy supply to ion transport across the plasmalemma of a leaf cell. Energy delivery by ATP or $NADH_2$ may depend on the ability to switch from photosynthetic to respiratory energy production and on feedback from ion transport (*1*); *2* photoinhibition of respiration, *3* transfer of carbohydrates. *I* shuttles mediating the ATP and/or $NAD(P)H_2$ transport by chloroplasts; *II* mitochondrial adenylate carrier; *III* mitochondrial malate shuttle. Similar energy balances and feedbacks would pertain to a transport across the tonoplast

light intensities (JESCHKE, 1972c). Also in *Tradescantia,* the rate of Cl$^-$ uptake was only 1/10 to 1/30 of the rate of photosynthetic oxygen evolution (JOHANSEN and LÜTTGE, 1974).

Although inhibition by uncouplers or inhibitors of phosphorylation indicates that ATP production is needed for Cl$^-$ or K$^+$ transport in most leaves (see *4.2.3.4.4* and *4.2.3.4.5*), both ATP and NAD(P)H are included in Fig. *4.5* as possible energy sources as both are interrelated again by their delivery to the cytoplasm by metabolic shuttles.

3.4.4 Energetics of Ion Transport in Hydrophytes

3.4.4.1 K$^+$ and Cl$^-$ Uptake

In *Vallisneria,* the Cl$^-$ uptake was suggested to depend on oxidative or photophosphorylation as it was inhibited by arsenate, KCN and UO$_2^{2+}$ ions (ARISZ, 1953, 1958; VAN LOOKEREN CAMPAGNE, 1957b). KCN was thought to inhibit photophosphorylation (VAN LOOKEREN CAMPAGNE, 1957b) although in chloroplasts higher KCN concentrations are needed and KCN specifically inhibits photosynthesis at the site of RUDP carboxylase (LOSADA and ARNON, 1963; GOOD and IZAWA, 1973); in this respect it is remarkable that KCN did not inhibit at light-limiting but mainly at saturating intensities (VAN LOOKEREN CAMPAGNE, 1957b). In these early experiments DNP showed variable results (ARISZ, 1958, see *4.2.3.4.6*). However, low concentrations of CCCP applied under light-limiting-conditions inhibited the Cl$^-$ and Rb$^+$ influx equally and strongly (PRINS, 1974).

In *Elodea* DNP and CCCP inhibited the light-driven Cl$^-$ and K$^+$ influx (pH 5.2) most severely under light-limiting conditions indicating that the inhibitors affected the energy supply close to a light reaction and not the transport site and that photophosphorylation is needed for the light-dependent Cl$^-$ and K$^+$ influx (JESCHKE, 1967, 1972b, c). This was suggested also for *Limnophila* (PENTH and WEIGL, 1971) where CCCP decreased the Cl$^-$ uptake and the ATP-level to the same degree. An inhibition by oligomycin (WEIGL, 1967) or by Dio-9 and atebrin (JESCHKE, 1972a) could have confirmed this energetic linkage, but evidence is available that these compounds may act on the transport site at the membrane in addition to effects on oxidative or photophosphorylation (HODGES, 1966; JESCHKE, 1972a).

When the Cl$^-$ influx of *Elodea* and *Vallisneria* leaves was measured in the absence of CO$_2$ in N$_2$ or air, it was less sensitive to DCMU than photosynthesis (WEIGL, 1967; JESCHKE, 1967, Table *4.4b*; PRINS, 1970); in the presence of CO$_2$, the DCMU-sensitivity of the Cl$^-$ influx was greatly increased (JESCHKE and SIMONIS, 1967, Table *4.4b*; PRINS, 1974). Similar results were obtained for the K$^+$ influx (JESCHKE, 1972c) and for the Rb$^+$ influx of *Vallisneria* (PRINS, 1970, 1974) where the Rb$^+$ influx—contrary to the results with algae (Part A, Chap. 6)—tended to be more sensitive to DCMU than the Cl$^-$ influx.

Also without an inhibitor, CO$_2$ severely decreased the K$^+$ and Cl$^-$ influxes of *Elodea,* when they were measured under light-limiting conditions (Fig. *4.4*; JESCHKE and SIMONIS, 1969; JESCHKE, 1972c) indicating that photosynthesis or noncyclic electron flow compete with the ion fluxes for energy. The effect of CO$_2$ was confirmed for *Vallisneria* (PRINS, 1974) although higher CO$_2$ concentrations were needed to inhibit the Cl$^-$ and Rb$^+$ influx in this species. This difference was attributed to difficult access of CO_2 *via* intercellular spaces in *Vallisneria* as compared with *Elodea* leaves (PRINS, 1974). Both in *Elodea* and in *Vallisneria* the inhibition was due to CO_2 and not to HCO_3^- (JESCHKE and SIMONIS, 1969; PRINS, 1974). Only at higher pH HCO_3^- inhibits the Cl^- influx of *Vallisneria*, presumably by competition at the transport site (VAN LOOKEREN CAMPAGNE, 1957b).

These results could indicate a linkage to cyclic photophosphorylation, particularly since the Cl^- and K^+ influxes were supported by light also in N_2 (VAN LOOKEREN CAMPAGNE, 1957b; JESCHKE, 1967; WEIGL, 1967; PENTH and WEIGL, 1971; PRINS, 1974), and since a light-dependent Cl^- influx occurred in far-red light ($\lambda > 700$ nm, JESCHKE, 1967; PRINS, 1973). However, far-red light was almost as effective in supporting photosynthesis as in stimulating the Cl^- and K^+ influxes: the relative quantum yields at 717 nm were 5 for photosynthesis, 5.6 and 8.3 for K^+, and 5.9 and 8.9 for Cl^- influx in N_2 and air respectively (quantum yield at 683 nm $= 100$) (JESCHKE, 1972c). In *Vallisneria* the Cl^- influx was supported somewhat more effectively by far-red light than photosynthesis and Rb^+ uptake (PRINS, 1973, see Fig. *4*.3). A similar result was obtained with the moss *Hookeria*, where 700 nm light supported Cl^- but not K^+ uptake (SINCLAIR, 1968).

Before it can be stated that cyclic photophosphorylation provides energy for ion fluxes, its occurrence under normal conditions *in vivo* has to be proven, see SIMONIS and URBACH (1973) for a review. KAISER (1973) and KAISER and URBACH (1973) found evidence for cyclic photophosphorylation in intact chloroplasts, but this process was DCMU-sensitive. According to HEBER (1973a, b), however, cyclic electron flow is not possible in leaves under physiological conditions and the additional ATP needed in photosynthesis or other processes is formed by pseudocyclic electron flow. HEBER (1969) found a high affinity of the electron transport chain to CO_2 and O_2 which favors noncyclic or pseudocyclic electron flow. The severe inhibition of the ion fluxes by CO_2 (Fig. *4*.4) is consistent with these results. It has been suggested therefore, that pseudocyclic photophosphorylation provides the energy for Cl^- and K^+ influx of *Elodea* (JESCHKE, 1972c), while in N_2 there could be a contribution by cyclic photophosphorylation, but see ULLRICH (1971). PRINS (1974) investigated the effect of O_2 on ion fluxes and their sensitivity to DCMU in *Vallisneria* and found no proof of a participation of pseudocyclic photophosphorylation. He concluded that cyclic photophosphorylation provides the energy, but a contribution by PS II was evident also in this species.

So it appears that ATP and *both* photosystems are involved in the energy supply to the Cl^- and K^+ (or Rb^+) influx of hydrophyte leaves which would be consistent also with the implications of the intracellular energy transport, see *4*.2.3.4.3.

The similarities of the Cl^- and K^+ (or Rb^+) influx imply that at least parts of the cation fluxes are linked to the Cl^- influx in *Elodea* and *Vallisneria* (JESCHKE, 1972c; PRINS, 1974).

3.4.4.2 $H_2PO_4^-$ and SO_4^{2-} Uptake

A light-dependent $H_2PO_4^-$ uptake by *Elodea* leaves occurred also in the absence of CO_2 in air (BIANCHETTI, 1963; MARRÉ et al., 1963) and at a high rate in N_2 (WEIGL, 1967). As it was inhibited by oligomycin (WEIGL, 1967) and CCCP (GRÜNSFELDER, 1971; GRÜNSFELDER and SIMONIS, 1973) and as the ATP-level and the $H_2PO_4^-$ uptake were similarly depressed by isotonic mannitol concentrations

(WEIGL, 1967; cf. also the sulfate uptake in *Vallisneria*, KYLIN, 1960b), ATP appears to be required. Its source was suggested to be cyclic photophosphorylation, since DCMU failed to inhibit in an atmosphere of N_2 (WEIGL, 1967); however, at 5 μM $H_2PO_4^-$, which may be termed a physiological concentration (BIELESKI, 1973), the uptake was inhibited by DCMU in air+CO_2 (GRÜNSFELDER, 1971), indicating that noncyclic or pseudocyclic photophosphorylation can provide the energy.

The uptake and reduction of SO_4^{2-} and the sulfur incorporation into organic compounds in *Vallisneria* and leaves of some terrestial species was extensively studied by KYLIN (1960a–d). Uptake was not limited by sulfur incorporation into proteins and *vice versa*. The uptake was sensitive to DNP and KCN, more so in the dark than in light; SeO_4^{2-} inhibited competitively in light and dark. KYLIN inferred a dependence on phosphorylation and estimated the degree of photoinhibition of oxidative phosphorylation. At higher sulfate concentrations a mediation by reducing equivalents (salt respiration) was not excluded (KYLIN, 1960a, c). In *Limnophila* a similar sensitivity of the ATP-level and sulfate uptake argues for a dependence on ATP (PENTH and WEIGL, 1969b, 1971) which also appears to power the sulfate uptake in *Elodea* (WEIGL, 1967).

3.4.5 Energetics of Ion Transport in Leaves of Terrestial Angiosperms

A linkage to cyclic photophosphorylation or oxidative phosphorylation in light or dark was suggested for the influx of K^+ in maize (RAINS, 1967, 1968), Rb^+ in barley (KHOLDEBARIN and OERTLI, 1972) and of K^+ and Na^+ in *Citrus* leaves (SMITH and ROBINSON, 1971). In maize and barley DCMU was ineffective or somewhat inhibitory while CCCP and DNP inhibited in both species more strongly in the dark than in the light (Table *4.4,* d). The same was found for azide and KCN which are known to be less inhibitory in photophosphorylation than in respiration or oxidative phosphorylation (LOSADA and ARNON, 1963). The light-dependent K^+ or Rb^+ uptake in maize or barley was not abolished under anaerobic conditions (RAINS, 1968; KHOLDEBARIN and OERTLI, 1972).

Quite analogous results were obtained for the Cl^- uptake in barley (KHOLDEBARIN and OERTLI, 1972). In *Citrus* (ROBINSON and SMITH, 1970) and *Tradescantia* leaves (LÜTTGE and BALL, 1971; JOHANSEN and LÜTTGE, 1974) and in leaves of C_3-plants, *Atriplex hastata, Spinacia* and *Oenothera* (LÜTTGE, 1970; LÜTTGE et al., 1971b) the Cl^- influx was sensitive to uncouplers and also to oligomycin (JOHANSEN and LÜTTGE, 1974). As light in air was not or variably stimulatory, oxidative or photophosphorylation were suggested as alternative energy sources. Where light inhibited the Cl^- influx in air, DCMU relieved this inhibition (LÜTTGE, 1970; JOHANSEN and LÜTTGE, 1974). Under anaerobic conditions the Cl^- influx was light-dependent and DCMU- or CMU-sensitive but less than photosynthesis as in hydrophyte leaves (ROBINSON and SMITH, 1970; JOHANSEN and LÜTTGE, 1974).

In two cases in C_4-plants and pea leaves, other energetic linkages were found. Firstly, LÜTTGE et al. (1970, 1971b) suggested for the C_4-plants *Amaranthus, Zea mays* and particularly *Atriplex spongiosa* a) a light-independent and ATP-linked KCl uptake, which was accompanied by Na^+ loss and H^+ uptake, and additionally b) a light-dependent Cl^- and K^+ influx, which was independent of ATP and more closely linked to noncyclic electron flow. The arguments for b) were as follows.

The light-dependent influx was inhibited by DCMU (but cf. RAINS, 1968, for maize) and by far-red light and not inhibited or even stimulated by the uncoupler FCCP. So the C_4-plants could be an example of similar energetic linkages as suggested for some giant algae (Part A, Chap. 6).

However, the apparent *stimulation* by FCCP was due to the calculation of the light-driven influx as the difference of the light and dark flux and to a stronger inhibition in the dark compared to light. The subtraction would be justified only if the dark processes occurred unchanged in the light (see *4.2.3.4.3*). Both the ATP level and the Cl^- influx in C_4- *and* C_3-plants (with exception of *Spinacia*) were more sensitive to FCCP in the dark than in the light (LÜTTGE et al., 1971a, b) and there is a reasonable correlation between the FCCP concentrations [I_{50}] effecting a 50% decrease in the ATP-level and the Cl^- influx in light and dark (Table *4.4*, c and d). The ratios of [I_{50}] in light and dark correspond to that of photo- and oxidative phosphorylation *in vitro* (HEYTLER and PRICHARD, 1962), although the ratios appear to be somewhat larger for C_4- than C_3-plants, especially *Spinacia*, which behaved quite variably also with respect to light-stimulation.

The results of LÜTTGE et al. (1970, 1971a, b) could be therefore consistent with an energy supply by ATP both in the light and dark in C_4- as in C_3-plants, although there appear to be quantitative differences between C_3- and C_4-plants.

Secondly NOBEL (1969a) suggested that the net Cl^- uptake of chopped pea leaves was independent of ATP and more closely linked to noncyclic electron transport. It was less sensitive to FCCP and somewhat more sensitive to DCMU than the net K^+ uptake (Table *4.4*, e). The properties of ion transport in chopped pea leaves differed, however, from other tissues as did the experimental conditions. Fluxes were measured in the absence of Ca^{2+} and in isotonic sucrose solution. This solution prevented a K^+ loss in the dark and increased the uptake in the light. Contrary to other leaves (RAINS, 1968, see *4.2.3.4.1*), the K^+ uptake was not light-saturated even at 9000 Lx and was stimulated by weak acid anions (NOBEL, 1969a, 1970), most strongly by HCO_3^- (cf. JYUNG et al., 1965). In the presence of HCO_3^- and extraordinarily high uptake rate of 56 $\mu mol\ g^{-1}\ h^{-1}\ K^+$ was measured.

Obviously these conditions do not mimic the natural environment of leaf cells, which is hypotonic and rich in Ca^{2+}. However, the observations may be relevant to the *control* of ion uptake possibly by the cell turgor (cf. GUTKNECHT, 1968). The K^+ uptake produced a considerable increase in the osmolarity of the cell sap (NOBEL, 1970) which will restore the turgor and may thus limit further uptake (see CRAM, Part A, *11.3.1.3*).

Net uptake as measured by NOBEL occurs in young leaves more so than in mature leaves (HORTON and BRUCE, 1972). Young expanding leaves hence appear to reward further studies to find out whether the energetic linkage of the Cl^- uptake to electron transport or reducing equivalents, as found by NOBEL, is predominant here.

Cation influxes in growing bean leaves have been measured by JACOBY and DAGAN (1969) and JACOBY et al. (1973), see *4.2.3.3.2.1* and Fig. *4.2*. Although light stimulated only the K^+ and Rb^+ and not the Na^+ influx, they all were inhibited by CCCP (Table *4.4*, f). Close examination showed the Na^+ influx to be inhibited by all treatments that decreased the ATP-level (STEINITZ and JACOBY, 1974). It was suggested therefore that ATP powers the Na^+ influx, consistent with the light-independent ATP-level and Na^+ influx, but complicating the question for the energy source of the light-stimulated K^+ influx. Although the energy-dependence of the

Na^+ fluxes may be due to omission of K^+ and linkage to an anion influx, the results illustrate the difficulties encountered in relating ion fluxes to energy sources.

3.4.6 Localization of the Light-Dependent Fluxes

As Φ_{ov} was measured by most authors (Table 4.4) ϕ_{oc} as well as ϕ_{cv} may be light-dependent or the stimulation may even be due to an inhibition of ϕ_{co} see 4.2.3.2. In his early work, ARISZ (1947b, 1948, 1958) concluded from his ingenious investigations that the net plasmalemma and tonoplast Cl^- influxes are light-dependent in *Vallisneria*. Briefly, his arguments were as follows. Light increases the net uptake to the vacuoles, which points to a light-dependent tonoplast influx, and it increases the net uptake into the symplasm (see Chap. 2 for the details) which shows that ϕ_{oc} is light-dependent too. Although these results do not prove a direct effect of light on ϕ_{cv} or the net flux $\phi_{cv} - \phi_{vc}$, as the uptake to the vacuole may be limited by ϕ_{oc}, ARISZ also found different sensitivities to inhibitors. While UO_2^{2+}, arsenate and CN^- inhibited the uptake to the cytoplasm, the accumulation in the vacuoles was more sensitive to azide and sometimes to DNP.

In leaves of *Elodea* the plasmalemma and tonoplast Cl^- influx ϕ_{oc} and ϕ_{cv} as measured by compartmental analysis were light-stimulated and inhibitor-sensitive, confirming the results of ARISZ, a direct effect of light on ϕ_{cv} was suggested only in some experiments and it has yet to be established (JESCHKE, 1971). Light also stimulated the K^+ influxes ϕ_{oc} and ϕ_{cv} (Table 4.5), but here the stimulation of ϕ_{cv} was due to the increase in the cytoplasmic content indicating that only ϕ_{oc} was directly light-dependent. Measurement of Na^+ fluxes yielded indirect indications for a light-stimulation of ϕ_{co}, that would have been expected on energetic grounds. However, if only ϕ_{co} were light-stimulated, this stimulation must be transitory and would escape direct detection in steady state measurements.

In aerial leaves, preliminary compartmental measurements showed a stimulation of the vacuolar K^+ efflux by Na^+ in *Atriplex* (OSMOND, 1968), the effect of light was not studied. In barley leaf slices the plasmalemma influx of Cl^- ϕ_{oc} was insensitive to CCCP while Φ_{ov} was inhibited which points to a linkage of ϕ_{cv} to phosphorylation at 5 and 40 mM Cl^- (PITMAN et al., 1974b).

As stated in 4.2.3.1, chloroplasts maintain high ionic concentrations within the cytoplasm (Table 4.3) and an obvious site of light-dependent fluxes is their envelope, as ion fluxes of chloroplasts have been found *in vitro* (for a review see PACKER et al., 1970) and *in vivo* (NOBEL, 1969b). Recently DE PHILIPPIS and PALLAGHY (1973) reported on preliminary studies of ion fluxes of chloroplasts in living *Elodea* cells, showing substantial light-dependent Cl^-, Na^+ and K^+ effluxes at high rates (330, 210 and 90 nmol m^{-2} s^{-1}). By these effluxes 12, 3 or 0.5% of the total cellular Cl^-, Na^+ or K^+ contents were released to the cytoplasm in the light. Although the conditions were different—the leaves were suspended in water for at least 2 h in this work—the data are particularly relevant in connection with the increased cytoplasmic K^+ and Cl^- (and decreased Na^+) content in *Elodea* leaves in the light (JESCHKE, 1971, Table 4.5).

3.5 Effects of Light on the Membrane Potential and H^+ Fluxes

The membrane potential of green cells in general is strongly affected by light (for reviews see Schilde, 1968; Bentrup, 1971; Higinbotham, 1973b). In leaf cells two types of light response have been observed, transient and steady changes in the resting potential. Light-*on* often results in a more or less fast transient depolarization which can be followed by a slow hyperpolarization towards the steady resting potential $_L\psi_{vo}$ in the light, light-*off* results in the opposite sequence of transient hyperpolarizations followed by a slow decrease in the negative potential towards the dark resting potential $_D\psi_{vo}$. In some aerial leaves of mosses (Sinclair, 1968; Lüttge and Pallaghy, 1969; Pallaghy and Lüttge, 1970) and higher plants (Osmond et al., 1969; Lüttge and Pallaghy, 1969; Brinckmann, 1973) only the transients have been observed, while in hydrophyte leaves (Jeschke, 1970b; Bentrup et al., 1973; Spanswick, 1973; Prins, 1974), in submerged thalli of *Riccia* (Felle and Bentrup, 1974) and occasionally in aerial leaves (Lüttge and Pallaghy, 1969) stable changes in the resting potential have been detected.

The values of $_L\psi_{vo}$ clearly exceeded the passive diffusion potentials (Bentrup et al., 1973; Spanswick, 1973 who found -296 mV for *Elodea canadensis* in the light) suggesting an electrogenic pump or pumps as sources of $_L\psi_{vo}$. This agrees to the finding that $_L\psi_{vo}$ is depolarized reversibly by low concentrations of inhibitors as CCCP, DCMU, CN^- or azide (Jeschke, 1970; Spanswick, 1973) although in *Vallisneria* $_L\psi_{vo}$ was unaffected by DCMU or DNP (Bentrup et al., 1973) and less sensitive to CCCP than the Cl^- and Rb^+ influxes (Prins, 1974). However, an electrogenic flux of 4 μmol m^{-2} s^{-1} has been estimated to be required in *Vallisneria* which by far exceeds the observed rates of mineral ion fluxes in leaves. An electrogenic Cl^- efflux is unlikely since $_L\psi_{vo}$ is also reached in $CaSO_4$ solution although at reduced rate (Jeschke, 1970b; Bentrup et al., 1973). An electrogenic proton extrusion has therefore been suggested as the source of $_L\psi_{vo}$ (Spanswick, 1973; Bentrup et al., 1973). A model of electrogenic proton extrusion has been proposed by Kitasato (1968) and by Spanswick (1973) (see Part A, *3.5.4.2; 4.2.6; 6.6.6.1*).

The transient potential changes are dependent on metabolic reactions too and not due to direct effects of light on the plasma membrane. They were attributed to the operation of PS II alone in *Atriplex* (Lüttge and Pallaghy, 1969) as they were suppressed in far-red light and inhibited by DCMU. In *Vallisneria* the transients depended on both photosystems (Bentrup et al., 1973). This was confirmed elegantly by Brinckmann (1973) and Brinckmann and Lüttge (1974), who showed that transients comparable to those of *Atriplex* were present in normal *Oenothera* leaves but absent in PS II- as well as PS I-deficient mutants of this species.

The transients have been attributed to light-dependent changes in the cytoplasmic pH (Lüttge and Pallaghy, 1969), although the pH-changes may be too small to account for the changes in membrane potential (Higinbotham, 1973b; Jeschke, 1970b) and a K^+ efflux from the chloroplasts may counteract the pH-changes (Lüttge and Pallaghy, 1969; Higinbotham, 1973b). As shown by Pallaghy and Lüttge (1970) the light-induced transients correspond to changes in the proton fluxes (see below) in *Atriplex*. Bentrup et al. (1973) suggested a connection between the transients and alteration in the cytoplasmic ATP pool.

Transient changes of the membrane potential of epidermal cells of *Potamogeton* leaves in response to light-dark and dark-light changes are dependent on the presence of low concentrations of CO_2 or HCO_3^- and were interpreted in terms of active transport of HCO_3^- and OH^- ions (Denny and Weeks, 1970).

In aquatic leaves, light-dependent apparent proton fluxes had been observed by early investigators, usually as an alkalinization of the external medium in the light due to withdrawal of CO_2 or to uptake of HCO_3^- in exchange for OH^- and the reverse in the dark (for a review see STEEMANN NIELSEN, 1960). Light-dependent apparent net H^+ influxes into leaves or leaf slices have been investigated recently (PALLAGHY and LÜTTGE, 1970; BRINCKMANN and LÜTTGE, 1972; HOPE et al., 1972; LÜTTGE, 1973a). As the ratio of H^+ uptake/O_2 release during photosynthesis was 1 (BRINCKMANN and LÜTTGE, 1972; HOPE et al., 1972) the question was whether the proton influx was due to photosynthetic CO_2 consumption alone or whether it reflected in part the H^+ fluxes within the chloroplasts as suggested by PALLAGHY and LÜTTGE (1970). In the presence of CCCP or pBQ (p-benzoquinone), or in N_2 HOPE et al. (1972) observed relatively small residual proton influxes (6–12 μmol g^{-1} h^{-1}) which were largely independent of CO_2 fixation. Although the interpretation of the H^+ fluxes in the presence of CCCP and pBQ was complicated by effects on H^+ permeability (HOPE et al., 1972), it was suggested that the proton fluxes in N_2 and also part of the control fluxes are linked to H^+ fluxes in the chloroplasts. This was concluded also from results with etiolated barley leaf slices in which high rates of H^+ influx and O_2 evolution independent of CO_2-fixation occurred during the first hours of greening (LÜTTGE, 1973a). Whereas these light-dependent proton fluxes and the corresponding opposite fluxes occurring after a light-dark change could induce the transients of the electropotential (PALLAGHY and LÜTTGE, 1970) an efflux of protons would be needed as the source of the hyperpolarized resting potential $_L\psi_{vo}$ and also as the primary event in the active uptake of ions in the models of ROBERTSON (1968) and SMITH (1970) (see Part A, Chap. 12). Considerable light-dependent net proton effluxes were found in pea and Atriplex leaves (NOBEL, 1969a; LÜTTGE et al., 1970).

As was pointed out by SMITH and LUCAS (1973), in measurements of the bulk external pH a proton extrusion can be masked by other proton fluxes; this will be particularly true for aquatic leaves as of Elodea and Potamogeton in which a spatial separation of regions with net proton efflux at the abaxial surface and apparent proton influx at the adaxial surface has been observed (4.2.3.2.2, see also Part A, 12.5.2.1). The differences in proton fluxes may result in an electrical potential difference across the leaf when the solution adjacent to the two surfaces are separated (HELDER and BOERMA, 1973). Separate measurements of ion fluxes at the two surfaces of such leaves may give important results with respect to the model of SMITH (1970) and may provide information for an understanding of the light-dependent changes in the resting potential.

4. Conclusions

The energetics of ion fluxes in leaves, though not fully known, certainly are different from those found in algal cells (Part A, Chap. 6). This may appear unexpected, as an ion transport process which has developed during evolution should be present in most species. But firstly the evolutionary distance between the coenocytic algal and higher plant cells is large and secondly quantitative differences are probable. For this reason experiments should be conducted with the aim of investigating separately the anion (and linked cation) transport and the cation (K^+-Na^+, K^+-H^+) exchange transport also in higher plant cells. This includes conditions which favor proton extrusion or else salt uptake as in growing tissues.

For green cells the same transport mechanisms should be operative at the cell membrane in the light and dark. The various observed effects of physiological conditions and inhibitors on ion fluxes could be due not only to different sensitivities of the species studied but also to metabolic changes and regulation in the cell in response to changed conditions (Fig. *4.5*). This implies that effects of inhibitors on living cells do not necessarily give an insight into the energy sources, particularly as the partial metabolic reactions may be linked, as for instance by the transfer of energy out of the chloroplasts. It is critical therefore to have knowledge of the transport mechanisms at the cellular membranes.

The localization of the energy-dependent fluxes require further investigations. On the ground of cellular economy, pumps for each ion or ion pair at all cellular membranes appear unlikely. Instead, changes at one membrane or in a compartment will reflect on other membranes and an effect of altered energy supplies on ion fluxes does not necessarily imply a direct dependence on energy. Further information about the effects of ion fluxes between major compartments on other fluxes are needed.

The dependence of ion fluxes of leaf cells on the developmental state will reward further studies. Here important insights can be anticipated as a separation of different developmental stages is possible by using leaves of different age while in the root the stages of development are continuously present in adjacent zones and can be separated only by mechanical means.

Attention must be paid to the obvious differences between submerged and aerial leaves. It appears significant that stable potential changes and substantial light-stimulations of ion fluxes were found in hydrophytes, while aerial leaves even of mosses showed mainly transient potential changes and smaller light-stimulations. This appears to be related to the different ecological or physiological situations to which the cells of these two types of leaves are exposed. It is intriguing to wonder whether the cytoplasmic phase of the hydrophyte leaf shows altered ionic contents in light and dark—on grounds of minimum energy expenditure—whereas the cytoplasm is more homeostatic in the aerial leaf cells.

References

ARENS, K.: Zur Kenntnis der Karbonatassimilation der Wasserpflanzen. Planta **10**, 814–816 (1930).

ARENS, K.: Physiologisch polarisierter Massenaustausch und Photosynthese bei submersen Wasserpflanzen I. Planta **20**, 621–658 (1933).

ARENS, K.: Physiologisch polarisierter Massenaustausch und Photosynthese bei submersen Wasserpflanzen II. Die $Ca(HCO_3)_2$ Assimilation. Jahrb. Wiss. Bot. **83**, 513–560 (1936).

ARISZ, W.H.: Uptake and transport of chlorine by parenchymatic tissue of leaves of *Vallisneria spiralis* I. The active uptake of chlorine; II. Analysis of the transport of chlorine; III. Discussion of the transport and uptake. Vacuole secretion theory. Proc. Koninkl. Ned. Akad. Wetenschap. **50**, 1019–1032, 1235–1245 (1947a, b); **51**, 25–32 (1948).

ARISZ, W.H.: Active uptake, vacuole secretion and plasmatic transport of chloride ions in leaves of *Vallisneria spiralis*. Acta Botan. Neerl. **1**, 506–516 (1953).

ARISZ, W.H.: Influence of inhibitors on the uptake and the transport of chloride ions in leaves of *Vallisneria spiralis*. Acta Botan. Neerl. **7**, 1–32 (1958).

ARISZ, W.H.: Influx and efflux by leaves of *Vallisneria* I. Active uptake and permeability. Protoplasma **57**, 5–26 (1963).

ARISZ, W.H., SOL, H.H.: Influence of light and sucrose on the uptake and transport of chloride in *Vallisneria* leaves. Acta Botan. Neerl. **5**, 218–247 (1956).

ATKINSON, M.R., FINDLAY, G.P., HOPE, A.B., PITMAN, M.G., SADDLER, H.D.W., WEST, K.R.: Salt regulation in the mangroves *Rhizophora mucronata* Lam. and *Aegialitis annulata* R. Br. Australian J. Biol. Sci. **20**, 589–599 (1967).

BENTRUP, F.W.: Zellphysiologie, Elektrophysiologie der Zelle. Fortschr. Botan. **33**, 51–61 (1971).

BENTRUP, F.W., GRATZ, H.J., UNBEHAUN, H.: The membrane potential of *Vallisneria* leaf cells: Evidence for light-dependent proton permeability changes. In: Ion transport in plants (W.P. ANDERSON, ed.), p. 171–182. London-New York: Academic Press 1973.

BERNSTEIN, L.: Method for determining solutes in the cell wall of leaves. Plant Physiol. **47**, 361–365 (1971).

BIANCHETTI, R.: Azione della luce sull'assorbimento salino da parte di tessuti clorofilliani in ambiente privo di CO_2. Giorn. Botan. Ital. **70**, 321–328 (1963).

BIELESKI, R.L.: Phosphate pools, phosphate transport and phosphate availability. Ann. Rev. Plant Physiol. **24**, 225–252 (1973).

BOLLARD, E.G.: The use of tracheal sap in the study of apple tree nutrition. J. Exptl. Bot. **4**, 363–368 (1953).

BOWEN, J.E.: Borate absorption in excised sugarcane leaves. Plant Cell Physiol. **9**, 467–478 (1968).

BOWEN, J.E.: Absorption of copper, zinc and manganese by sugarcane leaf tissue. Plant Physiol. **44**, 255–261 (1969).

BRAUNER, L.: Die Beeinflussung des Stoffaustausches durch Licht. In: Encyclopedia of plant physiology (W. RUHLAND, ed.) vol. II, p. 381–397. Berlin-Heidelberg-New York: Springer 1956.

BRINCKMANN, E.: Zur Messung des Membranpotentials und dessen lichtabhängigen Änderungen an Blattzellen höherer Landpflanzen. Diss. Darmstadt (1973).

BRINCKMANN, E., LÜTTGE, U.: Vorübergehende pH-Änderungen im umgebenden Medium intakter grüner Zellen bei Beleuchtungswechsel. Z. Naturforsch. **27b**, 277–284 (1972).

BRINCKMANN, E., LÜTTGE, U.: Lichtabhängige Membranpotentialschwankungen und deren interzelluläre Weiterleitung bei panaschierten Photosynthese-Mutanten von *Oenothera*. Planta **119**, 47–57 (1974).

BRISTOW, I.M., WHITCOMBE, M.: The role of roots in the nutrition of aquatic vascular plants. Amer. J. Bot. **58**, 8–13 (1971).

COLLANDER, R.: Die Electrolyt-Permeabilität und Salzakkumulation pflanzlicher Zellen. In: Tabulae biologicae (H. DENZER, V.J. KONINGSBERGER, H.J. VONK, eds.), vol. 19/2, p. 313–333. Den Haag: Junk 1941.

CROWDY, S.H., TANTON, T.W.: Water pathways in higher plants I. Free space in wheat leaves. J. Exptl. Bot. **21**, 102–111 (1970).

DENNY, P., WEEKS, D.C.: Effects of light and bicarbonate on membrane potential in *Potamogeton schweinfurthii* (Benn.). Ann. Bot. (London), N.S. **34**, 483–496 (1970).

EPSTEIN, E.: The essential role of calcium in selective cation transport by plant cells. Plant Physiol. **36**, 437–444 (1961).

EPSTEIN, E.: Dual pattern of ion absorption by plant cells and by plants. Nature **212**, 1324–1327 (1966).

FELLE, H., BENTRUP, F.W.: Light-dependent changes of the membrane potential and conductance in *Riccia fluitans*. In: Membrane transport in plants (J. DAINTY, U. ZIMMERMANN, eds.). Berlin-Heidelberg-New York: Springer 1974.

FOOTE, B.D., HANSON, J.B.: Ion uptake by soybean root tissue depleted of calcium by ethylenediaminetetraacetic acid. Plant Physiol. **39**, 450–460 (1964).

FRANKE, W.: Mechanisms of foliar penetration of solutes. Ann. Rev. Plant Physiol. **18**, 281–300 (1967).

GESSNER, F.: Untersuchungen über die Nitrataufnahme der Wasserpflanzen. Intern. Rev. Ges. Hydrobiol. Hydrogr. **43**, 211–224 (1943).

Gimmler, H., Urbach, W., Jeschke, W.D., Simonis, W.: Die unterschiedliche Wirkung von Disalicylidenpropandiamin auf die cyclische und nichtcyclische Photophosphorylierung *in vivo* sowie auf die ^{14}C-Markierung einzelner Photosyntheseprodukte. Z. Pflanzenphysiol. **58**, 353–364 (1968).

Good, N.E., Izawa, S.: Inhibition of photosynthesis. In: Metabolic inhibitors (R.M. Hochster, J.H. Quastel, eds.), vol. IV, p. 179–214. New York: Academic Press 1973.

Greenway, H.: Plant responses to saline substrates IV. Chloride uptake by *Hordeum vulgare* as affected by inhibitors, transpiration and nutrients in the medium. Australian J. Biol. Sci. **18**, 249–268 (1965).

Greenway, H., Pitman, M.G.: Potassium retranslocation in seedlings of *Hordeum vulgare*. Australian J. Biol. Sci. **18**, 235–247 (1965).

Grube, K.H.: Über den Zusammenhang von Phosphathaushalt und Photosynthese bei *Elodea densa*. Planta **42**, 279–303 (1953).

Grünsfelder, M.: Die Kinetik der Phosphataufnahme bei *Elodea densa*. Diss. Würzburg (1971).

Grünsfelder, M., Simonis, W.: Aktive und inaktive Phosphataufnahme in Blattzellen von *Elodea densa* bei hohen Phosphat-Außenkonzentrationen. Planta **115**, 173–186 (1973).

Gutknecht, J.: Salt transport in Valonia: Inhibition of potassium uptake by small hydrostatic pressures. Science **160**, 68–70 (1968).

Healey, F.P., Myers, J.: The Kok effect in *Chlamydomonas reinhardi*. Plant Physiol. **47**, 373–379 (1971).

Heber, U.: Conformational changes of chloroplasts induced by illumination of leaves *in vivo*. Biochim. Biophys. Acta **180**, 302–319 (1969).

Heber, U.: Flow of metabolites and compartmentation phenomena in chloroplasts. In: Transport and distribution of matter in cells of higher plants (K. Mothes, E. Müller, A. Nelles, D. Neumann, eds.), vol. b, p. 151–184. Berlin: Akademieverlag 1970.

Heber, U.: Stoichiometry of reduction and phosphorylation during illumination of intact chloroplasts. Biochim. Biophys. Acta **305**, 140–152 (1973a).

Heber, U.: Elektronentransport zum Sauerstoff und ATP-Verbrauch in der Photosynthese. Ber. Deut. Botan. Ges. **86**, 187–196 (1973b).

Heber, U., Krause, G.: Transfer of carbon, phosphate energy, and reducing equivalents across the chloroplast envelope. In: Photosynthesis and photorespiration (M.D. Hatch, C.B. Osmond, R.O. Slatyer, eds.). New York-London-Sydney-Toronto: Wiley and Sons 1971.

Heber, U., Santarius, K.A.: Compartmentation and reduction of pyridine nucleotides in relation to photosynthesis. Biochim. Biophys. Acta **109**, 390–408 (1965).

Heber, U., Santarius, K.A.: Direct and indirect transfer of ATP and ADP across the chloroplast envelope. Z. Naturforsch. **25b**, 718–728 (1970).

Helder, R.J.: Translocation in *Vallisneria spiralis*. In: Encyclopedia of plant physiology (W. Ruhland, ed.), vol. XIII, p. 20–43. Berlin-Heidelberg-New York: Springer 1967.

Helder, R.J., Boerma, J.: Polar transport of labelled rubidium ions across the leaf of *Potamogeton lucens*. Acta Botan. Neerl. **21**, 211–218 (1972).

Helder, R.J., Boerma, J.: Exchange and polar transport of rubidium ions across the leaves of *Potamogeton*. Acta Botan. Neerl. **22**, 686–693 (1973).

Heldt, H.W.: Adenine nucleotide translocation in spinach chloroplasts. F.E.B.S. Letters **5**, 11–17 (1969).

Heldt, H.W., Rapley, L.: Specific transport of inorganic phosphate, 3-phosphoglycerate and dihydroxyacetonephosphate, and of dicarboxylates across the inner membrane of spinach chloroplasts. F.E.B.S. Letters **10**, 143–148 (1970).

Heller, R., Crignon, C., Scheidecker, D.: Study of the efflux and influx of potassium in cell suspensions of *Acer pseudoplatanus* and leaf fragments of *Hedera helix*. In: Ion transport in plants (W.P. Anderson, ed.). London-New York: Academic Press 1973.

Heytler, P.G., Prichard, W.W.: A new class of uncoupling agents—carbonyl cyanide phenylhydrazones. Biochem. Biophys. Res. Commun. **7**, 272–275 (1962).

Higinbotham, N.: The mineral absorption process in plants. Botan. Rev. **39**, 15–69 (1973a).

Higinbotham, N.: Electropotentials of plant cells. Ann. Rev. Plant Physiol. **24**, 25–46 (1973b).

Higinbotham, N., Etherton, B., Foster, R.J.: Mineral ion contents and cell transmembrane electropotentials of pea and oat seedling tissue. Plant Physiol. **42**, 37–46 (1967).

Hill, B.S., Hill, A.E.: ATP-driven chloride pumping and ATPase activity in the *Limonium* salt gland. J. Membrane Biol. **12**, 145–158 (1973a).

HILL, B.S., HILL, A.E.: Enzymatic approaches to the chloride transport in the *Limonium* salt gland. In: Ion transport in plants (W.P. ANDERSON, ed.), p. 379–384. London-New York: Academic Press 1973 b.

HOCH, G.E., OWENS, O.V.H., KOK, B.: Photosynthesis and respiration. Arch. Biochem. Biophys. **101**, 171–180 (1963).

HODGES, T.K.: Oligomycin inhibition of ion transport in plant roots. Nature **209**, 425–426 (1966).

HOPE, A.B., LÜTTGE, U., BALL, E.: Photosynthesis and apparent proton fluxes in *Elodea canadensis*. Z. Pflanzenphysiol. **68**, 73–81 (1972).

HORTON, R.F., BRUCE, K.R.: Inhibition by abscisic acid of the light and dark uptake of potassium by slices of *Vicia faba* leaves. Canad. J. Bot. **50**, 1915–1917 (1972).

ILAN, I., GILAD, T., REINHOLD, L.: Specific effects of kinetin on the uptake of monovalent cations by sunflower cotyledons. Physiol. Plantarum **24**, 337–341 (1971).

INGOLD, C.T.: The effect of light on the absorption of salts by *Elodea canadensis*. New Phytologist **35**, 132–141 (1936).

IZAWA, S., WINGET, G.D., GOOD, N.E.: Phlorizin, a specific inhibitor of photophosphorylation and phosphorylation-coupled electron transport in chloroplasts. Biochem. Biophys. Res. Commun. **22**, 223–226 (1966).

JACOBY, B., ABAS, S., STEINITZ, B.: Rubidium and potassium absorption by bean-leaf slices compared to sodium absorption. Physiol. Plantarum **28**, 209–214 (1973).

JACOBY, B., DAGAN, J.: A comparison of two methods of investigating sodium uptake by bean-leaf cells and the vitality of isolated leaf cells. Protoplasma **64**, 325–329 (1967).

JACOBY, B., DAGAN, J.: Effects of age on sodium fluxes in primary bean leaves. Physiol. Plantarum **22**, 29–36 (1969).

JACOBY, B., PLESSNER, ORA, E.: Some aspects of chloride absorption by bean leaf tissue. Ann. Bot. (London), N.S. **34**, 177–182 (1970).

JESCHKE, W.D.: Die cyclische und die nichtcyclische Photophosphorylierung als Energiequellen der lichtabhängigen Chloridionenaufnahme bei *Elodea*. Planta **73**, 161–174 (1967).

JESCHKE, W.D.: Der Influx von Kaliumionen bei Blättern von *Elodea densa*, Abhängigkeit vom Licht, von der Kaliumkonzentration und von der Temperatur. Planta **91**, 111–128 (1970a).

JESCHKE, W.D.: Lichtabhängige Veränderungen des Membranpotentials bei Blattzellen von *Elodea densa*. Z. Pflanzenphysiol. **62**, 158–172 (1970b).

JESCHKE, W.D.: Über die Verwendung von ^{86}Rb als Indikator für Kalium, Untersuchungen am lichtgeförderten ^{42}K/K und ^{86}Rb/Rb-Influx bei *Elodea densa*. Z. Naturforsch. **25b**, 624–630 (1970c).

JESCHKE, W.D.: Energetic linkages of individual ion fluxes in leaf cells of *Elodea densa*. In: First European Biophysics Congress (E. BRODA, A. LOCKER, H. SPRINGER-LEDERER, eds.). Wien: Verlag Wiener Med. Akad. 1971.

JESCHKE, W.D.: The effect of the inhibitor of photophosphorylation Dio-9 and the uncoupler atebrin on the light-dependent Cl$^-$ influx of *Elodea densa*: direct inhibition of membrane transport? Z. Pflanzenphysiol. **66**, 379–408 (1972a).

JESCHKE, W.D.: The effect of DNP and CCCP on photosynthesis and light-dependent Cl$^-$ influx in *Elodea densa*. Z. Pflanzenphysiol. **66**, 409–419 (1972b).

JESCHKE, W.D.: Über den licht-geförderten Influx von Ionen in Blättern von *Elodea densa*. Vergleich der Influxe von K$^+$- und Cl$^-$-Ionen. Planta **103**, 164–180 (1972c).

JESCHKE, W.D., SIMONIS, W.: Über die Aufnahme von Phosphat- und Sulfationen durch Blätter von *Elodea densa* und ihre Beeinflussung durch Licht, Temperatur und Außenkonzentration. Planta **67**, 6–32 (1965).

JESCHKE, W.D., SIMONIS, W.: Effect of CO$_2$ on photophosphorylation *in vivo* as revealed by the light-dependent Cl$^-$ uptake in *Elodea densa*. Z. Naturforsch. **22b**, 873–876 (1967).

JESCHKE, W.D., SIMONIS, W.: Über die Wirkung von CO$_2$ auf die lichtabhängige Cl$^-$-Aufnahme bei *Elodea densa*: Regulation zwischen nichtcyclischer und cyclischer Photophosphorylierung. Planta **88**, 157–171 (1969).

JOHANSEN, C., LÜTTGE, U.: Respiration and photosynthesis as alternative energy sources for chloride uptake by *Tradescantia albiflora* leaf cells. Z. Pflanzenphysiol. **71**, 189–199 (1974).

JONES, H.G., OSMOND, C.B.: Photosynthesis by thin leaf slices in solution. Australian J. Biol. Sci. **26**, 15–24 (1973).

JYUNG, W.H., WITTWER, S.H., BUKOVAC, M.J.: Ion uptake by cells enzymatically isolated from green tobacco leaves. Plant Physiol. **40**, 410–414 (1965).

KAISER, W.: Endogene Photophosphorylierung in isolierten Chloroplasten. Diss. Würzburg (1973).

KAISER, W., URBACH, W.: Endogene cyclische Photophosphorylierung in isolierten Chloroplasten. Ber. Deut. Botan. Ges. **86**, 213–226 (1973).

KANNAN, S.: Course of cation accumulation by leaf tissue in *Phaseolus vulgaris* L. Experientia **26**, 552 (1970).

KANNAN, S., WITTWER, S.H.: Absorption of iron by enzymatically isolated leaf cells. Physiol. Plantarum **20**, 911–919 (1967).

KAPPEN, L., ULLRICH, W.: Verteilung von Chlorid und Zuckern in Blattzellen halophiler Pflanzen bei verschieden hoher Frostresistenz. Ber. Deut. Botan. Ges. **83**, 265–275 (1970).

KHOLDEBARIN, B., OERTLI, J.J.: Changes of organic acids during salt uptake by barley leaf tissues under light and dark conditions. Z. Pflanzenphysiol. **62**, 237–242 (1970).

KHOLDEBARIN, B., OERTLI, J.J.: Effects of metabolic inhibitors on salt uptake and organic acid synthesis by leaf tissues in the light and in the dark. Z. Pflanzenphysiol. **66**, 352–358 (1972).

KITASATO, H.: The influence of H^+ on the membrane potential and ion fluxes in *Nitella*. J. Gen. Physiol. **52**, 60–87 (1968).

KLEPPER, B., KAUFMANN, M.R.: Removal of salt from the xylem sap by leaves and stems of guttating plants. Plant Physiol. **41**, 1743–1747 (1966).

KOK, B.: The interrelation of respiration and photosynthesis in green plants. Biochim. Biophys. Acta **3**, 625–631 (1949).

KRAMER, P.J.: Plant and soil water relationships—a modern synthesis. New York: McGraw-Hill 1969.

KYLIN, A.: The apparent free space of *Vallisneria leaves*. Physiol. Plantarum **10**, 732–740 (1957).

KYLIN, A.: The accumulation of sulphate in isolated leaves as affected by light and darkness. Botan. Notiser **113**, 49–81 (1960a).

KYLIN, A.: The influence of the external osmotic conditions upon the accumulation of sulphate in leaves. Physiol. Plantarum **13**, 148–154 (1960b).

KYLIN, A.: The incorporation of radio-sulphur from external sulphate into different sulphur fractions of isolated leaves. Physiol. Plantarum **13**, 366–379 (1960c).

KYLIN, A.: The apparent free space of green tissues. Physiol. Plantarum **13**, 385–397 (1960d).

LÄUCHLI, A., LÜTTGE, U.: Untersuchung zur Kinetik der Ionenaufnahme in das Cytoplasma von *Mnium*-Blattzellen mit Hilfe der Mikroradioautographie und der Röntgenmikrosonde. Planta **83**, 80–98 (1968).

LARKUM, A.W.D.: Ionic relations of chloroplasts *in vivo*. Nature **218**, 447–449 (1968).

LARKUM, A.W.D., HILL, A.E.: Ion and water transport in *Limonium* V. The ionic status of chloroplasts in the leaf of *Limonium vulgare* in relation to the activity of salt glands. Biochim. Biophys. Acta **203**, 133–138 (1970).

LOOKEREN CAMPAGNE, R.N. VAN: The action spectrum of the influence of light on chloride absorption in *Vallisneria* leaves. Proc. Koninkl. Ned. Akad. Wetenshap. C **60**, 70–76 (1957a).

LOOKEREN CAMPAGNE, R.N. VAN: Light-dependent chloride absorption in *Vallisneria* leaves. Acta Botan. Neerl. **6**, 543–582 (1957b).

LOSADA, M., ARNON, D.I.: Selective inhibitors of photosynthesis. In: Metabolic inhibitors (R.M. HOCHSTER, J.H. QUASTEL, eds.), vol. II, p. 559–593. New York-London: Academic Press 1963.

LOWENHAUPT, B.: The transport of calcium and other cations in submerged aquatic plants. Biol. Rev. **31**, 371–395 (1956).

LOWENHAUPT, B.: Active cation transport in submerged aquatic angiosperms I. The effect of light upon the absorption and excretion of calcium by *Potamogeton crispus* L. leaves. J. Cell. Comp. Physiol. **51**, 199–208 (1958a).

LOWENHAUPT, B.: Active cation transport in submerged aquatic angiosperms II. Effect of aeration upon the equilibrium content of calcium in *Potamogeton crispus* L. leaves. J. Cell. Comp. Physiol. **51**, 209–219 (1958b).

LUNDEGÅRDH, H.: Salt and respiration. Nature **185**, 70–74 (1960).

LÜTTGE, U.: Aktiver Transport (Kurzstreckentransport) bei Pflanzen. Protoplasmatologia, vol. VIII, 7b. Wien-New York: Springer 1969.

LÜTTGE, U.: Die photosynthese-abhängige Ionenaufnahme durch die grünen Zellen von Luft-Blättern höherer Pflanzen. Ber. Deut. Botan. Ges. **83**, 473–479 (1970).

LÜTTGE, U.: Proton and chloride uptake in relation to the development of photosynthetic capacity in greening etiolated barley leaves. In: Ion transport in plants (W.P. ANDERSON, ed.), p. 205–221. London-New York: Academic Press 1973a.

LÜTTGE, U.: Stofftransport der Pflanzen. Berlin-Heidelberg-New York: Springer 1973b.

LÜTTGE, U., BALL, E.: Light-independent uncoupler-sensitive ion uptake by green and by pale cells of variegated leaves of higher plants in relation to protein content and chloroplast integrity. Z. Naturforsch. **26b**, 158–161 (1971).

LÜTTGE, U., BALL, E.: Ion uptake by slices from greening etiolated barley and maize leaves. Plant Science Letters **1**, 275–280 (1973).

LÜTTGE, U., BALL, E., WILLERT, K. VON: Gas exchange and ATP levels of green cells of leaves of higher plants as affected by FCCP and DCMU in *in vitro* experiments. Z. Pflanzenphysiol. **65**, 326–335 (1971a).

LÜTTGE, U., BALL, E., WILLERT, K. VON: A comparative study of the coupling of ion uptake to light reactions in leaves of higher plant species having the C_3- and C_4-pathway of photosynthesis. Z. Pflanzenphysiol. **64**, 336–350 (1971b).

LÜTTGE, U., BAUER, K.: Evaluation of ion uptake isotherms and analysis of individual fluxes of ions. Planta **80**, 52–64 (1968).

LÜTTGE, U., PALLAGHY, C.K.: Light-triggered transient changes of membrane potentials in green cells in relation to photosynthetic electron transport. Z. Pflanzenphysiol. **61**, 58–67 (1969).

LÜTTGE, U., PALLAGHY, C.K., OSMOND, C.B.: Coupling of ion transport in green cells of *Atriplex spongiosa* leaves to energy sources in the light and in the dark. J. Membrane Biol. **2**, 17–30 (1970).

LÜTTGE, U., SCHÖCH, E.V., BALL, E.: Can externally applied ATP supply energy to active ion transport mechanisms of intact plant cells? Australian J. Plant Physiol. **1**, 211–220 (1974).

MACDONALD, I.R.: Effect of vacuum infiltration on photosynthetic gas exchange in leaf tissue. Plant Physiol. **56**, 109–112 (1975).

MACDONALD, I.R., MACKLON, A.E.S.: Anion absorption by etiolated wheat leaves after vacuum infiltration. Plant Physiol. **49**, 303–306 (1972).

MACDONALD, I.R., MACKLON, A.E.S.: Light-enhanced chloride uptake by wheat laminae. A comparison of chopped and vacuum-infiltrated tissue. Plant Physiol. **56**, 105–108 (1975).

MACROBBIE, E.A.C.: The nature of the coupling between light energy and active ion transport in *Nitella translucens*. Biochim. Biophys. Acta **94**, 64–73 (1965).

MACROBBIE, E.A.C.: The active transport of ions in plant cells. Quart. Rev. Biophys. **3**, 251–294 (1970).

MARRÉ, E., FORTI, G., BIANCHIETTI, R., PARISI, B.: Utilization of photosynthetic chemical energy for metabolic processes different from CO_2 fixation. In: La Photosynthese **119**, 557–570 (1963).

MCCARTHY, R.J., GUILLORY, R.J., RACKER, E.: Dio-9, an inhibitor of coupled electron transport and phosphorylation in chloroplasts. J. Biol. Chem. **240**, 4822–4823 (1965).

MCFARLANE, J.C., BERRY, W.L.: Cation penetration through isolated leaf cuticles. Plant Physiol. **53**, 723–727 (1974).

MORROD, R.S.: A new method for measuring the permeability of plant cell membranes using epidermis-free leaf discs. J. Exptl. Bot. **25**, 521–533 (1974).

NETTER, H.: Theoretische Biochemie. Berlin-Göttingen-Heidelberg: Springer 1959.

NEUMANN, J., JAGENDORF, A.T.: Dinitrophenol as an uncoupler of photosynthetic phosphorylation. Biochem. Biophys. Res. Commun. **16**, 562–567 (1964).

NISSEN, P.: Uptake of sulfate by roots and leaf slices of barley: Mediated by single, multiphasic mechanisms. Physiol. Plantarum **24**, 315–324 (1971).

NOBEL, P.S.: Light-dependent potassium absorption by *Pisum sativum* leaf fragments. Plant Cell Physiol. **10**, 597–605 (1969a).

NOBEL, P.S.: Light-induced changes in the ionic content of chloroplasts in *Pisum sativum*. Biochim. Biophys. Acta **172**, 134–143 (1969b).

NOBEL, P.S.: Relation of light-dependent potassium uptake by pea leaf fragments to the pK of the accompanying organic acid. Plant Physiol. **46**, 491–493 (1970).

OSMOND, C.B.: Ion absorption in *Atriplex* leaf tissue I. Absorption by mesophyll cells. Australian J. Biol. Sci. **21**, 1119–1130 (1968).

Osmond, C.B., Lüttge, U., West, K.R., Pallaghy, C.K., Shacher-Hill, B.: Ion absorption in *Atriplex* leaf tissue II. Secretion of ions to epidermal bladders. Australian J. Biol. Sci. **22**, 797–814 (1969).

Packer, L., Murakami, S., Mehard, C.W.: Ion transport in chloroplasts and plant mitochondria. Ann. Rev. Plant Physiol. **21**, 271–304 (1970).

Pallaghy, C.K.: Electron probe microanalysis of potassium and chloride in freeze-substituted leaf sections of *Zea mays*. Australian J. Biol. Sci. **26**, 1015–1034 (1973).

Pallaghy, C.K., Lüttge, U.: Light-induced H^+-ion fluxes and bioelectric phenomena in mesophyll cells of *Atriplex spongiosa*. Z. Pflanzenphysiol. **62**, 417–425 (1970).

Penth, B., Weigl, J.: Unterschiedliche Wirkung von Licht auf die Aufnahme von Chlorid und Sulfat in *Limnophila*. Abhängigkeit der Lichtwirkung von der Konzentration der Anionen. Z. Naturforsch. **24b**, 342–348 (1969a).

Penth, B., Weigl, J.: Wirkung von CCCP auf Anionenaufnahme und ATP-Spiegel in *Limnophila gratioloides* und *Chara foetida*. Z. Naturforsch. **24b**, 1668–1669 (1969b).

Penth, B., Weigl, J.: Anionen-Influx, ATP-Spiegel und CO_2-Fixierung in *Limnophila gratioloides* und *Chara foetida*. Planta **96**, 212–223 (1971).

Perrin, A.: Contribution a l' étude de l'organisation et du fonctionnement des hydathodes: Recherches anatomiques, ultrastructurales et physiologiques. Thesis, Lyon 1972.

Philippis, L.F. de, Pallaghy, C.K.: Effect of light on the volume and ion relations of chloroplasts in detached leaves of *Elodea densa*. Australian J. Biol. Sci. **26**, 1251–1265 (1973).

Pierce, W.S., Higinbotham, N.: Compartments and fluxes of K^+, Na^+, and Cl^- in *Avena* coleoptile cells. Plant Physiol. **46**, 666–673 (1970).

Pitman, M.G.: Sodium and potassium uptake by seedlings of *Hordeum vulgare*. Australian J. Biol. Sci. **18**, 10–24 (1965).

Pitman, M.G., Courtice, A.C., Lee, B.: Comparison of potassium and sodium uptake by barley roots at high and low salt status. Australian J. Biol. Sci. **21**, 871–881 (1968).

Pitman, M.G., Lüttge, U., Kramer, D., Ball, E.: Free space characteristics of barley leaf slices. Australian J. Plant Physiol. **1**, 65–75 (1974a).

Pitman, M.G., Lüttge, U., Läuchli, A., Ball, E.: Ion uptake to slices of barley leaves, and regulation of K content in cells of the leaves. Z. Pflanzenphysiol. **72**, 75–88 (1974b).

Pratt, M.J., Matthews, R.E.F.: Non-uniformities in the metabolism of excised leaves and leaf discs. Planta **99**, 21–36 (1971).

Prins, H.B.A.: The effect of DCMU on ion uptake and photosynthesis in leaves of *Vallisneria spiralis* L. Acta Botan. Neerl. **19**, 813–820 (1970).

Prins, H.B.A.: The action spectrum of photosynthesis and the rubidiumchloride uptake by leaves of *Vallisneria spiralis*. Proc. Koninkl. Ned. Akad. Wetenschap. C **76**, 495–499 (1973).

Prins, H.B.A.: Photosynthesis and ion uptake in leaves of *Vallisneria spiralis* L., Thesis Groningen (1974).

Rains, D.W.: Light-enhanced potassium absorption by corn leaf tissue. Science **156**, 3780 (1967).

Rains, D.W.: Kinetics and energetics of light-enhanced potassium absorption by corn leaf tissue. Plant Physiol. **43**, 394–400 (1968).

Rains, D.W., Epstein, E.: Preferential absorption of potassium by leaf tissue of the mangrove *Avicennia marina*: an aspect of halophytic competence in coping with salt. Australian J. Biol. Sci. **20**, 847–857 (1967).

Ramati, A., Eshel, A., Liphschitz, N., Waisel, Y.: Localization of ions in cells of *Potamogeton lucens* L. Experientia **29**, 497–501 (1973).

Raven, J.A.: Photosynthesis and light-stimulated ion transport in *Hydrodictyon africanum*. In: Transport and distribution of matter in cells of higher plants (K. Mothes, E. Müller, A. Nelles, D. Neumann, eds.), vol. a, p. 145–152. Berlin: Akademieverlag 1968.

Robertson, R.N.: Ion transport and respiration. Biol. Rev. **35**, 231–264 (1960).

Robertson, R.N.: Protons, electrons, phosphorylation and active transport. Cambridge: Cambridge University Press 1968.

Robinson, J.B.: Salinity and the whole plant. In: Salinity and water use (T. Talsma, J.R. Philip, eds.), p. 193–206 (1971).

Robinson, J.B., Smith, F.A.: Chloride influx into *Citrus* leaf slices. Australian J. Biol. Sci. **23**, 953–960 (1970).

SCHEIDECKER, D., JAQUES, R., ANDREOPOULOS-RENAUD, U., CONNAN, A.: Effet de la lumière sur l'absorption simultanée du calcium par des fragments des feuilles de lierre panaché. Compt. Rend. D **272**, 3142–3145 (1971).

SCHILDE, C.: Zellphysiologie, Elektrophysiologie der Zelle. Fortschr. Botan. **30**, 43–56 (1968).

SCHÖCH, E.V., LÜTTGE, U.: Zur Entstehung einer lichtabhängigen Komponente der Kationenaufnahme bei Blattgewebestreifen mit zunehmendem Zeitabstand von der Präparation. Biochem. Physiol. Pflanzen **165**, 345–350 (1974).

SCHOLANDER, P.F., BRADSTREET, E.D., HAMMEL, H.T., HEMMINGSON, E.A.: Sap concentrations in halophytes and some other plants. Plant Physiol. **41**, 529–532 (1966).

SCULTHORPE, C.D.: The biology of aquatic vascular plants. London: Arnold 1967.

SHTARKSHALL, R.A., REINHOLD, L., HASSADAH, H.: Transport of amino acids in barley leaf tissue I. Evidence for a specific uptake mechanism and the influence of "ageing" on accumulation capacity. J. Exptl. Bot. **21**, 915–925 (1970).

SIMONIS, W., GRUBE, K.H.: Untersuchungen über den Zusammenhang von Phosphathaushalt und Photosynthese. Z. Naturforsch. **7b**, 194 (1952).

SIMONIS, W., URBACH, W.: Photophosphorylation in vivo. Ann. Rev. Plant Physiol. **24**, 89–114 (1973).

SINCLAIR, J.: Nernst potential measurements on the leaf cells of the moss, *Hookeria lucens*. J. Exptl. Bot. **18**, 594–599 (1967).

SINCLAIR, J.: The influence of light on the ion fluxes and electrical potential of the leaf cells of the moss, *Hookeria lucens*. J. Exptl. Bot. **19**, 254–263 (1968).

SMITH, C.R., EPSTEIN, E.: Ion absorption by shoot tissue: technique and first findings with excised leaf tissue of corn. Plant Physiol. **39**, 338–341 (1964a).

SMITH, C.R., EPSTEIN, E.: Ion absorption by shoot tissue: kinetics of potassium and rubidium absorption by corn leaf tissue. Plant Physiol. **39**, 992–996 (1964b).

SMITH, F.A.: The mechanism of chloride transport in *Characean* cells. New Phytologist **69**, 903–917 (1970).

SMITH, F.A., LUCAS, W.J.: The role of H^+ and OH^- fluxes in the ionic relations of *Characean* cells. In: Ion transport in plants (W.P. ANDERSON, ed.). London-Heidelberg-New York: Springer 1973.

SMITH, F.A., ROBINSON, J.B.: Sodium and potassium influx into *Citrus* leaf slices. Australian J. Biol. Sci. **24**, 861–871 (1971).

SMITH, F.A., WEST, K.R.: A comparison of the effects of metabolic inhibitors on chloride uptake and photosynthesis in *Chara australis*. Australian J. Biol. Sci. **22**, 351–363 (1969).

SOL, H.H.: Pretreatment and chloride uptake in *Vallisneria leaves*. Acta Botan. Neerl. **7**, 131–173 (1958).

SPANSWICK, R.M.: Electrogenesis in photosynthetic tissues. In: Ion transport in plants (W.P. ANDERSON, ed.), p. 113–128. London-New York: Academic Press 1973.

STEEMANN NIELSEN, E.: Passive and active transport during photosynthesis in water plants. Physiol. Plantarum **4**, 189–198 (1951).

STEEMANN NIELSEN, E.: Uptake of CO_2 by the plant. In: Encyclopedia of plant physiology (W. RUHLAND, ed.), vol. V, pt. 1, p. 70–84. Berlin-Heidelberg-New York: Springer 1960.

STEINITZ, B., JACOBY, B.: Energetics of $^{22}Na^+$ absorption by bean-leaf slices. Ann. Bot. (London), N.S. **38**, 453–457 (1974).

STENLID, G.: Salt losses and redistribution of salts in higher plants. In: Encyclopedia of plant physiology, vol. IV, p. 615–637. Berlin-Heidelberg-New York: Springer 1958.

STEVENINCK, R.F.M. VAN, CHENOWETH, A.R.F.: Ultrastructural localization of ions. I. Effect of high external sodium chloride concentrations on the apparent distribution of chloride in leaf parenchyma cells of barley seedlings. Australian J. Biol. Sci. **25**, 499–516 (1972).

STEWARD, F.C., MOTT, R.L.: Cells, solutes, and growth: salt accumulation in plants reexamined. Intern. Rev. Cytol. **28**, 275–370 (1970).

STEWARD, F.C., SUTCLIFFE, J.F.: Plants in relation to inorganic salts. In: Plant physiology (F.C. STEWARD, ed.), vol. II, p. 253–478. New York-London: Academic Press 1959.

STOCKING, C.R., LARSON, S.: A chloroplast cytoplasmic shuttle and the reduction of extraplastic NAD. Biochem. Biophys. Res. Commun. **37**, 278–282 (1969).

STOCKING, C.R., ONGUN, A.: The intracellular distribution of some metallic elements in leaves. Amer. J. Bot. **29**, 284–289 (1962).

Stokes, M.D., Walker, D.A.: Relative inpermeability of the intact chloroplast envelope to ATP. In: Photosynthesis and photorespiration (M.D. Hatch, C.B. Osmond, R.O. Slatyer, eds.), New York-London-Sydney-Toronto: Wiley and Sons 1971.

Sutcliffe, J.F.: Mineral absorption in plants. Oxford-London-New York-Paris: Pergamon Press 1962.

Torii, K., Laties, G.G.: Dual mechanisms of ion uptake in relation to the vacuolation in corn roots. Plant Physiol. **41**, 863–870 (1966).

Tukey, H.B., Jr.: The leaching of substances from plants. Ann. Rev. Plant Physiol. **21**, 305–324 (1970).

Ullrich, W.R.: Nitratabhängige nichtcyclische Photophosphorylierung bei *Ankistrodesmus braunii* in Abwesenheit von CO_2 und O_2. Planta **100**, 18–30 (1971).

Ullrich, W.R., Urbach, W., Santarius, K.A., Heber, U.: Die Verteilung von Orthophosphat auf Plastiden, Cytoplasma und Vacuole in der Blattzelle und ihre Veränderung im Licht-Dunkel-Wechsel. Z. Naturforsch. **20b**, 905–910 (1965).

Waisel, Y., Eshel, A.: Localization of ions in the mesophyll cells of the succulent halophyte *Suaeda maritima* Forssk. by X-ray microanalysis. Experientia **27**, 230–232 (1971).

Waisel, Y., Shapira, Z.: Functions performed by roots of some submerged hydrophytes. Israel J. Botany **20**, 69–77 (1971).

Weigl, J.: Beweis für die Beteiligung von beweglichen Transportstrukturen (Trägern) beim Ionentransport durch pflanzliche Membranen und die Kinetik des Anionentransportes bei *Elodea* im Licht und Dunkeln. Planta **75**, 327–342 (1967).

Wittwer, S.H., Bukovac, M.J.: The uptake of nutrients through leaf surfaces. In: Handbuch der Pflanzenernährung und Düngung (H. Linser, ed.), vol. I, p. 235–261. Wien-New York: Springer 1969.

Young, M., Sims, A.P.: The potassium relation of *Lemna minor* L. I. Potassium uptake and plant growth. J. Exptl. Bot. **77**, 958–969 (1972).

Young, M., Sims, A.P.: II. The mechanism of potassium uptake. J. Exptl. Bot. **79**, 317–327 (1973).

4.3 Stomatal Ion Transport

T.C. Hsiao

1. Introduction

Stomata, the leaf ports which regulate both the inward passage of CO_2 for photosynthesis and the unavoidably associated outward dissipation of water, have held the attention of plant physiologists continuously since the classic work of Hugo von Mohl, in the middle of the 19th century. The literature on the subject in the past hundred years has been voluminous, yet our understanding of the processes underlying stomatal behavior seems to have only begun. Current indications are that the mechanism involved is exceedingly complex and under regulation by a diverse variety of internal and external factors. This Chapter emphasizes ion transport as related to stomatal movement (opening and closing). For more comprehensive coverages of general stomatal physiology, pertinent references should be consulted (Meidner and Mansfield, 1968; Zelitch, 1969; Raschke, 1975).

Stomata of many species normally open during the day and close at night. Opening is the end result of solute accumulation in guard cells. The consequent water influx causes turgor pressure in guard cells to build up in excess of that in the surrounding epidermal cells, stretching the guard-cell pair so that they part in the middle to form the stomatal pore. Stomatal closing is accomplished by the reverse process: a decline in guard-cell solutes and consequent loss of guard-cell turgor. During the last two decades, controversies have centered on the identity of the solutes and on the nature of the build-up process. Explanations in the early years were dominated by the "starch-sugar" hypothesis, conceived by Lloyd in 1908 and modified by Sayre in 1926 (see Meidner and Mansfield, 1968) which holds that sugar is the key solute and that its concentration is determined by starch hydrolysis or formation, which in turn is triggered by pH changes arising from fluctuations in CO_2 levels. Levitt (1967) proposed the latest variation on this scheme. The starch-sugar hypothesis has been severely criticized (Meidner and Mansfield, 1965) and some of its major points are at variance with current understanding of starch-sugar metabolism. A different means of building up solutes is for guard cells to take them up externally. Extensive evidence was obtained recently to warrant a firm conclusion that the uptake of K^+ by guard cells accounts for a major part of the solute build-up underlying stomatal opening. First, a brief review of the history of this development.

2. Early Work on Stomatal Ion Relations

Although the sustained research that led directly to elucidation of the role of K^+ transport in stomatal movement started only in the middle 1960's, ions had been linked to stomata much earlier.

Some 50 years ago, Iljin and others studied the effects of various salts on stomata. Even earlier, at the turn of the century, K^+ was detected histochemically by Macallum (1905) in tulip guard cells. Iljin (1922a, b) found the effectiveness of Group I elements in promoting stomatal opening often to be $Li^+ > Na^+ > K^+$ when present at high concentrations in the medium. He presumed that those ions stimulated the hydrolysis of starch to sugar in guard cells. Although those findings were regarded as most important at that time (Maximov, 1929: p. 193), interest apparently waned. Stomatal ion relations were not given concerted attention again in the West until recently. In Japan, however, work continued sporadically.

In an extensive study, Imamura (1943) found stainable K^+ in guard cells to be well correlated with stomatal aperture and guard-cell osmotic potential[1] under a variety of conditions. He found $K^+ > Na^+ > Li^+$ in effectiveness in promoting opening when supplied externally, and concluded that K^+ probably caused swelling of cellular colloids. Subsequent studies on the effect of K^+ deficiency (Yamashita, 1952) and with isolated epidermal strips (Fujino, e.g., 1959) implicating K^+ were published in Japanese and went unnoticed in the West, where reviews on stomata for that period largely ignored ionic relations and none considered ions as possible osmotica (e.g., Stålfelt, 1956; Heath, 1959; Ketellapper, 1963; Meidner and Mansfield, 1965; Levitt, 1967). Only when Fujino (1967) expanded his work and published in English did those findings become known to Western workers. Fujino concluded that the active transport of K^+ in and out of guard cells is responsible for stomatal movement.

Unaware of the studies in Japan, which relied on the stain of Macallum (1905) for K^+, Fischer (1967) found that stomata in epidermal strips of *Vicia faba* open readily in light only when floated on solutions containing K^+. More importantly, he estimated K^+ uptake with $^{86}Rb^+$ and found it to be sufficient to account for the observed change in guard-cell osmotic potential if a counter ion for K^+ is assumed (Fischer, 1968a; Fischer and Hsiao, 1968). Thus some evidence was finally obtained indicating a direct role for K^+ as a major osmotic agent, in contrast to the nonosmotic role envisioned by Iljin and Imamura. Stimulated by publications of Fujino and Fischer, diverse and extensive evidence was quickly accumulated by various laboratories to establish K^+ transport as a fundamental stomatal mechanism.

3. Plant Material and Methodology

Perhaps more than anything else, progress in stomatal studies is hindered by the small size and dearth of guard cells in the leaf. The difficulty is compounded by the fact that guard cells are notoriously individualistic (Heath, 1959; Hsiao et al., 1973), as evinced by the wide range of osmotic potential in any guard-cell population (Fischer, 1968b; Mansfield and Jones, 1971). Further, stomata in

---

[1] The total potential of water (Ψ) in the cell is made up largely of two components, that due to the hydrostatic pressure (turgor pressure) P and that due to the colligative effect of solutes (osmotic potential) π. In the literature on water relations the former component potential of water is referred to also as pressure potential (Ψ_p) and the latter also as solute potential (Ψ_s). (These symbols will not be used here.) See Part A, 2.3.1 for further discussion of water potential in plant cells.

Before rolling   After rolling

Neutral
red
uptake

Stained
for K⁺
From
leaves
in light

Stained
for K⁺
From
leaves
in dark

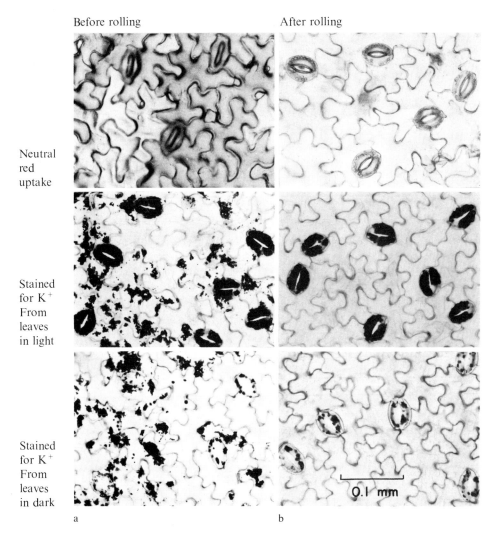

a   b

Fig. 4.6a and b. Quasi-isolation of guard cells of *V. faba* by rolling epidermal strips: Effects on neutral red uptake and K^+ location. Quasi-isolated guard cells (see text) depicted on the right were stained after rolling. Stomata in the leaves were open in light and closed in darkness; but staining for K^+ caused all stomata to close. The stain for K^+ was that of MACALLUM (1905) [$Co(NO_2)_3 \cdot 6H_2O$, 4 g; $NaNO_2$, 7 g; acetic acid, 13.3% by volume, 15 ml] and was darkened by reacting with 1% $(NH_4)_2S$. FISCHER (1971) has shown for epidermal strips a linear relationship between K^+ content and the proportion of guard-cell area stained. (a) Viable epidermal cells and guard cells took up the vital stain, neutral red (dark gray), and retained their K^+ (black precipitate). Note that some epidermal cells were broken by the stripping process and did not take up neutral red nor retain their K^+. (b) After rolling, only guard cells took up neutral red and retained their K^+. Rearranged from ALLAWAY and HSIAO (1973)

batches of plant material grown under similar or supposedly identical conditions sometimes behave differently (HSIAO et al., 1973; PALLAGHY and FISCHER, 1974; SQUIRE and MANSFIELD, 1974). This probably reflects our ignorance of all the subtle factors that influence stomata.

With guard cells constituting roughly 1% of the leaf volume and not having ready access to the external solution, early studies of ions in guard cells had to employ histochemical techniques or epidermal strips. Contributing importantly to elucidation of the role of K^+ in stomatal movement was the success in obtaining epidermal strips from some species (FISCHER, 1968b; FUJINO, 1967) containing guard cells that function similarly to those in intact leaves. Unfortunately, many species do not yield functional epidermal strips, though success was had with hard-to-strip maize leaves (PALLAGHY, 1971). Also, the relevance of some strip systems has been questioned (WILLMER and MANSFIELD, 1969a).

Even in epidermal strips, the tissue volume is occupied mostly by cells other than guard cells, making, for example, chemical analysis of the whole strip (e.g., PALLAS and WRIGHT, 1973) almost irrelevant as far as guard-cell contents go. This difficulty is overcome by a method (ALLAWAY and HSIAO, 1973) of rolling *Vicia* epidermis to break all epidermal cells while leaving the guard cells intact and apparently functional (Fig. *4.6*). The quasi-isolated guard cells thus obtained should greatly facilitate biochemical and other stomatal studies.

The use of electron microprobes to determine guard cell ion contents (SAWHNEY and ZELITCH, 1969) recently proved very fruitful, yielding the first definitive set of quantitative data (HUMBLE and RASCHKE, 1971; see Fig. *4.7*), though calibration can still be a problem. Also exciting is the reported success with microelectrodes in measuring guard cell membrane potential as well as K^+ activity (PENNY and BOWLING, 1974; see Fig. *4.13*). Electropotential data for guard cells were first reported by PALLAGHY (1968).

4. Evidence on the Osmotic Role of K^+

The initial work of IMAMURA strongly implicated K^+ as a specific ion in stomatal movement. Further work by YAMASHITA and FUJINO led to the conclusion that the transport of K^+ is a key process underlying stomatal movement. Those workers stained for guard-cell K^+, but they were not able to estimate the quantities involved. FISCHER was the first to conclude that the role of K^+ is directly osmotic. For an increase of 7 µm in stomatal aperture in *Vicia* epidermis induced by light and CO_2-free air, guard cell K^+ concentration increased some 300 mol m^{-3} as estimated from the uptake of ^{86}Rb-labeled K^+. This increase would amount to a 12-bar change in osmotic potential if K^+ was present as the Cl^- salt (FISCHER, 1968a; FISCHER and HSIAO, 1968). The measured change in the guard cells was about 10 bars for that change in aperture. This claim of massive transport of K^+ was soon substantiated by results SAWHNEY and ZELITCH (1969) obtained with an electron microprobe on tobacco guard cells, and was further confirmed by meticulous work of HUMBLE and RASCHKE (1971) on *Vicia* with the same instrument (Fig. *4.7*). More detailed work with ^{86}Rb$^+$ and ^{42}K$^+$ yielded similar results (FISCHER, 1972). A study with an electron microprobe of maize stomata also showed a comparable amount of K^+ transported, except that K^+ in this case is shuttled back and forth between the guard cells and the specialized subsidiary cells (PALLAGHY, 1971; RASCHKE and FELLOWS, 1971; see Fig. *4.10*). Additional confirmation came from

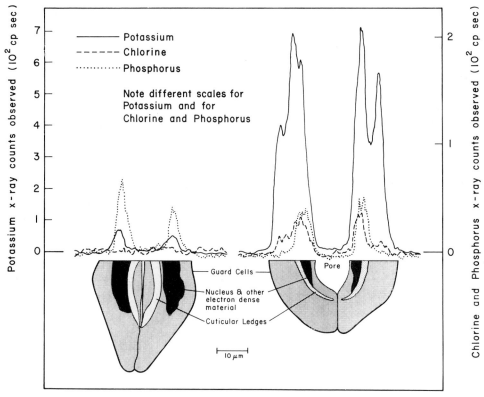

Fig. *4.7*. Profiles of relative amounts of K, Cl, and P across an open and a closed stoma of *Vicia*, as determined with an electron microprobe. The traces are the results of scanning across the stomata, shown diagrammatically below the traces, at the lines where they have been cut off. Reproduced from HUMBLE and RASCHKE (1971)

yet another technique, the analysis of quasi-isolated guard cells of *Vicia* for K⁺ by standard flame photometry (ALLAWAY and HSIAO, 1973). As shown in Fig. *4.6*, in such a preparation other cells are broken and only guard cells retain their K⁺. Still another totally independent technique, that of measuring guard cells with K⁺ microelectrodes, also yielded data (see Fig. *4.13*) indicating K⁺ transport of similar magnitude in stomata of *Commelina communis* (PENNY and BOWLING, 1974).

The changes in K⁺ estimated by the aforementioned investigators are compared in Table *4.6*, along with estimated rates of transport and associated parameters such as osmotic potential. In examining the table it is useful to keep in mind that for the dicots studied, relationships among guard-cell K⁺ content, osmotic potential and stomatal aperture are roughly linear (SAWHNEY and ZELITCH, 1969; FISCHER, 1972, 1973; ALLAWAY and HSIAO, 1973).

Comparing the measured changes in osmotic potential with the computed changes in K⁺ concentration in guard cells associated with stomatal opening, HUMBLE and RASCHKE (1971) deduced that K⁺ probably accounts for a major part of the osmotic-potential change, whereas anions balancing the positive charge of K⁺ account for only a minor part and hence should be di- or polyvalent. As shown in Table *4.6* (compare observed and measured $\Delta\pi$), though measurements made by ALLAWAY

Table 4.6. Changes in guard-cell K^+ and related parameters associated with stomatal opening

Method of K⁺ determination	Species	K⁺ in guard cells		ΔK⁺/Δ aperture		Δπ/Δ aperture		Rate of net K⁺ influx	Ref.
		Closed stomata (pmol cell⁻¹ [a])	Open stomata (pmol cell⁻¹ [a])	K⁺ as content per cell (pmol cell⁻¹ μm⁻¹)	K⁺ in conc. units (mol m⁻³ μm⁻¹)	Observed (bar μm⁻¹)	Calc. from ΔK⁺, assume K⁺ as KCl (bar μm⁻¹)		
Radioactive label	*Vicia faba*	–	–	0.22	43	1.4	1.7	90 nmol m⁻² s⁻¹ (cell surface)[b]	Fischer and Hsiao (1968)
Radioactive label	*V. faba*	0.25	1.5	0.21	41	2.0	1.9	100 to 150 nmol m⁻² s⁻¹ (cell surface)	Fischer (1972)
Electron microprobe	*V. faba*	0.10	2.1	0.20	–	1.6	2.8	200 μmol g$_{FW}^{-1}$ h⁻¹ (cell fresh weight)	Humble and Raschke (1971)
Extract of quasi-isolated guard cells	*V. faba*	0.28	1.4	0.25	50	1.2	2.2	–	Allaway and Hsiao (1973)
		(mol m⁻³)							
Electron microprobe	Tobacco	200	500	–	41	–	–	–	Sawhney and Zelitch (1969)
Electron microprobe	Maize	–	400	–	–	–	–	150 μmol g$_{FW}^{-1}$ h⁻¹ (cell fresh weight)	Raschke and Fellows (1971)
Micro-electrodes	*Commelina communis*	100	450	–	–	–	–	–	Penny and Bowling (1974)

[a] Guard-cell volumes were 5,000 μm³ for open stomata (Fischer and Hsiao, 1968), 2,400 μm³ and 5,000 μm³ respectively for open and closed stomata (Humble and Raschke, 1971), and 5,000 μm³ at incipient plasmolysis (Allaway and Hsiao, 1973).

[b] This corresponds to 120 μmol g$_{FW}^{-1}$ h⁻¹ on a cell fresh weight basis.

and HSIAO (1973) are consistent with this proposal, those made by FISCHER (1972) were in disagreement. There is a need for information on the activity coefficient for K^+ in these cells to confirm that most of the K^+ is uncomplexed as expected.

Other studies, while not providing absolute values for K^+, nevertheless indicate that massive K^+ transport is the general mechanism underlying stomatal movement. A fungal toxin (from *Helminthosporium maydis,* the causal agent of southern corn leaf blight) that causes rapid stomata closure only in the susceptible maize strain also causes a loss of stainable K^+ from guard cells only in the susceptible strain (ARNTZEN et al., 1973). A toxin (fusicoccin, see Chap. 8) that induces abnormally large stomatal opening also causes excessive K^+ accumulation (TURNER, 1973; SQUIRE and MANSFIELD, 1974). Additional evidence comes from the reduced stomatal opening associated with K^+ deficiency (YAMASHITA, 1952; GRAHAM and ULRICH, 1972). Finally, there is the fact that in almost all instances, changes in stomatal opening are intimately associated with changes in guard cell K^+ (or Rb^+, if it is used instead), regardless of the parameters that effected the change, be they light intensity or quality (Fig. 4.8), CO_2 concentration (PALLAGHY, 1971), water stress (HSIAO, 1973 b), or various metabolic inhibitors (FISCHER, 1972). As for species, pronounced changes in guard cell K^+ have been found to be associated with stomatal movement in a large number of diverse species (WILLMER and PALLAS, 1973; DAYANANDAN and KAUFMAN, 1973).

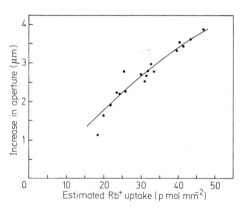

Fig. 4.8. Opening of aperture as related to Rb^+ uptake by quasi-isolated guard cells from *Vicia* under different light conditions. Conditions were light of different wave lengths, ranging from 400 nm to 680 nm and 20 nm apart, at a quantum flux of $38 \cdot 10^{18}$ quanta m^{-2} s^{-1} as well as darkness and "white light". Based on the data of HSIAO et al. (1973)

5. Other Cations

As will be recalled, ILJIN (1922a, b) considered several other monovalent cations to be more effective than K^+ in promoting stomatal opening. What, then, is the justification for emphasizing K^+ in the preceding discussion? It turned out that other alkali metal ions are promotive only when present at concentrations much higher than that expected in the apoplast of intact mesophytic plants. Many early studies on ionic effects focused on the concentrations required to plasmolyze guard cells, so solutions of perhaps 0.1 to 1 M were used to float epidermal strips. Employing KCl in a more dilute range, 1 to 300 mM, FISCHER and HSIAO (1968) determined that light markedly reduced the concentration required to effect full stomatal open-

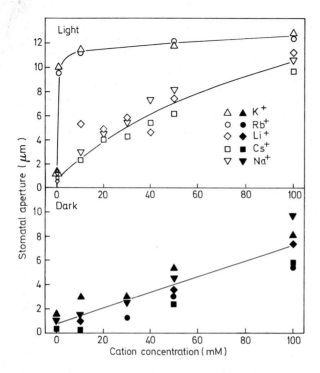

Fig. *4.9*. Stomatal opening in *Vicia* epidermal strips as dependent on monovalent cations in the floating medium. Strips with closed or nearly closed stomata were floated on buffered solutions of the Cl^- salt of the various cations and under CO_2-free air for 3 h, either under light or in darkness, before measuring aperture. Data for light and dark and for the different cations in many instances were obtained in different experiments but are presented together to show only the general patterns. Redrawn from Humble and Hsiao (1969)

ing. Humble and Hsiao (1969) later found that all alkali metal ions induced stomatal opening in light, though the concentration required for full opening differed markedly among ions. As shown in Fig. *4.9*, K^+ and Rb^+ were very effective at 0.1 to 1 mM, whereas comparable opening on Li^+, Na^+, and Cs^+ required concentrations some two orders of magnitude higher. In darkness, however, stomata opened substantially only at high concentrations and there was no marked difference among the ions. Thus, for light-activated stomatal opening, there appears to be a very specific requirement for K^+ (or Rb^+) if ions are supplied at physiological concentrations. Total concentration of ions in leaf apoplast has been estimated at a few mM (Chap. *4.2*; Bernstein, 1971). In contrast to the pronounced light-activated uptake of K^+, there is little or no light-activated $^{22}Na^+$ uptake by *Vicia* guard cells (Humble and Hsiao, 1970). Hence, the lack of light-activated stomatal opening on Na^+ is attributable to a lack of Na^+ transport. In electron microprobe studies of stomata in leaves, Na^+ was found in only small amounts in guard cells of *Vicia* (Humble and Raschke, 1971) and maize (Raschke and Fellows, 1971), and Na^+ content changed very little with stomatal opening. Pallaghy (1970) showed that the selectivity of K^+ over Na^+ was evident only when some Ca^{2+} was present.

Willmer and Mansfield (1969b) reported NaCl to be somewhat more effective than KCl for opening in *Commelina* epidermis and suggested that Na^+ may be an important ion for stomatal movement. Their work had been criticized (Humble and Hsiao, 1970; Humble and Raschke, 1971) for the use of concentrated solutions (100 mM and higher) and for not including some Ca^{2+} to preserve membrane integrity. Ca^{2+}, however, is inhibitory in the *Commelina* epidermis system. Possibly there is a still more fundamental reason for their failure to observe a specific response to exogenous K^+. Recent results indicate K^+ transport to be confined largely to the stomatal complex in *Commelina* and that this complex is probably self-sufficient in K^+ (see *4.3.10.2*). Consequently, response to exogenous ions may well not be indicative of the ionic species needed for stomatal opening in intact leaves.

Taken together, the evidence described establishes that K^+ (and Rb^+, if present) is the osmotic cation for stomatal movement in most natural situations. Some halophytes probably are an exception, however. A recent but limited study with electron microprobe (ESHEL et al., 1974) suggests that in the facultative halophyte *Cakile maritima*, stomatal movement involves K^+ under non-saline conditions but Na^+ under saline conditions.

Regarding cations outside Group I, Ca^{2+} is strongly inhibitory to stomatal opening at moderate concentrations (ILJIN, 1922a; IMAMURA, 1943; FISCHER, 1968b) and was thought in one study (FISCHER, 1972) to be slightly inhibitory even at low concentrations (0.1 to 1.0 mM). Mg^{2+} at a range of concentrations does not promote opening (HUMBLE and HSIAO, 1969). Interestingly, NH_4^+, though taken up as rapidly as K^+ by roots (TROMP, 1962), did not promote opening in light at any concentration; nor was it inhibitory to K^+-effected opening (HUMBLE and HSIAO, 1969).

The fact that divalent cations and one monovalent cation (NH_4^+) do not promote opening at any concentration points to a selective transport system even at high concentrations. Presumably something like the mechanism II of roots and leaves (Chap. *3.2*), as well as mechanism I with a very low affinity for Li^+, Na^+, and Cs^+, are operating in guard cells. According to the arguments given in *4.3.8.1*, the transport at high external concentration is also energy-requiring and probably active.

6. Anions and Charge Balance

With the positively charged K^+ transported in such large amounts, a charge balance must be maintained *via* the concomitant transport of anions in the same direction or of other cations in the opposite direction.

6.1 Inorganic Anions

Cl^-, Br^-, NO_3^- salts of K^+ all permitted good light-activated stomatal opening in *Vicia* epidermal strips, whereas K_2SO_4 was slightly less effective at high concentrations (HUMBLE and HSIAO, 1969). Analyses with the electron microprobe showed that, in guard cells *in situ*, Cl^- content correlated well with opening in maize (RASCHKE and FELLOWS, 1971) but not significantly in *Vicia* (HUMBLE and RASCHKE, 1971). This change in Cl^- in maize was also demonstrated histochemically (Fig. 4.10). In maize the equivalents of Cl^- transported were about half as many as the equivalents of K^+ transported, thus suggesting an important osmotic role for Cl^- in this species. In *Vicia* the content of P and S also did not vary materially with stomatal movement. When supplied with KCl labeled with $^{42}K^+$ and $^{36}Cl^-$,

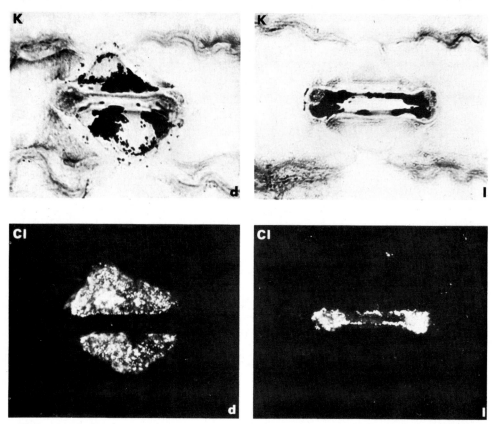

Fig. *4*.10. Distribution of K^+ and Cl^- within the stomatal complex of maize as affected by light. Epidermal strip was taken from leaves kept in darkness with closed stomata (*d*) or in light with open stomata (*l*). K^+ was stained with Macallum's stain (top) and Cl^- with $AgNO_3$ (bottom). Metallic silver precipitate, formed by photoreduction of AgCl, was photographed in incident light. Reproduced from RASCHKE and FELLOWS (1971)

epidermal strips of *V. faba* with intact guard cells but mostly broken epidermal cells took up Cl^- to only about 0.2 of the equivalent increase of K^+ during stomatal opening (PALLAGHY and FISCHER, 1974). These results, along with the fact that plants generally need only a very small amount of Cl^- for growth (0.01% of dry weight; EPSTEIN, 1972), indicate that electrical neutrality is not maintained solely by the uptake of inorganic anions. Even in maize, Cl^- accounts for only half of the needed negative charge. There is a faint possibility, however, that NO_3^- may contribute materially. KNO_3 permitted as much opening in epidermal strips as KCl. The uptake of NO_3^- by guard cells has not been studied. Uptake of organic anions from surrounding cells by guard cells *in situ* remains a possibility but apparently is not obligatory since quasi-isolated guard cells take up Rb^+ and their apertures open on RbCl solutions devoid of organic anions (HSIAO et al., 1973).

6.2 K$^+$-H$^+$ Exchange

The other alternative is to exchange cations. K$^+$ taken up could be exchanged for H$^+$ in guard cells, a continuous supply of the latter being provided by the synthesis of organic acids (Part A, Chaps. *12, 13*). Exchange of K$^+$ with cations such as Na$^+$ would be totally ineffective in altering the guard cell osmotic potential since there would not be a net gain of solutes. The importance of pH regulation and H$^+$ and OH$^-$ transport and exchange in cells and biological processes are discussed in Part A, Chap. *12*. SMITH and RAVEN (Part A, *12.5.2.3*) also speculate on the possible importance of pH regulation as part of the mechanism of stomatal opening.

RASCHKE and HUMBLE (1973) recently reported data supportive of a K$^+$-H$^+$ exchange mechanism for stomatal opening. Stomata in *Vicia* epidermal strips opened regardless of whether K$^+$ was supplied as the salt of absorbable or presumably nonabsorbable anions. The pH of the external solution declined in positive correlation with stomatal opening; the amount of H$^+$ released was within the known range of equivalents of K$^+$ transported. One puzzling aspect of their data is that floating the strips on 50 mM instead of 10 mM K$^+$ increased the H$^+$ released three-to-five-fold but did not alter stomatal opening noticeably. Nonetheless, it is reasonable to conclude that H$^+$ release (or OH$^-$ uptake) in exchange for K$^+$ is a mechanism for maintaining charge balance. As is well known since the starch-sugar hypothesis, guard-cell pH is apparently higher with stomata open than with stomata closed (e.g., FUJINO, 1967). These data are open to criticism, however, since pH was estimated from indicators taken up by the cells. K$^+$-H$^+$ exchange systems in other tissues are discussed in Chap. *3.3*, and Part A, Chaps. *6, 7, 8, 12* and *13*. Very recently PENNY and BOWLING (1975) measured *in situ* with micro-electrodes a vacuolar pH about 0.5 higher in guard cells of open stomata than in those of closed stomata.

6.3 Malic Acid

Malic acid, internally produced, is well known for providing H$^+$ for exchange and maintaining charge balance in other plant tissues when cation absorption is in excess (Part A, Chap. *13*). Epidermal strips have been analyzed for various organic acids in attempts to relate them to stomatal movement and K$^+$ transport. Unfortunately, the results of some of those studies (e.g., PALLAS and WRIGHT, 1973) are of little value, since epidermal cells were not necessarily broken and their contents were not taken into account. In the only meaningful analysis to date, ALLAWAY (1973) determined malate in quasi-isolated guard cells and found that when stomata opened in *Vicia* leaves, malate increased pronouncedly in guard cells although the increase was only 0.5 of the equivalent increase in K$^+$. In spite of quantitative uncertainties in the only H$^+$ exchange study to date (RASCHKE and HUMBLE, 1973), the malate data permit one to surmise that about half of the K$^+$ equivalents transported were probably exchanged for H$^+$. Yet to be studied is whether the remainder of the positive charge imported by K$^+$ is balanced by the uptake of anions other than OH$^-$ or by exchange with H$^+$ from sources other than malic acid (other organic acids or elevations in pH of a highly buffered cell interior).

7. Starch and CO$_2$

Most workers found stainable starch in guard cells to be inversely correlated with stomatal aperture (e.g., FISCHER and HSIAO, 1968), a relationship that led to formulation of the starch-sugar hypothesis. In spite of the elucidation of the role of K$^+$, the function of starch remains in question. Indeed, the opinion has been expressed (MANSFIELD and JONES, 1971) that changes in starch could be as primary a process as K$^+$ transport. Considerable data argue against that, however. In various instances (YAMASHITA, 1952; MEIDNER and MANSFIELD, 1968; MOURAVIEFF, 1972a) there is a lack of close correlation between starch and aperture. Particularly IMAMURA (1943) noted that upon changing conditions, changes in K$^+$ paralleled changes in osmotic potential in guard cells whereas changes in starch often lagged behind. In CO$_2$-free air, darkening can close stomata substantially without significant concurrent changes in starch (Table *4.7*).

Table *4.7*. Changes in stomatal aperture and guard-cell starch under CO$_2$-free air as affected by darkness following light. *Vicia* epidermal strips on 2 mM KCl. Strips were taken from leaves in darkness and floated briefly (initial values) before exposing to light. Starch was stained with I-KI-phenol and scored according to FISCHER (1968b). Original data of HSIAO

			3 h light followed by	
	Initial	3 h light	20 min dark	90 min dark
Aperture (μm)	1.5	10.0	5.0	3.6
Starch score	4.7	1.6	0.9	1.9

Nevertheless, in view of its general relationship with aperture and its unique behavior (MEIDNER and MANSFIELD, 1968), guard-cell starch almost certainly has a function in stomatal movement. It could be the ultimate source of carbon skeleton for the anions synthesized to balance K$^+$ (MANSFIELD and JONES, 1971) or a source of energy for opening (MOURAVIEFF, 1972a), particularly in the dark or in light of very low intensity (*4.3.8.3*).

Perhaps the most enigmatic aspect of stomatal mechanisms is that underlying the effect of CO$_2$, one of the key environmental variables that elicit K$^+$ transport and stomatal movement (PALLAGHY, 1971; FISCHER, 1972). Commonly, increases in CO$_2$ close stomata. In light, CO$_2$ fixation may compete with guard-cell transport for energy (KETELLAPPER, 1963; HSIAO et al., 1973). Whether a similar explanation could also be invoked for the induction of opening in the dark by low CO$_2$ is debatable. Guard cells can fix some CO$_2$ in the dark (SHAW and MACLACHLAN, 1954) and may be high in PEP carboxylase (WILLMER et al., 1973). The fact that guard-cell chloroplasts in C$_3$ species resemble the chloroplasts in C$_4$ plants (ALLAWAY and SETTERFIELD, 1972; PALLAS and MOLLENHAUER, 1972) also points to the possibility of high PEP carboxylase activity and dark CO$_2$ fixation. Although competition for energy may possibly play a part in stomatal response to CO$_2$, there perhaps exists still another causal link. The response of maize stomata to changes in CO$_2$ is in terms of seconds (RASCHKE, 1972), which seems to be too rapid to be explained solely on the basis of diminution in energy due to competition. Membrane permeability could be affected by CO_2/HCO_3^- (HOPE, 1965).

Another link suggested for CO_2 is that its fixation into organic acids provides counter-ions for K^+ (WILLMER et al., 1973). The problem is that, according to this idea, high CO_2 should open stomata, not close them. Still another proposed possibility is that a high content of organic acids such as malic acid would lower pH sufficiently to lead to the polymerization of sugar into starch, thus causing stomatal closure partly in accordance with the starch-sugar hypothesis (LEVITT, 1967; MEIDNER and MANSFIELD, 1968: p. 139). This idea also appears untenable because malate rises sharply in guard cells during the *opening* process (ALLAWAY, 1973).

8. Sources of Energy

8.1 General

The increase in guard-cell turgor (P) underlying stomatal opening is brought about by a corresponding change in guard-cell osmotic potential (π). This relationship is evident from Part A, Eq. (2.8), if water potential (Ψ) of the cell is assumed to be constant. It follows that the activity of solutes within the guard cell must rise sufficiently to account for the observed change in osmotic potential. Hence, the uptake of the major osmotic ion, K^+, is against an activity gradient after guard-cell turgor becomes positive relative to the surrounding. That transport would require metabolic energy regardless of whether it is active (against a gradient of electrochemical potential, not just of activity, see Part A, Chap. 3) or passive. In the passive case, energy would be needed to maintain an appropriate guard-cell membrane potential (negative inside). PENNY and BOWLING (1974) reported that the electropotential between adjacent cells of the stomatal complex of *Commelina* is near zero and that K^+ transport should be active for both the opening and closing processes. More electrophysiological data are needed.

What may be the metabolic source of energy for stomatal ion transport? K^+ uptake and stomatal opening can occur in light in an oxygen-free atmosphere (and with DCMU to inhibit internal oxygen release from photosynthesis; HUMBLE and HSIAO, 1970), or in total darkness when ambient CO_2 is reduced. Apparently either photosynthetic or respiratory energy can be used. Mitochondria are numerous in guard cells, especially in comparison with chloroplasts (THOMSON and DE JOUNETT, 1970; ALLAWAY and SETTERFIELD, 1972); so respiration seems a feasible source of energy. Useful evidence on respiratory energy is scant, however, because of the difficulties inherent in studying a microscopic apparatus so well dispersed among other cells and hard to isolate. Almost all early studies used inhibitors on leaves or leaf pieces (ZELITCH, 1965); thus, the results are complicated by indirect effects *via* leaf mesophyll. This discussion and that to follow make clear that stomata must have more than one source of energy for ion transport. A concept of a general pool of energy for transport in cells drawing from various sources is elaborated on in *4.2.3.4* and Part A, Chap. *9*.

8.2 Photosynthesis

The starch-sugar hypothesis attributes to light only a role of lowering the CO_2 level in guard cells and their environs. CO_2 is a key factor in stomatal behavior although its specific link to guard cell solutes is obscure (*4.3.7*). A more direct link of ion transport to light would be *via* photosynthetic electron flow or photophosphorylation. Though guard-cell chloroplasts are small, not well developed, and not abundant (ALLAWAY and SETTERFIELD, 1972), some speculative calculations, though involving many assumptions, do suggest that the energy requirements for K^+ transport can be met by photophosphorylation (PALLAS and DILLEY, 1972). The action spectrum for preventing stomatal closing in *Senecio* epidermis was found (KUIPER, 1964) to be similar to the absorption spectrum of chloroplasts. Action spectra (HSIAO et al., 1973) for stomatal opening in *Vicia* leaf discs (Fig. *4.11*), as well as of quasi-isolated guard cells under a CO_2-free atmosphere, also suggest action of a photosynthetic type, though only at a high quantum flux density. A nearly identical action spectrum was observed for guard-cell Rb^+ uptake (as can be inferred from Fig. *4.8*). The use of quasi-isolated guard cells (hence, removing other CO_2 sink and source) and CO_2-free air most likely eliminated any indirect effect operating *via* changes in CO_2 levels. The conclusion that under these conditions photosynthesis probably has a rather direct part in stomatal ion transport is not belaboring the obvious. Because of the emphasis on the indirect light effect of reducing intercellular CO_2, at times in the past a more direct photosynthetic light effect was taken as unlikely or of minimal importance (MEIDNER and MANSFIELD, 1965).

To examine the specific photosynthetic processes providing energy, various inhibitors for studying ion transport in green cells in general (Chap. *4.2* and Part A, Chap. *6*) were also tried on stomata in epidermal strips. Eliminating suspected energy sources one by one with stepwise addition of inhibiting chemicals or conditions, HUMBLE and HSIAO (1970) concluded that the energy for K^+ transport and stomatal opening in *Vicia* can be derived from photosynthetic cyclic electron flow and in large part does not require photosynthetic noncyclic electron flow or respiration. This conclusion was supported by independent evidence that *Vicia* stomata opened fully in light of > 700 nm, a condition that presumably permits only cyclic electron flow. RAGHAVENDRA and DAS (1972) reached a similar conclusion with *Commelina*. Particularly they found that opening was stimulated by PMS, a catalyst for cyclic photophosphorylation. Generally stomatal opening or K^+ uptake in epidermal strips is sensitive to inhibitors of cyclic electron flow (CCCP, FCCP, and salicylaldoxime) and relatively insensitive to inhibitors of noncyclic electron flow of the DCMU type (HUMBLE and HSIAO, 1970; WILLMER and MANSFIELD, 1970; RAGHAVENDRA and DAS, 1972; TURNER, 1973; PALLAGHY and FISCHER, 1974). KUIPER (1964) did report a marked stomatal closure caused by DCMU, but since he used air containing CO_2 the effect might be *via* an increase in CO_2 level. Though for light-dependent stomatal opening, cyclic electron flow appears to be an important energy source, probably obligatory under some experimental conditions, noncyclic electron flow as a possible source cannot be ruled out.

A word of caution is in order regarding the use of inhibitors. Stomata are readily inhibited by a very large number of diverse chemicals, some of them not known as inhibitors of energy production or ion transport. An extreme example is the inhibition by actinomycin D (FISCHER, 1967), an inhibitor of DNA-dependent RNA synthesis. It would appear that, depending on the biased selection of inhibitors and their concentrations, data may be obtained to support a number of hypotheses, each in direct conflict with the other. Therefore, with stomata more than anything else, inferences drawn from inhibitor studies should be confirmed with independent evidence.

8.3 Blue Light Effect

Light effects on stomata are not confined to those of photosynthesis. This point is heightened by a recent report of stomatal response of normal magnitudes to light in species which apparently contain no detectable chloroplasts in guard cells (NELSON and MAYO, 1975). It has been known for some time that there is a blue-light stimulation of stomatal opening which is independent of intercellular CO_2 (MEIDNER and MANSFIELD, 1968). Recent work on the action spectrum showed that, at a low quantum flux, stomatal opening (Fig. 4.11) and Rb^+ uptake by guard cells (HSIAO et al., 1973) are confined to the blue region almost completely, with little or no action in the

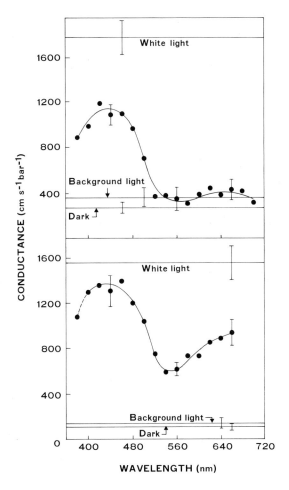

Fig. 4.11. Spectral dependence at equal quantum flux of abaxial stomatal opening in *Vicia* leaf discs as indicated by conductance to the mass flow of air. Mass-flow conductance is a curvilinear function of stomatal aperture (HSIAO et al., 1973); it is nearly proportional to aperture, however, if the change in aperture is small. Top: Low quantum flux ($7 \cdot 10^{18}$ quanta m^{-2} s^{-1}). Bottom: High quantum flux ($38 \cdot 10^{18}$ quanta m^{-2} s^{-1} except for 380 nm, which received $32.5 \cdot 10^{18}$ quanta m^{-2} s^{-1}). Experiments for the low and high quantum fluxes were conducted separately and exposure to light was for 3 h. Samples receiving background light (< 28 μcal cm^{-2} $min^{-1} = 19$ mW m^{-2}) were placed near the spectrograph light beam, but outside the area directly lit by the beam. "White" light was at about 0.25 cal cm^{-2} $min^{-1} = 174$ Wm^{-2}. Vertical bars represent twice the standard error of the mean. Modified from HSIAO et al. (1973)

red. Only at a much higher quantum flux did red become effective and the action spectrum more like that for photosynthesis (Fig. *4*.11). The low quantum flux was probably too low for light to be an important source of energy. More likely, light acts as a modulator of another process which supplies the energy needed for guard-cell ion uptake. KOWALLIK (1967) reported a comparable blue action spectrum for enhancement of respiration in *Chlorella*. The maximal effect at 455 nm was approached at only $1 \cdot 10^{18}$ quanta m^{-2} s^{-1}. Whether respiration is the source of energy for blue-activated K$^+$ uptake by guard cells remains to be investigated.

8.4 ATPase, ATP, and Glycolate

ATPase has been postulated to mediate stomatal ion transport, but the evidence is indirect and confusing. FUJINO (1967) reported a marked stomatal opening and increase in guard-cell K$^+$ in epidermal strips in response to exogenous ATP and a heavy ATPase-stain reaction in closed guard cells but not open guard cells. FISCHER (1967), on the other hand, observed no significant response to ATP; and WILLMER and MANSFIELD (1970) detected no difference between open and closed stomata in ATPase stain. Evaluation of any response to exogenous ATP is complicated by the known ability of ATP to act as an ion chelator (LÜTTGE et al., 1974). In a hypothesis that is apparently at variance with current concepts of the role of ATPase in ion transport (Part A, Chap. *10*), FUJINO supposed that ATPase is involved only in the *efflux* of K$^+$ whereas ATP, through some mechanism other than ATPase, supplies the energy for K$^+$ influx and stomatal opening.

ATPase was implicated also in some (but not all) studies with ouabain, an inhibitor of ATPase-linked K$^+$-Na$^+$ exchange in animal and some plant systems (Part A, Chap. *10*). THOMAS (1970b) reported a readily reversible inhibition by ouabain of stomatal opening in epidermis, and postulated an ATPase-mediated K$^+$ transport in guard cells[2]. Others, however, found ouabain of various concentrations to have little or no effect on stomatal opening in leaves (MOURAVIEFF, 1972b) or on opening and K$^+$ uptake in epidermal strips (PALLAGHY and FISCHER, 1974), in accordance with the general insensitivity of higher plant ATPases to ouabain. On the other hand, TURNER (1973) reported an inhibition of opening in light but not of fusicoccin-induced opening in darkness.

A discussion on guard-cell energetics cannot be complete without mentioning glycolate metabolism. ZELITCH made the initial observation that the bisulfite compound α-hydroxysulfonate, a competitive inhibitor of glycolate oxidase, inhibited stomatal opening in light. Finding that stomatal aperture correlated well with glycolate formation in leaf discs, ZELITCH (1965) postulated a key role for glycolate in stomatal movement. He proposed that glyoxylate, the product of glycolate oxidation, may be a precursor of carbohydrates needed as osmotic agents, or that its reduction to glycolate may be coupled to the oxidation of NADPH to regenerate the NADP needed for noncyclic photophosphorylation. ATP produced in the latter process is then used for guard-cell transport (ZELITCH, 1969). A crucial piece of evidence is that the inhibition of stomatal opening by α-hydroxysulfonate could be reversed by glycolate, the substrate of the enzyme for which the inhibitor is supposedly specific. Other investigators, however, were not able to get this reversal (MOURAVIEFF, 1965; FISCHER, 1967). More importantly, bisulfite compounds in general have been shown to be nonspecific, inhibiting photosynthesis, ion transport, as well as other membrane-dependent processes (LÜTTGE et al., 1972). The scheme also seems to involve an unnecessary waste of reducing power. A more likely means of regenerating NADP would be a system such as the PGA-triose-phosphate shuttle transporting reducing power and energy from noncyclic electron flow in the chloroplasts to the cytoplasm for ion uptake at the plasmalemma (see Vol. 3). Still another point is that current concepts place glycolate oxidase within microbodies (peroxisomes), a paucity of which has been noted in guard cells (e.g., ALLAWAY and SETTERFIELD, 1972). Apparently if glycolate has a part in stomatal mechanisms, it would not be as proposed in ZELTICH's scheme.

---

[2] THOMAS used a unique liquid-flow porometer (THOMAS, 1970a) to monitor stomata in epidermal peels while feeding chemicals and ions through the same aqueous liquid. In spite of two months of serious efforts, I could not make his porometer work effectively. Further, using tobacco epidermis from the same variety and seed source that he used, I failed to obtain a degree of stomatal movement any where approaching what he reported.

9. Stomatal Closure as Related to Opening

Stomatal closure depends on an efflux of K^+ from guard cells (Fig. *4.12*). A most intriguing aspect of stomatal ion transport is how the direction of net flux is determined and regulated. More specifically, how is stomatal closure brought about in a manner that is consistent with the opening process and with the maintenance of a zero net flux when a steady-state aperture is reached (Fig. *4.12*). The direction of net flux must be reversed frequently when stomata open and close repeatedly under rapidly fluctuating environmental conditions. Assuming that at least the influx of K^+ is an active process, one may consider a few of the possibilities. Speculations on changes in membrane permeability to water (e.g., ZELITCH, 1965: p. 477) can be discarded since they stem from a failure to recognize that water inside the guard cell is essentially at equilibrium with the water outside (e.g., FISCHER, 1973) and that changes in permeability to a passive flux such as that of water would affect rates only, not the direction of equilibrium. Part A, Chap. *11* discusses the regulatory means of attaining and maintaining steady-state ion concentrations in cells. Active influx balanced against a passive efflux, and active influx coupled with a passive efflux and negative feedback on the influx, are the two cases considered in some detail. The emphasis there is on unidirectional net flux. With stomata, on the other hand, emphasis must also be on reversing the direction of net flux.

One possibility is changes in passive permeability of guard-cell membrane to K^+. An increase in permeability would bring about at least partial closure of originally open stomata. Relevant observations are very limited, however. When *Vicia* epidermal strips with stomata opened on ${}^{86}Rb^+$-labeled K^+ by light were transferred to unlabeled K^+ and darkness, stomata closed partly and a small portion of the label taken

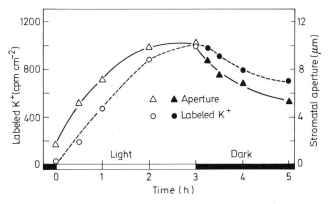

Fig. *4.12*. Stomatal movement and fluxes of ${}^{86}Rb^+$-labeled K^+ in *Vicia* epidermal strips in response to consecutive light ($\triangle$, $\circ$) and darkness ($\blacktriangle$, $\bullet$). Strips with mainly broken epidermal cells were placed under light and CO_2-free air on 1 mM KCl labeled with ${}^{86}Rb^+$ at time zero and remained under CO_2-free air throughout. The amount of labeled K^+ retained by strips after desorption and washing of the free space is given as cpm per cm^2 of strip area. Other data in this study indicated that K^+ was substantially localized in guard cells. The loss of each radioactive count during closing represents more K^+ moved than the gain of each count during opening, because the label was diluted by the original K^+ content of the cells. Reproduced from HUMBLE and HSIAO (1970)

up in light was lost to the medium in 2 h (Humble and Hsiao, 1970). The retention of a major portion of the label against exchange indicates a low permeability to Rb^+ and K^+ even under conditions inducive to stomatal closing. Under light, the loss of label on unlabeled medium was also slight (Fischer and Hsiao, 1968). Fischer (1972) estimated the influx and efflux of K^+ for open stomata at a steady state to be much smaller than influx at the beginning of the opening process. Those results suggest a minimal role for permeability change and indicate a feedback modulation of the influx pump. Another possibility is the presence of an efflux pump as well as an influx pump, or even of a pump that stops or reverses its direction *via* feedback control, as postulated for the marine alga *Valonia* (Zimmermann and Steudle, 1974). An efflux pump is indicated by the results of Penny and Bowling (1974), mentioned earlier, which suggest an active net efflux for stomatal closing in *Commelina*. Although those limited observations need to be expanded and substantiated, they nevertheless suggest that the mechanism for reversing the direction of K^+ flux is very complex.

10. Path of Transport and Reversible Source—Sink Relationship

The amount of K^+ transported is massive when stomata open or close. And stomatal modulation of gas diffusion in leaves is so dynamic that net K^+ flux may reverse directions frequently. Where does the large amount of K^+ come from, and where does it go? It would seem advantageous to have the reversible source-sink relationship localized to guard cells and only a few adjacent cells. The presence of specialized subsidiary cells in many species suggests such a possibility.

10.1 Shuttle for K^+ and Cl^- in Maize

Work of Pallaghy (1971) and Raschke and Fellows (1971) clearly established not only a localized source-sink relationship between guard cells and subsidiary cells in maize, but also a coupled turgor relation between them that makes particularly effective the stomatal apparatus of maize (and probably other Gramineae). In what is termed a K^+ and Cl^- shuttle system (Fig. *4*.10), for stomatal opening K^+ and Cl^- are transported into guard cells, apparently mostly from the two subsidiary cells; for closing, K^+ and Cl^- are moved back into the subsidiary cells. The advantages here are two-fold. For one, a large amount of K^+ is readily at hand and needs to be transported for only a minimal distance. For another, the concurrent turgor change in the opposite direction in guard cells and subsidiary cells during stomatal movement greatly increases the effectiveness of the ions in creating turgor differences between them. Stomatal aperture is dependent on the balance of turgor pressure between guard cells and adjacent cells. It is therefore not surprising that maize stomatal movement is one of the fastest known (Rascke and Fellows, 1971).

10.2 Stomatal Complex of *Commelina*

Another plant with subsidiary cells, *C. communis,* seems to have a source-sink arrangement that is only partially localized. Stomata were observed to open, and guard cells accumulated K^+, in *Commelina* epidermal strips on K^+-free medium when treated with fusicoccin. This led SQUIRE and MANSFIELD (1974) to conclude that subsidiary cells are the endogenous source of K^+. Further evidence came from the microelectrode work of PENNY and BOWLING (1974). With open stomata in leaves, K^+ activity decreased from the guard cell to the inner subsidiary cell to the outer subsidiary cell to the epidermal cell. With stomata closed, the gradient of K^+ activity was completely reversed. Changes in K^+ activity were as great in the epidermal cell next to the subsidiary cell as they were in the guard cell, but in the opposite direction (Fig. *4.13*). MANS-

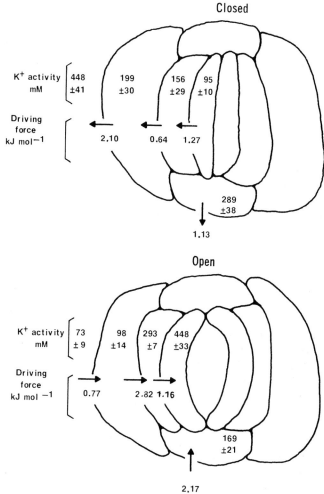

Fig. *4.13*. K^+ activity in cells of the stomatal complex of *Commelina communis*, with the stoma either closed or open, and the calculated driving force on K^+ required to maintain the observed differences in cellular K^+ activity. Activities were measured with K^+-sensitive microelectrodes calibrated against standard KCl solutions, and hence are expressed as equivalent KCl concentration. Driving forces were calculated from the Nernst potentials for K^+ and the electrical-potential differences between cells measured with microelectrodes. Arrows indicate the direction of active transport. Activities are the means ($\pm$ standard error of mean) of at least 5 estimates and those listed to the left of the stomatal complex are of epidermal cells. Adapted from PENNY and BOWLING (1974)

FIELD and JONES (1971), however, did not detect any difference in osmotic potential of subsidiary and epidermal cells between *Commelina* epidermal strips with open stomata and those with stomata which were kept closed by the presence of ABA. It seems that this arrangement of K^+ transport across several cells is rather inefficient although the activity gradients against transport between any two adjacent cells are kept small. (For recent electron microprobe data on the stomatal complex of *Commelina* see WILLMER and PALLAS, 1974.)

10.3 Species without Subsidiary Cells

Guard cells of many dicots such as *V. faba* and some monocots do not have subsidiary cells and are surrounded by cells which are indistinguishable in appearance from other epidermal cells. For stomata in *Vicia* epidermal strips to open wide, ions must be provided in the medium. The fact that *Vicia* stomata can effectively use K^+ in the external medium points to the possibility that, *in situ*, the source of K^+ for opening and sink for closing is the general leaf apoplast, including that in the leaf mesophyll. On the other hand, early data of STÅLFELT (1963, 1966) demonstrated that solutes (unidentified) can accumulate substantially in epidermal cells of *Vicia* leaves under some conditions, leading to back pressure on guard cells and stomatal closure. Those results would be more consistent with the idea that the source-sink arrangement is localized and similar to that in *C. communis*. TURNER (1973) was unable to detect any significant change in K^+ of epidermal cells in leaves of *Phaseolus vulgaris* associated with stomatal movement.

10.4 Symplastic *vs.* Trans-Membrane Transport

If the source and sink for ions of guard cells is the symplast of the surrounding cells rather than the general apoplast of the leaf, the specific path taken by the ions presents another problem. Symplastic transport between cells, as contrasted to trans-membrane flux, though consistent with the fast movements observed, is not the likely path as speculated on by some workers (e.g., FUJINO and JINNO, 1972). One reason is that many studies observed no plasmodesmata, the symplastic links, between mature guard cells and adjacent cells (THOMSON and DE JOURNETT, 1970; ALLAWAY and SETTERFIELD, 1972; SINGH and SRIVASTAVA, 1973). In other studies plasmodesmata were reported but were either not in abundance (PALLAS and MOLLENHAUER, 1972) or poorly resolved (LITZ and KIMMINS, 1968; FUJINO and JINNO, 1972). The K^+ and Cl^- shuttle represents a particularly problematic case because plasmodesmata between mature guard cells and subsidiary cells were not found in maize (SRIVASTAVA and SINGH, 1972) or other grass species (BROWN and JOHNSON, 1962). The other reason for doubting symplastic transport is that this path is presumed to be passive, i.e., down the gradient of electrochemical potential (Chap. 2). Since the symplastic network among cells is not believed to be separated by membranes (LÄUCHLI, 1972), electrical potential should be rather uniform throughout and the gradient for movement is essentially merely the activity gradient. Yet there is no question that K^+ accumulation by guard cells for stomatal opening is against a K^+ activity gradient. Even closing seems to involve active transport (PENNY and BOWLING, 1974). Hence, if plasmodesmata were present in adequate numbers, there still remains the difficulty of movement in the wrong direction. In fact, only in the absence of numerous plasmodesmata is the observed massive K^+ movement feasible. Otherwise the maintenance of large differences in ion activity between guard and adjacent cells would consume too much energy.

The alternative and apparently more plausible path is to move K^+ (and Cl^- when appropriate) between cells across the cell wall and the plasmalemma. Once in the wall, however, the ion may spread by diffusion through a large volume of apoplast. Viewed in a straightforward way, this seems to be inconsistent with the localized source-sink arrangement in maize. The very thin wall (ca. 0.1 μm) between guard and subsidiary cells of grasses (BROWN and JOHNSON, 1962), however, should facilitate trans-wall and trans-membrane flux and could aid in limiting apoplastic spreading by diffusion. Electron micrographs of *Opuntia* guard cells have been interpreted to be indicative of a highly active plasmalemma (THOMSON and DE JOURNETT, 1970). One troublesome point in this scheme is the apparent need for means of regulating and reversing the net flux also in subsidiary cells. Fluxes of about 100 mmol m^{-2} s^{-1} are required (Table 4.6).

11. ABA and Phytochrome

ABA is currently receiving much attention for one of its postulated roles, that of a regulator of plant behavior under stress conditions. This interest started with the first findings that ABA is a potent inhibitor of stomata opening (MITTELHEUSER and VAN STEVENINCK, 1969) and that ABA content increases sharply and rapidly in leaves subjected to mild to moderate water deficits (WRIGHT, 1969). Enforced by data from subsequent studies, the concept of ABA being a key endogenous modulator that mediates interactions between environmental factors and stomata soon gained wide popularity. Elaborated on elsewhere are the general aspects of stomatal physiology as related to ABA (RASCHKE, 1975), of water deficits as they affect endogenous ABA and stomata (HSIAO, 1973a), of possible interactions of ABA with other hormones in mediating plant adaptation to stress (LIVNE and VAADIA, 1972), and of ion transport as influenced by ABA (Chap. 7).

The discussion here is confined to the effects of ABA on stomata in terms of ion movements and to a comment on the extent of ABA participation in modulating stomatal changes. When fed or applied to leaves, ABA causes stomatal closure within minutes (Fig. 4.14; ARNTZEN et al., 1973). In epidermal strips ABA inhibits stomatal opening and K^+ uptake into guard cells (MANSFIELD and JONES, 1971; HORTON and MORAN, 1972; ARNTZEN et al., 1973). The inhibition is readily reversible (Fig. 4.14; HORTON, 1971). The evidence discussed below suggests that the inhibition of stomata is not a simple matter of increasing the membrane permeability to K^+ and thus enhancing passive efflux. Generally ABA does not seem to promote leakage of ions from cells and in fact, can stimulate ion uptake by root cells (Chap. 7). At a concentration that prevented opening, ABA neither caused substantial closing of open stomata in epidermal strips (HORTON, 1971; TURNER, 1973) nor prevented the opening elicited by fusicoccin (TURNER, 1973). In the presence of ABA but under conditions otherwise inductive to stomatal opening, K^+ was detected in subsidiary cells but not in guard cells of maize epidermal strips (ARNTZEN et al., 1973). In contrast, in the presence of decenylsuccinic acid, a compound known to induce leakage from cells, K^+

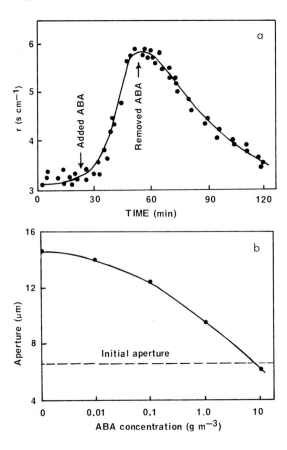

Fig. 4.14a and b. Effects of ABA on stomatal opening under light. (a) Changes in diffusion resistance (r) of an excised barley leaf after the water bathing the cut leaf base was changed to 10^{-7} M ABA and later after replacing the ABA with water. Diffusion resistance increases as stomata close and decreases as they open. Adapted from Cummins et al. (1971). (b) Opening in *Vicia* epidermal strips as affected by ABA concentration. Strips were taken from leaves in the dark and floated on buffered 25 mM KCl with various amounts of ABA for 3 h under light and CO_2-free air before measuring aperture. Initial aperture refers to that before the light period. Based on Horton (1971)

was not detected in either cell type. Thus, if ABA affects the passive permeability to K^+, it must be very selective regarding cell type. Together, these data point to uptake by guard cells as a site of ABA action. There is inconsistency in this interpretation, however. The most troublesome is how an inhibition of K^+ uptake can account for the rapid closure in leaves caused by ABA (Fig. 4.14). It would seem that guard-cell plasmalemma must be very permeable after ABA application to accomodate the rapid K^+ efflux underlying the rapid closing. An equally tenuous alternative is for ABA to activate an efflux pump.

As for the known increase in cell permeability to water resulting from applied ABA (Glinka and Reinhold, 1971), it has already been pointed out that such changes should not affect steady-state stomatal aperture or the direction of stomatal movement(4.3.9).

Returning now to the concept of ABA mediating interactions between stomata and environmental factors, there seems fair evidence that this does occur (Raschke, 1975), especially in affecting the sensitivity of stomata to opening or closing stimuli. However, as discussed elsewhere (Hsiao, 1973a), some definitive evidence is yet to be obtained for the postulated role of ABA in initiating stomatal closure at the onset of water stress and in limiting opening in the later stage of recovery from stress. The tendency to ascribe to ABA an all-pervasive role in stomatal modulation (Hiron and Wright, 1973; Milborrow, 1974), even if only for conditions of water stress, seems to be unjustified since there are several known instances of disparity between tissue ABA level and stomatal opening (e.g., Loveys and Kriedemann, 1973).

Phytochrome has been shown to be involved in the nyctinastic closure of *Albizzia*. That movement depends on K^+ transport in the pulvinar cells (SATTER and GALSTON, 1971), as stomatal movement depends on guard-cell K^+ transport. In a careful comparative study, however, EVANS and ALLAWAY (1972) showed that although the opening of *Albizzia* pinnules is also induced by blue light, there is a clear difference between the two systems in that phytochrome apparently is not involved in stomatal closure in *Vicia*. HABERMANN (1973), on the other hand, has interpreted her stomatal data as indicative of phytochrome participation, though at a light level much lower than customary.

12. Concluding Remarks

From the discussion above, limited largely to ion transport, it is clear that the mechanism underlying stomatal movement is very complex indeed. This is not totally unexpected in view of the regulatory function of stomata. To facilitate CO_2 assimilation, to prevent unnecessary water loss when photosynthesis is low, and to exert proportional feedback control so as to maintain plant water status within a narrow limit, stomata must respond to a host of environmental and internal factors in an integrated manner. These factors include such diverse parameters as light, CO_2 level, air humidity, temperature, internal water status, and leaf ontogeny. So far, indications are that most of these factors are somehow sensed and transduced into signals for K^+ transport and the associated process. Stomatal aperture at any time would then reflect the integrated response of ion transport and allied processes to all these signals. Thus, hypotheses on stomatal operations must not overemphasize a part (e.g., CO_2-response, ABA, etc.) as to lose sight of the whole.

What are some of the means to accelerate progress in the field of stomatal ion transport? It seems that detailed electrophysiological studies of stomatal movement in selected species may answer many of the questions raised in this article. Also high on the priority list is the finding of species that would yield epidermal strips with guard cells functioning more consistently and with less anomalies than the present strip systems. Finally, extension and improvement of the technique for obtaining quasi-isolated guard cells or even isolated functional guard cells would greatly facilitate the elucidation of guard-cell metabolic controls and energetics.

References

ALLAWAY, W.G.: Accumulation of malate in guard cells of *Vicia faba* during stomatal opening. Planta **110**, 63–70 (1973).
ALLAWAY, W.G., HSIAO, T.C.: Preparation of rolled epidermis of *Vicia faba* L. so that stomata are the only viable cells: Analysis of guard cell potassium by flame photometry. Australian J. Biol. Sci. **26**, 309–318 (1973).
ALLAWAY, W.G., SETTERFIELD, G.: Ultrastructural observations on guard cells of *Vicia faba* and *Allium porrum*. Canad. J. Bot. **50**, 1405–1413 (1972).

ARNTZEN, C.J., HAUGH, M.F., BOBICK, S.: Induction of stomatal closure by *Helminthosporium maydis* pathotoxin. Plant Physiol. **52**, 569–574 (1973).
BERNSTEIN, L.: Method for determining solutes in the cell walls of leaves. Plant Physiol. **47**, 361–365 (1971).
BROWN, W.V., JOHNSON, SR.C.: The fine structure of the grass guard cell. Amer. J. Bot. **49**, 110–115 (1962).
CUMMINS, W.R., KENDE, H., RASCHKE, K.: Specificity and reversibility of the rapid stomatal response to abscisic acid. Planta **99**, 347–351 (1971).
DAYANANDAN, P., KAUFMAN, P.B.: Stomata in *Equisetum*. Canad. J. Bot. **51**, 1555–1564 (1973).
EPSTEIN, E.: Mineral nutrition of plants: principles and perspectives. New York-London: Wiley and Sons 1972.
ESHEL, A., WAISEL, Y., RAMATI, A.: The role of sodium in stomatal movements of a halophyte: A study by X-ray microanalysis. In: 7th Intl. Colloquium on Plant Analysis and Fertilizer Problems (J. WEHRMAN, ed.). Hannover, Germany: German Soc. Plant Nutrition 1974.
EVANS, L.T., ALLAWAY, W.G.: Action spectrum for the opening of *Albizzia julibrissin* pinnules, and the role of phytochrome in the closing movements of pinnules and of stomata of *Vicia faba*. Australian J. Biol. Sci. **25**, 885–893 (1972).
FISCHER, R.A.: Stomatal physiology with particular reference to the after-effect of water stress and to behavior in epidermal strips. Ph.D. Dissertation, University of California Davis (1967).
FISCHER, R.A.: Stomatal opening: role of potassium uptake by guard cells. Science **160**, 784–785 (1968a).
FISCHER, R.A.: Stomatal opening in isolated epidermal strips of *Vicia faba*. I. Response to light and to CO_2-free air. Plant Physiol. **43**, 1947–1952 (1968b).
FISCHER, R.A.: Role of potassium in stomatal opening in the leaf of *Vicia faba*. Plant Physiol. **47**, 555–558 (1971).
FISCHER, R.A.: Aspects of potassium accumulation by stomata of *Vicia faba*. Australian J. Biol. Sci. **25**, 1107–1123 (1972).
FISCHER, R.A.: The relationship of stomatal aperture and guard-cell turgor pressure in *Vicia faba*. J. Exptl. Bot. **24**, 387–399 (1973).
FISCHER, R.A., HSIAO, T.C.: Stomatal opening in isolated epidermal strips of *Vicia faba*. II. Response to KCl concentration and the role of potassium absorption. Plant Physiol. **43**, 1953–1958 (1968).
FUJINO, M.: Stomatal movement and active migration of potassium. (In Japanese.) Kaguku **29**, 660–661 (1959).
FUJINO, M.: Role of adenosinetriphosphate and adenosinetriphosphatase in stomatal movement. Sci. Bull. Fac. Educ. Nagasaki Univ. **18**, 1–47 (1967).
FUJINO, M., JINNO, N.: The fine structure of the guard cell of *Commelina communis* L. Sci. Bull. Fac. Educ. Nagasaki Univ. **23**, 101–111 (1972).
GLINKA, Z., REINHOLD, L.: Abscissic acid raises the permeability of plant cells to water. Plant Physiol. **48**, 103–105 (1971).
GRAHAM, R.D., ULRICH, A.: Potassium deficiency-induced changes in stomatal behavior, leaf water potentials, and root system permeability in *Beta vulgaris* L. Plant Physiol. **49**, 105–109 (1972).
HABERMANN, H.M.: Evidence for two photoreactions and possible involvement of phytochrome in light-dependent stomatal opening. Plant Physiol. **51**, 543–548 (1973).
HEATH, O.V.S.: The water relations of stomatal cells and the mechanisms of stomatal movement. In: Plant physiology, a treatise (F.C. STEWARD, ed.), vol. II, p. 193–250. New York-London: Academic Press 1959.
HIRON, R.W.P., WRIGHT, S.T.C.: The role of endogenous abscisic acid in the response of plants to stress. J. Exptl. Bot. **24**, 769–781 (1973).
HOPE, A.B.: Ionic relations of cells of *Chara australis*. X. Effects of bicarbonate ions on electrical properties. Australian J. Biol. Sci. **18**, 789–801 (1965).
HORTON, R.F.: Stomatal opening: the role of abscissic acid. Can. J. Botany **49**, 583–585 (1971).
HORTON, R.F., MORAN, L.: Abscissic acid inhibition of potassium influx into stomatal guard cells. Z. Pflanzenphysiol. **66**, 193–196 (1972).
HSIAO, T.C.: Plant responses to water stress. Ann. Rev. Plant Physiol. **24**, 519–570 (1973a).
HSIAO, T.C.: Effects of water deficit on guard cell potassium and stomatal movement. Plant Physiol. **51**, Suppl., 9 (1973b).

HSIAO, T.C., ALLAWAY, W.G., EVANS, L.T.: Action spectra for guard cell Rb$^+$ uptake and stomatal opening in *Vicia faba*. Plant Physiol. **51**, 82–88 (1973).

HUMBLE, G.D., HSIAO, T.C.: Specific requirement of potassium for light-activated opening of stomata in epidermal strips. Plant Physiol. **44**, 230–234 (1969).

HUMBLE, G.D., HSIAO, T.C.: Light-dependent influx and efflux of potassium of guard cells during stomatal opening and closing. Plant Physiol. **46**, 483–487 (1970).

HUMBLE, G.D., RASCHKE, K.: Stomatal opening quantitatively related to potassium transport. Evidence from electron probe analysis. Plant Physiol. **48**, 447–453 (1971).

ILJIN, W.S.: Wirkung der Kationen von Salzen auf den Zerfall und die Bildung von Stärke in der Pflanze. Biochem. Z. **132**, 494–510 (1922a).

ILJIN, W.S.: Physiologischer Pflanzenschutz gegen schädliche Wirkung. Biochem. Z. **132**, 526–542 (1922b).

IMAMURA, S.: Untersuchungen über den Mechanismus der Turgorschwankung der Spaltöffnungsschließzellen. Jap. J. Bot. **12**, 251–346 (1943).

KETALLAPPER, H.J.: Stomatal physiology. Ann. Rev. Plant Physiol. **14**, 249–270 (1963).

KOWALLIK, W.: Action spectrum for an enhancement of endogenous respiration by light in *Chlorella*. Plant Physiol. **42**, 672–676 (1967).

KUIPER, P.J.C.: Dependence upon wavelength of stomatal movement in epidermal tissue of *Senecio odoris*. Plant Physiol. **39**, 952–955 (1964).

LÄUCHLI, A.: Translocation of inorganic solutes. Ann. Rev. Plant Physiol. **23**, 197–218 (1972).

LEVITT, J.: The mechanism of stomatal action. Planta **74**, 101–118 (1967).

LITZ, R.E., KIMMINS, W.C.: Plasmodesmata between guard cells and accessory cells. Canad. J. Bot. **46**, 1603–1605 (1968).

LIVNE, A., VAADIA, Y.: Water deficits and hormone relations. In: Water deficits and plant growth (T.T. KOZLOWSKI, ed.), vol. III, p. 255–271. New York-London: Academic Press 1972.

LOVEYS, B.R., KRIEDEMANN, P.E.: Rapid changes in abscissic acid-like inhibitors following alterations in vine leaf water potential. Physiol. Plantarum **28**, 476–479 (1973).

LÜTTGE, U., OSMOND, B., BALL, E., BRINCKMANN, E., KINZE, G.: Bisulfite compounds as metabolic inhibitors: nonspecific effects on membranes. Plant Cell Physiol. **13**, 505–514 (1972).

LÜTTGE, U., SCHÖCH, E.V., BALL, E.: Can externally applied ATP supply energy to active ion uptake mechanisms of intact plant cells? Australian J. Plant Physiol. **1**, 211–220 (1974).

MACALLUM, A.B.: On the distribution of potassium in animal and vegetable cells. J. Physiol. **32**, 95–118 (1905).

MANSFIELD, T.A., JONES, R.J.: Effects of abscissic acid on potassium uptake and starch content of stomatal guard cells. Planta **101**, 147–158 (1971).

MAXIMOV, N.A.: The plant in relation to water. A study of the physiological basis of drought resistance. London: Allen and Univin 1929.

MEIDNER, H., MANSFIELD, T.A.: Stomatal responses to illumination. Biol. Rev. **40**, 483–509 (1965).

MEIDNER, H., MANSFIELD, T.A.: Physiology of stomata. New York-London: McGraw-Hill 1968.

MILBORROW, B.V.: The chemistry and physiology of abscissic acid. Ann. Rev. Plant Physiol. **25**, 259–307 (1974).

MITTELHEUSER, C.J., VAN STEVENINCK, R.F.M.: Stomatal closure and inhibition of transpiration induced by (RS)-abscissic acid. Nature **221**, 281–282 (1969).

MOURAVIEFF, I.: Sur les réactions de l'appareil stomatique à l'acide α-hydroxy-2-pyridinemethane sulphonique, inhibiteur de la glycolique oxydase. Compt. Rend. **261**, 4487–4489 (1965).

MOURAVIEFF, I.: Microphotométrie des fluctuations de la teneur en amidon des stomates en rapport avec l'ouverture de l'ostiole à la lumière en présence ou en absence de gaz carbonique. Ann. Sci. Nat. Botan. Biol. Végétale **13**, 361–368 (1972a).

MOURAVIEFF, I.: Action des solutions d'inhibiteurs du transport actif des ions: l'ouabaine, le salicylaldoxime et le carbonyl cyanide *m*-chlorophényl hydrazone, sur le mouvement d'ouverture des stomates à la lumière en présence ou en absence de gaz carbonique. Physiol. Vég. **10**, 547–551 (1972b).

NELSON, S.D., MAYO, J.M.: The occurrence of functional non-chlorophyllous guard cells in *Paphiopedilum* spp. Canad. J. Bot. **53**, 1–7 (1975).

PALLAGHY, C.K.: Electrophysiological studies in guard cells of tobacco. Planta **80**, 147–153 (1968).

PALLAGHY, C.K.: The effect of Ca^{++} on the ion specificity of stomatal opening in epidermal strips of *Vicia faba*. Z. Pflanzenphysiol. **62**, 58–62 (1970).

PALLAGHY, C.K.: Stomatal movement and potassium transport in epidermal strips of *Zea mays*: the effect of CO_2. Planta **101**, 287–295 (1971).

PALLAGHY, C.K., FISCHER, R.A.: Metabolic aspects of stomatal opening and ion accumulation by guard cells in *Vicia faba*. Z. Pflanzenphysiol. **71**, 332–344 (1974).

PALLAS, J.E., JR., DILLEY, R.A.: Photophosphorylation can provide sufficient adenosine 5'-triphosphate to drive K^+ movements during stomatal opening. Plant Physiol. **49**, 649–650 (1972).

PALLAS, J.E., JR., MOLLENHAUER, H.H.: Physiological implications of *Vicia faba* and *Nicotiana tabacum* guard-cell ultrastructure. Amer. J. Bot. **59**, 504–514 (1972).

PALLAS, J.E., WRIGHT, B.G.: Organic acid changes in the epidermis of *Vicia faba* and their implication in stomatal movement. Plant Physiol. **51**, 588–590 (1973).

PENNY, M.G., BOWLING, D.J.F.: A study of potassium gradients in the epidermis of intact leaves of *Commelina communis* in relation to stomatal opening. Planta **119**, 17–25 (1974).

PENNY, M.G., BOWLING, D.J.F.: Direct determination of pH in the stomatal complex of *Commelina*. Planta **122**, 209–212 (1975).

RAGHAVENDRA, A.S., DAS, V.S.R.: Control of stomatal opening by cyclic photophosphorylation. Current Sci. (India) **41**, 150–151 (1972).

RASCHKE, K.: Saturation kinetics of the velocity of stomatal closing in response to CO_2. Plant Physiol. **49**, 229–234 (1972).

RASCHKE, K.: Stomatal Action. Ann. Rev. Plant Physiol. **26**, in press (1975).

RASCHKE, K., FELLOWS, M.P.: Stomatal movement in *Zea mays*: Shuttle of potassium and chloride between guard cells and subsidiary cells. Planta **101**, 296–316 (1971).

RASCHKE, K., HUMBLE, G.D.: No uptake of anions required by opening stomata of *Vicia faba*: Guard cells release hydrogen ions. Planta **115**, 47–57 (1973).

SATTER, R.L., GALSTON, A.W.: Potassium flux: a common feature of *Albizzia* leaflet movement controlled by phytochrome or endogenous rhythm. Science **174**, 518–520 (1971).

SAWHNEY, B.L., ZELITCH, I.: Direct determination of potassium ion accumulation in guard cells in relation to stomatal opening in light. Plant Physiol. **44**, 1350–1354 (1969).

SHAW, M., MACLACHLAN, G.A.: The physiology of stomata. I. Canad. J. Bot. **32**, 784–794 (1954).

SINGH, A.P., SRIVASTAVA, L.M.: The fine structure of pea stomata. Protoplasma **76**, 61–82 (1973).

SQUIRE, G.R., MANSFIELD, T.A.: The action of fusicoccin on stomatal guard cells and subsidiary cells. New Phytologist **73**, 433–440 (1974).

SRIVASTAVA, L.M., SINGH, A.P.: Stomatal structure in corn leaves. J. Ultrastruct. Res. **39**, 345–363 (1972).

STÅLFELT, M.G.: Die stomatäre Transpiration und die Physiologie der Spaltöffnungen. In: Encyclopedia of Plant Physiology (W. RUHLAND, ed.), vol. III, p. 351–426. Berlin-Heidelberg-New York: Springer 1956.

STÅLFELT, M.G.: Die Abhängigkeit des osmotischen Potentials der Stomatazellen vom Wasserzustand der Pflanze. Protoplasma **57**, 719–729 (1963).

STÅLFELT, M.G.: The role of the epidermal cells in the stomatal movements. Physiol. Plantarum **19**, 241–256 (1966).

THOMAS, D.A.: The regulation of stomatal aperture in tobacco leaf epidermal strips. I. The effect of ions. Australian J. Biol. Sci. **23**, 961–979 (1970a).

THOMAS, D.A.: The regulation of stomatal aperture in tobacco leaf epidermal strips. II. The effect of ouabain. Australian J. Biol. Sci. **23**, 981–989 (1970b).

THOMSON, W.W., JOURNETT, R., DE: Studies on the ultrastructure of the guard cells of *Opuntia*. Amer. J. Bot. **57**, 309–316 (1970).

TROMP, J.: Interactions in the absorption of ammonium, potassium, and sodium ions by wheat roots. Acta Botan. Neerl. **11**, 147–192 (1962).

TURNER, N.C.: Action of fusicoccin on the potassium balance of guard cells of *Phaseolus vulgaris*. Amer. J. Bot. **60**, 717–725 (1973).

WILLMER, C.M., KANAI, R., PALLAS, J.E., JR., BLACK, C.C., JR.: Detection of high levels of phosphoenolpyruvate carboxylase in leaf epidermal tissue and its significance in stomatal movements. Life Sci. **12**, 151–155 (1973).

WILLMER, C.M., MANSFIELD, T.A.: A critical examination of the use of detached epidermis in studies of stomatal physiology. New Phytologist **68**, 363–375 (1969a).

WILLMER, C.M., MANSFIELD, T.A.: Active cation transport and stomatal opening: A possible physiological role of sodium ions. Z. Pflanzenphysiol. **61**, 398–400 (1969b).

WILLMER, C.M., MANSFIELD, T.A.: Effects of some metabolic inhibitors and temperature on ion-stimulated stomatal opening in detached epidermis. New Phytologist **69**, 983–992 (1970).

WILLMER, C.M., PALLAS, J.E., JR.: A survey of stomatal movements and associated potassium fluxes in the plant kingdom. Canad. J. Bot. **51**, 37–42 (1973).

WILLMER, C.M., PALLAS, J.E., JR.: Stomatal movements and ion fluxes within epidermis of *Commelina communis* L. Nature **252**, 126–127 (1974).

WRIGHT, S.T.C.: An increase in the "Inhibitor-β" content of detached wheat leaves following a period of wilting. Planta **86**, 10–20 (1969).

YAMASHITA, T.: Influence of potassium supply upon properties and movement of the guard cell. Sieboldia Acta Biol. **1**, 51–70 (1952).

ZELITCH, I.: Environmental and biochemical control of stomatal movement in leaves. Biol. Rev. **40**, 463–482 (1965).

ZELITCH, I.: Stomatal control. Ann. Rev. Plant Physiol. **20**, 329–350 (1969).

ZIMMERMANN, U., STEUDLE, E.: The pressure-dependence of the hydraulic conductivity, the membrane resistance and membrane potential during turgor pressure regulation in *Valonia utricularis*. J. Membrane Biol. **16**, 331–352 (1974).

5. Elimination Processes by Glands

5.1 General Introduction

U. LÜTTGE and M.G. PITMAN

The term *elimination* as conceived by FREY-WYSSLING (1935, 1972) encompasses a vast range of processes which all have in common that matter is released from the metabolically active parts of the plant, i.e. from the "living cytoplasm", and that this release naturally involves *transport*. Nevertheless, it would be far beyond the scope of this Volume, to describe all of these elimination processes, which comprise disposal of substances in the cell vacuoles and other disposal compartments (intracellular transport), deposits of substances in the cell walls (cellulose, pectins, lignins, cutins, waxes and suberins, mineral constituents), and specialized systems such as storage of latex in lacticifers. In many cases single cells or groups of cells have developed that facilitate elimination. We call these cells and tissues glands when they display cytological specialization, i.e. usually a dense cytoplasm with evidently active and numerous organelles and membrane systems, and physiological specialization, i.e. more or less specific elimination processes. A discussion of a few selected cases of elimination by typical glands is attempted in the following Chaps. *5.2* and *5.3*.

To describe elimination processes by glands the terms *secretion* and *excretion* are frequently used, and various attempts have been made to define them, resulting

i) in a *teleological* or *ecological definition,* which is largely adopted from animal physiology, and which uses the term excretion for elimination of end products of metabolism serving no further purposes, and the term secretion for the elimination of substances having a function in establishing relations to the environment;

ii) in a *morphological definition* (HABERLANDT, 1884/1918), where excretion is an elimination out of the plant and secretion into compartments within the plant, and

iii) in a dynamic *physiological* or *biochemical-metabolic definition* (FREY-WYSSLING 1935, 1972), where secretion is the elimination of substances which have been subject to assimilative metabolism after they were taken up by the plant (resorption), where excretion is the elimination of substances which have been subject to dissimilative metabolism after resorption and assimilation, and where the additional term *recretion* is used for substances which are eliminated unmetabolized, i.e. as they were resorbed.

All of these definitions are subject to criticism. The ecological importance of an elimination is often not clear, and research continuously reveals new functions, especially in the elimination of the so-called secondary plant constituents; nectar secretion of floral nectaries has an obvious ecological function, whereas the chemically similar product of extrafloral nectaries has not (definition (i)). Similar or even identical materials are frequently eliminated both to the outside and to the inside (definition (ii)). Products of dissimilative metabolism are often shown to be reintroduced into intermediary metabolism, and it becomes increasingly difficult to maintain the term "secondary" plant constituents; in the case of eliminated water molecules and to some extent also of mineral ions (e.g. PO_4^{3-}, NO_3^-, SO_4^{2-}) it is impossible to say whether they were involved in metabolic reactions before elimination (secretion or excretion) or not (recretion) (definition (iii)). Furthermore, the elimination products of many glands are a mixture of various substances of different chemical nature, although there is frequently a high specificity in that a particular compound or group of compounds prevails in the eliminate (LÜTTGE, 1971: p. 23–24; LÜTTGE, 1973: p. 140 on).

The definition and choice between the terms secretion or excretion becomes a matter of personal predilection. In conclusion, the large diversity of gland-mediated elimination phenomena observed in nature does not allow clearcut definition, and it is no surprise that the terms secretion and excretion are used rather loosely by plant physiologists. We do not regret this, because we feel that it is more important to understand cytological and physiological mechanisms of particular cases than to establish lists of phenomena. Accordingly, the following treatments do not aim to give a comprehensive classification of observations, which space would not allow anyway, but endeavour to describe a few selected cases which are particularly well investigated.

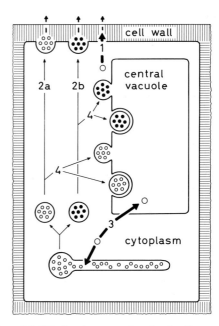

Fig. 5.1. Elimination processes: general cytological model. Elimination out of a gland cell can occur by membrane transport of individual particles, i.e. by *eccrine* secretion or excretion (*1*), or by transport of numerous particles in small vesicles, i.e. by *granulocrine* secretion or excretion (*2*). Granulocrine elimination may or may not include metabolic modification of the particles ((*2b*)+closed circles and (*2a*)+open circles respectively) which are accumulated in the vesicles or in the intracellular compartments generating them. Elimination by disposal into compartments within the cell (e.g. into the central vacuole) may also occur by molecular (*3*) or by vesicular (*4*) transport (see Encyclopedia of Plant Physiology. New Series, Vol. 3). *Lysigenous* (*holocrine*) elimination would involve loss of cell integrity and export of total cell content to the outside. Thick arrows: membrane transport of individual particles (ions or molecules). Thin arrows: transport in compartments (vesicles). Dashed arrows: transport out of the cell wall of glands, which often functions as a free space. Thick lines: membranes. Open circles: a given species of particles to be eliminated from the cytoplasm. Solid circles: metabolized particles

The subdivision of this section is purely operational, but it is justified especially because metabolic implications are very different in the elimination of *mineral ions* and *organic substances* respectively. A very difficult case is the *elimination of water.* Water is eliminated in bulk by transpiration (long distance transport). Of course, this does not involve glands. Special structures are serving water elimination where hydathodes are involved. However, the hydathodes are an extremely heterogenous group of structures both cytologically and functionally (Perrin, 1972). Cytologically they range from simple openings in the epidermis (often non-functional stomata) to typical glands. Functionally it is conceivable that water molecules are never transported actively in a strict thermodynamic sense (see Part A, Chap. *2*), and that the transport of water across membranes is always driven by gradients established by the transport of some sort of solutes, which can range from rather simple mineral ions to relatively complicated organic substances. Hence, gland-like hydathodes eliminate water as a consequence of the elimination of ions and/or hydrophilic organic molecules, or *vice versa,* the elimination of solutes is always accompanied by the elimination of water (e.g. salt glands, nectary glands). In sugar secretion for example, depending on the degree of anatomical and physiological specialization of the structures involved, one observes a gradual variation from hydathodes (dilute secretion) to nectary glands (concentrated secretion: nectar) (Frey-Wyssling and Häusermann, 1960). Therefore, it appears justified that no special section is devoted to water excretion.

In order to conceive mechanistic models of particular cases of elimination processes in plant glands, it is useful to introduce a few specific terms (cf. Schnepf, 1969). This is done by the scheme depicted in Fig. 5.1. This scheme may be considered as a *general model of elimination* which is to be replaced below by models of the particular cases described (Figs. 5.12 and 5.16).

References

Frey-Wyssling, A.: Die Stoffausscheidungen der höheren Pflanzen. Berlin: Springer 1935.
Frey-Wyssling, A.: Elimination processes in higher plants. Saussurea **3**, 79–90 (1972).
Frey-Wyssling, A., Häusermann, E.: Deutung der gestaltlosen Nektarien. Ber. Schweiz. Botan. Ges. **70**, 150–162 (1960).
Haberlandt, G.: Physiologische Pflanzenanatomie. 5th ed. Leipzig: W. Engelmann 1918 (1st ed. 1884).
Lüttge, U.: Aktiver Transport. Kurzstreckentransport bei Pflanzen. Protoplasmatologia. Vol. VIII/7 b. Wien-New York: Springer 1969.
Lüttge, U.: Stofftransport der Pflanzen. Berlin-Heidelberg-New York: Springer 1973.
Perrin, A.: Contribution a l'étude de l'organisation et du fonctionnement des hydathodes: recherches anatomiques, ultrastructurales et physiologiques. Thèse Docteur des Sciences Naturelles, Lyon (1972).
Schnepf, E.: Sekretion und Exkretion bei Pflanzen. Protoplasmatologia, Vol. VIII/8. Wien-New York: Springer 1969.

5.2 Mineral Ions

A.E. HILL and B.S. HILL

1. Introduction

The transport of mineral ions by glands or gland-type cells in plants that have received study by physiologists may conveniently be classed into three groups. First there are the salt glands of halophytes whose function is that of secreting excess ions acquired by the plant during its growth on media of high salinity, often seawater. These glands all possess a similar cellular organisation and excrete both salts and water from the mesophyll of the leaf to the exterior, where the water is lost by evaporation and the salts are removed by subsequent dissolution; in drier habitats they usually form crystalline deposits on the cuticle. This process is therefore one which involves continual loss of water by the plant, and consequently plants possessing these glands are usually found in maritime or salt marsh habitats.

A second group comprises those plants which develop "salt hairs" on the leaf. A salt hair is typically a two-celled structure in which the upper cell accumulates salts and water throughout its life, a process accompanied by considerable swelling, and which ultimately dies. The accumulated salt is thus shed with the hair, which has functioned during its life as a reservoir for salt excreted from the leaf mesophyll. The water loss by leaves with salt hairs is therefore minimal, and throughout the life of the hair its internal water is available to the mesophyll. As a consequence of this, many plants with such hairs can colonize regions both arid and saline.

These two systems are basically for desalination of the leaf parenchyma, and they are of comparable magnitude. Simple calculations show that the salt glands of *Limonium*, which are about as numerous as the bladders of *Atriplex* per unit leaf surface, secrete as much salt as is found in a chenopod bladder in about 2–3 weeks. This is about equal to the average life-time of a bladder when the parent plant is growing in media of high salinity, bladders being continually regrown. It is not easy to measure the sap concentration in the xylem of these halophytes, but such a study has been made on mangroves by SCHOLANDER et al. (1962) and by ATKINSON et al. (1967). It appears that in mangrove species which possess glands the rate of salt input to the leaf *via* the transpiration stream is about equal to its rate of secretion. There is marked diurnal variation in the secretion rate which may therefore be attributed to the diurnal variation in transpiration and hence salt supply. Comparison with non-glandular species shows that although the transpiration rate is very similar, the xylem sap is much less concentrated, being about $^1/_{30}$ that of seawater as compared to about $^1/_4$ seawater in species with glands. ATKINSON et al. showed that in *Rhizophora* the increase in leaf size kept pace with the salt accumulation and so the mean salt concentration in the leaf remained constant, a situation previously found in *Iva oraria* by STEINER (1939). Plants with glands thus seem to lack a powerful salt exclusion mechanism in their root systems but directly control the concentration in the leaf by active excretion.

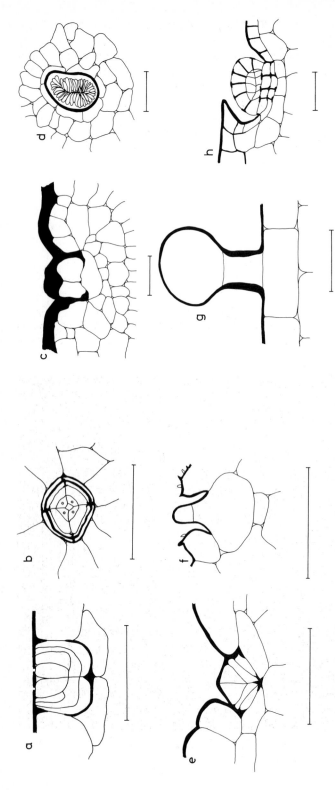

Fig. 5.2a–h. Diagrams of the major gland types discussed in the text. The cuticular insulation is shown as a heavy black continuous line. In *Limonium* and *Tamarix* pores can be seen through the upper gland cuticle, but in *Aegiceras* and *Spartina* the exudate seems to percolate through a cuticle of uncertain structure. A 50 μm line is drawn beneath every section. (a) The *Limonium* salt gland in longitudinal section. (b) The same in transverse section, showing the four exit pores. (c) The *Aegiceras* salt gland in longitudinal section. (d) The same in transverse section, showing the numerous small secretory cells. (e) The *Tamarix* salt gland in longitudinal section. (f) The *Tamarix* salt gland in longitudinal section, showing the small upper and bottle-shaped lower cell. (g) The *Atriplex* salt hair in longitudinal section, showing a young bladder cell and the stalk cell beneath. (h) A secretory cell of *Nepenthes* sited on the inner pitcher surface

The third group represents those glands which produce ionic and proteinaceous secretions, of which only those of *Nepenthes* (the pitcher plant) have been studied (see *5.3.2*). This type of gland is found in many carnivorous plants, and serves not only to regulate the ionic concentration and composition of the digestive fluid, but presumably also to move water as well. In some systems, such as the bladder mechanism of *Utricularia* (SYDENHAM and FINDLAY, 1973), high rates of water transport are observed, which may be due to very efficient coupling of salt and water flows. The three gland systems are here treated separately with respect to their ability to transport or accumulate mineral ions from the mesophyll. In no system is the gland complex supplied directly with vascular tissue.

2. Salt Glands

Salt glands are embedded in the surface of leaves, and their frequency is quite high, from 300–1,000 per cm^2. In size most glands measure from 15 to $30 \cdot 10^{-6}$ metres across, approximately the size of a stomatal complex (Fig. *5.3*). Their precise ontogeny is unknown. What is generally called the "gland cell complex" is an association of several cells encased in a cuticular envelope about 30 micrometres across; as many as 20 cells can sometimes comprise such a gland, and consequently

Fig. *5.3*. A scanning electron microscope picture of the surface of a leaf of *Limonium vulgare*. A gland and three stomates can be seen. Magnification × 400

the cells are very small. The exceptions seem to be salt glands of the Gramineae (the genera *Spartina, Aeluropus* and *Chloris*) where the complex is two-celled, one cell sited above the other (Fig. 5.2f).

2.1 Taxonomic Distribution

Plants with salt glands seem to be of wide general distribution and the following genera have been recognized to date: Plumbaginaceae-*Limonium, Limoniastrum, Acantholimon, Plumbago, Armeria* and *Aegialitis**; Frankeniaceae-*Frankenia, Hypericopsis;* Primulaceae-*Glaux*; Tamaricaceae-*Tamarix, Reaumuria*; Acanthaceae-*Acanthus**, *Neuracanthus*; Avicenniaceae-*Avicennia**; Verbenaceae-*Clerodendron*; Convolvulaceae-*Ipomoea, Cressa*; Combretaceae-*Laguncularia**; Myrsinaceae-*Aegiceras**; Gramineae-*Spartina, Aeluropus, Chloris* (* denotes mangrove spp.).

This list is almost certainly incomplete, and there may be many plants with salt glands still to be recognized.

2.2 Cellular Aspects of the Secretory Process

2.2.1 Cell Structure

Amongst the glands which have received structural study are *Limonium* (RUHLAND, 1915; ZIEGLER and LÜTTGE, 1966; HILL and HILL, 1973c), *Tamarix* (CAMPBELL and STRONG, 1964; THOMSON and LIU, 1967; SHIMONY and FAHN, 1968), *Aegialitis* (ATKINSON et al., 1967), *Aegiceras* (CARDALE and FIELD, 1971) and *Spartina* (SKELDING and WINTERBOTHAM, 1939; LEVERING and THOMSON, 1971). The gland cells are avacuolate, but contain many small quasi-vacuolar spaces which are usually described as vesicles. The nuclei are very prominent, often with more than one nucleolus and the cytoplasm is rich in endoplasmic reticulum and mitochondria. Plastids are without well-developed lamellae and lack grana. A striking feature of most salt glands is that the wall at certain places shows protuberances, which are outgrowths of wall material into the cell. These outgrowths are often quite extensive and usually branched, and the plasma membrane extends around them, thereby increasing its area quite substantially (Figs. 5.4–5.6). In transporting gland cells there often appears to be a clear space between the wall protuberance and the membrane and the latter appears distended.

With the exception of *Spartina* it appears that the cuticular envelope extends continuously around the gland cell complex except for the presence of specific gaps, termed by THOMSON and LIU "transfusion areas". In *Limonium* and *Tamarix* these areas are lateral but in *Aegialitis* and *Aegiceras* there is one basal area (Fig. 5.2). Through these areas the secretory cells of the gland make contact with the surrounding leaf mesophyll by numerous plasmodesmata; in addition, the cells of the gland complex can be seen to be internally connected by plasmodesmata (Fig. 5.4). On the surface the cuticle is perforated by several pores through which the exudate leaves the gland. In the salt glands of dicotyledons therefore, the secretory cells are completely encased in cuticle which opens only into the mesophyll to allow direct symplastic connection with the leaf cytoplasm, and any apoplastic

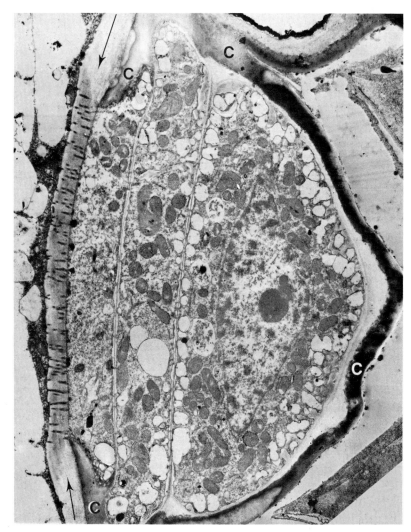

Fig. 5.4. A longitudinal section through a secretory cell of the *Tamarix* salt gland. The transfusion area can be clearly seen on the left (arrows) where the plasmodesmata cross the wall from the adjacent mesophyll. The apparent vesicles or membrane-bounded spaces can be seen in profusion against the inner cell membrane to the right. *C* cuticular insulation. Magnification × 26,400. Reproduced with kind permission from THOMSON and LIU (1967)

route of salt supply can only be obliquely maintained *via* a very small area of uncutinized cell wall and secretory cell plasma membrane into the gland cytoplasm.

In *Spartina*, the gland is composed of two cells (Fig. 5.2f) in which the lower is quite large and again symplastically connected with the cytoplasm of the mesophyll. The cuticular "transfusion area" is very large indeed, comprizing most of the surface of the lower gland cell; this is balanced however by the fact that the transport of salt appears to be occurring at the top of this cell where there is quite a thick lining of cuticle (Fig. 5.2f). The function of the top cell is not quite clear, but presumably it too *(also)* is a secreting gland cell. There is a considerable space between this cell and the cuticular cap to the gland, which seems to be filled with exudate, and pores can be found across the cuticle here. The wall protuberances in this gland are found only on the upper wall of the lower gland cell and the membrane invaginations associated with them are very large and run deeply into the lower cell cytoplasm.

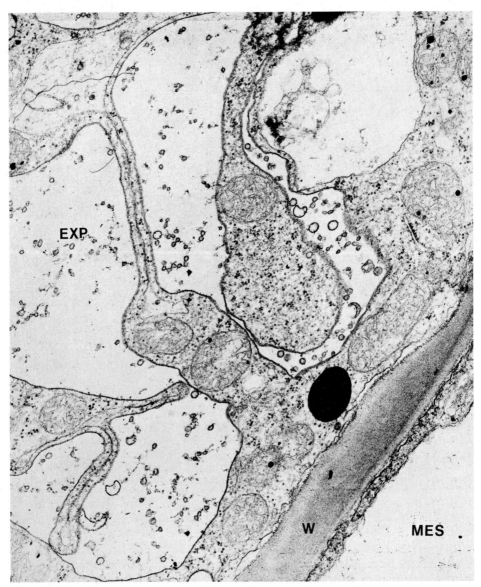

Fig. 5.5. A portion of a secretory cell of the *Spartina* salt gland. The expanded spaces in the cytoplasm (*EXP*) appear as extensions of the plasma membrane, and these extracytoplasmic spaces are distended when the gland is transferring salts and water. *W* wall, *MES* mesophyll cell. Magnification × 28,000. Reproduced with kind permission from Levering and Thomson (1971)

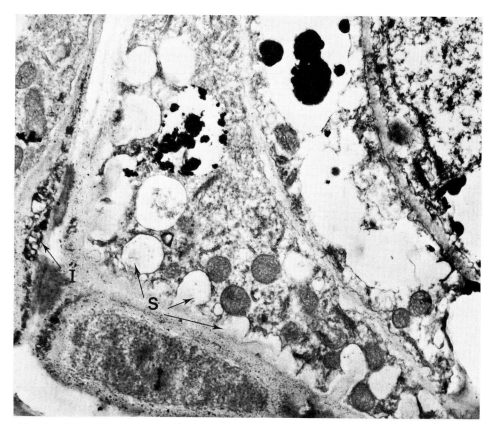

Fig. 5.6. Part of a secretory cell of the *Limonium* salt gland. The extended quasi-vesicular spaces are seen as distension of the plasma membrane, usually built upon wall protuberances (*S*). A skimming section (*I*) shows numerous membrane infoldings. Magnification × 9,000. Reproduced from HILL and HILL (1973c)

Electrophysiological studies have also been made in order to compare various conductance pathways for ions. In *Limonium* it has been shown that the gland cell complex represents the major conductance pathway for ions across the leaf surface, and that the fraction of the surface conductance contributed by the cuticle is very small (HILL and HILL, 1973c).

On the basis of a microelectrode insertion into the *Aegiceras* salt gland BOSTROM and FIELD (1973) have adduced that the active transport of ions is most likely to be associated with the basal membrane of gland cells as it is here that the potential step seems to be found on advancing a microelectrode through the tissue; the cellular potential of the gland cells seems to be identical to the secretory potential, about −40 to −100 millivolts referred to the basal medium. It should be noted however that the cells of the gland are so small that extensive damage must have occurred, and there is every reason to suspect that the potential observed was due to continued electrical contact with the surface phase, which would be at the secretory potential, through the damaged gland surface. In addition, the tissue is so elastic that the position of the microelectrode tip cannot be ascertained with any real precision. In another study BILLARD and FIELD (1974) determined the resistances between various phases of the gland complex by injecting current and noting the potential responses with the same microelectrode, but this is only possible when the resistance of the electrode can be independently monitored (see for example ANDERSON et al., 1974) as it changes upon insertion. Thus it seems that these insertion studies need to be pursued on a higher technical level before any conclusions can be firmly drawn from them.

2.2.2 Coupling to the Mesophyll

A direct method of measuring the coupling between the gland cells and various leaf compartments is provided by compartmental analysis, and to this end ATKINSON et al. (1967) measured the specific activity of radiochloride in the leaf mesophyll and gland exudate of *Aegialitis* as a function of time, when high salt tissue was fed with label through the transpiration stream. The specific activity of the exuded chloride rose to a plateau of 40–50% that of the basal medium within 5 h, whilst that of the tissue chloride ranged from 20–30%. They interpreted this in terms of a direct chloride pathway to the gland cells *via* the apoplasm, but the slow time course of the rise suggests a correlation with the cytoplasm rather than the apoplasm. The results could more legitimately be interpreted as radiochloride entering the cytoplasm and then passing to the gland cells and the vacuole in parallel; the large chloride pool of the vacuole would then keep the mean specific activity of the leaf cells well below that of the glandular exudate.

In *Limonium* it can be shown by non-aqueous extraction of chloroplasts that their ion content, and hence that of the cytoplasm, is directly controlled by the salt gland (LARKUM and HILL, 1970), and an analysis of ionic compartments in the leaf by efflux analysis has revealed four spaces for chloride and three for sodium (HILL, 1970a). Three of these for each ion corresponded to the classical three compartments of plant tissue, free-space, cytoplasm and vacuole. Those assigned to the cytoplasm had halftimes of about 10 min for chloride and 30 min for sodium. Appearance of labelled ions in the exudate when these were added to the medium was asymptotic, and characterized by transit-half-times which agreed closely with the cytoplasmic half-times for each ion. The compartment with the shortest half-time was identified as the free-space as its loading could be predicted from a knowledge of the thickness of the lamina and the ionic mobilities, and it showed no specificity. The free-space half-time is so small compared with the transit half-time, that any direct coupling with the extrusion process is impossible. This indicates that the cytoplasm is freely coupled to the salt transport mechanism of the gland cell, and that the apoplastic route is negligible.

A similar analysis could well be extended to other genera, especially as some authors have assumed that the glands are supplied wholly *via* the tissue free space (POLLACK and WAISEL, 1970).

2.2.3 Transport and the Gland Cell Membranes

A theory of salt transport across the gland has come from electron microscopical examination of the secretory cell cytoplasm. Although gland cells contain no large central vacuole they do contain many apparent vesicles, and this has led ZIEGLER and LÜTTGE (1966, 1967) to suggest that in *Limonium* salt is transported over the plasmalemma by a process of reverse pinocytosis, the vesicles enclosing salt and water in the cytoplasm and discharging into a glandular free-space by membrane fusion (i.e. the secretion is granulocrine, Chap. 5.1, Fig. 5.1). This scheme has also been proposed by THOMSON et al. (1969) to explain the apparent presence of rubidium in vesicles of *Tamarix* gland cytoplasm when plants were fed this ion in their basal medium; the metal ion appeared as electron-dense material which was absent from rubidium-free plants. In the anatomical study of the *Tamarix* gland by THOMSON and LIU (1967) the vesicles can be seen clustered closely up against the plasma membrane (Fig. 5.4).

Such a proposition must however involve the following points: (i) the membrane of the vesicles must possess selective transport systems, as unlike a simple pinocytotic mechanism, the exudate does not represent a sequestration of the cytoplasmic milieu with all its solutes and soluble proteins, (ii) if the vesicles are fusing with the gland cell plasmalemma at a high rate, as they should do to accommodate the observable fluid transfer, then this membrane must contain the same transport systems and be pumping the same ions, (iii) if vesicles of the size seen in *Limonium* and *Tamarix* (0.5–1.0 μm) are transferring the exudate at the rates observed, then simple calculation shows that in the secretory cell the membrane turnover is as high as if the plasmalemma were recycling about 30–40 times an hour (HILL and HILL, 1973c). If the vesicles are smaller, then the turnover rate is much higher, but if they contain concentrated salt solution the rate could be lower. It is difficult to see, though, how the vesicle with its enormous surface-to-volume ratio could resist such an osmotic swelling force. In *Limonium* it is clear that most of the "vesicles" are merely cross-sections of plasma membrane invaginations, and it certainly seems simpler that, as in most animal systems, the salt transport involves a direct membrane transfer.

A more serious objection to the vesicular concept has come from a consideration of the effects of voltage clamps on the active ion fluxes, described above (HILL and HILL, 1973a). The active fluxes of both sodium and chloride ions and their apparent stoichiometry were found to depend strongly upon the clamping current. This result was explained by a model involving electrogenic pumping into a channel; it would be impossible, however, for the clamp current to flow unilaterally across the membranes of cytoplasmic vesicles, as these are distinct entities electrically uncoupled from the plasma membrane. They should be unaffected by clamping and function like a neutral salt pump. In the *Spartina* gland the quasi-vacuolar spaces in the lower secretory cell are large and irregular in shape (Fig. 5.5), resembling those in the avian salt gland (KOMNICK, 1963). When secretion is not occurring the cytoplasm is traversed by folded membranes and it is quite clear that these distend themselves during secretion to accommodate the secreted solution, being continuous invaginations of the plasma membrane (LEVERING and THOMSON, 1971). In the absence of any direct experiments involving vesicles in the salt transfer process, it would seem that in *Limonium* at least, the exudate is produced by ionic pumping at the plasma membrane, and that the vesicles are merely cross sections of plasma membrane invaginations (Fig. 5.6).

Table 5.1. The rate and composition of the exudate of several salt gland systems

Species	Exudation (M)	Rate ($mm^3\ cm_{leaf}^{-2}\ h^{-1}$)	Ref.
Limonium vulgare	medium: 0.1–0.7 NaCl exudate: 0.35–0.5 Cl$^-$	0.03	Arisz et al. (1955)
Limonium gmelinii	—	0.28–0.86	Ruhland (1915)
Aegialitis annulata	medium: seawater exudate: 1.0 NaCl	0.3	Atkinson et al. (1967)
Aegialitis annulata	medium: seawater exudate: 0.3–0.8 NaCl	0.03	Scholander et al. (1962)
Aegiceras corniculatum	medium: seawater exudate: 0.2–0.5 NaCl	0.004–0.01	Scholander et al. (1962)
Avicennia marina	medium: seawater exudate: 0.7 NaCl	0.01–0.02	Scholander et al. (1962)
Aeluropus litoralis	medium: 0.3 NaCl exudate: 0.4–1.2 Na$^+$	—	Pollack and Waisel (1970)
Spartina townsendii	medium: seawater exudate: 0.5 NaCl	—	Skelding and Winterbotham (1939)

Notes: Most authors found that the osmotic pressure (O.P.) of the exudate was accounted for mostly by NaCl; in some cases e.g. Ruhland (1915), the O.P. was often some 20% higher than that predicted on an NaCl basis. Berry (1970) collected *Tamarix* exudate, but his presentation is rather complex and should be consulted for details.

2.3 Ion Transport Studies

The rate of excretion and the composition of the exudate of several salt gland systems are summarized in Table 5.1. Although Schtscherback (1910) observed the movement of various salts out of the *Limonium* salt gland, Ruhland (1915) was the first to measure the salt concentration in the exudate. He came to the conclusion that little osmotic work is performed by the glands, and that the exudation is achieved by hydrostatic pressure. Arisz et al. (1955) showed that not only was there osmotic work done, but that chloride secretion was dependent upon oxygen and partially upon light. The secretion was increased by raising the temperature, but ultimately reduced by several respiratory and glycolytic inhibitors at high enough concentration. In almost all treatments the chloride concentration in the exudate and the rate of fluid secretion were similarly affected, so that the concentration remained unaffected. Only an increase in the osmotic concentration of the basal medium diminished the rate of secretion and raised the exudate concentration, leaving the rate of chloride transport little affected. Arisz et al. concluded that chloride salts were accumulated in the gland cells by an active mechanism and the resulting hydrostatic pressure forced out the exudate, i.e. that the glands represented a *local osmotic space*. This can be shown to be incorrect, however, as the glands will still transport fluid at a low rate when plasmolyzed in high salt concentrations.

Electrochemical potentials of sodium, potassium and chloride ions were subsequently measured in the *Limonium* exudate, and these indicated that all three ions were actively transported into the fluid (HILL, 1967a). Voltage clamp experiments on *Limonium* have shown that a stable negative short-circuit current can be drawn from the gland on chloride media, and that this current can be correlated with the net ion fluxes (HILL, 1967b). The current is increased by progressively higher salt loading, and is of 2–3 μA magnitude in the light and dark. The charge flow out of the gland cells, like the secretory potential, is always negative. The current is strongly dependent upon the presence of chloride (halide) ions, although its magnitude depends to some extent upon the accompanying cation. Measurement of the net fluxes of sodium and chloride ions using tracers has shown that both ions are actively extruded from the gland cells under short-circuit conditions, and that the clamp current accounts for the difference between the net fluxes (HILL, 1967b). In addition, most alkali metal and halide ions show flux ratios which depart strongly from unity under short-circuit conditions, indicating that they too are actively extruded.

It is possible, however, that cation and anion fluxes could be electrically coupled in localized regions within the gland complex, where, owing to the complicated geometry of the membranes, the potential has not been reduced to zero by the short-circuit technique. To test this, a study was made of the effect of clamping the transglandular potential at various positive and negative values on the sodium and chloride fluxes, with and without inhibitors. The results were interpreted in terms of electrogenic chloride pumping into a glandular lumen accompanied by passive sodium transfer, a model suggested by the structure of the system (HILL and HILL, 1973a). This would explain not only the strong dependence of the secretion on chloride ions, but also the fact that organic cations can also be transported by these cells.

In the mangrove *Aegialitis* ATKINSON et al. have also shown that chloride is the principal anion in the exudate, and that the depression of the freezing point can be predicted from the chloride titre. Apart from their study of the specific activity of the extruded chloride described in a previous section, they also showed that the rate of glandular secretion is controlled by the delivery of salt to the leaves by the transpiration stream. The concentration of exudate from leaves *in situ* was almost 1.0 M, and this was balanced by a mean xylem water potential of -4.7 MPa, contributed almost entirely by the hydrostatic component as determined by the pressure-bomb technique. Results of a similar nature were obtained by SCHOLANDER et al. (1962) who showed that in *Aegialitis, Avicennia* and *Aegiceras* chloride was also the major anion in the exudate, and that where large negative hydrostatic pressures were found in the xylem, these were offset by high osmotic pressure of the leaf sap enabling water to move into the parenchyma (SCHOLANDER, 1968).

The composition of the exudate from the *Tamarix* gland has been shown to contain varying amounts of monovalent and divalent cations (BERRY and THOMSON, 1967; WAISEL, 1961), but it has also been claimed that when growing on dilute Hoagland solution, and when salt-loaded with sodium chloride, chloride ions were at low concentration in the exudate which contained large amounts of sulfate, bicarbonate and micronutrients (BERRY, 1970). This peculiar result is at odds with most other determinations of exudate composition, and in view of the very drastic treatment to which the leaves were subjected, should be treated with reserve. The *Aeluropus* gland will also transport various amounts of potassium ions when these are added to the basal medium (Pollack and Waisel, 1970).

2.4 Energetics and Metabolism

The secretory cell cytoplasm of salt glands is supplied with mitochondria but does not contain apparently functional chloroplasts; it would thus seem that respiratory energy should be directly available to drive the ion transport but that light energy could be provided only by diffusion of high energy compounds into the cells from the surrounding mesophyll. Arisz et al. found that the exudation of chloride in *Limonium* was stimulated several-fold by light, and that inhibitors of glycolysis, respiration and phosphorylation all progressively abolished the transport as their concentration was raised, although most of them showed stimulations at intermediate concentrations. Anaerobiosis was very effective in suppressing exudation, but there is no evidence whether this was conducted under light or dark conditions. In the *Aegialitis* system Atkinson et al. similarly showed that the osmotic work performed by the glands could be abolished by the uncoupler CCCP. Using the short-circuit current as a direct measure of the electrogenic chloride pump in *Limonium* tissue, Hill and Hill (1973b) investigated the effects of anaerobiosis, and found that CO_2-free nitrogen abolished the current completely in the dark but not in the light. Inhibitors of photosystem II had little effect, but CCCP was very effective. The results have been interpreted as ATP being the energy source for chloride pumping in this system. In support of this, the tissue has been shown to contain a Cl^--stimulated ATPase, which behaves in many ways similarly to the chloride transport system, the only anion-specific ATPase yet found in a plant cell. Whether this ATPase is in fact the chloride pump has yet to be confirmed, but it seems that in the *Limonium* gland, the chloride transport is not driven by electron transport or reducing power as may be the case in other tissues (MacRobbie, 1970; see *4.2.3.4*; and Part A, *6.6.2.5*; *6.6.3.3*).

The effect of salt loading on the transport system has been followed in the *Limonium* gland by recording the secretory potential, zero clamp current and ion fluxes as a function of time after transfer of low-salt leaf tissue from water to sodium chloride solution (Hill, 1970b). The time course is similar for all the parameters of transport, and cannot be correlated with the filling of compartments in the leaf cells as can the transit flux of label in the steady state. The time course is sigmoid, rising to a plateau after 3 h, and is sensitive to inhibitors of protein synthesis and low temperature (Shachar-Hill and Hill, 1970); it has therefore been argued that this represents a phase of induction of chloride transport by the salt load, and that the chloride pump is here being synthesized and assembled (see *9.3.3.4*). A further observation is that the transglandular conductance rises during the phase of induction following a salt load in close parallel with the rise in short-circuit current, suggesting that part of the conductance may be that of an inducible charge transfer process in the gland cells. This proposition has also been strongly supported by the fact that the Cl^--ATPase activity rises markedly after similar salt loading, and its rise can be blocked by inhibitors (Hill and Hill, 1973b, c). These authors have claimed that such an induction of ion pumps is a necessary part of a homeostatic mechanism which allows the plant to adapt physiologically to a wide range of salinities.

In a briefly reported study of aspects of lipid metabolism in salt glands BENSON and ATKINSON (1967) found that sulfate and phosphate were rapidly and extensively incorporated into choline esters in salt gland plants but not in others, and that the glands were rich in phospholiphase-D activity; they speculated that the phosphatidic acid cycle could be operative in these salt-transporting cells as has been shown for those of avian salt glands (HOKIN and HOKIN, 1960).

2.5 Water Transport

Nothing is really known about the water-salt coupling. Apart from the suggestion of ARISZ et al. that fluid flow is produced by hydrostatic pressure, it has also been proposed that water is withdrawn osmotically after the salt, presumably within a local osmotic space (SUTCLIFFE, 1962). This gains support from the demonstration by ATKINSON et al. that the water potential of the exudate is virtually that of the xylem fluid, and from the experiments of ARISZ et al. in which the rate of exudation was decreased by increases in the osmotic pressure of the basal medium; electro-osmosis has not been ruled out however.

3. Salt Hairs

The salt hair that has been extensively studied is that of *Atriplex* (Chenopodiaceae) and the following discussion is almost entirely confined to this system (Fig. 5.7).

3.1 Structure and Supply

The chenopod hair is a two-celled structure sited upon the epidermis (Fig. 5.2g). The lower cell is a stalk cell containing extensive endoplasmic reticulum, poorly developed plastids, and no central vacuole. It resembles the secretory cells of salt glands. The upper cell termed the "bladder cell" is up to 200 µm in diameter and has a similar cytoplasm but an enormous vacuole which can occupy 95% of the cell volume. The stalk cell is connected by plasmodesmata to the epidermal cell and the bladder cell. It seems that the salt is accumulated in the bladder cell vacuole, and that it can move to the bladder cell through the symplastic connections between the cells. Autoradiographs show concentrations of radiosulfate and radio-chloride in the hairs, but no specifically in the vacuoles (OSMOND et al., 1969). The whole structure is covered with a waxy material which reduces transpiration, and eventually the hair dies, depositing its salt as a scale on the surface; in some species depending also upon the habitat, a leaf will grow several populations of bladders during its lifetime.

Fig. 5.7. A scanning electron microscope picture of *Atriplex spongiosa* epidermal bladder and stalk cells (with kind permission from TROUGHTON and DONALDSON, 1972). Magnification × 50

3.2 Salt Accumulation

In the extreme halophyte *Atriplex halimus* the accumulation is so great as to pose problems (MOZAFAR and GOODIN, 1970). On salt loading (0.1 M NaCl + 0.1 M KCl) bladders contained 60 times as much salt as the leaf cells per unit water, and the equivalent of 5–6 M Na$^+$ and K$^+$ ions, the chloride concentration reaching 9.2 M with large amounts of oxalate. Vapor phase osmometry showed that the bladder water potential was less than −50 MPa and that this more than matched the osmotic pressure. In the steady state the water potential in the bladders must match that of the mesophyll cells, but it is difficult to see how the discrepancy can be made up by turgor pressure if the salt is accumulated in the vacuole of the salt hair. If pressure differences are ruled out then two alternatives remain. If the bladder salt is all in solution (and phase contrast microscopy revealed no crystals *in vivo*) then the leaf sap might match the osmotic pressure with an organic osmotic effector of its own; it would have to be present in the cells at enormous concentration. Another alternative is that the bladder vacuoles can actively eliminate water, which is there at a permanently lower potential than in the lamina. Whatever the explanation, the osmotic relations of this system are very interesting and as yet unsolved.

3.3 Ion Transport Studies

Microelectrode studies have shown that the bladder cell vacuole is 80 to 100 mV negative with respect to the external solution, and that the mesophyll cell vacuoles also have a high negative potential (LÜTTGE and PALLAGHY, 1969; OSMOND et al., 1969). Transitions from light to dark produce abrupt transients in the membrane potential, which LÜTTGE and PALLAGHY showed is due to light absorption in the photosynthetic wavelength region. In *Chenopodium*, the epidermis was stripped off with its hairs, and those that seemed undamaged no longer showed transients: thus it seems that either the effect emanates from the chlorenchyma and that light energy must find its way into the hair from there, or that there is good electrical coupling and the transient represents an electrical signal in the chlorenchyma (see 2.5.1). It is not evident that light causes any consistent hyperpolarization as might be expected from a negative charge transfer process.

OSMOND et al. measured the chloride concentration both in the bladders, which were removed mechanically after freezing, and in the leaf cells. The Nernst equilibrium potential for chloride differed from the measured vacuolar potential by about 100 mV in bladders and 130 mV in leaf cells, with respect to the basal medium (5 mM KCl). It appears that chloride must be actively accumulated into the bladder cell vacuole as into the mesophyll cells, but the results do not indicate that the bladders are in fact functioning as salt-accumulating systems with respect to the mesophyll. The fluxes of radiochloride into bladders and lamina are also very similar, except that the input to the bladders shows a lag phase. It is only at higher sodium chloride concentrations in the medium that the salt content of the bladders reveals that accumulation is in fact occurring; on 250 mM NaCl solution they contain 4–5 times as much NaCl as the leaf cells per unit water. Although a vacuolar uptake mechanism seems probable as the basis of salt accumulation in this system, it has no experimental support as yet.

3.4 Energetics and Metabolism

It is impossible to study the uptake of chloride directly into the isolated bladder cell of *Atriplex*, and the fluxes into the hair are therefore dependent upon the initial transport into the mesophyll and the chloride concentration there. In a study of the overall process, LÜTTGE and OSMOND (1970) found that chloride uptake was stimulated by light, but in the dark uncouplers of phosphorylation were very effective inhibitors. In the light, the inhibitor of photosystem II, DCMU, was partially effective whilst the remaining uptake was abolished by uncouplers (FCCP and CCCP). CO_2-fixation was mainly confined to the chlorenchyma, the bladders accounting for about 10% of the total carbon fixed. These results have been interpreted as showing that ATP is the energy source in the dark and photosynthetic electron transport or reducing power that in the light; the effect of DCMU is to reduce the light flux to the dark level. The assumptions upon which this rests are (i) that respiratory metabolism is unaffected by photosynthesis, the two processes being simply additive; this is not supported by recent work. (ii) That there are no diffusion problems encountered in using the inhibitors in leaf-disc experiments. It has been previously shown that diffusion problems in such discs may indeed be present (OSMOND, 1968). It is difficult to see how a simple transport system could use two such different energy sources, and this may be due to the fact that the uptake studied here is really a complex overall process involving several steps.

4. Secretory Glands of Carnivorous Plants

Of this large group the only one that has received any study where ion transport is concerned is the *Nepenthes* gland. The glands are found on the inside of the pitcher (a modified leaf), and probably fulfil the functions of fluid transfer, enzyme secretion, ion transport, and solute reabsorption. Isolated pitcher discs with their outer cuticle removed will secrete both ions and water from glands on the inner surface.

4.1 Structure and Supply

The glands have been described by STERN (1917) in considerable detail, using the light microscope. They are composed of roughly isodiametric cells with thin walls, and the outer cells are cutinized on their lateral but not on their transverse walls (see Fig. 5.2h). As the passage of ions and water is also in a transverse direction, these "transfusion areas" are probably where symplastic connections are made between cells, although this cannot be seen by light microscopy. The gland cell cytoplasm is very dense without any obvious vacuole. A vascular trace usually runs near the gland, but not directly to it.

4.2 Ion Transport Studies

The pitcher fluid contains Na^+, K^+, Mg^{2+} and Ca^{2+} ions, the principal anion being Cl^- at a concentration of about 20 mM (LÜTTGE, 1966a; MORRISSEY, 1955; NEMČEK et al., 1966). There are also appreciable amounts of hydrogen ions at some stages, as the pH can reach 2.0, but these have been little considered, and the fluid from unopened pitchers is much nearer to neutrality.

In a series of investigations by LÜTTGE (1964, 1965, 1966a, 1966b) the transport of chloride ions has been studied, using preparations of pitcher wall with the external cuticle removed. In an uptake study using ^{36}Cl it was found that when tissue had taken up the label, it could be detected in the glandular exudate against a chemical potential difference, and the gland flux was sensitive to cyanide, showing an initial stimulation and then severe inhibition above 1 mM cyanide concentration (cf. similar effect, ARISZ et al., 1955). The back flux from tissue to medium was little affected. The approximate K_m for through-transport of chloride to the exudate was 30 mM. The inhibitors cyanide and arsenate both reduced the secretion rate, but left the chloride concentration of the fluid unaffected. Histoautoradiography of the tissue shows that the gland cells are not sites of chloride accumulation, but radiosulfate can be preferentially localized in them when this is being absorbed from the pitcher fluid.

The potential difference between the pitcher fluid and the vacuoles of pitcher cells has been determined (NEMČEK et al., 1966) and has a mean value of -35 mV referred to the fluid. It is not really clear whether any true gland cells were among those whose potential was measured. When tissue discs were mounted between short-circuit chambers filled with Ringer's solution, the secretory potential was $+50$ mV and the clamp current $+10$–15 μA. Changing to pure sodium sulfate or nitrate solutions at 10–20 mM concentration made little difference to the current, but a similar concentration of potassium salts or choline chloride reversed the potential and the current to about -5 μA.

As a result of an analysis of the tissue salt concentration, NEMČEK et al. calculated the Nernst equilibrium potentials for Na^+, K^+ and Cl^- ions across the inner pitcher wall, and comparing it with the measured potential, deduced that Na^+ and Cl^- ions were actively pumped from fluid to cells, whilst K^+ was in passive equilibrium. The short-circuit data (cited above) indicated that this was probably true except for K^+ ions, which seemed to be actively extruded from the cells. They suggested that there might be a large K^+ ion shunt which would have the effect of reducing the Nernst potential for this ion, or that the K^+ ions may be pumped at a deeper tissue site. The former situation would not lead to any apparent K^+-clamp current however, and is unlikely.

Unfortunately the authors have misinterpreted the current data, and the results indicate a different polarity for the transport processes. If the secretory potential is positive in sodium-only media for example, then the clamp current must be negative, and the electrogenic process indicated is a sodium transport into the pitcher fluid. It would thus seem that on the basis of the data alone, Na^+ and Cl^- are pumped into the lumen presumably by the glands, and K^+ is absorbed; the direction of chloride transport then accords with the data of LÜTTGE that transport of Cl^- into the pitcher fluid is an active process. It should also be noted that (i) the potential difference measured between the epidermal cells and the fluid may not be that between the gland cells and fluid, where the active processes are presumably located, (ii) the potential is certainly between the vacuole of the epidermal cells and fluid, and this must therefore include potential steps (e.g. that of the tonoplast) which are unrelated to the secretion; the internal tissue concentrations are also those of whole cells, comprising several compartments, (iii) changes in clamp current are difficult to interpret when the bathing solutions are changed abruptly, as plant tissue is full of salts and has a slow diffusion time-constant. Nevertheless, this work makes an interesting starting point for future studies on the electrophysiology of this tissue.

References

ANDERSON, W.P., HENDRIX, D.L., HIGINBOTHAM, N.: Higher plant cell membrane resistance by an intracellular microelectrode method. Plant Physiol. **53**, 122–124 (1974).

ARISZ, W.H., CAMPHUIS, I.J., HEIKENS, H., VAN TOOREN, A.J.: The secretion of the salt glands of *Limonium latifolium* Ktze. Acta Botan. Neerl. **4**, 322–338 (1955).

ATKINSON, M.R., FINDLAY, G.P., HOPE, A.B., PITMAN, M.G., SADDLER, H.D.W., WEST, K.R.: Salt regulation in the mangroves *Rhizophora mucronata* Lam. and *Aegialitis annulata* R. Br. Australian J. Biol. Sci. **20**, 589–599 (1967).

BENSON, A.A., ATKINSON, M.R.: Choline sulphate and choline phosphate in salt excreting plants. Federation Proc. **26**, 394 (1967).

BERRY, W.L.: Characteristics of salts secreted by *Tamarix aphylla*. Amer. J. Bot. **57**, 1226–1230 (1970).

BERRY, W.L., THOMSON, W.W.: Composition of salt secreted by salt glands of *Tamarix aphylla*. Canad. J. Bot. **45**, 1774–1775 (1967).

BILLARD, B., FIELD, C.D.: Electrical properties of the salt gland of *Aegiceras*. Planta **115**, 285–296 (1974).

BOSTROM, T.E., FIELD, C.D.: Electrical potentials in the salt gland of *Aegiceras*. In: Ion transport in plants. (W.P. ANDERSON, ed.). London-New York: Academic Press 1973.

CAMPBELL, C.J., STRONG, J.E.: Salt gland anatomy in *Tamarix pentandra* (Tamaricaceae). S. West. Nat. **9**, 232–238 (1964).

CARDALE, S., FIELD, C.D.: The structure of the salt gland of *Aegiceras corniculatum*. Planta **99**, 183–191 (1971).

HILL, A.E.: Ion and water transport in *Limonium*. I. Active transport by the leaf gland cells. Biochim. Biophys. Acta **135**, 454–460 (1967a).

HILL, A.E.: Ion and water transport in *Limonium*. II. Short-circuit analysis. Biochim. Biophys. Acta **135**, 461–465 (1967b).

HILL, A.E.: Ion and water transport in *Limonium*. III. Time constants of the transport system. Biochim. Biophys. Acta **196**, 66–72 (1970a).

HILL, A.E.: Ion and water transport in *Limonium*. IV. Delay effects in the transport process. Biochim. Biophys. Acta **196**, 73–79 (1970b).

HILL, A.E., HILL, B.S.: The electrogenic chloride pump of the *Limonium* salt gland. J. Membrane Biol. **12**, 129–144 (1973a).

HILL, A.E., HILL, B.S.: The *Limonium* salt gland: a biophysical and structural study. Intern. Rev. Cytol. **35**, 229–320 (1973c).

HILL, B.S., HILL, A.E.: ATP-driven chloride pumping and ATP-ase activity in the *Limonium* salt gland. J. Membrane Biol. **12**, 145–158 (1973b).

HOKIN, L.E., HOKIN, M.R.: Studies on the carrier function of phosphatidic acid in sodium transport. J. Gen. Physiol. **44**, 61–85 (1960).

KOMNICK, H.: Elektronenmikroskopische Untersuchungen zur funktionellen Morphologie des Ionentransportes in der Salzdrüse von *Larus argentatus*. Protoplasma **56**, 274–314 (1963).

LARKUM, A.W.D., HILL, A.E.: Ion and water transport in *Limonium*. V. The ionic status of chloroplasts in the leaf of *Limonium vulgare* in relation to the activity of salt glands. Biochim. Biophys. Acta **203**, 133–138 (1970).

LEVERING, C.A., THOMSON, W.W.: The ultrastructure of the salt gland of *Spartina foliosa*. Planta **97**, 183–196 (1971).

LÜTTGE, U.: Untersuchungen zur Physiologie der Carnivoren-Drüsen. Ber. Deut. Botan. Ges. **77**, 181–187 Suppl. (1964).

LÜTTGE, U.: Untersuchungen zur Physiologie der Carnivoren-Drüsen. II. Über die Resorption verschiedener Substanzen. Planta **66**, 331–344 (1965).

LÜTTGE, U.: Untersuchungen zur Physiologie der Carnivoren-Drüsen. IV. Die Kinetik der Chloridsekretion durch das Drüsengewebe von *Nepenthes*. Planta **68**, 44–56 (1966a).

LÜTTGE, U.: Untersuchungen zur Physiologie der Carnivoren-Drüsen. V. Mikroautoradiographische Untersuchung der Chloridsekretion durch das Drüsengewebe von *Nepenthes*. Planta **68**, 269–285 (1966b).

LÜTTGE, U., OSMOND, C.B.: Ion absorption in *Atriplex* leaf tissue. III. Site of metabolic control of light-dependent chloride secretion to epidermal bladders. Australian J. Biol. Sci. **23**, 17–25 (1970).

LÜTTGE, U., PALLAGHY, C.K.: Light triggered transient changes of membrane potentials in green cells in relation to photosynthetic electron transport. Z. Pflanzenphysiol. **61**, 58–67 (1969).

MACROBBIE, E.A.C.: The active transport of ions in plant cells. Quart. Rev. Biophys. **3**, 251–294 (1970).

MORRISSEY, S.: Chloride ions in the secretion of the pitcher plant. Nature **176**, 1220–1221 (1955).

MOZAFAR, A., GOODIN, R.J.: Vesiculated hairs: a mechanism for salt tolerance in *Atriplex halimus*. Plant Physiol. **45**, 62–65 (1970).

NEMČEK, O., SIGLER, K., KLEINZELLER, A.: Ion transport in the pitcher of *Nepenthes henryana*. Biochim. Biophys. Acta **126**, 73–80 (1966).

OSMOND, C.B.: Ion absorption in *Atriplex* leaf tissue. I. Absorption by mesophyll cells. Australian J. Biol. Sci. **21**, 1119–1130 (1968).

OSMOND, C.B., LÜTTGE, U., WEST, K.R., PALLAGHY, C.K., SHACHAR-HILL, B.: Ion absorption in *Atriplex* leaf tissue. II. Secretion of ions to epidermal bladders. Australian J. Biol. Sci. **22**, 797–814 (1969).

POLLACK, G., WAISEL, Y.: Salt secretion in *Aeluropus litoralis*. Ann. Botany **34**, 879–888 (1970).

RUHLAND, W.: Untersuchungen über die Hautdrüsen der Plumbaginaceen. Ein Beitrag zur Biologie der Halophyten. Jahrb. Wiss. Botany **55**, 409–498 (1915).

SCHOLANDER, P.F.: How mangroves desalinate seawater. Physiol. Plantarum **21**, 251–261 (1968).

SCHOLANDER, P.F., HAMMEL, H.T., HEMMINGSEN, E., GAREY, W.: Salt balance in mangroves. Plant Physiol. **37**, 722–729 (1962).

SCHTSCHERBACK, J.: Über die Salzausscheidung durch die Blätter von *Statice gmelini* (Vorläufige Mitteilung). Ber. Deut. Botan. Ges. **28**, 30–34 (1910).

SHACHAR-HILL, B., HILL, A.E.: Ion and water transport in *Limonium*. VI. The induction of chloride pumping. Biochim. Biophys. Acta **211**, 313–317 (1970).

SHIMONY, C., FAHN, A.: Light- and electron-microscopical studies on the structure of salt glands of *Tamarix aphylla* L. J. Linn. Soc. (Bot.) **60**, 283–288 (1968).

SKELDING, A.D., WINTERBOTHAM, J.: The structure and development of the hydathodes of *Spartina townsendii* Groves. New Phytologist **38**, 69–79 (1939).

STEINER, M.: Die Zusammensetzung des Zellsaftes bei höheren Pflanzen in ihrer ökologischen Bedeutung. Ergeb. Biol. **17**, 151–254 (1939).

STERN, K.: Beiträge zur Kenntnis der Nepenthaceen. Flora (Jena) **109**, 213–282 (1917).

SUTCLIFFE, J.F.: Mineral salt absorption in plants, p. 159–162. London: Pergamon Press 1962.

SYDENHAM, P.H., FINDLAY, G.P.: Solute and water transport in the bladders of *Utricularia*. In: Ion transport in plants (W.P. ANDERSON, ed.). London-New York: Academic Press 1973.

THOMSON, W.W., BERRY, W.L., LIU, L.L.: Localization and secretion of salt by the salt glands of *Tamarix aphylla*. Proc. Natl. Acad. Sci. U.S. **63**, 310–317 (1969).

THOMSON, W.W., LIU, L.L.: Ultrastructural features of the salt gland of *Tamarix aphylla* L. Planta **73**, 201–220 (1967).

TROUGHTON, J., DONALDSON, L.A.: Probing plant structure. Wellington-Auckland-Sydney-Melbourne: A.H. and A.W. Reed 1972.

WAISEL, Y.: Ecological studies on *Tamarix aphylla*. III. The salt economy. Plant Soil **13**, 356–364 (1961).

ZIEGLER, H., LÜTTGE, U.: Die Salzdrüsen von *Limonium vulgare*. I. Die Feinstruktur. Planta **70**, 193–206 (1966).

ZIEGLER, H., LÜTTGE, U.: Die Salzdrüsen von *Limonium vulgare*. II. Die Lokalisierung des Chlorids. Planta **74**, 1–17 (1967).

5.3 Organic Substances

U. Lüttge and E. Schnepf

1. Carbohydrates

1.1 Mechanisms of Membrane Transport of Sugars in Eucaryotic Plant Cells

Considerable fluxes of sugars across lipo-protein membranes are a salient feature of the physiological activity of glands secreting carbohydrates. In the case of eccrine sugar secretion membrane transport may be involved at various locations in an anatomically complexly compartmented system (see 5.3.1.2.4, Fig. 5.12). In the case of granulocrine elimination of polysaccharides, membrane transport of precursors (monomers) also may be involved at various locations in a cytologically complexly compartmented system (see 5.3.1.3, Fig. 5.16). Membrane transport of sugars is of similar importance in other higher plant systems of comparable structural and functional complexity; e.g. in "vein loading" and "unloading" associated with the translocation process in the phloem (see Encyclopedia of Plant Physiology, New Series, Vol. 1, Chap. 17), in radial transport in rays (see Encyclopedia of Plant Physiology, New Series, Vol. Chap. 18) and in carbohydrate storage and mobilization in grains and in storage tissues. However, although considerable achievements were made in investigating the mechanisms of sugar membrane transport in erythrocytes, bacteria and yeasts, surprisingly little is known about the molecular mechanisms of sugar fluxes in higher plants, even though rates of transport are extremely large (cf. nectary secretion; 5.3.1.2.3.1). The only detailed information available for a eucaryotic green plant system is that regarding the inducible hexose transport mechanism of *Chlorella* (see 5.3.1.1.2.2, 3.3.1, Fig. 3.16, 9.3.3.2 and Part A, 6.7.2.3).

It has been noted repeatedly that sugar transport in plant cells may be directed against a gross concentration gradient (Bieleski, 1960; Grant and Beevers, 1964; Laties, 1964; Hancock, 1970; Komor and Tanner, 1971; Whitesell and Humphreys, 1972; Linask and Laties, 1973) and hence is a *bona fide* active process. The mediation of such sugar transport by carrier-like molecules is invoked by kinetic data (Linask and Laties, 1973). Carrier-dependent exchange diffusion or counter flow alone, however, appear to be often insufficient to explain the observed uphill transport of sugars (e.g. Glasziou and Gayler, 1972), inasmuch as these processes linking uphill transport of one species of solutes with downhill transport of another species would only temporarily lead to concentration gradients for a given sugar species (cf. Lüttge, 1973: pp. 30–38). A coupling of sugar fluxes to sodium fluxes as in animal tissues and possibly in fungi (see Part A, 7.10.2) has not been observed in plants (Gayler and Glasziou, 1972; Komor et al., 1972).

Metabolic reactions actively driving uphill transport of sugars may involve the sugar molecules or the carrier entities mediating sugar transport or both (see 5.3.1.1.1.3). Metabolism of the sugars may cause a pull (sink) for sugar transport, but this is considered trivial, i.e. not genuine metabolically mediated membrane transport. Non-metabolized sugar derivatives like 3-0-methyl-glucose (3-0-MG) are frequently used to eliminate this possibility. Metabolic reactions providing energy to drive sugar membrane transport may be phosphorylating or non-phosphorylating.

1.1.1 Phosphorylating Mechanisms

1.1.1.1 The Acid Phosphatase: A Cytological Structure-Function Correlation

For a long while, much emphasis rested on the most striking correlation that in the cytochemical Gomori-test (cf. GLICK, 1949) massive acid phosphatase activities are displayed by cells that are strategically situated in higher plant systems at locations where massive short-distance sugar transport occurs (WANNER, 1952; ZIEGLER, 1956).

Examples for this coincidence are the companion cells of angiosperm phloem (ZIEGLER, 1956; FIGIER, 1968), the Strasburger cells of gymnosperm phloem (ZIEGLER and HUBER, 1960), rays and wood parenchyma cells (ZIEGLER, 1956; SAUTER, 1966, 1972; SAUTER and MARQUARDT, 1965; SAUTER et al., 1973; see Encyclopedia of Plant Physiology, New Serie, Vol. 1, especially Chaps. 17 and 18) but also the tips of conifer nucelli which lysigenously secrete small droplets of pollination fluid (ZIEGLER, 1959).

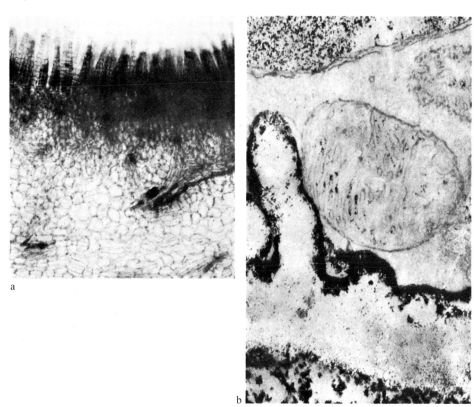

Fig. 5.8a and b. Acid phosphatase in sugar transporting cells. (a) Calyx nectary of *Abutilon striatum*. The black precipitation indicates phosphatase activity in the nectary trichomes, in the parenchyma just below them and in the vascular tissue supplying the nectary. With kind permission from ZIEGLER (1956) (× 80) (b) Cell-wall protuberances from a companion cell of sieve tubes supplying the extrafloral nectary of *Vicia faba*. The black precipitation indicates phosphatase activity associated with the plasmalemma. With kind permission from FIGIER (1968) (× 46,000)

In nectary glands, acid phosphatase has been considered as a prerequisite for genuine secretory tissue (Fig. *5.8*a; ZIEGLER, 1956; VIS, 1958; FREY-WYSSLING and HÄUSERMANN, 1960; LÜTTGE, 1966). FIGIER (1968, 1971, 1972) has shown acid phosphatase to be associated with the plasmalemma in the cell-wall labyrinths of sugar transporting transfer cells (Fig. *5.8*b; see also *5.3.1.2.2.4*). This correlation appears to be more than a fortuitous result of the simplicity and hence the ready use of the Gomori-test. Nevertheless, the acid phosphatase is a very unspecific enzyme and the message retained in the observed correlation remains obscure, i.e. whether sugar or carrier phosphorylations and de-phosphorylations are involved in the carbohydrate movements across membranes, or whether simply the general intermediary metabolism of the considered cells is particularly active (see *5.3.1.2.3.2*).

1.1.1.2 Sucrose-Phosphate

Specific evidence that sucrose-phosphate is the transport from of sugar moving across the tonoplast into the vacuoles ("storage compartments") of sugar cane stalk cells comes from feeding specifically labeled precursors and from compartmentation experiments. Sucrose is shown to be split into hexoses in the free space by outer space invertase (GLASZIOU and GAYLER, 1972; BOWEN and HUNTER, 1972). The hexoses are thought to be taken up unphosphorylated (a conclusion largely based on the competitive effect of 3-0-MG, but see below), phosphorylated in the cytoplasm ("metabolic compartment") yielding eventually sucrose-phosphate (from uridinediphosphate-glucose + fructose-phosphate), which "is the direct precursor of sucrose in the storage compartment", and the hydrolysis of which "is an integral part of the transfer process" (review: GLASZIOU and GAYLER, 1972). The crucial experiment appears to be that label from the fructose moiety of specifically labeled sucrose is randomized in the hexose pool when labeled sucrose itself is fed to the stalk tissue, but transferred unaltered to the vacuoles, when labeled sucrose-phosphate is administered (HATCH, 1964; but see BOWEN and HUNTER, 1972, for a different interpretation).

A similar mechanism appears to be involved in sucrose transport at the tonoplast of maize scutellum cells (HUMPHREYS, 1973). Disagreement falls, however, on the mechanism of uptake from the free space. In the scutellum HUMPHREYS and GARRARD (1968) and WHITESELL and HUMPHREYS (1972) find sucrose to be taken up without prior inversion. In sugar cane, by contrast to GLASZIOU and coworkers (see above) MARETZKI and THOM (1972) interpret the very rapid phosphorylation of sugars and the finding that L-arabinose or L-ribose do not compete for uptake to suggest that carbon 6 must be available for phosphorylation. Since both galactose and 3-0-MG compete, these authors conclude that carbon 3 and 4 are not important, but that the OH group of carbon 2 plays a large role, especially because mannose does not compete with glucose.

1.1.1.3 The Phosphoenolpyruvate-Glucose-Phosphotransferase System

In bacteria, a molecular mechanism of sugar membrane transport has been described, which in principle is a vectorial group transfer reaction. It involves the transfer of a phosphate group from PEP (phosphoenolpyruvate) generated in glycolysis, *via* a heat-stable, low molecular-weight protein (HPr), which is part of the carrier system, onto the transported sugar molecule according to the following equations:

$$PEP + HPr \xrightleftharpoons[Mg^{2+}]{Enzyme\ I} pyruvate + \text{\textcircled{P}} - HPr,$$

$$\text{\textcircled{P}} - HPr + sugar_{out} \xrightarrow[Mg^{2+}]{Enzyme\ II} sugar - \text{\textcircled{P}}_{in} + HPr,$$

where the indices "out" and "in" refer to the outside and inside respectively. Mg^{2+} is required for the reactions. The carrier system is composed of HPr and enzymes I and II. Enzyme II is a membrane-bound component, enzyme I and HPr are soluble factors but must be attached to enzyme II during catalysis of the transmembrane transport of sugar (reviews: KABACK, 1970a and b, with a model on p. 91 and p. 575 respectively). WHITESELL and HUMPHREYS (1972; see also GARRARD and HUMPHREYS, 1969) argue that this mechanism may also operate in sugar uptake by scutellum cells, but all the evidence they have is a correlation of uptake with glycolysis. Interestingly, BARNES (1972) has suggested recently that the phosphoenolpyruvate-glucose-phosphotransferase system is only operative in the facultative anaerobes among the bacteria (see *5.3.1.1.2*).

1.1.2 Non-Phosphorylating Mechanisms

1.1.2.1 Redox Reactions of Membrane-Bound Systems Mediating Sugar Transport

In obligatorily aerobic bacteria, hexose uptake appears to be clearly independent of the supply of $\sim P$ from ATP or PEP but coupled to a flavine-adenine-dinucleotide-(FAD)-linked malate dehydrogenase reaction in the membrane (BARNES and KABACK, 1970; BARNES, 1972). This redox-driven transport has been also considered as an interesting model case for eucaryotic cells.

A membrane-bound glucose dehydrogenase system mediating uptake of gluconate after extracellular oxidation of glucose has been described for *Pseudomonas aeruginosa* (MIDGLEY and DAWES, 1973).

1.1.2.2 The Hexose Uptake System of *Chlorella*

As argued above (*5.3.1.1.1.2*) the use of 3-0-MG alone does not rule out the phosphorylation of the sugar molecule transported. KOMOR and TANNER (1971) used a larger spectrum of sugar derivatives to exclude phosphorylation at all possible carbon atoms, showing that no specific carbon atom was required for phosphorylation of the sugars transported by the substrate-inducible hexose transport system of *Chlorella*. It was concluded that the sugar molecules transported are not phosphorylated since the results of KOMOR and TANNER would have required a different C-atom to be phosphorylated for each derivative, which would imply a most unspecific phosphorylating system (see also Part A, *7.8.4*).

The energy necessary for active sugar transport is equivalent to 1.18 ATP per hexose transported, which can be provided in the dark by respiration or in anaerobic solution in the light by cyclic photophosphorylation (TANNER, 1969; DECKER and TANNER, 1972; KOMOR et al., 1973). The ATP is most likely used neither for substrate phosphorylation (see above) nor for carrier phosphorylation (activation), but may reflect the need of metabolic maintenance of a pH-gradient at the cell membrane (see below; 5.3.1.1.2.3). Efflux of glucose from the cells can generate energy (i.e. reversal of energy consuming influx) which can be utilized by influx, so that the apparent energy requirement of steady-state influx is an equivalent of 0.5–0.6 ATP as compared with 1.18 ATP required for initial influx where concomitant efflux is low (DECKER and TANNER, 1972). (The interesting hexose transport system of *Chlorella* is described in more detail in 9.3.3.2 and Part A, 6.7.2.3.)

1.1.2.3 H⁺ Transport Coupled Sugar Uptake

In experiments with some bacteria it has been found that uptake of sugar (β-galactoside) is linked with H^+ influx (sugar-H^+-symport or cotransport) or OH^- efflux (sugar-OH^--antiport or countertransport) with a stoichiometry of 1 sugar molecule: $1 H^+$ or OH^- (*Escherichia coli:* WEST, 1970, WEST and MITCHELL, 1972, 1973; *Streptococcus lactis:* KASHKET and WILSON, 1973; see also Part A, Chap. 7.10). KOMOR (1973) has found the same stoichiometry for hexose-H^+-symport (or hexose-OH^--antiport) in *Chlorella vulgaris*. Thus, hexose uptake by the eucaryotic microalga cells may in principle follow a similar mechanism as lactose uptake by *E.coli*. In such systems, energy for membrane transport of sugar molecules appears to be drawn from the electrochemical proton potential at the membrane, i.e. sugar transport against a concentration gradient is driven by H^+ or OH^- transport down an electrochemical gradient (WEST, 1970; KOMOR and TANNER, 1974). The inhibitory effect of uncoupling agents is thought to be due to increasing H^+ permeability of the membrane.

1.1.2.4 The Glucosidic Bond

The inversion of sucrose is thought to participate in maintenance of a gradient for sugar secretion in nectaries (see 5.3.1.2, Fig. 5.12). More specifically, the energy stored in the glucosidic bond has been considered to make di-saccharides superior forms of sugar translocates. This case has been elaborated regarding the overwhelming role of sucrose (ZIEGLER, 1956; ZIMMERMANN, 1958). It has been also conceived for trehalose in analogy to some animal systems, where a membrane-bound trehalase appears to play a large role in trehalose-mediated glucose transport (SACKTOR, 1968). An intermediary role of trehalose as a translocate for glucose would explain the presence of trehalose-metabolizing enzymes in sugar cane (GLASZIOU and GAYLER, 1969) and in germinating pollen (GUSSIN et al., 1969), while appreciable amounts of trehalose are not detectable. A proposal of a role of trehalose in hexose efflux from sugar cane (GLASZIOU and GAYLER, 1969) has been revoked, however, more recently (GAYLER and GLASZIOU, 1972; GLASZIOU and GAYLER, 1972).

Vectorial group transfer reactions mediated by membrane-bound enzyme (here = carrier) systems are necessarily accompanied by transport. KABACK (1970a, p. 96–97) has pointed out that such reactions may be a general principle of catalyzed membrane transport, whereas the specific reactions observed may vary. Examples quoted by KABACK are phosphoryl-transfer, oxidation-reduction, and acetylation-deacetylation reactions. Formation-hydrolysis of the glucosidic bond in sucrose or trehalose as suggested above may be yet another example.

1.2 Elimination of Sugars

The nectary glands, which are the glands specialized for the secretion of sugars, are widely distributed among angiosperms (Fig. 5.9). Floral nectaries (for a review of early literature see FELDHOFEN, 1933; SPERLICH, 1939) provide sugars as food for pollinating animals. However, so-called extrafloral nectaries are also frequently found on the leaves and stems of angiosperms (ZIMMERMANN, 1932). Extrafloral nectaries serve no obvious function. It has been suggested though, from correlations observed between nectar secretion and developmental stage of the organs carrying the glands, that nectaries serve as "*sap valves*" eliminating a surplus of sugar supply which is not used by the organ (FREY-WYSSLING, 1935; ZIMMERMANN, 1953; MATILE, 1956). However, this problem has never been investigated by quantitatively determining carbon and nitrogen budgets. Nectary glands have not been reported in gymnosperms, but some fern leaves carry "extrafloral" nectaries (FREY-WYSSLING and HÄUSERMANN, 1960; LÜTTGE, 1961). This suggests that the role of nectar in pollination developed phylogenetically later than the nectar secretion process itself.

1.2.1 Chemical Composition of Nectars

In most nectars, sugars make up close to 100% of the dry weight, the predominant sugars being sucrose, glucose, and fructose (MAURIZIO, 1959; LÜTTGE, 1961; PERCIVAL, 1961; VAN HANDEL et al., 1972). Oligosaccharides composed of glucose and fructose units are also frequently found. These oligosaccharides are intermediate products of reactions catalyzed by extracellular trans-glucosidases and trans-fructosidases in the nectar, which eventually invert the sucrose to its hexose monomers (ZIMMERMANN, 1952, 1953, 1954; FREY-WYSSLING et al., 1954; MATILE, 1956). The ratio of total glucose to total fructose units in the nectar is very often close to unity (LÜTTGE, 1961; VAN HANDEL et al., 1972). Occasionally, other sugars (e.g. galactose) are found in small amounts either as free hexoses or as units in oligosaccharides (MAURIZIO, 1959; PERCIVAL, 1961).

Substances other than sugars are found only in very low concentrations in typical nectar. As the following list shows there is a wide variety of these substances; the figures give the amount of the respective substances per 1 g of sugar in the nectar:

i) *amino acids* 0.5–19.6 mg/g (LÜTTGE, 1961, 1966),
ii) *mineral ions* in µmoles/g (LÜTTGE, 1962b, 1966)

K^+	Na^+	Ca^{2+}	Mg^{2+}	PO_4^{3-}
3–310	1–35	1–100	trace–14	0.6–64,

iii) traces of *di- and tri-carboxylic acids* (LÜTTGE, 1961, 1966),
iv) traces of various *vitamins* (ZIEGLER et al., 1964; LÜTTGE, 1966).

The following tabulation gives average quantitative estimations of various substances in the nectar of banana flowers (*Musa sapientum*) on a weight/weight basis (LÜTTGE, 1961, 1966):

water	68%	amino acids	0.014%
glucose	8%	protein	0.28–0.59%
fructose	8%	inorganic phosphate	0.032%
sucrose	16%	phosphate in organic compounds	0.013%

Fig. 5.9a – j. Legend see opposite page

The protein content of this nectar is most probably not due to secretion of protein but to secondary contamination with pollen and also with micro-organisms. This also raises the question whether some of the other substances accompanying the sugars in the nectar might originate from bacterial metabolism. Another secondary modification of the nectar after secretion regards the water content which is altered by equilibration with the atmosphere (BEUTLER, 1953; HUBER, 1956). However, the nectar concentration is also affected by anatomical (see 5.3.1.2.2.1) and physiological factors (REED et al., 1971).

1.2.2 Secretory Pathways

1.2.2.1 Supply of Sugar to the Gland Tissue

Storage of polysaccharides in the gland tissue (e.g. starch in the leucoplasts of *Gasteria* nectaries, SCHNEPF, 1964) appears able to provide sugars secreted in the nectar to only a limited extent. Instead, the analytical results discussed above suggest that the bulk of the sugars found in the nectar originates from sucrose, which is the predominant form in which carbohydrate is translocated in the phloem (ZIEGLER, 1956; ARNOLD, 1968). Furthermore, anatomically most nectary glands are specifically supplied with phloem strands, and the concentration of sugar in the nectar, i.e. the relative amounts of sugars and water in the nectar, depends on the relative amounts of phloem and xylem elements in the gland parenchyma (FREY-WYSSLING and AGTHE, 1950; AGTHE, 1951; FREI, 1955; FREY-WYSSLING, 1955). The supply of sugar to the gland tissue must be maintained during the secretion process. In *Abutilon*, the total amount of sugar secreted is at least seven times the sugar present in the gland tissue at the onset of secretion (FINDLAY et al., 1971). Using marker substances, it has been clearly shown that the phloem translocate is the source of the nectar fluid (AGTHE, 1951; ZIMMERMANN, 1952, 1953, 1954; FREY-WYSSLING et al., 1954; MATILE, 1956).

1.2.2.2 Differentiation of the Gland Tissue

The secretory pathway from the sieve tubes to the actual site of elimination on the gland surface may vary, and of course is related to the particular anatomy of the gland, which also determines to what extent the phloem translocate is subject to modification during the secretion process (LÜTTGE, 1961). The degree of specialization at the anatomical level can be correlated with the chemical specificity of the nectar; i.e. it is inversely correlated to the amounts of non-sugar compounds in the nectar (e.g. amino acids: Fig. 5.10). This suggests an important regulatory role of the gland tissue.

◁ ─────────────────────────────────────

Fig. 5.9a–j. Nectar secretion. Arrows indicate nectary glands or nectar droplets secreted. (a) *Pteridium aquilinum*, extrafloral nectaries. (b) *Sanseveria ceylanica*, nectar droplets on the bracts which function as extrafloral nectaries. (c) *Sambucus nigra*, nectar droplets on stipules which function as extrafloral nectaries. (d) *Hosta* hybr. septal nectaries with nectar droplets. (e) *Euphorbia pulcherrima*, cyathial nectaries with nectar. (f) *Musa sapientum*, nectar-filled flower. (g) *Dendrobium chrysotoxum*, extrafloral nectar secretion. (h) *Abutilon striatum*, nectar at the base of the calyx. (i) *Abutilon striatum*, longitudinal section of flower after removal of the petals, nectary trichomes at the base of the calyx. (j) *Hoya carnosa*, floral nectar secretion. (Fig. 5.9a from LÜTTGE, 1964c, Figs. 5.9 f, h, i, j from LÜTTGE, 1964b)

Plant Species	Nectary	Ratio: amino acids : sugars
Platycerium div. spec.	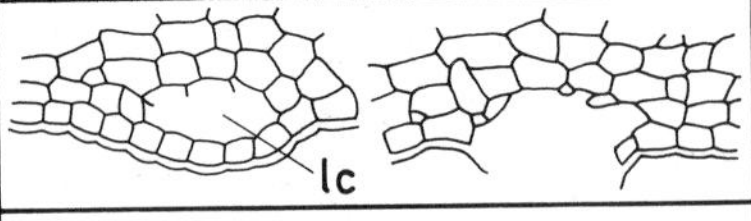 *(diagram: lc)*	$20\,000 \cdot 10^{-6}$
Sambucus nigra L.	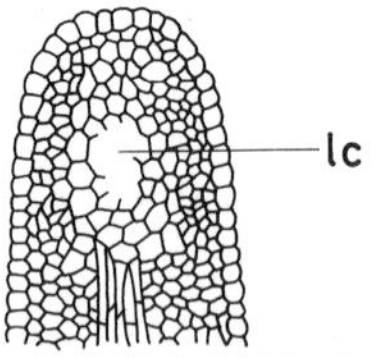 *(diagram: lc)*	$5\,000 \cdot 10^{-6}$
Sambucus racemosa L.	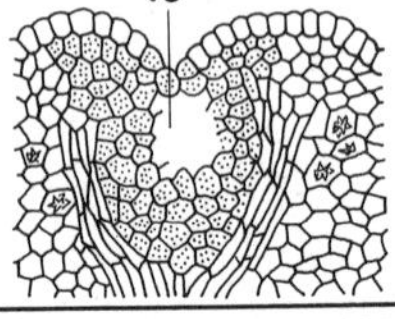 *(diagram: lc)*	$4\,000 \cdot 10^{-6}$
Pteridium aquilinum Kuhn	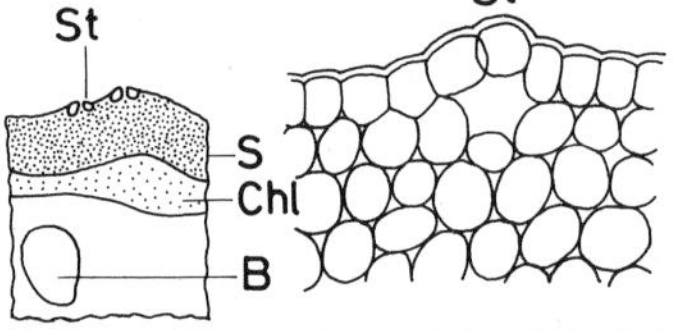 *(diagram: St, S, Chl, B)*	$100 \cdot 10^{-6}$
Abutilon striatum Dicks. Robinia pseudo-acacia L. Hoya carnosa R. Br.	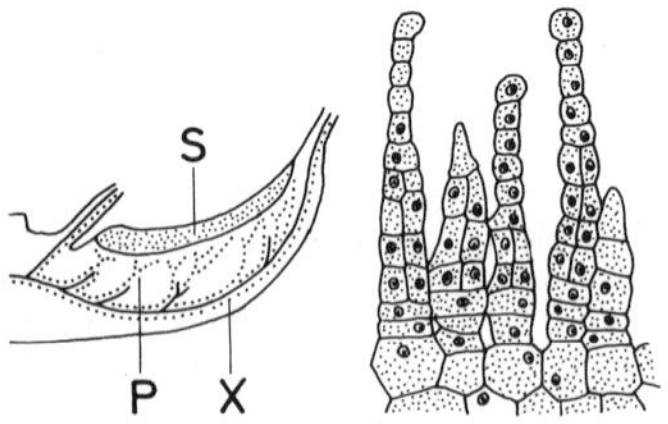 *(diagram: S, P, X)*	$5 \cdot 10^{-6}$

Fig. *5*.10. Relation of anatomical gland organization to chemical specificity of nectar (e.g., decreasing ratio amino acids: sugars with increasing structural specialization). The degree of structural spezialization increases from top to bottom, i.e. from lysigenous secretion (*Platycerium, Sambucus* spp.), *via* glands where secretion occurs through modified stomata (*Pteridium*), to glands with highly specialized glandular tissues and cells (the example drawn shows *Abutilon* calyx nectaries). Lines 1–4 extrafloral, line 5 floral nectaries. From Lüttge 1973, Fig. 7. 13, p. 260. *B* bundle, *Chl* chlorenchyma, *lc* lysigenous cavity, *S* secretory tissue, *St* modified stomata, *P* phloem elements, *X* xylem elements

1.2.2.3 Apoplasmic or Symplasmic Transport in the Gland Tissue?

It appears to be difficult to conceive a controlling function of the gland-cell metabolism if the sugars secreted were moving predominantly in the apoplast as suggested by Vasiliev (1969b, 1971). In some nectaries, a continuous apoplasmic pathway for the secreted fluid is blocked by specific cell-wall incrustations (*1*.2.2.1) consisting of hydrophobic material (Fig. *5*.12) (Schrödter, 1926; Frey-Wyssling, 1935; Sper-

LICH, 1939; LÜTTGE, 1969; SCHNEPF, 1969a; FINDLAY and MERCER, 1971b). Physiological observations on nectar secretion stress that at least at one stage in the secretory pathway metabolically controlled membrane transport is involved (5.3.1.2.3 and 5.3.1.2.4). Various possibilities for the site are depicted in the structural model discussed below (Fig. 5.12), but it is difficult to decide where this step might be located. Symplasmic transport between the sieve tubes, companion cells, parenchyma cells and gland cells is possible. All of these cells are usually connected by plasmodesmata, the number of which is often particularly high between gland cells and also between sieve tubes and companion cells.

EYMÉ (1966, 1967) suggested that ER-vesicles are involved in a sort of a granulocrine transport across the plasmalemmas from cell to cell in the gland parenchyma.

1.2.2.4 Cell-Wall Protuberances and Increased Plasmalemma Surface

It is possible that the outer plasmalemma of the gland complex plays an important role in sugar elimination. The outer plasmalemma surface of secreting nectary cells is often considerably enlarged by cell-wall protuberances (WRISCHER, 1962; SCHNEPF, 1964; FAHN and RACHMILEVITZ, 1970), similar to those found in many other cells mediating extensive short distance transport (GUNNING and PATE, 1969; PATE and GUNNING, 1972: *"transfer cells"*; see 1.4.2.2.2). Although protuberances are not consistently found in secreting cells of nectaries (SCHNEPF, 1969a, 1974), in some species the culmination in the formation of a wall labyrinth is clearly correlated with the most active period of nectar secretion (Fig. 5.11, SCHNEPF, 1964, 1974). The enlarged plasmalemma surface may serve secretion either by providing more space for membrane-bound carrier mechanisms or by facilitating passive elimination by diffusion (LÜTTGE, 1971; PATE and GUNNING, 1972). Hence this cytological phenomenon does not allow conclusions on the secretion mechanism.

1.2.2.5 Granulocrine or Eccrine Secretion?

It has been argued that nectar elimination might occur by granulocrine secretion (MERCER and RATHGEBER, 1962; EYMÉ, 1966; FAHN and RACHMILEVITZ, 1970; FINDLAY and MERCER, 1971b; RACHMILEVITZ and FAHN, 1973). However, the demonstration of an abundance of endoplasmic reticulum, of increased numbers of vesicles, of active dictyosomes and of irregular plasmalemma surfaces (invaginations?) in gland cells appears to be only rather circumstantial evidence. So far, only in the exceptional case of the elimination of polysaccharide slimes, has electron microscopy provided evidence for granulocrine secretion by plant glands, that can stand rigorously critical interpretation (see 5.3.1.3). Furthermore, a number of specific molecular mechanisms of active membrane transport of sugars have evolved in various organisms (see 5.3.1.1), and it is likely that one or more of them serve membrane transport of sugar in nectar secretion (i.e. eccrine secretion). Future research may reveal that both eccrine and granulocrine nectar secretion is possible, i.e. that due to the diversity of various types of nectaries, different secretion mechanisms may occur.

1.2.2.6 Elimination across the Cuticle

After elimination from the cytoplasm into the cell-wall free space, the nectar must pass the cuticle and sometimes cutinized cell-wall layers. The cuticle itself is often

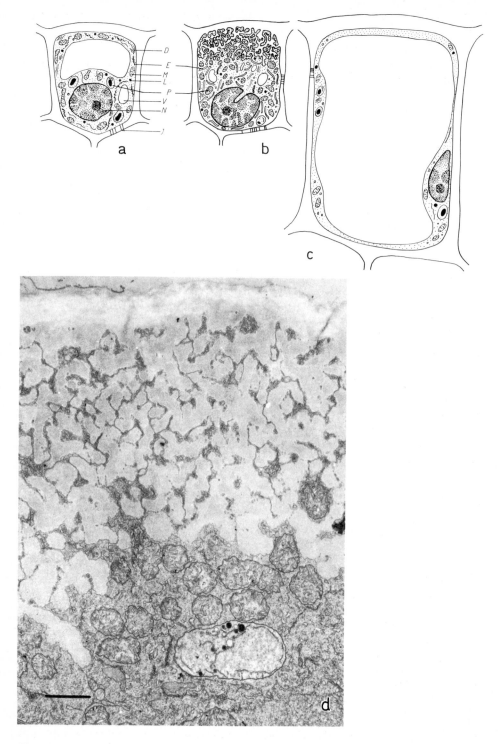

Fig. 5.11a – d. Legend see opposite page

very thin, or it may be ruptured during secretion (cf. SPERLICH, 1939; SCHNEPF, 1969 a).

In the trichome nectaries of *Abutilon* the nectar accumulates apically in a subcuticular space between the cell wall and the cuticle. The hydrostatic pressure building up causes opening of fine pores in the cuticle, and concomitant release of a nectar droplet with a sudden decrease in the hydrostatic pressure, whereupon the pores close and the periodic process starts again (FINDLAY and MERCER, 1971 a).

1.2.3 Physiology of the Secretion Process

1.2.3.1 Nectar Secretion: Specific, against a Concentration Gradient, at High Flux Rates

A discussion of the physiology of the secretion process must be based on the following experimental findings:

i) As suggested by the preponderance of sugars (5.3.1.2.1), the nectary secretion product is chemically very specific.

ii) Sugar concentration in the phloem may be quite high (i.e. in the range of 18 up to 38%: KLUGE, 1970); and in some nectars sugar concentration is lower (e.g. 6% in *Fritillaria,* BEUTLER, 1929/30). Hydrolysis of sucrose by invertases (5.3.1.2.1), which contributes to maintenance of water potential gradients driving the elimination of nectar fluid, further obscures measurements of sugar concentration gradients (REED et al., 1971). Nevertheless, sugar concentrations of nectar of 60% (and even up to saturated solutions with crystallizing sugar) have been recorded (cf. BEUTLER, 1953). It is, therefore, evident, that in most nectaries sugar at least at one stage in the system (Fig. 5.12) must be transported actively against a concentration gradient. In the nectaries of *Abutilon* this step may occur when the sugar enters the trichomes *via* their stalk cells, since the average sugar content in the nectary tissue is highest within the trichomes (REED et al., 1971).

iii) The rates of transmembrane sugar fluxes in nectaries appear to be unusually large.

The rate of sugar efflux from glucose-storing onion cells into a sugar-free solution is 28 nmol $m^{-2} s^{-1}$ (STEINBRECHER and LÜTTGE, 1969); the fluxes at the plasmalemma and tonoplast of scutellum cells in germinating seeds are 122 and 2.2–4.5 nmol $m^{-2} s^{-1}$ respectively (HUMPHREYS, 1973). If sugar secretion in banana flowers were proceeding at similar rates, the effective secretory area would have to be 0.25–0.5 m^2 per flower (LÜTTGE, 1971). Taking the upper half of the spherical apical cell of *Abutilon* trichomes as the secretory surface, the dimensions together with the estimated rate of exudation of $3 \cdot 10^3$ μm^3 per trichome and a sugar concentration

$\triangleleft$

Fig. 5.11 a–d. Epithelium cells of the septal nectary glands in the ovaries of *Gasteria trigona.* (a) Gland cell in a young flower bud resembling normal epidermal cells. (b) Actively secreting cell with reduced vacuole and dense cytoplasm, numerous mitochondria, cell-wall protuberances, the nucleus having invaginations. (c) Postsecretory stage: the protuberances have disappeared, the cell is elongating and forming a large central vacuole, the cytoplasm is restricted to a thin peripheral layer, no cytological features are reminiscent of the original secretory function. (d) Electron micrograph of the outer part of a secretory cell of an active gland showing the wall protuberances and many mitochondria. Fixation: OsO_4; marker bar: 1 μm.
D dictyosome, *E* Cisternae and vesicles of the endoplasmic reticulum, *J* plasmodesmata, *L* lipoid droplets, *M* mitochondria, *N* nucleus, *P* leucoplasts (starch if present indicated black), *V* vacuole. From SCHNEPF (1973)

in the nectar (expressed as sucrose) of 0.4 M given by Findlay and Mercer (1971a: Figs. 5–7 and page 651) suggest a transmembrane flux of 66,500 nmol $m^{-2} s^{-1}$. (For comparison the mass transfer of sugar in the phloem is 20–$200 \cdot 10^6$ nmol $m^{-2}\ s^{-1}$, MacRobbie, 1971.) If the surface of the total trichome were actively secreting, this figure would be smaller by a factor of about $^1/_{20}$, and would be then in the order of magnitude of the salt fluxes reported for mangrove glands (*Aegialitis:* 4,200 nmol $m^{-2} s^{-1}$ Atkinson et al., 1967). If in addition it is considered in the above estimation that the trichome in its basal parts is more than one cell wide, the surface will still be larger, i.e. the apparent rate will be smaller. In cases where the plasmalemma surface is increased by the presence of cell-wall protuberances, taking this into account, might bring the rate of secretion down to quite normal figures. Hence the apparently enormous rates of secretion by glands may to a fair extent be due to a structural-cytological achievement rather than to a physiological one.

This summary emphazises that the carbohydrate metabolism of the gland cells modifying and/or supplying the sugars secreted, and the possible driving forces for active sugar transport must be discussed.

1.2.3.2 Sugar Metabolism and Phosphorylation in the Nectary Tissue

Respiratory O_2 uptake by the gland tissue is increased during active nectar secretion, and is larger than in adjacent non-glandular tissue (Ziegler, 1956). Nectary gland cells often contain a very dense cytoplasm with numerous mitochondria (Schnepf, 1969a; Findlay and Mercer, 1971b). The abundance of mitochondria may be related to both requirements of increased carbohydrate metabolism and of increased energy transfer in the gland cells. Nectar secretion is temperature-dependent; it is also inhibited by anaerobiosis, by uncouplers of phosphorylation, and by inhibitors of respiratory electron flow (Ziegler, 1956; Matile, 1956; Lüttge, 1966, 1969; Findlay et al., 1971).

In view of the particular metabolic activity of nectary tissue, the finding that most enzymes of intermediary carbohydrate metabolism are active (Ziegler, 1965; Fekete et al., 1967) is not surprising. Feeding experiments have shown that sugars are indeed subject to metabolic modification during the secretion process, when they pass through the gland tissue (Zimmermann, 1952, 1953; Frey-Wyssling et al., 1954; Matile, 1956; Shuel, 1956). However, this modification appears to be limited. Sugars other than sucrose, its hexose monomers, and oligosaccharides composed of them are secreted only in minimal amounts (5.3.1.2.1). Excised *Abutilon* nectaries supplied with ^{14}C-sucrose labeled in the glucose moiety but not in the fructose moiety, secreted nectar sugars with 72% of the radioactivity still in the glucose and 28% in the fructose (Ziegler, 1965).

The high amounts of acid phosphatase found in the nectary tissues and particularly in association with the plasmalemma of the gland cells (see 5.3.1.1.1.1, Fig. 5.8), and perhaps favorable Mg^{2+}/Ca^{2+} ratios in the gland tissue (Lüttge, 1962a) suggest that energetically phosphorylation reactions may play a key role in sugar elimination from the gland cells (5.3.1.1.1.1). However, these are only circumstantial correlations, and Findlay et al. (1971) have calculated that the ratio of $\sim$P available from oxidative phosphorylation to the amount of sugar transported is only 1–$5 \sim$P/ sugar molecule, which does not take into account other $\sim$P-consuming processes in the cells. Hence, a direct correlation of sugar phosphorylation and sugar transport is not readily supported by the evidence.

1.2.3.3 Hormonal Control, Action Potentials

In addition to metabolism, hormonal control systems obviously take part in the regulation of nectar secretion (MATILE, 1956; SHUEL, 1959, 1964, 1967). *In vivo* the reproductive organs appear to be involved in such control of the secretion by floral nectaries (SHUEL, 1961). Action potentials in the phloem and in the nectary cells are correlated with secretion in the floral nectaries of *Tilia* which is triggered by mechanical stimulation (MOLOTOK et al., 1968).

1.2.4 Possible Mechanisms of Nectar Secretion: Discussion of Models

Possible mechanisms and sites of active processes are summarized in Fig. 5.12.
Nectar sugars originate largely from sucrose translocated in the phloem (5.3.1.2.2.1). Hence the first possible site of active transport in the scheme of Fig. 5.12 is unloading of the phloem. However, in many cases isolated glands also secrete nectar. Therefore, the active step supplying sugar to the gland tissue may occur farther within the gland. The data of FINDLAY et al. (1971) and REED et al. (1971) suggest that such a mechanism concentrating the sugar in the trichomes of *Abutilon* nectaries (5.3.1.2.3.1 (ii)) establishes the water potential gradient required for secretion. In this case, the plasmalemma in the apical cells of the trichomes would have to be much more permeable for the sugars than the plasmalemma of the stalk cell. Fluid would be eliminated by pressure filtration (FREY-WYSSLING and HÄUSERMANN, 1960; ZIEGLER, 1965; REED et al., 1971). Similarly, leakage of sucrose from maize scutella has been explained by a pressure-flow mechanism depending on turgor pressure in the scutellar parenchyma cells and extending through the phloem (HUMPHREYS and GARRARD, 1971; HUMPHREYS, 1972a, b). However, to explain the chemical specificity of the nectary secretion, specific reabsorption processes would be required in addition to pressure filtration (ZIEGLER, 1965; VASILIEV, 1969b). It has been shown that reabsorption occurs concomitantly to elimination (ZIEGLER and LÜTTGE, 1959; SHUEL, 1961; LÜTTGE, 1962b), and it is conceivable that like other plant cells, gland cells have various kinds of uptake systems. However,

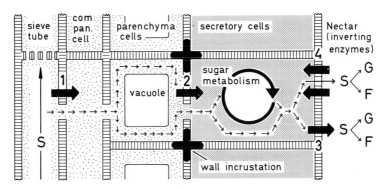

Fig. 5.12. Model of nectar secretion. Possible sites of active sugar membrane transport: *1* active sieve-tube unloading, *2* pump concentrating sugars in the secretory cells, *3* active secretion. *4* passive leakage (pressure filtration) of nectar fluid accompanied by specific active reabsorption processes. *S* sucrose, *G* glucose, *F* fructose. Small arrows: transport in plasmodesmata and symplasmic transport. Thick arrows: metabolically controlled membrane transport

if the active mechanism were restricted to the basal plasmalemma of the stalk cells of *Abutilon* trichomes, during steady-state nectar secretion the active flux at this membrane would have to be of the enormous rate of about 110 μmol m^{-2} s^{-1} (see *5.3.1.2.3.1* (iii)).

Alternative or possibly additional driving forces may be the modification of sugars by gland cell metabolism (although in *Abutilon* this will not affect more than 30% of the sugar transported (*5.3.1.2.3.2*); or by the invertases at the gland surface or in the excreted fluid itself (*5.3.1.2.1*) maintaining water potential and/or sugar concentration gradients as already mentioned above (*5.3.1.2.3.1* (ii)).

Another important driving force may be active secretion at the outer gland plasmalemma. This model of active eccrine nectar secretion has often been favored in the literature (Lüttge, 1962b; Shuel, 1967; Schnepf, 1969a). Reed et al. (1971) discard it on the basis of osmotic considerations.

1.3 Elimination of Polysaccharides

1.3.1 Types of Glands

Polysaccharides are secreted by every plant cell, at least during cell-wall formation. Specialized secretory cells occur in mucilage glands, for example in carnivorous plants (Schnepf, 1969a), in young shoots or buds (Schnepf, 1968; Rougier, 1972; Wollenweber et al., 1971) or as epithelia of mucilage ducts (Schnepf, 1963c; Evans et al., 1973). The slime-producing outer root cap (Mollenhauer et al., 1961; Northcote and Pickett-Heaps, 1966; Jones et al., 1966; Morré et al., 1967) may also be considered as a slime gland. These glands belong to the merocrine type though they may be rather short-lived, as in the root cap. Holocrine secretion is observed in archegonia of ferns and mosses (Vian et al., 1970). Cells (mucilage cells) storing the secretion product in the vacuole or, more often, between plasmalemma and cell wall also occur (Mollenhauer, 1967; Bouchet and Deysson, 1973).

1.3.2 Chemical Composition of Secretion Products

The secreted polysaccharides may be bound to or mixed with proteins (Jones and Morré, 1967) or not (Schnepf, 1963a). They usually contain a considerable amount of non-glucose sugars and sugar acids and are highly hydrated. In the hydrolyzed trapping slime of the insectivorous plant *Drosophyllum*, galactose, arabinose, xylose, rhamnose, and gluconic acid were identified; the secreted fluid is also rich in ascorbic acid (Schnepf, 1963a). The *Rheum* slime glands secrete a polysaccharide composed of galactose (55%), arabinose (20%), xylose (10%), mannose (10%), and glucose (5%) (Kling, 1961). The polysaccharide of the maize root cap slime contains 22% glucose, 39% galactose, 12% galaturonic acid, 6% mannose, 8% fucose, 1% rhamnose, 7% arabinose, and 5% xylose (corrected data, from Jones and Morré, 1973). The mucilage cells of *Hibiscus* contain a slime with 80% galactose, 10% rhamnose and 6% galacturonic acid (MOLLENHAUER, 1967). It is possible that in all these examples the macromolecules have a rather heterogeneous composition.

In algae, secreted polysaccharides often are sulfated including those which are eliminated by gland cells (EVANS et al., 1973).

1.3.3 Secretory Pathways

The secretory pathways are very similar in the different types of glands (see 5.3.1.3.1). The process of slime elimination was studied extensively in the glands of *Drosophyllum* (Fig. 5.13) and in the root cap cells of maize. It resembles the secretion of the non-cellulosic matrix of the cell wall (MOLLENHAUER and MORRÉ, 1966a; SCHNEPF, 1969a), with the Golgi apparatus playing the most important role in the synthesis and the extrusion. Thus many well-developed dictyosomes and Golgi apparatus-derived vesicles are the predominant cell organelles (Fig. 5.14). These vesicles originate in peripheral regions of the dictyosomes, where the Golgi cisternae usually are more or less reticulate (MOLLEN-HAUER and MORRÉ, 1966b). The Golgi vesicles still associated with the Golgi cisternae reveal the polarity of the dictyosome; their content increases in density from the regeneration, or forming face to the secretory, or maturing face (MOLLENHAUER and WHALEY, 1963). Sometimes the last cisterna at the secretion face is transformed completely into a vesicle (SCHNEPF, 1968). These vesicles turned out to be secretion vesicles because of:

    i) the staining character of their content (SCHNEPF, 1963a),

    ii) cytochemical reactions (SCHNEPF, 1968; ROUGIER, 1972),

    iii) autoradiographic evidence (NORTHCOTE and PICKETT-HEAPS, 1966),

    iv) the membrane of the vesicles becoming increasingly similar in structure to the plasmalemma with which it fuses eventually (SCHNEPF, 1965),

    v) the correlation of cytological and physiological activity (SCHNEPF, 1961b) (Fig. 5.15).

Fig. 5.13. *Drosophyllum lusitanicum*. Leaf with numerous stalked glands secreting the trapping slime

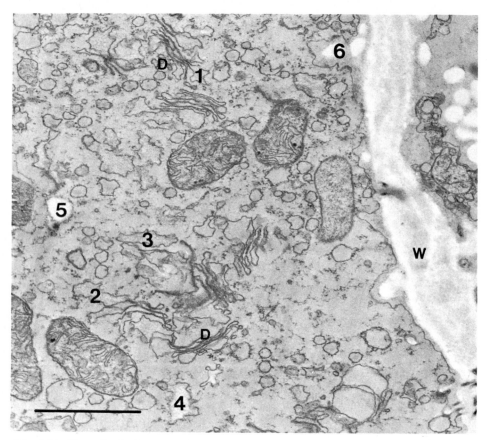

Fig. 5.14. Slime gland of *Drosophyllum*. The slime is produced by the Golgi apparatus. The dictyosomes (*D*) form secretion vesicles which look empty in the beginning (*1*), then they become filled with content (*2*), and detach from the dictyosomes (*3*); their content becomes denser (*4, 5*), and finally is extruded into the space between the plasmalemma and the cell wall (*W*) (*6*). Fixation: $OsO_4 + K_2Cr_2O_7$, without poststaining; marker bar: 1 μm. From Schnepf (1973)

Thus overwhelming evidence from simultaneous physiological and cytological investigations proves that in this case elimination is granulocrine, by contrast to other examples in higher plants where this has been only assumed (see 5.3.1.2.2.5). The secretion products, slime or similar substances, cannot be recognized within the cytoplasm or in cell organelles other than Golgi vesicles suggesting that they are synthesized in the Golgi apparatus. In cases where the polysaccharides are sulfated, such as in some algae, this reaction also appears to occur within the dictyosomes (Evans et al., 1973). If this is right—and there is no reason to doubt it—the dictyosomes and/or the Golgi vesicles have to be supplied:

i) with the sugar (or sugars) and, if necessary, with SO_4^{2-}, which are the precursors of the polysaccharide,

ii) with the enzymes involved in the synthesis, and

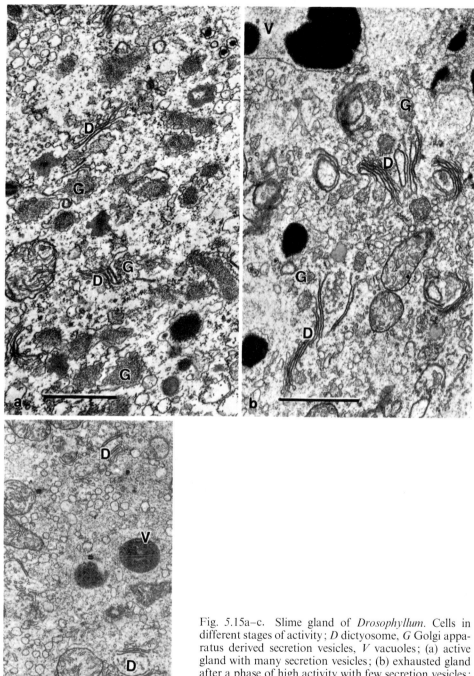

Fig. 5.15a–c. Slime gland of *Drosophyllum*. Cells in different stages of activity; *D* dictyosome, *G* Golgi apparatus derived secretion vesicles, *V* vacuoles; (a) active gland with many secretion vesicles; (b) exhausted gland after a phase of high activity with few secretion vesicles; (c) inactive gland in N_2-atmosphere. Fixation: $OsO_4 + K_2Cr_2O_7$, poststaining with lead citrate; marker bar: 1 µm. In part from SCHNEPF (1969a), modified

iii) with the membrane material necessary to compensate for the consumption of membranes by the formation of the secretion vesicles and their disappearance during discharge.

In lower plants and in animal cells secreting glycoproteids there are small vesicles which originate from smooth elements of the ER and fuse in the forming face of the dictyosomes or with peripheral elements. These vesicles may fulfil the before-mentioned three requirements. In the slime glands of higher plants such small transitory vesicles usually do not occur. (For an exception see ROUGIER (1965): slime glands, squamulae, of *Elodea*.) Hence there are only few indications that an incorporation of ER-derived elements into the dictyosomes is realized to a considerable degree (SCHNEPF, 1969a). Yet, the synthesizing enzymes may originate directly or indirectly from the ER whereas the membrane passage of the sugars takes place mainly at the level of the Golgi membranes. This transmembrane transport must precede the synthesis but could also be combined with it. Autoradiographic studies suggest binding of externally applied labeled sugars to ER elements involved in cell-wall formation in the root tip (PICKETT-HEAPS, 1967) but not in slime-secreting root cap cells.

The contents of the Golgi vesicles are modified further, even after the detachment from the dictyosomes; usually they become denser (Fig. 5.14) (SCHNEPF, 1963a). This may indicate a condensation of the synthesized material as well as a continuation of the synthesis (VAN DER WOUDE et al., 1971). During the maturation of the secretion vesicles the structure and composition of their membranes are modified; finally they are very similar to the plasmalemma (SCHNEPF, 1965; MORRÉ et al., 1971).

The vesicles extrude their contents by a typical exocytosis: the vesicle opens by fusion of its membrane with the plasmalemma; the slime droplets are deposited thereby between the plasmalemma (with the incorporated vesicle membranes) and the cell wall. In some slime glands, this extrusion seems to be directed to a special wall (SCHNEPF, 1963a; MORRÉ et al., 1967), but this was questioned for the root-cap cells by JUNIPER and PASK (1973). Nevertheless, there are intracellular mechanisms transporting and guiding the vesicles from the dictyosomes to the plasmalemma and restricting the fusion to mature vesicles. In pollen tubes the transport of the secretion vesicles is inhibited by cytochalasin B and, therefore, assumed to depend on protoplasmic streaming (FRANKE et al., 1972), whereas in slime glands protoplasmic streaming has not been observed. Colchicin has little or no influence on root-cap slime secretion (MORRÉ et al., 1967), thus microtubules seem not to be involved in the transport.

In the last step of the secretion process, the extruded polysaccharide passes the outer cell wall. If the exocytosis occurred near an inner wall this wall is the pathway towards the surface of the gland (SCHNEPF, 1961a). In some slime glands, e.g. of carnivorous plants (SCHNEPF, 1961a; 1963a) and in *Rumex* (SCHNEPF, 1968), cell-wall incrustations in subepithelial cells form a barrier in the apoplast inhibiting a backflow of the secreted material, but they do not occur in the slime glands of mosses, in the root cap, and in mucilage ducts. Therefore this barrier is not a prerequisite for the secretion process. In *Drosophyllum* (SCHNEPF, 1965) and *Drosera* (RAGETLI et al., 1972) the cuticle is perforated by pores allowing the secretion product to pass or it is permeable enough even without pores (SCHNEPF, 1961a: *Pinguicula*). In some cases (HORNER and LERSTEN, 1968: secretory trichomes in

Psychotria) the glands have no cuticle. In the glands of *Rumex* and *Rheum* the slime is deposited transitorily in subcuticular spaces (SCHNEPF, 1968).

1.3.4 Physiology of the Secretion Process

The intensity of the secretion process is temperature dependent. The Q_{10} in *Drosophyllum* is about 2 (SCHNEPF, 1961b), in maize root cap cells >2 (MORRÉ et al., 1967). In the latter a temperature-independent periodicity of the secretory activity (3 h) was stated. In *Drosera* the secretion also seems to be discontinuous; different secretory cells within one gland appear in different stages of activity (DEXHEIMER, 1972).

The amount of polysaccharide secretion is influenced by supply of different sugars, in *Drosophyllum* (SCHNEPF, 1963a) as well as in the root cap (MORRÉ et al., 1967; JONES and MORRÉ, 1973); glucose, fucose, xylose, sucrose, maltose, lactose, fructose, and ribose stimulate, galactose, arabinose, and galacturonic acid inhibit the secretion in the latter. An extreme increase was achieved by mannose, this effect was partially reversed by inorganic phosphate and other sugars. It is assumed that the mannose effect is due to an alteration of the normal sugar metabolism.

Inhibitors of ATP production reduce the synthesis of the polysaccharide and, in consequence or independently, the formation of secretion vesicles in *Drosophyllum* (SCHNEPF, 1963b) and in root cap cells (WHALEY, 1966; MORRÉ et al., 1967). Secretion vesicles are not found under such conditions; their transport and discharge obviously are less influenced. After inhibition the dictyosomes usually are enlarged. They are composed of more and larger cisternae which often are cup-shaped. In contrast a period of stimulation results in smaller dictyosomes with fewer cisternae (SCHNEPF, 1961b, 1963b). Obviously the renewal of the dictyosomal membranes limits the production of secretion vesicles in periods of high secretory activity (SCHNEPF, 1969a). The enlargement of the dictyosomes under conditions of ATP deficiency suggests that the formation of new membranes does not require much energy and is, to some extent, independent of synthesis of new membrane material. This seems possible if a backflow of membranes is taken into account which compensates the increase of the plasmalemma in a non-growing cell by the incorporation of vesicle membranes (SCHNEPF, 1969a).

The idea of recycling membrane molecules is favored by the fact that puromycin has no effect on slime secretion (MORRÉ et al., 1967; SCHNEPF, 1972b) (interestingly cycloheximide has an effect, but its action seems to be complex, COCUCCI and MARRÉ, 1973). The data of SCHNEPF (1961b) indicate that an increase of temperature stimulates the synthesis of the mucilage in *Drosophyllum* and the formation of secretion vesicles more than the regeneration of the dictyosomes by recycling membrane material; this suggests that the latter is a physical rather than a chemical process.

The exudation of the slime after its extrusion into the space between plasmalemma and cell wall seems to depend mainly on the turgor pressure but to be a passive process with respect to the direct influence of ATP (SCHNEPF, 1961b; MORRÉ et al., 1967).

The enzymology of the polysaccharide secretion by glands is difficult to study in detail: it seems not possible to isolate the cell organelles involved and presumably only some of their enzymes can be identified cytochemically.

In hypersecretory outer root cap cells inosine- and other nucleoside-diphosphatases and thiamine pyrophosphatase were localized cytochemically in the Golgi apparatus by DAUWALDER et al. (1969) and GOFF (1973). Acid phosphatase was not found here (DAUWALDER et al., 1969) but in other cells of the root tip and even in the slime glands (squamulae) of *Elodea* (ROUGIER, 1972). Similar results concerning the distribution of these enzymes in many other cell types are summarized in DAUWALDER et al. (1972). The nucleosidediphosphatase activity may represent glycosyl synthetases or transferases (GOFF, 1973) and therefore reflect participation in the synthesis of the polysaccharide. Such or similar enzymes occurring in the Golgi apparatus fraction also were detected biochemically (see Encyclopedia of Plant Physiology, New Series, Vol. 3). The enzymes required for the transfer of active sulfate to synthesize sulfated polysaccharides seem also to be located in the Golgi apparatus (YOUNG, 1973). With respect to the membrane turnover it is important that enzymes involved in membrane lipid synthesis have a high activity in the Golgi apparatus (MORRÉ, 1970; MORRÉ et al., 1970; for details see Encyclopedia of Plant Physiology, New Series, Vol. 3). Morphometric analyses combined with measurements of the amount of slime secretion in *Drosophyllum* allowed a calculation of the kinetics of the secretion process. It was estimated that the time between the formation and the extrusion of a secretion vesicle is a few minutes (SCHNEPF, 1961 b). This is about the same kinetics as in Golgi apparatus-mediated elimination processes in hydathodes of *Monarda* (HEINRICH, 1973 c) and in some unicellular algae (SCHNEPF, 1969 a; BROWN, 1969) whereas in the root cap secretion the time is about 15–20 min according to autoradiographic data of NORTHCOTE and PICKETT-HEAPS (1966; cf. also MORRÉ et al., 1967). Pulse labeling of the production of extracellular polysaccharide in the Golgi apparatus in the red alga *Porphyridium* (RAMUS, 1972) and in animal cells secreting glycoproteids (e.g. NEUTRA and LEBLOND, 1966) also revealed longer times.

1.3.5 Conclusions

In Fig. 5.16. our knowledge of the secretion of polysaccharides by glands is summarized. (The numbers given in the following text refer to the processes depicted in this Figure.) In part the details shown are based on observations in non-specialized cells eliminating matrix materials of the cell wall. Cellulose secretion usually seems to be different (SCHNEPF, 1969 a; for an exception see BROWN et al., 1970). The way of sugar supply (1) to the secreting cells is not known, its mechanisms are discussed in 5.3.1.2.4. Further, it is not known where the different sugars constituting the polysaccharide are formed. They probably originate from glucose. The secretion process proper begins with the passage of a membrane, presumably a membrane of the Golgi apparatus (cisternae and vesicles (2a)), but a transport into an element of the ER (2b) cannot be excluded completely; then the ER compartments are modified and fuse to form Golgi cisternae and vesicles (3). The synthesis is localized in the compartments of the dictyosomes (4a) and their derivatives, i.e. the secretion vesicles (4b). At least some of the enzymes are membrane-bound. The vesicles

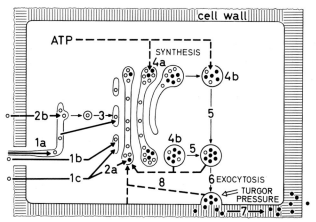

Fig. 5.16. Model of polysaccharide secretion *via* the Golgi apparatus. Thick solid arrows: transport of molecules, thin solid arrows: transport of compartments. *1* Loading of the secretory cell through plasmodesmata (*1a, 1b*) or by membrane passage (*1c*). *2* Membrane passage into the Golgi apparatus (*2a*) or the ER (*2b*). *3* Supply of the dictyosome by ER derived vesicles with membrane material, synthesizing enzymes, and sugars (open circles). *4* Polysaccharide (closed circles) synthesis in the dictyosome (*4a*) and in detached vesicles (*4b*), participation of ATP. *5* Transport of secretion vesicles. *6* Exocytosis. *7* Transport and exudation of the secretion product into the cell wall by turgor pressure (direction of turgor pressure: thick open arrow). *8* Backflow of membrane material from the plasmalemma with the incorporated vesicle membranes and from secretion vesicles to the Golgi apparatus. Thick dashed arrows indicate interactions other than transport of moleculus or transport of compartments in the secretory process (i.e. ATP supply, and backflow of membrane compounds). This model is speculative in several details

extrude their contents by exocytosis (6). There are indications of recycling membrane molecules or micelles (8).

The intensity of the synthesis regulates the activity in vesicle formation; in extreme cases the latter seems to be influenced also by the rate of membrane re-formation. The synthesis (4) depends strongly on the supply of energy and of sugars whereas transport of the vesicles (5), their discharge, the recycling of the membrane material and the re-formation of new membranes seems to be less directly linked to metabolic energy transfer. The extruded polysaccharide is transported passively through the cell wall (7) to the exterior, moved only by the turgor pressure of the cell.

2. Elimination of Proteins

The capability of plant cells to transport proteins across their plasmamembranes is evident from secretion of enzymes, i.e.

i) proteolytic enzymes detected in bacteriologically sterile fluid secreted by digestive glands of some carnivorous plants (e.g. *Nepenthes:* LÜTTGE, 1964a; STECKELBERG et al., 1967; *Dionaea:* SCALA et al. 1969),

ii) secretion of amylases and proteinases from scutellum and aleurone cells of germinating grass (cereal) seeds (e.g.: PALEG, 1960; GOODWIN and CARR, 1972; SUNDBLOM and MIKOLA, 1972), and

iii) in general, where extracellular enzymes of fungi and plants have been described.

Although a lot of work has been devoted to these systems, very little is known about the mechanism of protein transport, so that it may suffice here to mention the questions of granulocrine elimination and of hormonal control.

In analogy to the granulocrine secretion of proteins (probably glycoproteins) *via* the Golgi apparatus in animal gland cells (Sievers, 1973) one would expect that in plant cells, protein secretion also involves ER or Golgi vesicles. However, cytological investigations of active aleurone and scutellum cells (Jones, 1969) have not yet provided evidence for granulocrine elimination (Jones, 1972), which is as clear as that obtained for secretion of polysaccharide slime (see 5.3.1.3.3). A more advanced aspect in the investigation of protein secretion of scutellum and aleurone cells is hormonal control. Gibberellic acid appears to be the key hormone, whose action may be supported or modified by other hormones, such as cytokinins, abscisic acid, cAMP, and ethylene (cf. Carr, 1972).

3. Elimination of Secondary Plant Products

3.1 Secretion Products

A large diversity of secondary plant products is eliminated by glands. Usually they have a lipophilic character and occur in mixtures, even together with hydrophilic material such as slimes. Most of them are terpenoids (resins, essential oils), besides these, flavonoid aglycons, fatty oils, and quinones are common. In Table 5.2 different glands and their characters are listed.

3.2 Cytological Types of Gland Cells

The eliminatory cells are epidermal structures (often hairs) or surrounding intercellular spaces. They are long-lived if the gland belongs to the merocrine type. In holocrine or lysigenous glands the excretion product is usually discharged by the desintegration of groups of internal cells. Excretion cells accumulate the excretion products, but they do not transport them to the exterior. Sometimes there are transitions between these types (Sperlich, 1939; Schnepf, 1969a). With respect to the fine structure of the cytoplasm many glands show common features (see also Table 5.2 for References).

3.2.1 Plastidal Type

In the glandular epithelium of the excretion ducts of conifers and *Hedera helix* as well as in the holocrine glands of *Rutaceae* leucoplasts are the predominant organelles (Table 5.2). In active glands these leucoplasts contain osmiophilic materials, most probably the terpenoids or their precursors. In the conifers, these products were also found between the membranes of the plastidal envelope, in a periplastidal cisterna of the ER, and in the cytoplasm. It is assumed that the bulk of the terpenes is synthesized in the plastids, but the ER and the ground cytoplasm may also participate in the synthetic processes. It must be mentioned that the smooth ER usually is well developed. The transport of the excretion product into the duct is not fully understood. Electron micrographs (Wooding and Northcote, 1965) suggest a passage of the complete lipophilic excretion droplets through the—lipophilic—membranes, especially the plasmalemma. The changes in the molecular architecture of the membrane necessarily involved in this process are not explained. The material, then, is suggested to percolate the cell wall in the form of tiny droplets.

3.2.2 ER Type

Another group of glands is characterized by an extended, smooth ER, mainly in the form of tubular elements (Table 5.2; Fig. 5.17). This, for example, is also a property of the cells which eliminate the stigmatic fluid of *Petunia* (KONAR and LINSKENS, 1966) or the scent in the Aracea, *Typhonium* (E. SCHNEPF, unpublished). In this respect the glands resemble animal cells active in steroid synthesis, which has many features in common with the biosynthesis of terpenes. Several enzymes of the biosynthetic pathway of steroids have been localized in the ER fraction (see WOODING and NORTHCOTE, 1965). In many of these glands the leucoplasts also have a sheath of a periplastidal cisterna of the ER which may play a role in intracellular transport of some substances involved. In these glands the excretion products or their precursors have rarely been localized within the cell. In some cases (Table 5.2) osmiophilic materials were observed in the vacuole, in the ER, or in the ground cytoplasm, presumably at least in part identical with certain components of the extruded excretion. Yet it is an open question whether they have been synthesized in these compartments or been fixed during the extrusion process, or whether they are immobile deposits as in holocrine glands or excretion cells. As in the glands of the plastidal type, extrusion vesicles containing the excretion products have not been identified with certainty. An enlargement of the plasmalemma by wall protuberances does not occur. Therefore one is inclined to assume that

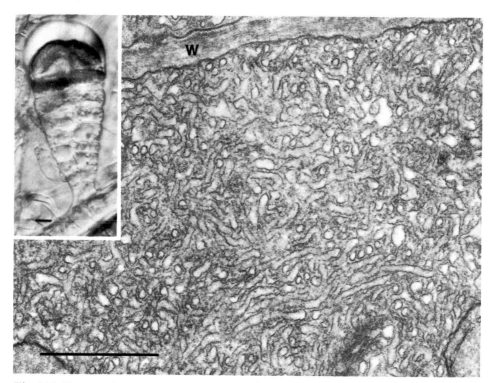

Fig. 5.17. Electron micrograph of an oil gland of *Arctium* with extended, tubular, smooth-surfaced ER; *W* cell wall. Fixation: $OsO_4 + K_2Cr_2O_7$, marker bar: 1 μm. Inset: Gland, photographed in the light microscope with Nomarski interference contrast, marker bar: 10 μm

Table 5.2. Structure of glands eliminating secondary products

Species	Excretion products	Anatomical gland type[a]	Develop. op. gland type[b]	Development of c			
				apo-plast barrier	plas-tids	tubu-lular s ER	r E
Pinus ssp. ⎱ *Picea abies* ⎰	resins	duct	mer	−	+	+	
Heracleum sp. ⎱ *Aegopodium podagraria* ⎱ *Dorema ammoniacum* ⎰	essential oils slimes, gums	duct	mer	−		+ +	
Hedera helix	terpenes	duct	mer	−	+	+	
Solidago canadensis	oils	duct	mer	−		+	
Mentha piperita	essential oils	p. hair	m/h	+		+	
Mentha piperita	ess. oils, slimes?[d]	h. hair	mer	+		+	+
Salvia ssp.	oils in emulsion[e]	p. hair	mer	+		+ +	
Salvia ssp.	(fatty?) oils[e]	p. hair	mer	+		+ +	
Monarda fistulosa	essential oils	p. hair	mer	+ ?		+	
Monarda fistulosa	essential oils	h. hair	mer	+ ?	+		
Arctium lappa	oils	hair	mer	−		+ +	
Ribes nigrum	oils	hair	mer	−		+ +	
Ledum palustre	oils	hair	m/h	+		+ +	
Lonicera periclymenum	oils	hair	mer	−		+ +	
Ononis repens	oils	hair	mer	−		+ +	
Senecio viscosus	oils	hair	mer	−		+ +	
Viscaria vulgaris	oils/resins	hair	mer	−		+	+
Calceolaria sp.	fatty oils	hair	mer	+	+	+	
Primula ssp.	flavon., quinones	hair	mer	+		+ +	
Pityrogramma chrysoconia	flavonoids	hair	mer	−		+ +	+
Populus ssp.	terpen., flavon.	epithel.	mer	−	+	+	
Alnus ssp.	terpen., flavon., slime	hair	mer	−		+ +	+
Ficus benjamina	fats, "wax"	epithel.	mer	−		+ +	
Cleome speciosa	essential oils	hair	hol	− ?		+ +	
Ruta graveolens	methylketones	cavity	hol				
Citrus ssp. ⎱ *Poncirus trifoliata* ⎰	sesqui-, monoterpenes	cavity	hol		+		
Dictamnus albus	essential oils	cavity	hol		+		

[a] p. hair = peltate hair, h. hair = hair with a glandular head.
[b] mer = merocrine, hol = holocrine, m/h merocrine/holocrine.
[c] presumably carbohydrates.
[d] G. Heinrich (personal communication) has stated that there are two types of glandular heads (as in *Monarda* and *Salvia*), one is a "hydathode" or slime gland with a well developed Golgi apparatus, the other an oil gland with a subcuticular space.

structures			Excretion products found in						Ref.
peri-plast. ER	Golgi appa-ratus	plas-tids	ground plasm	vacu-ole	ER	Golgi appa-ratus	extra-plasm. space	sub-cuti-cul. space	
+		+	+	+	+	−	+		Wooding and Northcote (1965) / Vasiliev (1970)
+	+	−	−	−	−	+[c]	+		Vasiliev (1969a) / Schnepf (1969e)
+		+	+	+	+	−	+		Vasiliev (1970)
+		−	+?	−	−	−	+?		Schnepf (1969d)
−		−	+?	+	+?	−	+	+	Amelunxen (1965, 1973, pers. comm.)
−	++	−	−	+	−	+[c]	−	+?	Amelunxen et al. (1969) / Amelunxen (1964)
−		−	−	−	−	−	−	+	Schnepf (1972a)
−		−	−	−	−	−	−	+	Schnepf (1972a)
−?		−	+	−	−	−	−	+	Heinrich (1973a, b)
−		−	−[f]	−	−	−	−	+	Heinrich (1973, pers. comm.)
−		−	−	−	−	−	+	+	Schnepf (1969b)
−		−	−	?	−	−	+	+	Vasiliev (1970)
−		−	+	−	−	−	−	+[g]	Schnepf (1972a)
−		−	−	?	−	−	−	+	Schnepf and Klasová (1972)
−		−	−	?	−	−	−	+	Schnepf and Klasová (1972)
−		−	−	?	−	−	−	+	Schnepf and Klasová (1972)
−		−	−	−	−	−	−	+	Tsekos and Schnepf (1974)
−		−	−	−	−	−	−	+	Schnepf (1969c)
−		−	−	−	−	−	−	−	Wollenweber and Schnepf (1970)
−		−	−	−	−	−	−	−	Schnepf and Klasová (1972)
	+?		−	+?	−	−	+?	+?	Vasiliev (1970); Charrière-Ladreix (1973)
+	−		−	−	−	+[c]	−	+	Wollenweber et al. (1971)
−			−	−	−	−	−	−	Schnepf (1972a)
−			−	+	−	−	−	−	Amelunxen and Arbeiter (1969)
		−	+?	+?	−	−	−	+[g]	Heinrich (1969)
	++		+	−	−	−	−	−	Heinrich (1966, 1969, 1970)
	++		+	−	−	−	−	−	Amelunxen and Arbeiter (1967)

[e] WOLLENWEBER (1974) has identified some flavonoids in the excretion of *Salvia glutinosa*.
[f] osmiophilic substances are found in mitochondria.
[g] in intercellular spaces.

the plasmalemma is passed in the eccrine way and that the discharge does not depend on a membrane pump.

3.2.3 Further Types

Several glands do not fit well in these categories (Table 5.2), among them are those with the rough ER as the predominating structure. Here the role of the ribosomes is not intelligible whereas the membranes might participate in the synthesis of the lipophilic excretions as assumed for the smooth membranes. Some glands have a well-developed Golgi apparatus: in addition to the lipophilic substances they eliminate carbohydrates, i.e. either cell wall materials during their growth or slimes which become mixed with terpenoids and flavonoids to form sticky masses at the surface of scales and leaflets in certain buds (WOLLENWEBER et al., 1971).

In exposed positions glandular hairs with a relatively permeable cuticle often have a barrier in the apoplast formed by hydrophobic wall incrustations in the stalk (Table 5.2). Excretion products are found also in the cells surrounding the glands (CHARRIÈRE-LADREIX, 1973; cf. AMELUNXEN, 1967).

In excretion cells, secondary plant products are often deposited in the vacuole. Sometimes they are subjected to further biochemical changes (FREY-WYSSLING, 1972). The capacity of a cell to accumulate them, naturally, is restricted and the excretion process runs slowly. Therefore the cells usually show no distinct features which allow conclusions about the secretory pathway; the eliminated material is more obvious than the eliminating structures. The transport processes are intracellular (see: Encyclopedia of Plant Physiology, New Series, Vol. 3).

References

AGTHE, C.: Über die physiologische Herkunft des Pflanzennektars. Ber. Schweiz. Botan. Ges. **61**, 240–274 (1951).

AMELUNXEN, F.: Elektronenmikroskopische Untersuchungen an den Drüsenhaaren von *Mentha piperita* L. Planta Med. **12**, 121–139 (1964).

AMELUNXEN, F.: Elektronenmikroskopische Untersuchungen an den Drüsenschuppen von *Mentha piperita* L. Planta Med. **13**, 457–473 (1965).

AMELUNXEN, F.: Einige Beobachtungen in den Blattzellen von *Mentha piperita* L. Planta Med. **15**, 32–34 (1967).

AMELUNXEN, F., ARBEITER, H.: Untersuchungen an den Spritzdrüsen von *Dictamnus albus* L. Z. Pflanzenphysiol. **58**, 49–69 (1967).

AMELUNXEN, F., ARBEITER, H.: Untersuchungen an den Drüsenhaaren von *Cleome spinosa* L. Z. Pflanzenphysiol. **61**, 73–80 (1969).

AMELUNXEN, F., WAHLIG, T., ARBEITER, H.: Über den Nachweis des ätherischen Öls in isolierten Drüsenhaaren und Drüsenschuppen von *Mentha piperita* L. Z. Pflanzenphysiol. **61**, 68–72 (1969).

ARNOLD, W.N.: The selection of sucrose as the translocate of higher plants. J. Theoret. Biol. **21**, 13–20 (1968).

ATKINSON, M.R., FINDLAY, G.P., HOPE, A.B., PITMAN, M.G., SADDLER, H.D.W., WEST, K.R.: Salt regulation in the mangroves *Rhizophora mucronata* Lam. and *Aegialitis annulata* R. Br. Australian J. Biol. Sci. **20**, 589–599 (1967).

BARNES, E.M. JR.: Respiration-coupled glucose transport in membrane vesicles from *Azotobacter vinelandii*. Arch. Biochem. Biophys. **152**, 795–799 (1972).

BARNES, E.M., KABACK, H.R.: β-galactoside transport in bacterial membrane preparations: energy coupling *via* membrane-bound D-lactic dehydrogenase. Proc. Natl. Acad. Sci. U.S. **66**, 1190–1198 (1970).

BEUTLER, R.: Biologische Beobachtungen über die Zusammensetzung des Blütennektars. Sitzber. Ges. Morphol. und Physiol. München **39**, 41–48 (1929/30).

BEUTLER, R.: Nectar. Bee World **34**, 106–116; 128–136; 156–162 (1953).

BIELESKI, R.L.: The physiology of sugar-cane. III. Characteristics of sugar uptake in slices of mature and immature storage tissue. Australian J. Biol. Sci. **13**, 203–220 (1960).

BOUCHET, P., DEYSSON, G.: Application de la technique à l'acide périodique-thiocarbohydrazide-protéinate d'argent à l'étude des cellules à mucilage chez diverses espèces végétales. Compt. Rend. D **276**, 2167–2170 (1973).

BOWEN, J.E., HUNTER, J.E.: Sugar transport in immature internodal tissue of sugarcane. II Mechanism of sucrose transport. Plant Physiol. **49**, 789–793 (1972).

BROWN, R.M., JR.: Observations on the relationship of the Golgi apparatus to wall formation in the marine chrysophycean alga, *Pleurochrysis scherffelii* Pringsheim. J. Cell Biol. **41**, 109–123 (1969).

BROWN, R.M., JR., FRANKE, W.W., KLEINIG, H., FALK, H., SITTE, P.: Scale formation in chrysophycean algae. I. Cellulosic and noncellulosic wall components made by the Golgi apparatus. J. Cell Biol. **45**, 246–271 (1970).

CARR, D.J. (ed.): Proc. 7th Intern. Conf. Plant Growth Substances, Canberra 1972.

CHARRIÈRE-LADREIX, Y.: Étude de la sécrétion flavonoidique des bourgeons de *Populus nigra* L. var. *italica*: cinétique du phénomène sécrétoire; ultrastructure et évolution du tissu glandulaire. J. Microscopie **17**, 299–316 (1973).

COCUCCI, M.C., MARRÉ, E.: The effects of cycloheximide in respiration, protein synthesis and adenosine nucleotide levels in *Rhodotorula gracilis*. Plant Sci. Letters **1**, 293–301 (1973).

DAUWALDER, M., WHALEY, W.G., KEPHART, J.E.: Phosphatases and differentiation of the Golgi apparatus. J. Cell Sci. **4**, 455–497 (1969).

DAUWALDER, M., WHALEY, W.G., KEPHART, J.E.: Functional aspects of the Golgi apparatus. Sub-cell. Biochem. **1**, 225–275 (1972).

DECKER, M., TANNER, W.: Respiratory increase and active hexose uptake of *Chlorella vulgaris*. Biochim. Biophys. Acta **266**, 661–669 (1972).

DEXHEIMER, J.: Quelques aspects ultrastructuraux de la sécrétion de mucilage par les glandes digestives de *Drosera rotundifolia* L. Compt. Rend. D **275**, 1983–1986 (1972).

EVANS, L.V., SIMPSON, M., CALLOW, M.E.: Sulphated polysaccharide synthesis in brown algae. Planta **110**, 237–252 (1973).

EYMÉ, J.: Infrastructure des cellules nectarigènes de *Diplotaxis erucoides* D.C., *Helleborus niger* L. et *H. foetidus*. Compt. Rend. D **262**, 1629–1632 (1966).

EYMÉ, J.: Nouvelles observations sur l'infrastructure de tissus nectarigènes floraux. Le Botaniste **50**, 169–183 (1967).

FAHN, A., RACHMILEVITZ, T.: Ultrastructure and nectar secretion in *Lonicera japonica*. Botan. J. Linn. Soc. **63**, Suppl. 1, 51–56 (1970).

FEKETE, M.A.R. DE, ZIEGLER, H., WOLF, R.: Enzyme des Kohlenhydratstoffwechsels in Nektarien. Planta **75**, 125–138 (1967).

FELDHOFEN, E.: Beiträge zur physiologischen Anatomie der nuptialen Nektarien aus den Reihen der Dikotylen. Beih. Botan. Zbl. **50**, 459–634 (1933).

FIGIER, J.: Localisation infrastructurale de la phosphomonoestérase acide dans la stipule de *Vicia faba* L. au niveau du nectaire. Rôles possibles de cet enzyme dans les mécanismes de la sécrétion. Planta **83**, 60–79 (1968).

FIGIER, J.: Étude infrastructurale de la stipule de *Vicia faba* L. au niveau du nectaire. Planta **98**, 31–49 (1971).

FIGIER, J.: Localisation infrastructurale de la phosphatase acide dans les glandes pétiolaires d'*Impatiens holstii*: rôles possibles de cette enzyme au cours des processus sécrétoires. Planta **108**, 215–226 (1972).

FINDLAY, N., MERCER, F.V.: Nectar production in *Abutilon*. I. Movement of nectar through the cuticle. Australian J. Biol. Sci. **24**, 647–656 (1971a).

FINDLAY, N., MERCER, F.V.: Nectar production in *Abutilon*. II. Submicroscopic structure of the nectary. Australian J. Biol. Sci. **24**, 657–664 (1971b).

FINDLAY, N., REED, M.L., MERCER, F.V.: Nectar production in *Abutilon*. III. Sugar secretion. Australian J. Biol. Sci. **24**, 665–675 (1971).

Franke, W.W., Herth, W., Woude, W.J., van der, Morré, D.J.: Tubular and filamentous structures in pollen tubes: possible involvement as guide elements in protoplasmic streaming and vectorial migration of secretory vesicles. Planta **105**, 317–341 (1972).

Frei, E.: Die Innervierung der floralen Nektarien dikotyler Pflanzenfamilien. Ber. Schweiz. Botan. Ges. **65**, 60–114 (1955).

Frey-Wyssling, A.: Die Stoffausscheidung der höheren Pflanzen. Berlin: Springer 1935.

Frey-Wyssling, A.: The phloem supply to the nectaries. Acta Botan. Neerl. **4**, 358–369 (1955).

Frey-Wyssling, A.: Elimination processes in higher plants. Saussurea **3**, 79–90 (1972).

Frey-Wyssling, A., Agthe, C.: Nektar ist ausgeschiedener Phloemsaft. Verhandl. Schweiz. Naturforsch. Ges., p. 175–176 (1950).

Frey-Wyssling, A., Häusermann, E.: Deutung der gestaltlosen Nektarien. Ber. Schweiz. Botan. Ges. **70**, 150–162 (1960).

Frey-Wyssling, A., Zimmermann, M., Maurizio, A.: Über den enzymatischen Zuckerumbau in Nektarien. Experientia **10**, 490-492 (1954).

Garrard, L.A., Humphreys, T.E.: The effect of D-mannose on sucrose storage in the corn scutellum: evidence for two sucrose transport mechanisms. Phytochem. **8**, 1065–1077 (1969).

Gayler, K.R., Glasziou, K.T.: Sugar accumulation in sugarcane. Carrier-mediated active transport of glucose. Plant Physiol. **49**, 563–568 (1972).

Glasziou, K.T., Gayler, K.R.: Sugar transport: occurrence of trehalase activity in sugar cane. Planta **85**, 299–302 (1969).

Glasziou, K.T., Gayler, K.R.: Storage of sugars in stalks of sugar cane. Botan. Rev. **38**, 471–490 (1972).

Glick, D.: Techniques of histo- and cytochemistry. New York-London: Interscience Publ. 1949.

Goff, C.W.: Localization of nucleoside diphosphatase in the onion root tip. Protoplasma **78**, 397–416 (1973).

Goodwin, P.B., Carr, D.J.: Actinomycin D and the hormonal induction of amylase synthesis in barley aleurone layers. Planta **106**, 1–12 (1972).

Grant, B.R., Beevers, H.: Absorption of sugars by plant tissues. Plant Physiol. **39**, 78–85 (1964).

Gunning, B.E.S., Pate, J.S.: Transfer cells. Plant cells with wall ingrowths, specialized in relation to short distance transport of solutes—their occurrence, structure and development. Protoplasma **68**, 107–133 (1969).

Gussin, A.E.S., McCormack, J.H., Waung, L.Y.-L., Gluckin, D.S.: Trehalase: a new pollen enzyme. Plant Physiol. **44**, 1163–1168 (1969).

Hancock, J.G.: Properties and formation of the squash high-affinity glucose transport system. Canad. J. Bot. **48**, 1515–1520 (1970).

Hatch, M.D.: Sugar accumulation by sugar-cane storage tissue: the role of sucrose phosphate. Biochem. J. **93**, 521–526 (1964).

Heinrich, G.: Licht- und elektronenmikroskopische Untersuchungen zur Genese der Exkrete in den lysigenen Exkreträumen von *Citrus medica*. Flora (Jena) A **156**, 451–456 (1966).

Heinrich, G.: Elektronenmikroskopische Beobachtungen zur Entstehungsweise der Exkret-behälter von *Ruta graveolens, Citrus limon* und *Poncirus trifoliata*. Österr. Botan. Z. **117**, 397–403 (1969).

Heinrich, G.: Elektronenmikroskopische Beobachtungen an den Drüsenzellen von *Poncirus trifoliata;* zugleich ein Beitrag zur Wirkung ätherischer Öle auf Pflanzenzellen und eine Methode zur Unterscheidung flüchtiger von nichtflüchtigen lipophilen Komponenten. Protoplasma **69**, 15–36 (1970).

Heinrich, G.: Entwicklung, Feinbau und Ölgehalt der Drüsenschuppen von *Monarda fistulosa*. Planta Med. **23**, 154–166 (1973a).

Heinrich, G.: Über das ätherische Öl von *Monarda fistulosa* und den Einbau von markiertem CO_2 in dessen Komponenten. Planta Med. **23**, 201–212 (1973b).

Heinrich, G.: Die Feinstruktur der Trichom-Hydathoden von *Monarda fistulosa*. Protoplasma **77**, 271–278 (1973c).

Horner, H.T. Jr., Lersten, N.R.: Development, structure and function of secretory trichomes in *Psychotria bacteriophila (Rubiaceae)*. Amer. J. Bot. **55**, 1089–1099 (1968).

Huber, H.: Die Abhängigkeit der Nektarsekretion von Temperatur, Luft- und Bodenfeuchtigkeit. Planta **48**, 47–98 (1956).

HUMPHREYS, T.E.: Leakage of hexose phosphates from the maize scutellum. Phytochem. **11**, 541–545 (1972a).

HUMPHREYS, T.E.: Sucrose leakage from excised maize scutella. Phytochem. **11**, 1311–1320 (1972b).

HUMPHREYS, T.E.: Sucrose transport at the tonoplast. Phytochem. **12**, 1211–1219 (1973).

HUMPHREYS, T.E., GARRARD, L.A.: The storage of exogeneous sucrose by corn scutellum slices. Phytochem. **7**, 701–713 (1968).

HUMPHREYS, T.E., GARRARD, L.A.: Sucrose leakage from the maize scutellum. Phytochem. **10**, 2891–2904 (1971).

JONES, D.D., MORRÉ, D.J.: Golgi apparatus mediated polysaccharide secretion by outer root cap cells of *Zea mays*. II. Isolation and characterization of the secretory product. Z. Pflanzenphysiol. **56**, 166–169 (1967).

JONES, D.D., MORRÉ, D.J.: Golgi apparatus mediated polysaccharide secretion by outer root cap cells of *Zea mays*. III. Control by exogenous sugars. Physiol. Plantarum **29**, 68–75 (1973).

JONES, D.D., MORRÉ, D.J., MOLLENHAUER, H.H.: Slime secretion by outer root cap cells of *Zea mays*. Amer. J. Bot. **53**, 621 (1966).

JONES, R.L.: Gibberellic acid and the fine structure of barley aleurone cells. I. Changes during the lag-phase of α-amylase synthesis. Planta **87**, 119–133 (1969).

JONES, R.L.: Fractionation of the enzymes of the barley aleurone layer: evidence for a soluble mode of enzyme release. Planta **103**, 95–109 (1972).

JUNIPER, B.E., PASK, G.: Directional secretion by the Golgi bodies in maize root cells. Planta **109**, 225–231 (1973).

KABACK, H.R.: The transport of sugars across isolated bacterial membranes. In: Current topics in membranes and transport (F. BRONNER, A. KLEINZELLER, eds.), vol. I. New York-London: Academic Press 1970a.

KABACK, H.R.: Transport. Ann. Rev. Biochem. **39**, 561–598 (1970b).

KASHKET, E.R., WILSON, T.H.: Proton-coupled accumulation of galactoside in *Streptococcus lactis* 7962. Proc. Natl. Acad. Sci. U.S. **70**, 2866–2869 (1973).

KLING, L.: Knospenbau und Sproßentwicklung von *Rheum palmatum*. L. Diss. Tübingen 1961.

KLUGE, H.: Jahresperiodische Schwankungen des Kohlenhydratgehaltes in Siebröhrensäften, Blättern und Wurzeln einiger Holzgewächse. Biochem. Physiol. Pflanzen **161**, 142–165 (1970).

KOMOR, E.: Proton-coupled hexose transport in *Chlorella vulgaris*. F.E.B.S.-Letters **38**, 16–18 (1973).

KOMOR, E., HAASS, D., TANNER, W.: Unusual features of the active hexose uptake system of *Chlorella vulgaris*. Biochim. Biophys. Acta **266**, 649–660 (1972).

KOMOR, E., LOOS, E., TANNER, W.: A confirmation of the proposed model for the hexose uptake system of *Chlorella vulgaris*. Anaerobic studies in the light and in the dark. J. Membrane Biol. **12**, 89–99 (1973).

KOMOR, E., TANNER, W.: Characterisation of the active hexose transport system of *Chlorella vulgaris*. Biochim. Biophys. Acta **241**, 170–179 (1971).

KOMOR, E., TANNER, W.: The hexose-proton cotransport system of *Chlorella*. pH-dependent change in K_m values and translocation constants of the uptake system. J. Gen. Physiol. **64**, 568–581 (1974).

KONAR, R.N., LINSKENS, H.F.: The morphology and anatomy of the stigma of *Petunia hybrida*. Planta **71**, 356–371 (1966).

LATIES, G.G.: Relation of glucose absorption to respiration in potato slices. Plant Physiol. **39**, 391–397 (1964).

LINASK, J., LATIES, G.G.: Multiphasic absorption of glucose and 3-0-methylglucose by aged potato slices. Plant Physiol. **51**, 289–294 (1973).

LÜTTGE, U.: Über die Zusammensetzung des Nektars und den Mechanismus seiner Sekretion. I. Planta **56**, 189–212 (1961).

LÜTTGE, U.: Über die Zusammensetzung des Nektars und den Mechanismus seiner Sekretion. II. Mitteilung. Der Kationengehalt des Nektars und die Bedeutung des Verhältnisses Mg^{++}/Ca^{++} im Drüsengewebe für die Sekretion. Planta **59**, 108–114 (1962a).

LÜTTGE, U.: Über die Zusammensetzung des Nektars und den Mechanismus seiner Sekretion. III. Mitteilung. Die Rolle der Rückresorption und der spezifischen Zuckersekretion. Planta **59**, 175–194 (1962b).

Lüttge, U.: Untersuchungen zur Physiologie der Carnivoren-Drüsen. I. Mitteilung. Die an den Verdauungsvorgängen beteiligten Enzyme. Planta **63**, 103–117 (1964a).

Lüttge, U.: Die Gewinnung und die chemische Analyse des Nektars. Deut. Bienenwirtsch. **15**, 101–105 (1964b).

Lüttge, U.: Die Nektarsekretion. Deut. Bienenwirtsch. **15**, 238–243 (1964c).

Lüttge, U.: Funktion und Struktur pflanzlicher Drüsen. Naturwissenschaften **53**, 96–103 (1966).

Lüttge, U.: Aktiver Transport. Kurzstreckentransport bei Pflanzen. Protoplasmatologia. Vol. VIII/7 b. Wien: Springer 1969.

Lüttge, U.: Structure and function of plant glands. Ann. Rev. Plant Physiol. **22**, 23–44 (1971).

Lüttge, U.: Stofftransport der Pflanzen. Berlin-Heidelberg-New York: Springer 1973.

MacRobbie, E.A.C.: Phloem translocation. Facts and mechanisms: a comparative survey. Biol. Rev. **46**, 429–481 (1971).

Maretzki, A., Thom, M.: Membrane transport of sugars in cell suspensions of sugarcane. I. Evidence for sites and specificity. Plant Physiol. **49**, 177–182 (1972).

Matile, P.: Über den Stoffwechsel und die Auxinabhängigkeit der Nektarsekretion. Ber. Schweiz. Botan. Ges. **66**, 237–266 (1956).

Maurizio, A.: Papierchromatographische Untersuchungen an Blütenhonigen und Nektar. Ann. Abeille **2**, 291–341 (1959).

Mercer, F.V., Rathgeber, N.: Nectar secretion and cell membranes. Proc. 5. Intl. Congr. Electron Microsc. Philadelphia, vol. II, WW-11. New York-London: Academic Press 1962.

Midgley, M., Dawes, E.A.: The regulation of transport of glucose and methyl α-glucoside in *Pseudomonas aeruginosa.* Biochem. J. **132**, 141–154 (1973).

Mollenhauer, H.H.: The fine structure of mucilage secreting cells of *Hibiscus esculentus* pods. Protoplasma **63**, 353–362 (1967).

Mollenhauer, H.H., Morré, D.J.: Golgi apparatus and plant secretion. Ann. Rev. Plant Physiol. **17**, 27–46 (1966a).

Mollenhauer, H.H., Morré, D.J.: Tubular connections between dictyosomes and forming secretory vesicles in plant Golgi apparatus. J. Cell Biol. **29**, 373–376 (1966b).

Mollenhauer, H.H., Whaley, W.G.: An observation on the functioning of the Golgi apparatus. J. Cell Biol. **17**, 222–225 (1963).

Mollenhauer, H.H., Whaley, W.G., Leech, J.H.: A function of the Golgi apparatus in outer rootcap cells. J. Ultrastruct. Res. **5**, 193–200 (1961).

Molotok, G.P., Britikov, E.A., Sinyukhin, A.M.: Electrophysiological and functional activity of the nectaries of lime on mechanical stimulation. Dokl. Akad. Nauk SSSR. **181**, 750–753. Engl. transl. p. 122–125 (1968).

Morré, D.J.: *In vivo* incorporation of radioactive metabolites by Golgi apparatus and other cell fractions of onion stem. Plant Physiol. **45**, 791–799 (1970).

Morré, D.J., Jones, D.D., Mollenhauer, H.H.: Golgi apparatus mediated polysaccharide secretion by outer root cap cells of *Zea mays.* I. Kinetics and secretory pathway. Planta **74**, 286–301 (1967).

Morré, D.J., Mollenhauer, H.H., Bracker, C.E.: Origin and continuity of Golgi apparatus. In: Results and problems in cell differentiation. Vol. II. Origin and continuity of cell organelles (J. Reinert, H. Ursprung, eds.), p. 82–126. Berlin-Heidelberg-New York: Springer 1971.

Morré, D.J., Nyquist, S., Rivera, E.: Lecithin biosynthetic enzymes of onion stem and the distribution of phosphorylcholine-cytidyl transferase among cell fractions. Plant Physiol. **45**, 800–804 (1970).

Neutra, M., Leblond, C.P.: Synthesis of the carbohydrate of mucus in the Golgi complex as shown by electron microscope radioautography of goblet cells from rats injected with glucose-H^3. J. Cell Biol. **30**, 119–136 (1966).

Northcote, D.H., Pickett-Heaps, J.D.: A function of the Golgi apparatus in polysaccharide synthesis and transport in the rootcap cells of wheat. Biochem. J. **98**, 159–167 (1966).

Paleg, L.: Physiological effects of gibberellic acid. II. On starch hydrolyzing enzymes of barley endosperm. Plant Physiol. **35**, 902–906 (1960).

Pate, J.S., Gunning, B.E.S.: Transfer Cells. Ann. Rev. Plant Physiol. **23**, 173–196 (1972).

Percival, M.S.: Types of nectar in angiosperms. New Phytologist **60**, 235–281 (1961).

Pickett-Heaps, J.D.: The use of radioautography for investigating wall secretion in plant cells. Protoplasma **64**, 49–66 (1967).

RACHMILEVITZ, T., FAHN, A.: Ultrastructure of nectaries of *Vinca rosea* L., *Vinca major* L. and *Citrus sinensis* Osbeck cv. Valencia and its relation to the mechanism of nectar secretion. Ann. Bot. **37**, 1–9 (1973).

RAGETLI, H.W.J., WEINTRAUB, M., LO, E.: Characteristics of *Drosera* tentacles: I. Anatomical and cytological detail. Canad. J. Bot. **50**, 159–168 (1972).

RAMUS, J.: The production of extracellular polysaccharide by the unicellular red alga *Porphyridium aerugineum*. J. Phycol. **8**, 97–111 (1972).

REED, M.L., FINDLAY, N., MERCER, F.V.: Nectar production in *Abutilon*. IV. Water and solute relations. Australian J. Biol. Sci. **24**, 677–688 (1971).

ROUGIER, M.: Ultrastructure des squamules d'*Elodea canadensis* (Hydrocharitacée) et *Potamogeton perfoliatus* (Potamogétonacée). J. Microscopie **4**, 523–530 (1965).

ROUGIER, M.: Étude cytochimique des squamules d'*Elodea canadensis*. Mise en évidence de leur sécrétion polysaccharidique et de leur activité phosphatasique acide. Protoplasma **74**, 113–131 (1972).

SACKTOR, B.: Trehalase and the transport of glucose in the mammalian kidney and intestine. Proc. Natl. Acad. Sci. U.S. **60**, 1007–1014 (1968).

SAUTER, J.J.: Untersuchungen zur Physiologie der Pappelholzstrahlen. II. Jahresperiodische Änderungen der Phosphataseaktivität im Holzstrahlenparenchym und ihre mögliche Bedeutung für den Kohlenhydratstoffwechsel und den aktiven Assimilattransport. Z. Pflanzenphysiol. **55**, 349–362 (1966).

SAUTER, J.J.: Respiratory and phosphatase activities in contact cells of wood rays and their possible role in sugar secretion. Z. Pflanzenphysiol. **67**, 135–145 (1972).

SAUTER, J.J., ITEN, W., ZIMMERMANN, M.H.: Studies on the release of sugar into the vessels of sugar maple (*Acer saccharum*). Canad. J. Bot. **51**, 1–8 (1973).

SAUTER, J.J., MARQUARDT, H.: Phosphatase-Aktivität und Stofftransport im Holzstrahlparenchym von *Populus*. Naturwissenschaften **52**, 61–68 (1965).

SCALA, J., IOTT, K., SCHWAB, D.W., SEMERSKY, F.E.: Digestive secretion of *Dionaea muscipula* (Venus's-flytrap). Plant Physiol. **44**, 367–371 (1969).

SCHNEPF, E.: Licht- und elektronenmikroskopische Beobachtungen an Insektivoren-Drüsen über die Sekretion des Fangschleimes. Flora (Jena) **151**, 73–87 (1961a).

SCHNEPF, E.: Quantitative Zusammenhänge zwischen der Sekretion des Fangschleimes und den Golgi-Strukturen bei *Drosophyllum lusitanicum*. Z. Naturforsch. **16b**, 605–610 (1961b).

SCHNEPF, E.: Zur Cytologie und Physiologie pflanzlicher Drüsen. 1. Teil. Über den Fangschleim der Insektivoren. Flora (Jena) **153**, 1–22 (1963a).

SCHNEPF, E.: Zur Cytologie und Physiologie pflanzlicher Drüsen. 2. Teil. Über die Wirkung von Sauerstoffentzug und von Atmungsinhibitoren auf die Sekretion des Fangschleimes von *Drosophyllum* und auf die Feinstruktur der Drüsenzellen. Flora (Jena) **153**, 23–48 (1963b).

SCHNEPF, E.: Golgi-Apparat und Sekretbildung in den Drüsenzellen der Schleimgänge von *Laminaria hyperborea*. Naturwissenschaften **50**, 674 (1963c).

SCHNEPF, E.: Zur Cytologie und Physiologie pflanzlicher Drüsen. 4. Teil. Licht- und elektronenmikroskopische Untersuchungen an Septalnektarien. Protoplasma **58**, 137–171 (1964).

SCHNEPF, E.: Die Morphologie der Sekretion in pflanzlichen Drüsen. Ber. Deut. Botan. Ges. **78**, 478–483 (1965).

SCHNEPF, E.: Zur Feinstruktur der schleimsezernierenden Drüsenhaare auf der Ochrea von *Rumex* und *Rheum*. Planta **79**, 22–34 (1968).

SCHNEPF, E.: Sekretion und Exkretion bei Pflanzen. Protoplasmatologia VIII/8. Wien-New York: Springer 1969a.

SCHNEPF, E.: Über den Feinbau von Öldrüsen. I. Die Drüsenhaare von *Arctium lappa*. Protoplasma **67**, 185–194 (1969b).

SCHNEPF, E.: Über den Feinbau von Öldrüsen. II. Die Drüsenhaare in *Calceolaria*-Blüten. Protoplasma **67**, 195–203 (1969c).

SCHNEPF, E.: Über den Feinbau von Öldrüsen. III. Die Ölgänge von *Solidago canadensis* und die Exkretschläuche von *Arctium lappa*. Protoplasma **67**, 205–212 (1969d).

SCHNEPF, E.: Über den Feinbau von Öldrüsen. IV. Die Ölgänge von Umbelliferen: *Heracleum sphondylium* und *Dorema ammoniacum*. Protoplasma **67**, 375–390 (1969e).

SCHNEPF, E.: Tubuläres endoplasmatisches Reticulum in Drüsen mit lipophilen Ausscheidungen von *Ficus*, *Ledum* und *Salvia*. Biochem. Physiol. Pfl. **163**, 113–125 (1972a).

SCHNEPF, E.: Über die Wirkung von Hemmstoffen der Proteinsynthese auf die Sekretion des Kohlenhydrat-Fangschleimes von *Drosophyllum lusitanicum*. Planta **103**, 334–339 (1972b).

Schnepf, E.: Sezernierende und exzernierende Zellen bei Pflanzen. In: Grundlagen der Cytologie (G.C. Hirsch, H. Ruska, P. Sitte, eds.), p. 461–477. Stuttgart: Fischer 1973.

Schnepf, E.: Gland cells. In: Dynamic aspects of plant ultrastructure (A.W. Robards, ed.), p. 331–357. Maidenhead: McGraw-Hill 1974.

Schnepf, E., Klasová, A.: Zur Feinstruktur von Öl- und Flavon-Drüsen. Ber. Deut. Botan. Ges. **85**, 249–258 (1972).

Schrödter, K.: Zur physiologischen Anatomie der Mittelzelle drüsiger Gebilde. Flora **120**, 19–86 (1926).

Shuel, R.W.: Studies of nectar secretion in excised flowers. I. The influence of cultural conditions on quantity and composition of nectar. Canad. J. Bot. **34**, 142–153 (1956).

Shuel, R.W.: Studies of nectar secretion in excised flowers. II. The influence of certain growth regulators and enzyme inhibitors. Canad. J. Bot. **37**, 1167–1180 (1959).

Shuel, R.W.: Influence of reproductive organs on secretion of sugars in flowers of *Streptosolen jamesonii*, Miers. Plant Physiol. **36**, 265–271 (1961).

Shuel, R.W.: Nectar secretion in excised flowers. III. The dual effect of indolyl-3-acetic acid. J. Apic. Res. **3**, 99–111 (1964).

Shuel, R.W.: Nectar secretion in excised flowers. IV. Selective transport of sucrose in the presence of other solutes. Canad. J. Bot. **45**, 1953–1961 (1967).

Sievers, A.: Golgi-Apparat. In: Grundlagen der Cytologie (G.C. Hirsch, H. Ruska, P. Sitte, eds.), p. 281–296. Stuttgart: Fischer 1973.

Sperlich, A.: Exkretionsgewebe. In: Handbuch der Pflanzenanatomie (K. Linsbauer, ed.), vol. IV: Das trophische Parenchym. Berlin: Borntraeger 1939.

Steckelberg, R., Lüttge, U., Weigl, J.: Reinigung der Proteinase aus *Nepenthes*-Kannensaft. Planta **76**, 238–241 (1967).

Steinbrecher, W., Lüttge, U.: Sugar and ion transport in isolated onion epidermis. Australian J. Biol. Sci. **22**, 1137–1143 (1969).

Sundblom, N.O., Mikola, J.: On the nature of the proteinases secreted by the aleurone layer of barley grain. Physiol. Plantarum **27**, 281–284 (1972).

Tanner, W.: Light-driven active uptake of 3-0-methylglucose via an inducible hexose uptake system of *Chlorella*. Biochem. biophys. Res. Commun. **36**, 278–283 (1969).

Tsekos, I., Schnepf, E.: Der Feinbau der Drüsen der Pechnelke, *Viscaria vulgaris*. Biochem. Physiol. Pfl. **165**, 265–270 (1974).

van der Woude, W.J., Morré, D.J., Bracker, C.E.: Isolation and characterization of secretory vesicles in germinated pollen of *Lilium longiflorum*. J. Cell Sci. **8**, 331–351 (1971).

van Handel, E., Haeger, J.S., Hansen, C.W.: The sugars of some Florida nectars. Amer. J. Bot. **59**, 1030–1032 (1972).

Vasiliev, A.E.: Some peculiarities of the endoplasmic reticulum in secretory cells of *Heracleum* sp. [Russian]. Akad. Nauk SSSR, Citologia **11**, 298–307 (1969a).

Vasiliev, A.E.: Submicroscopic morphology of nectary cells and problems of nectar secretion [Russian]. Akad. Nauk SSSR, Botan. J. **54**, 1023–1038 (1969b).

Vasiliev, A.E.: On the localization of the synthesis of terpenoids in plant cells [Russian]. Rastit. Resursy **6**, 29–45 (1970).

Vasiliev, A.E.: New data on the ultrastructure of the cells of flower nectary [Russian]. Akad. Nauk SSSR, Botan. J. **56**, 1292–1306 (1971).

Vian, B., Barbier, C., Reis-Crepin, D.: Données sur l'élaboration et l'évolution de mucilages dans le col de l'archégone de *Mnium undulatum*. Compt. Rend. D **271**, 1978–1981 (1970).

Vis, J.H.: The histochemical demonstration of acid phosphatase in nectaries. Acta Botan. Neerl. **7**, 124–130 (1958).

Wanner, H.: Phosphataseverteilung und Kohlenhydrattransport in der Pflanze. Planta **41**, 190–194 (1952).

West, I.C.: Lactose transport coupled to proton movements in *Escherichia coli*. Biochem. Biophys. Res. Commun. **41**, 655–661 (1970).

West, I.C., Mitchell, P.: Proton-coupled β-galactoside translocation in non-metabolizing *Escherichia coli*. J. Bioenerg. **3**, 445–462 (1972).

West, I.C., Mitchell, P.: Stoicheiometry of lactose-H$^+$ symport across the plasma membrane of *Escherichia coli*. Biochem. J. **132**, 587–592 (1973).

Whaley, W.G.: Proposals concerning replication of the Golgi apparatus. In: Funktionelle und morphologische Organisation der Zelle. Probleme der biologischen Reduplikation (P.

SITTE, ed.), 3. Wiss. Konf. Ges. dt. Naturf. u. Ärzte, p. 340–370. Berlin-Heidelberg-New York: Springer 1966.

WHITESELL, J.H., HUMPHREYS, T.E.: Sugar uptake in the maize scutellum. Phytochemistry 11, 2139–2147 (1972).

WOLLENWEBER, E.: Flavones and flavonols in exudate of *Salvia glutinosa*. Phytochemistry 13, 753 (1974).

WOLLENWEBER, E., EGGER, K., SCHNEPF, E.: Flavonoid-Aglykone in *Alnus*-Knospen und die Feinstruktur der Drüsenzellen. Biochem. Physiol. Pfl. 162, 193–201 (1971).

WOLLENWEBER, E., SCHNEPF, E.: Vergleichende Untersuchungen über die flavonoiden Exkrete von „Mehl"- und „Öl"-Drüsen bei Primeln und die Feinstruktur der Drüsenzellen. Z. Pflanzenphysiol. 62, 216–227 (1970).

WOODING, F.B.P., NORTHCOTE, D.H.: The fine structure of the mature resin canal cells of *Pinus pinea*. J. Ultrastruct. Res. 13, 233–244 (1965).

WRISCHER, M.: Elektronenmikroskopische Beobachtungen an extrafloralen Nektarien von *Vicia faba* L. Acta Botan. Croat. 20/21, 75–94 (1962).

YOUNG, R.W.: The role of the Golgi complex in sulfate metabolism. J. Cell Biol. 57, 175–189 (1973).

ZIEGLER, H.: Untersuchungen über die Leitung und Sekretion der Assimilate. Planta 47, 447–500 (1956).

ZIEGLER, H.: Über die Zusammensetzung des Bestäubungstropfens und den Mechanismus seiner Sekretion. Planta 52, 587–599 (1959).

ZIEGLER, H.: Die Physiologie pflanzlicher Drüsen. Ber. Deut. Botan. Ges. 78, 466–477 (1965).

ZIEGLER, H., HUBER, F.: Phosphataseaktivität in den „Strasburger-Zellen" der Koniferennadeln. Naturwissenschaften 47, 305 (1960).

ZIEGLER, H., LÜTTGE, U.: Über die Resorption von C^{14}-Glutaminsäure durch sezernierende Nektarien. Naturwissenschaften 46, 176–177 (1959).

ZIEGLER, H., LÜTTGE, U., LÜTTGE, U.: Die wasserlöslichen Vitamine des Nektars. Flora (Jena) 154, 215–229 (1964).

ZIMMERMANN, J.G.: Über die extrafloralen Nektarien der Angiospermen. Beih. Botan. Zbl. 1. Abt. 49, 99–196 (1932).

ZIMMERMANN, M.: Über ein neues Trisaccharid. Experientia 8, 424–425 (1952).

ZIMMERMANN, M.: Papierchromatographische Untersuchungen über die pflanzliche Zuckersekretion. Ber. Schweiz. Botan. Ges. 63, 402–429 (1953).

ZIMMERMANN, M.: Über die Sekretion saccharosespaltender Transglukosidasen im pflanzlichen Nektar. Experientia 10, 145–146 (1954).

ZIMMERMANN, M.H.: Translocation of organic substances in the phloem of trees. In: The physiology of forest trees (K.V. THIMANN, ed.), p. 381–400. New York: Ronald Press 1958.

6. Transport in Symbiotic Systems Fixing Nitrogen

J.S. PATE

1. Introduction

The fixation of nitrogen by associations involving plants is taxonomically widespread and of considerable importance ecologically or agriculturally in the provision of organically-bound nitrogen to other organisms. Seven main groupings of symbiotically paired organisms may be involved (Table 6.1), blue-green algae, Actinomycetes

Table 6.1. Symbiotic systems involving plants capable of fixing nitrogen

General classification of association	Genera of N-fixing microsymbiont	Genera of association or macrosymbiont
A. Lichen (lichen fungus, heterocystous blue-green alga) (± green alga)	*Nostoc, Anabaena, Calothrix*	*Collema*[a]; *Lichina*[b]; *Peltigera*[c]; *Placopsis*[d]; *Lobaria*[e]; *Stereocaulon*[f]; *Leptogium*[g]; *Ephebe*[h]
B. Liverwort—Blue-green alga	*Nostoc*	*Blasia*[i]; *Cavicularia*[j]
C. Fern—Blue-green alga	*Anabaena*	*Azolla*[k]
D. Cycad—Blue-green alga	*Nostoc, Anabaena*	*Macrozamia*[l]; *Encephalartos*[m]; *Ceratozamia*[n]
E. Angiosperm—Blue-green alga	*Nostoc*	*Gunnera*[o]
F. Angiosperm—Actinomycete	*Streptomyces*	*Coriaria*[p] (Coriariaceae); *Myrica*[q] (=*Gale*=*Comptonia*) (Myricaceae); *Alnus*[r] (Betulaceae); *Casuarina*[s] (Casuarinaceae); *Elaeagnus*[t], *Hippophae*[u], *Shepherdia*[v] (Elaeagnaceae); *Ceanothus*[w], *Discaria*[x] (Rhamnaceae); *Dryas*[y], *Purshia*[z], *Cercocarpus*[aa] (Rosaceae)
G. Angiosperm—Bacterium	*Rhizobium*	Many genera of legumes (Leguminosae); *Trema*[bb] (Ulmaceae)

Selected references providing or giving reference to experimental evidence of N_2 fixation (and/or C_2H_2 reduction) in the foregoing associations: [a] HENRIKSSON (1951); BOND and SCOTT (1955); ROGERS et al. (1966); HITCH and STEWART (1973). [b] HITCH and STEWART (1973). [c] SCOTT (1956); KERSHAW and MILLBANK (1970). [d] HITCH and STEWART (1973). [e] MILLBANK and KERSHAW (1970). [f] FOGG and STEWART (1968). [g] BOND and SCOTT (1955). [h] J. HALLIDAY, J.S. PATE and N.C. SAMMY (unpublished). [i] BOND and SCOTT (1955); PANKOW and MARTENS (1964). [j] STEWART (1968). [k] OES (1913); MISHUSTIN and SHIL'NIKOVA (1971); J. HALLIDAY, J.S. PATE (Unpublished). [l] BERGERSEN et al. (1965). [m] BOND (1967); GROBBELAAR et al. (1971). [n] BOND (1967). [o] SILVESTER and SMITH (1969); SILVESTER (1974). [p] HARRIS and MORRISON (1958). [q] BOND (1955, 1957a); SLOGER and SILVER (1965). [r] BOND (1955). [s] BOND (1957a). [t] BOND (1963). [u] BOND (1955). [v] GARDNER and BOND (1957); BOND (1957b). [w] BOND (1957b); DELWICHE et al. (1965); RUSSELL and EVANS (1966). [x] MORRISON (1961). [y] LAWRENCE et al. (1967); BOND (1971a). [z] WEBSTER et al. (1967). [aa] BOND (1971a). [bb] TRINICK (1973).

and *Rhizobium* functioning as the microsymbionts fixing nitrogen, plants representing any of the major taxa acting as the various non-fixing macrosymbiont. The list in Table 6.1 is by no means exhaustive. Only those symbioses in which it appears to have been demonstrated conclusively that fixation contributes materially to the normal nitrogen nutrition of the association are included. The "loose" associations of rhizosphere and phyllosphere between N-fixing bacteria and higher plants, and the casual endophytism of blue-green algae recorded for a variety of cryptogams (see, for example, LHOTSKY, 1946), are deemed to fall outside the scope of present discussion, despite their intrinsic interest in considerations of the evolution of symbiotic systems. Also, the various examples of angiosperm leaf nodules (*Pavetta, Psychotria, Ardisia*), mycorrhiza, arbuscular-vesicular mycorrhiza, proteoid roots of various *Proteaceae* and the nodules formed by fungi or Actinomycetes on certain Gymnosperms (*Podocarpus*) are omitted, since the evidence now seems to be that, if fixation occurs at all in these, it fails to reach physiologically significant proportions.

Unfortunately only relatively few of the associations listed in Table 6.1 have been examined in any detail with regard to transport between participants, with only certain lichens, the Actinomycete-*Alnus* association, and the *Rhizobium*-legume association providing studies in which both carbohydrate and nitrogen metabolism have been considered adequately. Making these admissions is tantamount to stating that it cannot yet be said with certainty whether some of the associations listed are truly mutualistic or merely casual or parasitic forms of symbiosis. For, in the opinion of several writers (see discussions of SCHAEDE, 1962; SCOTT, 1969; LEWIS, 1973) mutualism remains unproven in the absence of evidence of two-way exchange of nutrients between the partners. Assessed on this basis, the fern and liverwort associations with blue-green algae (Categories B and C, Table 6.1) fail the test for mutualism, lichens pass it only if one concedes that passage of water and inorganic nutrients from fungus to alga fulfils a "transport" requirement (see SMITH, 1974), whilst in many of the examples involving nodules (Categories D, E, F, G) mutualism is inferred on the logical basis that the microsymbiont is heterotrophic for carbon and must therefore acquire energy-yielding carbon-containing compounds from its host.

2. Morphological, Anatomical and Ultrastructural Attributes of the Symbioses

The proper description of transport phenomena demands understanding in depth of the basic structure and compartmentation of the organs, tissues and cells through which the flow of solutes is likely to occur, and this is indeed available for several of the systems to be considered here. For the sake of brevity, only the mature, fully functional symbiotic structures will be described in detail, and, for comparative purposes, discussion will commence with the simplest extracellular forms of symbiosis and end with the more complicated associations involving intimate, intracellular accommodation of the microsymbiont.

2.1 Symbioses Involving Blue-Green Algae

2.1.1 Liverworts, Ferns

The liverwort associations all show the algal component lying free in intercellular spaces or in relatively unspecialized cavities of the thallus (MOLISCH, 1925; LHOTSKY, 1946; WATANABE and KIYOHARA, 1963; STEWART, 1968).

There is no evidence of special modifications of host or algal tissues. The *Azolla-Anabaena* symbiosis is somewhat more specialized structurally, the algae being accommodated and proliferating rapidly in special ventral cavities on the dorsal lobe of each vegetative leaf (YU-FENG SHEN, 1960; SCHAEDE, 1962). Light microscope observations suggest no unusual host cell modification to the lining of this cavity, nor any unusual filament structure or heterocyst frequency in the contained alga (FRITSCH, 1952; YU-FENG SHEN, 1960).

2.1.2 Cycads

In the Cycad-blue-green alga symbiosis (Figs. 6.1–6.5) the alga is confined to special apogeotropic roots (Figs. 6.1 and 6.2) probably evolved initially as pneumatophores. Not all such roots are invaded, but, if they are, the algae come to lie within intercellular spaces of the root epidermis (Fig. 6.4), a tissue overlain by a persistent root cap. The only apparent effects of invasion are that the root loses its strictly apogeotropic properties and that its epidermal cells elongate radially (WATANABE, 1924; WITTMANN et al., 1965). Infected coralloid roots carry a periderm and develop prominent lenticels (ALLEN and ALLEN, 1964; see also Fig. 6.3). In old plants the contractile activities of the tap root may pull the shoot base and the infected and uninfected coralloid roots 0.2–0.5 m under the ground (observations on *Macrozamia riedlei* in sandy soils of Western Australia, B.B. LAMONT and J.S. PATE, unpublished). Algae in these buried coralloid roots are unlikely to receive significant amounts of light and even if located nearer the surface, light penetration through the opaque surface layers of the nodule is likely to be minimal. The mode of infection of Cycad roots by the blue-green alga, the ultrastructure of epidermis and adjacent algae, and the heterocyst frequency of the alga in comparison with free-living forms have all still to be investigated.

2.1.3 Lichens

The final extracellular association is that of the lichen, listed last because of the apparently obligate nature of the symbiosis as far as the fungus is concerned, and

---

Figs. 6.1–6.5. Symbiosis between the cycad *Macrozamia riedlei* and *Anabaena* spp.; Fig. 6.1. Seedlings showing a pair of young apogeotropic roots arising from the top of the radicle ($\times 1$); Fig. 6.2. Older seedling with apogeotropic roots much elongated and branched. Note swollen tips to these roots ($\times 1.5$); Fig. 6.3. Part of infected coralloid root from a mature plant. Note the dichotomous branching of the roots and the abundant lenticels ($\times 3.5$); Fig. 6.4. Surface view of transverse section through coralloid root showing algal zone, discernible as a dark ring in the cortical region of each section. ($\times 2$); Fig. 6.5. Cluster of coralloid roots on plant approximately 30 years of age. Continued apical growth has extended the root cluster above the crown of the plant. The uppermost coralloid roots were buried 20 cm below the sand surface ($\times 0.3$)

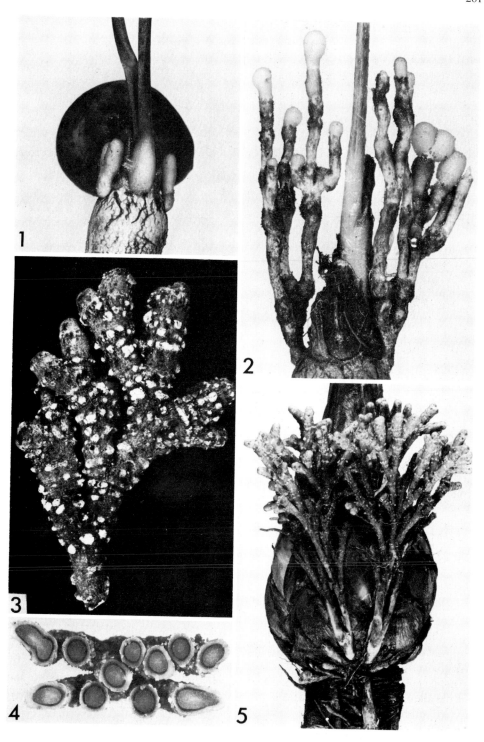

Figs. 6.1–6.5. Legend see opposite page

because of the quite highly developed ultra-structural and physiological modifications of the participants. The blue-green algae are confined to the surface layers of the lichen, either as a homogeneous layer (*Lichina, Collema, Cora,* most species of *Peltigera*) or confined to special structures, called cephalodia, situated either on the surface of (*Peltigera aphthosa*), or buried in (*Lobaria*) the thallus. If cephalodia are present, a second phycobiont (e.g. the green alga *Coccomyxa* in the case of *P. aphthosa*) may be present, presenting the interesting possibility of three-way exchange of nutrients. Collectively the phycobiont(s) comprise only a few percent of the thallus dry weight; clearly they are the microsymbionts.

Ultrastructural studies are available for several species of lichen containing N-fixing blue-green algae [PEAT, 1968 (*Peltigera polydactyla*); PEVELING 1969, 1973 (*Peltigera canina, Leptogium hildenbrandii, Lichina pygmaea*); ROSKIN, 1970 (*Cora pavonia*); GRIFFITHS et al., 1972 (*Peltigera canina*); SMITH, 1974 (*Peltigera canina*)]. The majority of algal cells are generally depicted as being healthy and not invaded by fungal hyphae, although the latter may be closely appressed to the algal component. Modifications to algal structure are relatively minor, compared with relevant free-living forms. The four-layered wall typical of the alga is present in the lichenized state but its outermost electron-dense layer and the underlying plasmamembrane may develop crenulations (*Lichina*). The mucilaginous cell sheath typical of *Nostoc* may be much reduced in symbiosis (*Peltigera canina*), and the proportion of heterocysts in the filaments (*P. canina*) is recorded to be much lower (3%) than in the free-living state (15%) (GRIFFITHS et al., 1972). However, in the cephalodia of *P. aphthosa* the heterocyst frequency is higher (18–25%) (J.W. MILLBANK, personal communication). Modification to the fungal component is more difficult to assess since the fungus may not be known outside symbiosis. However, the hyphae in the algal zone possess much more dense contents than in the lower regions of the thallus, and according to the observations of PEVELING (1969, 1973), the regions of plasmamembrane of the fungal hyphae in close contact with algal cells may develop tufts or infoldings. The latter are interpreted as amplifying the surface area available for absorption of nutrients released from the algal cells (PEVELING, 1969).

2.1.4 *Gunnera*

An example of intracellular symbiosis involving blue-green algae is afforded by the *Gunnera* symbiosis. The *Nostoc* invades glands (secretory domatia) on the nodes of the rhizomes near the leaf bases (HARDER, 1917; WINTER, 1935), travelling intracellularly into the host tissues and penetrating cells at a relatively early stage in the symbiosis (LHOTSKY, 1946; SCHAEDE, 1951). Proliferation of host cells is involved in nodule formation and the mature nodules are each supplied with two fine vascular strands arising from the main vascular network and ending blindly among the *Nostoc* cells (BATHAM, 1943). Ultrastructural investigations of *Gunnera* nodules by VON NEUMANN et al. (1970) and SILVESTER (1974) confirm the intracellular positioning of the algal cells, and show that the typical 4-layered algal wall and outer mucilage envelope are present just as in comparable free-living forms of *Nostoc*. Mature nodules possess algae with a heterocyst frequency of up to 50–60%—a much higher proportion than in the free-living state. Studying nodules of different age, SILVESTER (1974) found that ability to reduce acetylene, and, hence, presumed rate of fixation of nitrogen by the algal component, was positively correlated with the frequency of heterocysts. As nodules may be covered by leaf bases or soil and are likely to be shaded by leaves of the host, it has been concluded that they are likely to be dependent on host carbohydrate, just as in Cycad nodules (see SCHAEDE, 1962).

2.2 Symbioses Involving *Rhizobium* or an Actinomycete

Symbioses involving Angiosperms and Actinomycetes or *Rhizobium* involve the formation of root nodules, and in all cases investigated so far, root hair infection is an essential prerequisite for initiation of the nodule (e.g. see BECKING, 1970; ANGULO et al., 1974). The entity fixing nitrogen in Actinomycetes is believed to be the intracellular vesicle, a septate structure formed by the swelling of tips of the intracellular hyphae (MORRISON and HARRIS, 1958; SILVER, 1964; ALLEN et al., 1966; BECKING, 1970; LALONDE and FORTIN, 1973; ANGULO et al., 1974). The sub-units of each vesicle are surrounded by a common outer membrane and the whole vesicle is enclosed with a dense capsular material probably of host origin (GARDNER, 1965; GATNER and GARDNER, 1970; BECKING, 1968, 1970).

Actinomycete-based root nodules closely resemble the parent root in structure, though they are obviously of more stunted growth and they branch much more frequently (see FURMAN, 1959; BECKING, 1970). The nodule possesses a central stele enclosed by an endodermis and a concentric layer of infected tissues in its root cortex. A periderm and lenticels are usually present in surface layers of mature nodules. Nodules of *Casuarinaceae* and *Myricaceae* are exceptional in possessing a clothing of apogeotropic roots (BECKING, 1970), while those of *Coriaria* are unusual in showing the infected tissue occupying only certain regions of the cortex, so that the stele appears to occupy a somewhat peripheral position (BOND, 1962). The same pattern is described for *Trema* nodules by TRINICK (1973).

In the legume nodule the unit operative in nitrogen fixation is the bacteroid (see BERGERSEN, 1971), groups of these being enclosed in "infection vacuoles" bounded by membrane envelopes. The latter probably represent portions of host plasmamembrane which proliferate around the developing bacteroids after the endocytotic liberation of bacteria into the cells from the intercellular infection threads (GOODCHILD and BERGERSEN, 1966). The bacteroids are usually many times larger than the rod-shaped, non-fixing, bacterium from which they originate, and eventually they occupy a large fraction of the volume of each infected cell. As the infected cells develop, cytoplasmic organelles such as mitochondria become displaced to a peripheral position, being particularly frequent bordering the intercellular spaces running through the infected tissue (see GOODCHILD and BERGERSEN, 1966). The infected tissue of legume nodules is centrally placed, whilst the vascular strands serving it occupy an outer position (Fig. 6.6) at least in their ultimate branches (cf. actinomycete-non-legume nodules). Many strands may flank a single lobe of infected tissue whilst the xylem lies outside this and closer to the nodule surface (i.e. an orientation of conducting tissues reverse to that of a root (see Figs. 6.7 and 6.8)). Endodermal investments occur around each element of the vascular network (Fig. 6.7) (the so-called bundle endodermal layers), and a common nodule endodermis may also be present encompassing the outer cortex of the nodule. These

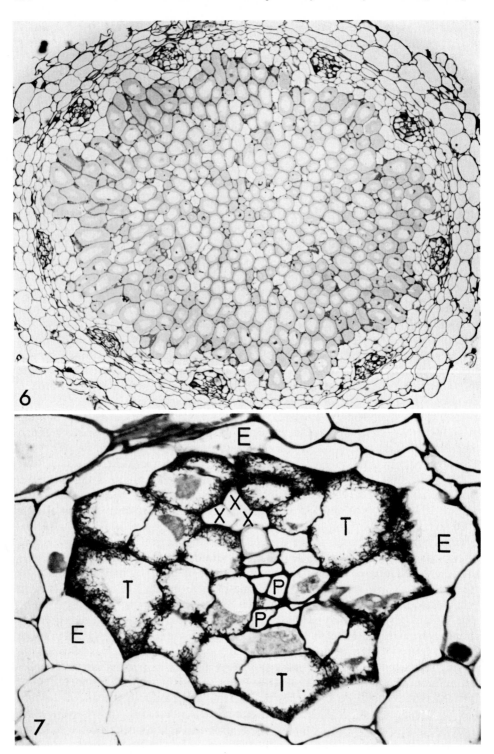

Figs. 6.6–6.7. Legend see opposite page

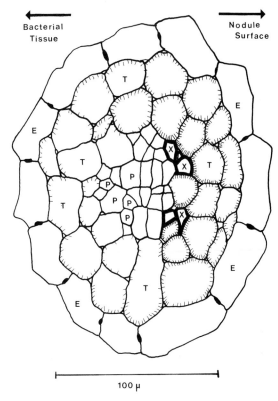

Fig. 6.8. Camera lucida drawing of a vascular strand of a root nodule of *Pisum sativum* showing the typical arrangement of endodermis (*E*) and pericycle transfer cells (*T*) and the orientation of elements of xylem (*X*) and phloem (*P*). Note the high density of wall ingrowths in transfer cells closest to the xylem. Estimates for the period 21–30 d after germination indicate that each cell of the infected tissue fixes 2.3 ng d^{-1} N, and consumes 23 ng d^{-1} sugar. Since there are approximately twice the number of infected cells as transfer cells in a nodule, each transfer cell on average ferries outwards through its symplast 46 ng sugar each day from phloem elements to the infected tissue and other parts of the nodule. Concurrently, each transfer cell releases to the free space of the nodule bundle 4.6 ng of fixed nitrogen in the form of amides and amino acids. Judging from composition of xylem sap collected from detached nodules, osmotic flushing of nitrogen from the vascular strand through its xylem elements requires only 0.3 cm^3 of water for each 1 mg of fixed nitrogen which is exported

◁ Figs. 6.6 and 6.7. Anatomical features of an effective root nodule of the legume *Pisum sativum*; Fig. 6.6. Transsection of mature region of nodule. Nine, peripherally-located vascular strands are visible. Note the uninfected cells dispersed amongst the larger, bacteroid-filled cells of the inner region of the nodule. Vacuoles are present in the infected cells ($\times 40$); Fig. 6.7. Section through a single vascular strand of nodule showing "transfer cells" (*T*) with wall protuberances surrounding the vascular elements of xylem (*X*) and phloem (*P*). An endodermis (*E*) invests the strand ($\times$ 850). Material embedded in glycol methacrylate, stained with toluidine blue and counterstained with periodic acid Schiff's reagent to delimit the wall ingrowths. (Material prepared in collaboration with B.E.S. GUNNING)

endodermal layers are attached at their proximal boundaries to the endodermis of the root. As in other classes of nodule, legume nodules often carry a periderm (Fig. *6.11*).

The large differences in size, shape and longevity of nodules of different legume genera, or of nodules formed on the one legume by different strains of *Rhizobium*, result primarily from the shape and activity of the nodule meristem (see Spratt, 1919; Allen and Allen, 1964; Corby, 1971). In *Lupinus luteus* (Fig. *6.9*) continued activity of a lateral meristem gives rise to a mushroom-shaped nodule which may eventually encircle the parent root, whilst in *Trifolium* and *Vicia* (Fig. *6.10*) the meristem, though often bifurcating, is strictly apical, so that nodules are of an essentially cylindrical construction.

In genera such as *Phaseolus, Glycine* and *Vigna,* the meristem is peripheral and short-lived and a spherical nodule therefore results, whose bacterial tissue consists of cells all of approximately the same age and state of development. In these "spherical"-type nodules, the periderm layer may be broken by lenticels, which presumably operate in ventilation of the nodule interior. In nodules with persistent meristems, water and oxygen uptake by the growing regions is likely to occur much more readily than in regions bounded by periderm and endodermis (Frazer, 1942). Several genera of legumes whose nodules possess a persistent meristem have specialized "transfer cells" in the pericycle of their vascular strands (Figs. *6.7–6.9* and see Pate et al., 1969). The supposed functioning of these is discussed later.

3. General Requirements for Transport between Participants

3.1 Gases: O_2 and N_2

The process of nitrogen fixation, catalyzed by the enzyme nitrogenase, is driven by energy in the form of ATP and requires the provision of a strong reductant for the formation of ammonia. In heterotrophic micro-organisms, whether living symbiotically or not, an external carbon source, such as pyruvate from glycolysis, is likely to provide both the ATP and source of electrons for the process, but in photoautotrophic nitrogen fixers it is conceivable that photosynthetic systems might satisfy either or both of these requirements.

Nitrogenase is irreversibly inactivated when exposed to oxygen and in at least the more highly evolved symbiotic systems means or mechanisms are in evidence which maintain near-anaerobic conditions in the immediate vicinity of the microsymbiont. Yet at the same time the heterotrophic nitrogen-fixing system must be supplied with sufficient oxygen for oxidative phosphorylation of host carbohydrate to occur,

---

▷

Figs. *6.9–6.11*. Variations in morphology and size of legume root nodules; Fig. *6.9*. *Lupinus luteus* showing collar-like arrangement of nodules encircling the top of the main root. Nodule growth takes place by means of a peripheral meristem ($\times$ 2.5); Fig. *6.10 Vicia faba* nodules showing branched apical growth. The nodule is basically of a cylindrical construction (see also nodule of *Pisum* Figs. *6.6* and *6.7*) ($\times$ 3); Fig. *6.11*. Large coralloid nodules of *Lupinus arboreus*. A periderm is well-developed and numerous lateral meristems are present. A single root nodule of this species may weigh up to 50 g ($\times$ 0.9)

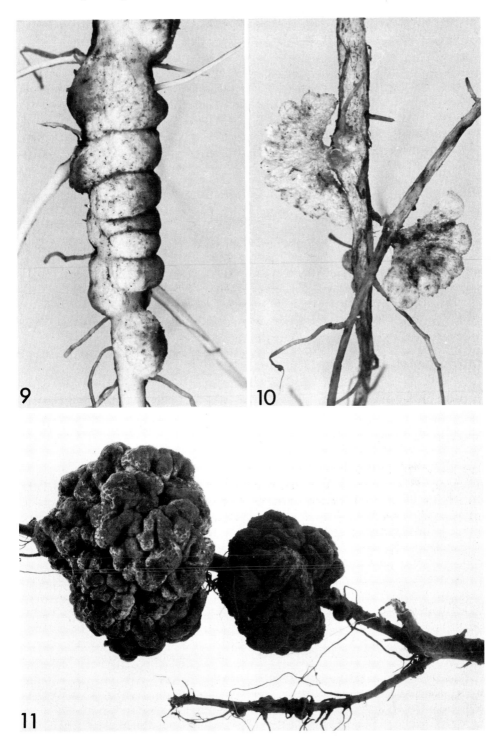

Figs. 6.9–6.11. Legend see opposite page

either in host cells near the microsymbiont, or in the microsymbiont itself. In the legume nodule a most sophisticated structural-physiological system for rationing of oxygen is evident (see Tjepkema, 1971; Tjepkema and Yocum, 1973), the endodermal layers probably acting as major barriers to inward diffusion of oxygen (Frazer, 1942), whilst the oxygen which does enter the central core of infected tissue, is likely to be distributed evenly and rapidly by means of the network of intercellular spaces and the leghaemoglobin solution in the infected cells (Bergersen, 1971). The membrane systems around the bacteroids may well act as the ultimate barrier excluding oxygen from the nitrogenase, and the somewhat similar compartmentation of microsymbiont in Actinomycete-based nodules may serve a similar function. Molecular nitrogen must of course reach the microsymbiont but its high partial pressure in air and in the nodule and its reasonably high solubility in water will probably allow adequate supplies to reach the nitrogenase (see Tjepkema, 1971). In blue-green algae, the organization of filaments into vegetative cells and heterocysts may favor the efficient operation of what is, in structural terms, a considerably less sophisticated symbiotic system. It is now widely believed that nitrogen fixation occurs in heterocysts, the thick walls of which (Dunn and Wolk, 1970), may exclude oxygen and thereby allow a reducing environment to be maintained at the nitrogenase (Stewart et al., 1969; Kale et al., 1970).

3.2 ATP and Reductant

The provision of these commodities to heterotrophic microsymbionts has already been discussed. In blue-green algae, the oxygen-producing photosystem of photosynthesis is absent from heterocysts, so that they are likely to have to depend on the complete photosynthetic system of adjacent vegetative cells for the provision of carbon and possibly also reductant (Wolk, 1968; Wolk and Wojciuch, 1971; Donze et al., 1972; Lex and Stewart, 1973). However, the heterocyst does appear to be able, when illuminated, to engage in cyclic photophosphorylation and thus supply at least part of the substantial requirements of the nitrogenase for ATP (Weare and Benemann, 1973; Lyne and Stewart, 1973). In the dark, when photosynthetic assistance to nitrogenase activity is no longer possible, fixation rates are generally much lower and the nitrogenase must be primed with ATP and reductant generated from cellular metabolism. Vegetative cells of the alga might then provide the necessary carbohydrate for fixation and, if living symbiotically, the alga might be supplied with carbohydrate from its host. Bothe (1970) suggests that $NADPH_2$ generated from glucose-6-phosphate by a dehydrogenase system functions as electron donor. Leach and Carr (1971) demonstrate activity of an ATP-activated pyruvate-ferredoxin oxidoreductase which might act similarly in generating reducing potential. Dark N_2 fixation of blue-green algae is markedly dependent on aerobic respiration, whereas the optimum atmospheric oxygen level for N_2 fixation in the light is much lower, presumably because photorespiratory losses will be discouraged at low oxygen tensions (Lyne and Stewart, 1973). Another feature which must be borne in mind is that, if partial pressures of oxygen are very low in the structures in which the alga is located, vegetative cells of the alga are likely to develop nitrogenase activity as well as heterocysts.

3.3 Carbon Acceptors, Ammonia and Amino Compounds

A final general point to be considered is the provision to one or other of the symbiotic partners of a suitable source of carbon acceptors to combine with the ammonia generated in fixation. It is still not known where the ammonia-assimilating enzymes are located in relation to the nitrogenase. One possibility is that free ammonia is the product translocated from the heterocyst, bacteroid, or Actinomycete vesicle and that it is the vegetative cells of the algal filament, or the host cells in the case of root nodules, which are responsible for synthesis of amino compounds. Alternatively, ammonia incorporation might be viewed as taking place within the microsymbiont, close to the nitrogenase, in which case amino compounds would be the form of nitrogen translocated to the non-fixing component of the system. In any event, studies on free-living blue-green algae (see HAYSTEAD et al., 1973), and the *Rhizobium* legume system (see DILWORTH, 1974), indicate that the enzymatic machinery exists for several types of assimilatory systems to operate, within or outside the microsymbiont. Once formed, the amino compounds may simply diffuse away from the microsymbiont, as may well occur in several examples of loosely organized, extracellular symbioses, but in those symbioses involving root nodules it is likely that specific conducting channels will be employed and an amount of water will be required to flush the sparingly-soluble fixation products into the host plant.

It is now time to look in more detail at transport processes in the various classes of symbiotic systems.

4. Transport in Symbioses Involving Blue-Green Algae

4.1 Liverworts and Ferns

Virtually nothing is known of the exchange of materials in these systems. In *Azolla,* nitrogen fixation, as assayed by acetylene reduction, occurs some eight times more rapidly in light than in dark, and the same order of difference is observed over a 24-h cycle between day- and night-time fixation. Transfer from light to dark or *vice-versa* elicits rapid changes in acetylene reducing activity (J. HALLIDAY and J.S. PATE, unpublished). These findings suggest that assistance to fixation from host-furnished carbohydrate is probably less important than the photosynthetic activities of the *Azolla* itself, a conclusion supported by the absence of any obvious conducting system close to the *Anabaena* cavity which might enable translocation of assimilates to the alga to be effected. More definitive experiments involving tracer studies are obviously called for. The reverse flow of nitrogen from *Azolla* to host must reach significant proportions, judging from rates of fern growth in nature or in nitrogen-free culture (see MISHUSTIN and SHIL'NIKOVA, 1971). VENKA-TARA-AMAN and SAXENA (1963) have shown that *Anabaena azollae* grown in free culture fixes atmospheric nitrogen and liberates some of this to the medium as aspartate, glutamate and alanine. These might be the forms in which nitrogen passes to the host in symbiosis. The observation that *Anabaena* grows much faster in *Azolla* leaf cavities than in free-living culture (YU-FENG SHEN, 1960) raises the question of whether the host might provide some factor stimulatory to algal growth. The same may also apply to the closely packed *Nostoc* colonies in thallus cavities of liverworts.

4.2 Cycads and *Gunnera*

In both of these symbioses the alga can apparently provide a substantial fraction of the nitrogen necessary for healthy growth of the host and for this to be achieved, considerable assistance from the host in providing carbohydrate must be required. Judging from a recent study on *Macrozamia riedlei* in Western Australia (J. Halliday and J.S. Pate, unpublished) a dense stand of these plants can fix some 20–30 kg N hectare^{-1} year^{-1}, a quite substantial contribution to the ecosystem with which these plants are associated. High rates of fixation in *Gunnera* associations in New Zealand have been recorded (Silvester and Smith, 1969). The marked (8-fold) stimulation of C_2H_2 reduction by light in *Gunnera* nodules (Silvester, 1974), and the observation that ^{15}N fixation in detached nodules of *Macrozamia* can be stimulated 2.7 fold by light (Bergersen et al., 1965) suggest that when the alga is suitably illuminated it can generate reductant, even if such a degree of illumination is not a normal event in nature. The finding that *Nostoc ex Gunnera* will grow and fix nitrogen in the dark on a carbohydrate source such as fructose, the unusually high (60%) proportion of heterocysts in filaments of symbiotic *Nostoc* in *Gunnera*, and the observation that light stimulation of $^{14}CO_2$ uptake in detached *Gunnera* nodules is not marked, all suggest that the host normally provides carbohydrate (Silvester, 1974). Moreover, the finding that C_2H_2 reducing activity/unit algal protein is 5–10 times greater in the *Gunnera* nodule than in free-living *Nostoc* (*ex Gunnera*), implies that provision of carbon to the intracellular *Nostoc* proceeds efficiently, and that other factors stimulatory to fixation might be released from the host.

$^{15}N_2$ experiments have shown that fixed nitrogen is released to aerial parts of the host of *Macrozamia* and *Gunnera* (Bergersen et al., 1965; Silvester and Smith, 1969). In *Gunnera*, fairly rapid incorporation of $^{15}N_2$ into soluble nitrogen of the leaf petiole and root occurs, indicating that products of fixation may be used directly in protein synthesis. The nitrogen content of nodules of *Cycas* and *Stangeria* is noticeably higher than that of other parts of the plant (Douin, 1953), so the possibility exists that the algal filaments may retain a proportion of the nitrogen which they fix. Citrulline and glutamine, the principal nitrogenous solutes of coralloid roots of *Macrozamia riedlei* are involved in transport of fixed nitrogen from alga to host in this association (J. Halliday and J.S. Pate, unpublished).

4.3 Lichens

In nitrogen-fixing lichens there is ample evidence that the algal component supplies the fungus both with fixed nitrogen and a source of carbohydrate (see Smith, 1974). The inhibition technique (see Hill and Smith, 1972), has indicated that glucose is the sugar released by the blue-green alga (Drew and Smith, 1967), although the $^{14}CO_2$ feeding studies of Hill (1972) suggested that glucan and sugar phosphates might also be forms in which carbohydrate moves to the fungus. In any event, the photosynthetically-fixed carbon moves rapidly to the lower fungal regions of the lichen (e.g. 40% of fixed ^{14}C transported in 4 h) at a turnover rate estimated to be commensurate with that of translocation in the phloem of higher plants (see SMITH et al., 1969; SMITH, 1974). The fungal filaments apparently transform the glucose to mannitol, an activity which may maintain steep diffusion gradients of glucose from alga to fungus (SMITH, 1974). Since *Nostoc* is apparently unable to utilize mannitol (DREW and SMITH, 1967), one-way flow of carbon to the fungus is assured. In *Peltigera aphthosa,* where *Nostoc* and the green alga *Coccomyxa* inhabit the same thallus, the fungal component receives glucose and fixed N from the *Nostoc* of the cephalodia and also obtains ribitol from the *Coccomyxa* in the general body of the thallus. The fungus apparently transforms both of these carbohydrates to mannitol. A similar nutritional situation is held to exist in the lichen *Lobaria amplissima* (RICHARDSON et al., 1967). Interchange of carbohydrates between the two algal partners has not been studied, but the partitioning of ^{15}N in *Peltigera aphthosa* thalli after feeding ^{15}N$_2$ (KERSHAW and MILLBANK, 1970) has shown that the *Coccomyxa* component sequesters a smaller share of the fixed nitrogen exported from the cephalodia than might be expected from its volume fraction in the thallus. In this species, nitrogen transfer to the non-cephalodial components occurs at almost as high a rate as that at which it is being fixed by the blue-green alga (MILLBANK and KERSHAW, 1969). Peptides are believed to feature in the transfer of fixed nitrogen (J.W. MILLBANK, personal communication).

There is substantial evidence that the physiological performance of the blue-green algal component of *Peltigera* spp. is somewhat different in the lichenized state than in free-living condition. DREW and SMITH (1967) found that *Nostoc* isolated directly from *P. polydactyla* continued at first to excrete large amounts of photosynthetically-fixed carbon into the medium (as glucose), but that this propensity became much reduced after two days of free-living culture. Conversely, if alga and lichen are artificially combined to form a lichen in laboratory culture (e.g. in *Cladonia*) it takes a considerable time before the alga becomes "entrained" to export photosynthate efficiently to its partner (HILL and AHMADJIAN, 1972). The abnormally high percentage of vegetative cells (97% versus 3% as heterocysts) in the filaments of the lichenized *Nostoc* is consistent with the concept of these cells having to provide carbon to the fungus as well as to the N-fixing heterocysts (cf. the situation for *Gunnera*).

SMITH and colleagues (SMITH, 1974) have suggested that high rates of carbon and nitrogen transfer are achieved in the lichen by the liberation from the fungus of factors inducing leakiness in the algal cells, but concrete evidence of this is not yet forthcoming. It is equally possible that in the lichenized state, the provision of essential inorganic nutrients from the fungus is so meagre that only very slow rates of multiplication of the alga are possible. In these circumstances an excess of photosynthetically-fixed carbon and fixed nitrogen is likely to be available continuously from the alga to the host fungus. Once removed from the lichen to a medium encouraging higher rates of growth, less photosynthate will be excreted from the alga since more will be utilized in growth (see ^{14}CO$_2$ feeding studies of HILL and AHMADJIAN, 1972). Evidence suggesting that the fungus maintains the alga in a relatively starved condition is available from experiments in which a lichen is provided with large amounts of nutrients (e.g. nitrogen), whereupon the alga outgrows the lichen and symbiosis becomes disrupted (see Scott, 1960). In an attempt to explain the low degree of retention of fixed nitrogen in the *Nostoc* of *Peltigera aphthosa,* Millbank and Kershaw (1969) suggest that fungal secretions might inhibit protein synthesis in the alga, leaving an excess of unused amino nitrogen available for the fungus. These authors have observed that the *Nostoc* in *Peltigera* fixes nitrogen at a faster rate than in comparable free-living forms, the rate of fixation being sufficient to permit algal cell division every 11 h, were the fixed nitrogen all retained within and utilized by the alga. The nature of this stimulation to fixation is not known, there being no evidence of the fungus providing the alga with carbohydrate or more specific stimulatory factors.

It is possible that the alga might also contribute growth substances to the fungus of a lichen. For instance, Bednar and Holm-Hansen (1964) report that *Coccomyxa* isolated from *Peltigera aphthosa* excretes biotin, and does this at a rate 17 times higher than in the free-living alga *Chlorella*. Thus, the lichen alga might possess a role analogous to that suggested for plant shoots in providing B-vitamin type growth factors for their roots.

There are apparently no reports of transfer of organic nutrients from fungus to alga in N-fixing lichens, so it remains to be seen whether the role of the fungus in symbiosis is anything more than a simple harborer of the alga and a provider of water and inorganic nutrients.

5. Transport in Symbioses Involving *Rhizobium* or an Actinomycete

5.1 Import of Photosynthetically Fixed Carbon to Root Nodules

Studies of the supply of photosynthetically-fixed carbon to root nodules have been carried out on a wide range of herbaceous legumes, and on the woody, non-legume genera *Myrica* and *Alnus*. There is general agreement that the availability to the nodule of recently-formed carbohydrate from shoot tissues is the most important factor regulating the rate of nitrogen fixation within the root nodules. Evidence for this comes from several quarters.

(i) Darkening of plants, removal of shoots, or detachment of nodules, results in a more or less rapid lowering of the rate of fixation (e.g. Virtanen et al., 1955; Wong and Evans, 1971; Wheeler, 1971; Mague and Burris, 1972).

(ii) Fixation is generally higher during the day when translocation to nodules is occurring more rapidly than during the night when translocation is likely to be reduced (Hardy et al., 1968; Bergersen, 1970; Wheeler, 1971).

(iii) $^{14}CO_2$ feeding of illuminated shoots of nodulated plants results in a rapid transfer of ^{14}C-photosynthate to the root nodules (Small and Leonard, 1969; Lawrie and Wheeler, 1973), some of this assimilate being used immediately in the nodules in the formation of amino compounds generated in nitrogen fixation (Pate, 1962).

(iv) The fixation rate of nodules will usually increase markedly and rapidly if the carbohydrate supply from shoot to root is improved; e.g. by raising light intensity; by spraying sugars on leaves (van Schreven, 1958); by increasing the carbon dioxide level around the photosynthesizing shoot (HARDY and HAVELKA, in press), or by removing sinks on the shoot (apices or fruits) likely to be competing with nodules for carbohydrate (J. HALLIDAY, unpublished; PATE, 1975).

Although so dependent in the long term on outside sources of carbohydrate, root nodules do contain in their host cells quite sizeable pools of starch and sugar (DANGEARD, 1926; WHEELER, 1969; MINCHIN and PATE, 1974). In *Alnus,* sugar levels are low when fixation is at its mid-day peak in activity (WHEELER, 1969, 1971). In *Pisum,* sugar and starch levels rise throughout the latter part of the photo-period and then fall progressively during the following night, implying that resident carbohydrates may supplement dwindling supplies of translocate during hours of darkness (MINCHIN and PATE, 1974). Removal of shoots of *Pisum* during daytime also prompts a rapid fall in sugar levels in the root nodules, and, whilst this sugar lasts, fixation is able to occur at almost as high a rate as in the intact, illuminated plant (MINCHIN and PATE, 1974).

Legumes with fleshy roots containing substantial amounts of reserve carbohy-drate (e.g. *Lupinus* spp.) are able to continue fixing at a high rate for several days after plants are darkened or decapitated (M.J. DILWORTH, personal communication), suggesting that in these there may be a great capacity for buffering the symbiotic system against starvation of assimilates from the shoot.

LAWRIE and WHEELER, 1973, studying *Pisum* and *Alnus* provide autoradiographic evidence that uninfected cells interspersed among the infected tissues of the root nodule (e.g. see Fig. 6.6) act as repositories for surplus ^{14}C-labeled photosynthates. Despite its abundance in the bacteroids of legume nodules poly-β-hydroxybutyrate is not regarded as a readily-utilizable source of carbon (WONG and EVANS, 1971), although it may function as a long-term reserve, such as in the support of basal respiration of overwintering nodules on deciduous perennials.

Judging from the results of $^{14}CO_2$ feeding studies (BACH et al., 1958; PATE, 1962; WHEELER, 1969), and from analysis of phloem sap of legumes (PATE, 1975), sugar, usually sucrose, is the major form in which carbon is translocated from shoot to nodule. Other assimilates, particularly amino compounds and organic acids are also likely to be carried to the nodule *via* the phloem, the ^{15}N feeding studies of OGHOGHORIE and PATE (1972) suggesting that considerable amounts of fixed nitrogen might cycle through the shoot and back again to the nodules. The possibility of phloem-borne amino acids exercising feedback control of nitrogen fixation has still to be tested, but, in any event, imported amino acids may contribute substantially to the growth of the nodule.

Once sucrose arrives in the legume nodule, it is assumed that it is not utilized as such by the bacteroid, but that the active host plant invertases first cleave it to hexose (see KIDBY, 1966; BERGERSEN, 1971; ROBERTSON and TAYLOR, 1973). The finding of KIDBY and PARKER (see BERGERSEN, 1971), that freshly prepared bacteroids require a period of adaptation before being able to utilize glucose, suggests that further metabolism of the hexose units from sucrose may take place in the host cytoplasm and that short-chain respiratory substrates (e.g. pyruvate or acetyl-coenzyme A) are what are normally passed to the bacteroid. At any rate, bacteroids can use TCA cycle intermediates such as succinate (BERGERSEN and TURNER, 1967), so that the TCA cycle may operate in providing ATP and reductant for fixation (BERGERSEN, 1971).

5.2 Export of N-Compounds from Root Nodules

The systems for forming amino compounds in root nodules are still not properly explored, but there is general agreement that NH_4^+ is the first stable product of fixation, and that from it a variety of compounds can form. $^{15}N_2$ feeding studies on *Glycine* (APRISON et al., 1954), *Ornithopus* (KENNEDY, 1966) and *Myrica* (LEAF et al., 1959) suggest that glutamate is the primary recipient of ammonia and that other compounds, e.g. aspartic acid and the amides, asparagine and glutamine, are formed from it secondarily. In *Alnus* the carbamyl nitrogen of citrulline is most heavily labeled in short-term ^{15}N feeding experiments, indicating that ammonia is incorporated *via* the carbamyl phosphate pathway, with ornithine as accepter molecule for the fixed nitrogen (LEAF et al., 1958). The intracellular localization of ammonia incorporating enzymes is still not properly evaluated. Any analysis must take into account the existence of glutamate dehydrogenases inside and outside the bacteroids (GRIMES and FOTTRELL, 1966), the presence of glutamine synthetase and glutamine (amide) 2-oxo-glutarate aminotransferase in nodule tissues (DUNN and KLUCAS, 1973; C.M. BROWN and M.J. DILWORTH, personal communication). The last two enzymes would together furnish an alternative pathway for glutamate synthesis in bacteroids.

The xylem is generally regarded as the eventual avenue of export of nitrogen released from root nodules, the best direct proof of this coming from $^{15}N_2$ experiments on *Alnus* (BOND, 1956) and *Hippophäe* (BOND, 1964). In both cases, removal of a ring of stem bark into the cambium did not impede upward transport of labeled nitrogen from the nodulated roots. The recovery of nitrogen compounds in xylem sap bleeding from root stumps, or extracted by vacuum from woody twigs of nodulated plants, provides confirmatory evidence of transport in the xylem, and allows the compounds moving from the roots and nodules to be identified. The compounds transported are species specific, most legumes and non-legumes carrying fixed nitrogen as amide (asparagine and/or glutamine), others utilizing citrulline (e.g. *Alnus, Albizzia*), methylene glutamine (*Arachis*) or allantoic acid (*Phaseolus* see reviews by PATE, 1971, 1973). These molecules carry much nitrogen to the shoot with the loss of a minimum amount of carbon from the nodule. Proof that fixed nitrogen is indeed carried in these xylem fluids comes from $^{15}N_2$ feeding studies on the non-legume *Coriaria* (SILVESTER, 1968), and the legume *Pisum* (OGHOGHORIE and PATE, 1972). In *Coriaria* it was found that 60% of the ^{15}N recovered from xylem sap collected over the period 4–16 h after feeding $^{15}N_2$ gas to nodulated roots was attached to the amide-nitrogen of glutamine. In *Pisum* the total amount of fixed ^{15}N recovered in the amino fraction of xylem sap collected over a period of three days from the root stumps was found to amount to 65% of the $^{15}N_2$ which had been fixed by the nodules of the root system.

The concept of a rapid upward export of fixed nitrogen from nodule to host in the xylem agrees well with earlier findings in which it was shown that during periods when fixation was occurring rapidly in well established nodules 80–90% of the nitrogen that they fixed passed immediately to the host, particularly to its shoot system (e.g. see BOND, 1936, for *Glycine*; PATE, 1958, for *Pisum*; STEWART, 1962, for *Alnus*).

There is evidence (see PATE et al., 1969; GUNNING et al., 1974) that there are certain similarities between the export system of the nodule and of the root. In both organs, various solutes are released from the symplast to the apoplast of a vascular strand, or strands, enclosed by an endodermis, and, in both, osmotic attraction of water to this solute-enriched apoplast can result in bleeding under pressure from the xylem, if the latter is cut. Generally, however, the solutes of xylem bleeding sap of nodules are many times more concentrated than those of comparable root bleeding sap (see PATE et al., 1969; MINCHIN, 1973). Export of fixed nitrogen from nodules and of ions and of organic solutes from roots then share the property of xylem mediated transport. In both therefore the potential exists for export to be aided by osmotic forces generated within the root or nodule, or by transpirational activity, generated within the attached shoot. This point is considered later.

The export of nitrogen from nodules of legumes is a selective process. Judging from the levels of certain amino compounds in the sap bleeding from the xylem of detached nodules, it would appear that the compounds selected for export leave the nodule at several times their concentration in the infected tissues, this differential certainly being higher than in the bleeding root (PATE et al., 1969; WONG and EVANS, 1971; MINCHIN, 1973). The species studied by these authors cover the genera, *Vicia, Glycine, Pisum, Lupinus,* and, in all, asparagine and glutamine represent the major compounds exporting nitrogen. These and aspartic acid are released from the nodule at from 3–15 times their concentration in donor tissues. By contrast, other compounds (e.g. glutamate in *Glycine max;* γ-amino butyric acid, glycine, valine in *Vicia faba*; lysine in *Pisum sativum*) appear to be actively excluded from export for, in each case, their level in the nodule is found to be much higher than in xylem sap bleeding from the nodule (PATE et al., 1969; MINCHIN, 1973). The pericycle cells of the nodule vascular bundles are suggested as the sites from which amino compounds are secreted to the extracellular space of the bundle (i.e. they are assumed to be analogous in role to the stele parenchyma of the root). Water, attracted osmotically across the bundle endodermis, then flushes these compounds out through the xylem (PATE, 1969; PATE and GUNNING, 1972). Where these pericycle cells become modified as "transfer cells" (see GUNNING and PATE, 1969), the enlarged plasmamembranes associated with the cellulose wall ingrowths of the transfer cells permit a particularly rapid flux of amino compounds to the nodule xylem (see Figs. 6.7 and 6.8 and GUNNING et al., 1974).

Assistance from transpiration to the export process of the nodule is suggested from the experiments of MINCHIN and PATE (1974) on plants of *Pisum sativum* grown in growth cabinets. During the photoperiod when transpiration is proceeding rapidly, the amides and amino acids of the nodule are maintained at relatively low levels, but during the night period these compounds increase greatly in concentration, indicating that the export process is then unable to keep pace with fixation. When transpiration commences at the beginning of the next photoperiod, this backlog of fixation products is rapidly cleared from the nodule. If plants are transferred to a high humidity during the photoperiod, the level of amino acids in their nodules rises, suggesting that reduced transpirational flow engenders a build-up in fixation products. Conversely, if plants are transferred to lower humidities than they have been normally receiving in the growth cabinets, the increase in transpiration rate which ensues is accompanied by a fall in the level of free amino compounds in the nodules. It would appear, therefore, that the nodule is able to adjust to quite wide fluctuations in host-plant water balance during the normal daily functioning of the plant and may indeed be able to exhibit even greater adaptability during periods of unusual environmental stress (MINCHIN 1973; MINCHIN and PATE, 1974).

5.3 Amounts and Density of Traffic of Different Solutes between Symbiotic Partners

Information on this topic is of considerable importance to an overall understanding of symbiosis and of the contributions and demands which it places on plant functioning. In an attempt to gather such information a series of physiological studies at whole plant, organ and cellular levels has been conducted recently on *Pisum sativum* (see MINCHIN and PATE, 1973, 1974; MINCHIN, 1973; GUNNING et al., 1974). All observations were restricted to a specific interval in the life cycle (21–30 days after sowing), just before flowering, and at the time when the root nodules reached peak efficiency in fixing nitrogen. The plants were totally dependent on nodules for their nitrogen.

Expressed on a whole plant basis the nodule population required during the nine-day study period 65 cm^3 O_2 for respiratory purposes and 22 cm^3 N_2, as substrate for fixation. These gases were presumably supplied through the surface of the nodule from the soil atmosphere. 112 mg of translocated carbon ($\equiv 280$ mg sucrose) were estimated to be consumed by the nodules, 16% of this carbon for nodule growth, 37% in respiration, while the remaining 47% was returned back to the shoot in the xylem as amino compounds produced in nitrogen fixation (see Fig. 6.12). Judging from the volumes of xylem fluid collected from detached nodules, 10 cm^3 of water were estimated to be required to export the 25 mg N found to be fixed by the nodules, although this was probably a conservative estimate since in actively transpiring plants water flow through nodules might be higher and xylem fluids leaving the nodule correspondingly more dilute. This volume of water (10 cm^3) was approximately 7% of the total water transpired by shoots of the plant during the experiment.

The importance of the exchange of carbohydrate and amino compounds in symbiosis is highlighted in the flow sheet of Fig. 6.12. In the pea plants studied, the nodules commanded in translocation 32 out of every 100 units net gain of carbon from the photosynthesizing shoot, while the shoot depended on the amino compounds formed in nodules for over one third of the gain in dry matter which it made during the period of study.

It has been possible to relate the above information to cellular activities within the nodule from a recent series of studies (MINCHIN, 1973; GUNNING et al., 1974), in which assessments were made of the various classes and numbers of cells in the nodule population, and the average volumes and surface areas of these cells. It was assumed that a symplastic route delivers sugar to the infected cells, and that the same route in reverse carries amino compounds to the pericycle transfer cells. The latter are envisaged as secreting amino compounds to the apoplast of the nodule vascular tissue and then to the xylem draining the nodule, and the rate at which this process occurs is estimated to be 240 nmol m^{-2} s^{-1}, a measurement which takes into account the increased surface area of plasmamembrane of the transfer cells due to their possession of wall ingrowths (see Figs. 6.7 and 6.8 and GUNNING et al., 1974). (The quantities of nitrogen—as amino compounds—and

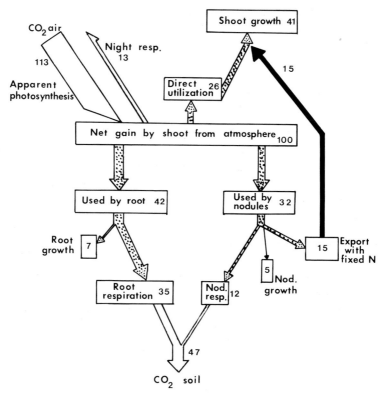

Fig. 6.12. Balance sheet for carbon in effectively nodulated pea plants grown in a medium deficient in mineral nitrogen. The data refer to the period 21–30 days after sowing and are expressed in terms of a net gain by the shoot of 100 units of carbon from the atmosphere. Clear regions of flow sheet—gaseous diffusion of CO_2; dotted regions—solute transport in symplast (principally in phloem); black regions—transport in xylem. Accompanying this flow of 100 units by weight of carbon there is an upward flow in the xylem of 13.5 units by weight of nitrogen (Fig. modified from Fig. 3 of MINCHIN and PATE, 1973)

sugar handled by each transfer cell are shown in the legend to Fig. 6.8.) The activity of these specialized cells may be the key to the successful operation of such high rates of transport within an organ poorly endowed with vascular tissue, and certainly ineffective in absorbing water rapidly from its external environment. By concentrating fixation products prior to export—indeed, concentrating them almost to their limit of solubility—the nodule achieves a high economy in water usage. It is thought possible that almost half of the water requirements of the nodule might be met by intake through mass flow in the phloem, and if the remainder were abstracted osmotically across endodermal layers of the root, a nodule growing in upper dry layers of a soil might be virtually independent of external sources of water for its functioning (MINCHIN and PATE, 1973). This suggestion has still to be critically tested.

6. Concluding Remarks

The many gaps in present knowledge of transport processes in symbiotic systems fixing nitrogen are all too obvious, and it is particularly frustrating not to be able to assemble comparatively physiological information for the classes of symbiosis studied here. Certainly the morphological and anatomical evidence displayed leads one to conclude that in evolutionary terms, increase in the complexity of structural organization of the symbiotic union has proceeded side by side with increases in the level of nutritional and biochemical integration of the activities of the associates. Even in the legume nodule, where interest in relation to agricultural potential has sponsored many detailed investigations, there is still no proper understanding of the general physiological and biochemical forces shaping symbiosis. Nor is there any real understanding of the biochemical events at ultrastructural level governing exchange reactions between the bacteroid and its unique cytoplasmic environment. Possibly the most frustrating aspect of the present situation is the almost total lack of information on physiological and ecological aspects of the majority of examples of non-leguminous symbiosis. This is particularly evident in relation to the contribution that these various systems make to the overall economy of nitrogen within the habitat with which they are associated, and the extent to which the macrosymbiont relies on symbiosis for its supply of nitrogen.

If it achieves little else it is hoped that this Chapter highlights the need in future research for an integrated, multi-purpose approach covering a wide, yet fully representative span of symbiotic systems at all levels of investigation.

References

Allen, E.K., Allen, O.N.: Nonleguminous plant symbiosis. Proc. Biol. Colloq. Microbiol. Soil Fertility. Oregon State University, Oregon, USA, 77–106 (1964).

Allen, J.D., Silvester, W.B., Kalin, M.: *Streptomyces* associated with root nodules of *Coriaria* in New Zealand. New Zealand J. Bot. **4**, 57–65 (1966).

Angulo, A.F., Van Dijk, C., Quispel, A.: Symbiotic interactions in non-leguminous root nodules with special reference to alder. In: Nitrogen fixation and the biosphere. Intern. Synthesis Meeting. Intern. Biological Programme. Cambridge: Cambridge University Press (in press).

Aprison, M.H., Magee, W.E., Burris, R.H.: Nitrogen fixation by excised soy-bean root nodules. J. Biol. Chem. **208**, 29–39 (1954).

Bach, M.K., Magee, W.E., Burris, R.H.: Translocation of photosynthetic products to soybean nodules and their role in nitrogen fixation. Plant Physiol. **33**, 118–124 (1958).

Batham, E.J.: Vascular anatomy of New Zealand species of *Gunnera*. Trans. Roy. Soc. New Zealand **73**, 209–216 (1943).

Becking, J.H.: Nitrogen fixation by non-leguminous plants. In: Symposium nitrogen in soil. Dutch Nitrogenous Fertilizer Review **12**, 47–74 (1968).

Becking, J.H.: Plant endophyte symbiosis in non-leguminous plants. Plant Soil **32**, 611–654 (1970).

Bednar, T.W., Holm-Hansen, O.: Biotin liberation by the alga *Coccomyxa* sp. and by *Chlorella pyrenoidosa*. Plant Cell Physiol. (Tokyo) **5**, 297–303 (1964).

Bergersen, F.J.: The quantitative relationship between nitrogen fixation and the acetylene reduction assay. Australian J. Biol. Sci. **23**, 1015–1025 (1970).

BERGERSEN, F.J.: Biochemistry of nitrogen fixation in legumes. Ann. Rev. Plant Physiol. **22**, 121–140 (1971).

BERGERSEN, F.J., KENNEDY, G.S., WITTMAN, W.: Nitrogen fixation in coralloid roots of *Macrozamia communis* L. Johnson. Australian J. Biol. Sci. **18**, 1135–1142 (1965).

BERGERSEN, F.J., TURNER, G.L.: Nitrogen fixation by the bacteroid fraction of breis of soybean root nodules. Biochim. Biophys. Acta **141**, 507–515 (1967).

BOND, G.: Quantitative observations on the fixation and transfer of nitrogen in the soya bean, with especial reference to the mechanism of transfer of fixed nitrogen from bacillus to host. Ann. Bot. (London) N.S. **50**, 559–578 (1936).

BOND, G.: An isotopic study of the fixation of nitrogen associated with nodulated plants of *Alnus, Myrica* and *Hippophäe*. J. Exptl. Bot. **6**, 303–311 (1955).

BOND, G.: Some aspects of translocation in root nodule plants. J. Exptl. Bot. **7**, 387–394 (1956).

BOND, G.: The development and significance of the root nodules of *Casuarina*. Ann. Bot. (London) N.S. **21**, 373–380 (1957a).

BOND, G.: Isotopic studies on nitrogen fixation in non-legume root nodules. Ann. Bot. (London) N.S. **21**, 513–521 (1957b).

BOND, G.: Fixation of nitrogen in *Coriaria myrtifolia*. Nature **193**, 1103–1104 (1962).

BOND, G.: The root nodules of non-leguminous angiosperms. Symp. Soc. Gen. Microbiol. **13**, 72–91 (1963).

BOND, G.: Isotopic investigations of nitrogen fixation in non-legume root nodules. Nature **204**, 600–601 (1964).

BOND, G.: Nitrogen fixation in some non-legume root nodules. Phyton (B. Aires) **24**, 57–66 (1967).

BOND, G.: Fixation of nitrogen by higher plants other than legumes. Ann. Rev. Plant Physiol. **18**, 107–126 (1967).

BOND, G.: Root-nodule formation in non-leguminous angiosperms. Plant Soil (special volume), 317–324 (1971a).

BOND, G., SCOTT, G.D.: An examination of some symbiotic systems for fixation of nitrogen. Ann. Bot. (London) N.S. **19**, 67–77 (1955).

BOTHE, H.: Photosynthetische Stickstoffixierung mit einem zellfreien Extrakt aus der Blaualge *Anabaena cylindrica*. Ber. Deut. Bot. Ges. **83**, 421–432 (1970).

CORBY, H.D.L.: The shape of leguminous nodules and the colour of leguminous roots. Plant Soil (special vol.) 305–314 (1971).

DANGEARD, P.A.: Recherches sur les tubercules radicaux des légumineuses. Le Botaniste, Ser. 16 (Paris) (1926).

DELWICHE, C.C., ZINKE, P.J., JOHNSON, C.M.: Nitrogen fixation by *Ceanothus*. Plant Physiol. **40**, 1045–1047 (1965).

DILWORTH, M.J.: Dinitrogen fixation. Ann. Rev. Plant Physiol. **25** (1974).

DONZE, M., HAVEMAN, J., SCHIERECK, P.: Absence of photosystem 2 in heterocysts of the blue-green alga *Anabaena*. Biochim. Biophys. Acta **256**, 157–161 (1972).

DOUIN, R.: Sur la fixation de l'azote libre par Myxophycees endophytes Cycadacees. Compt. Rend. **236**, 956–958 (1953).

DREW, E.A., SMITH, D.C.: Studies in the physiology of lichens. VII. The physiology of the *Nostoc* symbiont of *Peltigera polydactyla* compared with cultured and free living forms. New Phytologist **66**, 379–388 (1967).

DUNN, J.H., WOLK, C.P.: Composition of the cellular envelopes of *Anabaena cylindrica*. J. Bacteriol. **103**, 153–158 (1970).

DUNN, S.D., KLUCAS, R.V.: Studies on possible routes of ammonium assimilation in soybean root nodule bacteroids. Canad. J. Microbiol. **19**, 1493–1499 (1973).

FOGG, G.E., STEWART, W.D.P.: *In situ* determinations of biological nitrogen fixation in Antarctica. Br. Antarct. Surv. Bull. **15**, 39 (1968).

FRAZER, H.L.: The occurrence of endodermis in leguminous root nodules and its effect upon nodule function. Proc. Roy. Soc. Edinburgh B **61**, 328–343 (1942).

FRITSCH, F.E.: The structure and reproduction of the Algae, vol. 2. London: Cambridge University Press 1952.

FURMAN, T.: The structure of the root nodules of *Ceanothus sanguineus* with special reference to endophyte. Amer. J. Bot. **46**, 698–703 (1959).

GARDNER, I.C.: Observations on the fine structure of the endophyte of the root nodules of
  Alnus glutinosa Gaerten. Arch. Mikrobiol. **51**, 365–383 (1965).
GARDNER, I.C., BOND, G.: Observations on the root nodules of *Shepherdia*. Canad. J. Bot.
  35, 305–314 (1957).
GATNER, E.M.S., GARDNER, I.C.: Observations on the fine structure of the root nodule endophyte
  of *Hippophäe rhamnoides* L. Arch. Mikrobiol. **70**, 183–196 (1970).
GOODCHILD, D.J., BERGERSEN, F.J.: Electron microscopy of the infection and subsequent develop-
  ment of soybean nodule cells. J. Bacteriol. **92**, 204–213 (1966).
GRIFFITHS, H.B., GREENWOOD, A.D., MILLBANK, J.W.: The frequency of heterocysts in the
  Nostoc phycobiont of the lichen *Peltigera canina* Willd. New Phytologist **71**, 11–13
  (1972).
GRIMES, H., FOTTRELL, P.F.: Enzymes involved in glutamate metabolism in legume root nodules.
  Nature **212**, 295–296 (1966).
GROBBELAAR, N., STRAUSS, J.M., GROENEWALD, E.G.: Non-leguminous seed plants in Southern
  Africa which fix nitrogen symbiotically. Plant Soil (special volume) 325–334 (1971).
GUNNING, B.E.S., PATE, J.S.: Transfer cells. Plant cells with wall ingrowths, specialized in
  relation to short distance transport of solutes. Their occurrence, structure and development.
  Protoplasma **68**, 107–133 (1969).
GUNNING, B.E.S., PATE, J.S., MINCHIN, F.R., MARKS, I.: Quantitative aspects of transfer cell
  structure in relation to vein loading in leaves and solute transport in legume nodules. Symp.
  Soc. Exptl. Biol. **28**, 87–124 (1974).
HARDER, R.: Ernährungsphysiologische Untersuchungen an Cyanophyceen, hauptsächlich dem
  endophytischen *Nostoc punctiforme*. Z. Botan. **9**, 145–242 (1917).
HARDY, R.W.F., HAVELKA, U.D.: Photosynthate as a major factor limiting N_2 fixation by
  field-grown legumes with emphasis on soybeans. In: Nitrogen fixation and the biosphere.
  Intern. Synthesis Meeting, Intern. Biol. Programme. Cambridge: Cambridge University Press
  (in press).
HARDY, R.W.F., HOLSTEN, R.P., JACKSON, E.K., BURNS, R.C.: The acetylene-ethylene assay
  for nitrogen fixation; laboratory and field evaluation. Plant Physiol. **43**, 1185–1207 (1968).
HARRIS, G.P., MORRISON, T.M.: Fixation of nitrogen-15 by excised nodules of *Coriaria arborea*
  Lindsay. Nature **182**, 1812 (1958).
HAYSTEAD, A., DHARMAWARDENE, M.W.N., STEWART, W.D.P.: Ammonia assimilation in a
  nitrogen-fixing blue-green alga. Plant Sci. Letters **1**, 439–445 (1973).
HENRIKSSON, E.: Nitrogen fixation by a bacteria-free, symbiotic *Nostoc* strain isolated from
  Collema. Physiol. Plantarum **4**, 542–545 (1951).
HILL, D.J.: The movement of carbohydrate from the alga to the fungus in the lichen *Peltigera
  polydactyla*. New Phytologist **71**, 31–39 (1972).
HILL, D.J., AHMADJIAN, V.: Relationship between carbohydrate movement and the symbiosis
  in lichens with green algae. Planta **103**, 267–277 (1972).
HILL, D.J., SMITH, D.C.: Lichen physiology. XII. The inhibition technique. New Phytologist
  71, 15–30 (1972).
HITCH, C.J.B., STEWART, W.D.P.: Nitrogen fixation by lichens in Scotland. New Phytologist
  72, 509–524 (1973).
KALE, S.R., BAHAL, M., TALPASAYI, E.R.S.: Wall development and tetrazolium chloride reduc-
  tion in heterocysts of blue-green algae. Experientia **26**, 605–606 (1970).
KENNEDY, I.R.: Primary products of symbiotic nitrogen fixation. II. Pulselabelling of *Serradella*
  nodules with $^{15}N_2$. Biochim. Biophys. Acta **130**, 295–303 (1966).
KERSHAW, K.A., MILLBANK, J.W.: Nitrogen metabolism in lichens. II. The partition of cephalo-
  dial-fixed nitrogen between the mycobiont and phycobionts of *Peltigera aphthosa*. New
  Phytologist **69**, 75–79 (1970).
KIDBY, D.K.: Activation of a plant invertase by inorganic phosphate. Plant Physiol. **41**, 1139–
  1144 (1966).
LALONDE, M., FORTIN, J.A.: Microscopie photonique des nodules racinaires axéniques d'*Alnus
  crispa* var. *mollis*. Canad. J. Microbiol. **19**, 1115–1118 (1973).
LAWRENCE, D.B., SCHOENIKE, R.E., QUISPEL, A., BOND, G.: The role of *Dryas drummondii*
  in vegetation development following ice recession at Glacier Bay, Alaska, with special reference to its nitrogen fixation by root nodules. J. Ecol. **55**, 793–813 (1967).
LAWRIE, A.C., WHEELER, C.T.: The supply of photosynthetic assimilates to nodules of *Pisum
  sativum* L. in relation to the fixation of nitrogen. New Phytologist **72**, 1341–1348 (1973).

LEACH, C.K., CARR, N.G.: Pyruvate: ferredoxin oxidoreductase and its activation by ATP in the blue-green alga *Anabaena variabilis*. Biochim. Biophys. Acta **245**, 165–174 (1971).

LEAF, G., GARDNER, I.C., BOND, G.: Observations of the composition and metabolism of the nitrogen-fixing root nodules of *Alnus*. J. Exptl. Bot. **9**, 320–331 (1958).

LEAF, G., GARDNER, J.C., BOND, G.: Observation on the composition and metabolism of the nitrogen-fixing root nodules of *Myrica*. Biochem. J. **72**, 662 (1959).

LEWIS, D.H.: Concepts in fungal nutrition and the origin of biotrophy. Biol. Rev. **48**, 261–278 (1973).

LEX, M., STEWART, W.D.P.: Algal nitrogenase, reductant pools and photosystem I activity. Biochim. Biophys. Acta **292**, 436–443 (1973).

LHOTSKY, S.: The assimilation of free nitrogen in symbiotic Cyanophyceae. Studia bot. čsl. **7**, 20–35 (1946).

LYNE, R.L., STEWART, W.D.P.: Emerson enhancement of carbon fixation but not of acetylene reduction (nitrogenase activity) in *Anabaena cylindrica*. Planta **109**, 27–38 (1973).

MAGUE, T.H., BURRIS, R.H.: Reduction of acetylene and nitrogen by fieldgrown soybeans. New Phytologist **71**, 275–286 (1972).

MILLBANK, J.W., KERSHAW, K.A.: Nitrogen metabolism in lichens. I. Nitrogen fixation in the cephalodia of *Peltigera aphthosa*. New Phytologist **68**, 721–729 (1969).

MILLBANK, J.W., KERSHAW, K.A.: Nitrogen metabolism in lichens. III. Nitrogen fixation by internal cephalodia in *Lobaria pulmonaria*. New Phytologist **69**, 595–597 (1970).

MINCHIN, F.R.: Physiological functioning of the plant: nodule symbiotic system of garden pea (*Pisum sativum* L. cv. Meteor). Ph. D. Thesis. University of Belfast (1973).

MINCHIN, F.R., PATE, J.S.: The carbon balance of a legume and the functional economy of its root nodules. J. Exptl. Bot. **24**, 259–271 (1973).

MINCHIN, F.R., PATE, J.S.: Diurnal functioning of the legume root nodule. J. Exptl. Bot. **25**, 295–308 (1974).

MISHUSTIN, E.N., SHIL'NIKOVA, V.K.: Biological fixation of atmospheric nitrogen. London: Macmillan 1971.

MOLISCH, H.: Botanische Beobachtungen in Japan. IX. Über die Symbiose der beiden Lebermoose *Blasia pusilla* und *Cavicularia densa* mit *Nostoc*. Sci. Rep. Tôhoku Univ. (1925).

MORRISON, T.M.: Fixation of nitrogen-15 by excised nodules of *Discaria toumatou* Raoul Choix. Nature **189**, 945 (1961).

MORRISON, T.M., HARRIS, G.P.: Root nodules in *Discaria toumatou* Raoul Choix. Nature **182**, 1746–1747 (1958).

NEUMANN, D. VON, ACKERMANN, M., JACOB, F.: Zur Feinstruktur der endophytischen Cyanophyceen von *Gunnera chilensis* Lam. Biochemie und Physiologie der Pflanzen **161**, 483–498 (1970).

OES, A.: Über die Assimilation des freien Stickstoffs durch *Azolla*. Z. Botan. **5**, 145–163 (1913).

OGHOGHORIE, C.G.O., PATE, J.S.: Exploration of the nitrogen transport system of a legume using ^{15}N. Planta **104**, 35–49 (1972).

PANKOW, H., MARTENS, B.: Über *Nostoc sphaericum* Vauch. Arch. Mikrobiol. **48**, 203–212 (1964).

PATE, J.S.: Nodulation studies in legumes. I. The synchronization of host and symbiotic development in the field pea, *Pisum arvense* L. Australian J. Biol. Sci. **11**, 366–381 (1958).

PATE, J.S.: Root exudation studies on the exchange of ^{14}C-labelled organic substances between the roots and shoot of the nodulated legume. Plant Soil. **17**, 333–356 (1962).

PATE, J.S.: Movement of nitrogenous solutes in plants. Nitrogen-15 in soil plant studies. Intern. Atomic Energy Agency, Vienna. IAEA-Pl-341/ **13**, 165–187 (1971).

PATE, J.S.: Uptake, assimilation and transport of nitrogen compounds by plants. Soil Biol. Biochem. **5**, 109–119 (1973).

PATE, J.S.: Physiology of the reaction of nodulated legumes to environment. In: Nitrogen fixation and the biosphere. Intern. Synthesis Meeting, Intern. Biological Programme. Cambridge: Cambridge University Press (in press).

PATE, J.S.: Exchange of solutes between phloem and xylem and circulation in the whole plant. Encyclopedia of plant physiology. New Series. Vol. I Phloem transport, p. 451–473. Berlin-Heidelberg-New York: Springer 1975.

PATE, J.S., GUNNING, B.E.S.: Transfer cells. Ann. Rev. Plant Physiol. **23**, 173–196 (1972).

PATE, J.S., GUNNING, B.E.S., BRIARTY, L.G.: Ultrastructure and functioning of the transport system of the leguminous root nodule. Planta **85**, 11–34 (1969).

Peat, A.: Fine structure of the vegetative thallus of the lichen, *Peltigera polydactyla*. Arch. Mikrobiol. **61**, 212–222 (1968).

Peveling, E.: Elektronenoptische Untersuchungen an Flechten. IV. Die Feinstruktur einiger Flechten mit Cyanophyceen-Phycobionten. Protoplasma **68**, 209–222 (1969).

Peveling, E.: Vesicles in the phycobiont sheath as possible transfer structures between the symbionts in the lichen *Lichinia pygmaea*. New Phytologist **72**, 343–345 (1973).

Richardson, D.H.S., Smith, D.C., Lewis, D.H.: Carbohydrate movement between the symbionts of lichens. Nature **214**, 879–882 (1967).

Robertson, J.G., Taylor, M.P.: Acid and alkaline invertases in roots and nodules of *Lupinus angustifolius* infected with *Rhizobium lupini*. Planta **112**, 1–6 (1973).

Rogers, R.W., Lange, R.T., Nicholas, D.J.D.: Nitrogen fixation by lichens of arid soil crusts. Nature **209**, 96–97 (1966).

Roskin, P.A.: Ultrastructure of the host-parasite interaction in the basidiolichen *Cora pavonia* (Web.) E. Fries. Arch. Mikrobiol. **70**, 176–182 (1970).

Russell, S.A., Evans, H.J.: The nitrogen-fixing capacity of *Ceanothus velutinus*. Forest Sci. **12**, 164–169 (1966).

Schaede, R.: Über die Blaualgensymbiose von *Gunnera*. Planta **39**, 154–170 (1951).

Schaede, R.: Die pflanzlichen Symbiosen. Stuttgart: Gustav Fischer 1962.

Scott, G.D.: Further investigations of some lichens for fixation of nitrogen. New Phytologist **55**, 111–116 (1956).

Scott, G.D.: Studies in lichen symbiosis. I. The relationship between nutrition and moisture in the maintenance of the symbiotic state. New Phytologist **59**, 376–391 (1960).

Scott, G.D.: Plant symbiosis. London: Edward Arnold 1969.

Silver, W.S.: Root nodules symbiosis. I. Endophyte of *Myrica cerifera* L. J. Bacteriol. **87**, 416–421 (1964).

Silvester, W.: Nitrogen fixation by *Coriaria*. Ph. D. Thesis, University of Canterbury, New Zealand (1968).

Silvester, W.B.: Endophyte adaptation in *Gunnera-Nostoc* symbiosis. In: Nitrogen fixation and the biosphere. Intern. Synthesis Meeting, Intern. Biological Programme. Cambridge: Cambridge University Press (in press).

Silvester, W.B., Smith, D.R.: Nitrogen fixation by *Gunnera-Nostoc* symbiosis. Nature **224**, 1231 (1969).

Sloger, C., Silver, W.S.: Note on nitrogen fixation by excised root nodules and nodular homogenates of *Myrica cerifera* L. In: Non-heme iron proteins. Role in energy conservation. (San Pietro ed.). Symposium sponsored by the Charles F. Kettering Res. Lab. Yellow Springs, Ohio, p. 299–302. Yellow Springs, Ohio: Antioch Press 1965.

Small, J.G.C., Leonard, D.A.: Translocation of ^{14}C-labelled photosynthate in nodulated legumes as influenced by nitrate nitrogen. Amer. J. Bot. **56**, 187–194 (1969).

Smith, D.C.: Transport from symbiotic algae and symbiotic chloroplasts to host cells. Symp. Soc. Exptl. Biol. (1974).

Smith, D., Muscatine, L., Lewis, D.: Carbohydrate movement from autotrophs to heterotrophs in parasitic and mutualistic symbiosis. Biol. Rev. **44**, 17–90 (1969).

Spratt, E.R.: A comparative account of the root nodules of the Leguminosae. Ann. Bot. (London) **33**, 189–199 (1919).

Stewart, W.D.P.: A quantitative study of fixation and transfer of nitrogen in *Alnus*. J. Exptl. Bot. **13**, 250–256 (1962).

Stewart, W.D.P.: Nitrogen fixation in plants. London: Athlone Press 1968.

Stewart, W.D.P., Haystead, A., Pearson, H.W.: Nitrogenase activity in heterocysts of blue-green algae. Nature **224**, 226 (1969).

Tjepkema, J.D.: Oxygen transport in the soybean nodules and the function of haemoglobin. Ph. D. Thesis. University of Michigan (1971).

Tjepkema, J.D., Yocum, C.S.: Respiration and oxygen transport in soybean nodules. Planta **115**, 59–72 (1973).

Trinick, M.J.: Symbiosis between *Rhizobium* and the non-legume, *Trema aspera*. Nature **244**, 459–460 (1973).

Van Schreven, D.A.: Some factors affecting the uptake of nitrogen by legumes. In: Nutrition of the legumes, p. 137–163. London: Butterworths 1958.

Venkataraman, G.S., Saxena, H.K.: Studies on nitrogen fixation by blue-green algae. IV. Liberation of free amino acids in the medium. Indian J. Agr. Sci. **33**, 21–24 (1963).

VIRTANEN, A.I., MOISIO, T., BURRIS, R.H.: Fixation of nitrogen by nodules excised from illuminated and darkened pea plants. Acta Chem. Scand. **9**, 184–186 (1955).

WATANABE, K.: Studien über die Koralloide von *Cycas revoluta*. Bot. Mag. (Tokyo) **38**, 165–187 (1924).

WATANABE, A., KIYOHARA, T.: Symbiotic blue-green algae of lichens, liverworts and cycads. In: Studies on microalgae and photosynthetic bacteria, p. 189–196. Tokyo: Univ. Tokyo Press 1963.

WEARE, N.M., BENEMANN, J.R.: Nitrogen fixation by *Anabaena cylindrica*. Arch. Mikrobiol. **90**, 323–332 (1973).

WEBSTER, S.R., YOUNGBERG, C.T., WOLLUM, A.G.: Fixation of nitrogen by bitterbrush (*Purshia tridentata* (Pursh.) D.C.). Nature **216**, 392–393 (1967).

WHEELER, C.T.: The diurnal fluctuation in nitrogen fixation in the nodules of *Alnus glutinosa* and *Myrica gale*. New Phytologist **68**, 675–682 (1969).

WHEELER, C.T.: The causation of the diurnal changes in nitrogen fixation in the nodules of *Alnus glutinosa*. New Phytologist **70**, 487–495 (1971).

WINTER, G.: Über die Assimilation des Luftstickstoffs durch endophytische Blaualgen. Beitr. Biol. Pflanz. **23**, 295–335 (1935).

WITTMANN, W., BERGERSEN, F.J., KENNEDY, G.S.: The coralloid roots of *Macrozamia communis* L. Johnson. Australian J. Biol. Sci. **18**, 1129–1134 (1965).

WOLK, C.P.: Movement of carbon from vegetative cells to heterocysts in *Anabaena cylindrica*. J. Bacteriol. **96**, 2138–2143 (1968).

WOLK, C.P., WOJCIUCH, E.: Photoreduction of acetylene by heterocysts. Planta **97**, 126–134 (1971).

WONG, P.P., EVANS, H.J.: Poly-β-hydroxybutyrate utilization by soybean (*Glycine max* Merr.) nodules and assessment of its role in maintenance of nitrogenase activity. Plant Physiol. **47**, 750–756 (1971).

YU-FENG SHEN, E.: *Anabaena azollae* and its host *Azolla pinnata*. Taiwania **7**, 1–8 (1960).

III. Control and Regulation of Transport in Tissues and Integration in Whole Plants

7. Effect of Hormones and Related Substances on Ion Transport

R.F.M. VAN STEVENINCK

1. Introduction

It is now recognized that plant hormones play an important role in the regulation of ion transport and it may be safely predicted that recent observations will stimulate research into its mechanistic aspects. This applies especially to such topics as the effect of auxin on the movement of H^+ ions, the kinin-abscisic acid controlled mechanism of stomatal aperture, the stimulation of salt glands by kinetin, the abscisic acid-induced stimulation of Cl^- influx in storage tissues and the effects of abscisic acid on K^+ and Cl^- exudation from cut ends of barley and maize roots.

The topic of hormonal control of ion transport has been neglected since the Chapters by POHL (1961a, b) were published in the first Encyclopedia of Plant Physiology at a time when the concept of auxin-induced water uptake caused a considerable controversy. There have been some papers which provided useful introductions to the topic, but it is immediately apparent from a summary of the effects recorded to date (Table 7.1) that these observations are scattered over a wide range of experimental material, cover a variety of physiological phenomena and imply participation by practically all known plant growth hormones.

It will be necessary to limit this review to those aspects which are directly related to ion transport. A more complete treatment of the various theories of hormone action can be found in recent reviews by GALSTON and DAVIES (1969), AUDUS (1972), THIMANN (1972), KALDEWEY and VARDAR (1972), LIVNÈ and VAADIA (1972), EVANS (1974), MILBORROW (1974), and in the proceedings of the international conferences on plant growth substances (the most recent, CARR, 1972).

From a fundamental point of view it is useful to ask the following questions with respect to hormone action and ion transport:

a) does the hormone directly participate in the ion pump or carrier mechanisms (e.g. H^+, Cl^-, K^+ pumps)?;

b) does the hormone affect the induction or activity of enzymes which are directly related to ion transport (e.g. ATPases)?;

c) has the hormone any direct effects on membrane permeability or conductance through molecular interaction with the membrane structure?;

d) does the hormone indirectly affect the rate of synthesis or the composition of the cellular membrane systems through a control on protein and lipid synthesis?;

e) does the hormone affect metabolic processes which involve the establishment of ion gradients (e.g. proton gradients, see Part A, Chap. *12*), the production of organic ions (see Part A, Chap. *13*), the metabolic utilization of ions (phosphates, sulfates, nitrates, ammonium, etc.) or the supply of organic substrates in metabolism?

Since plant hormones are classified according to the major physiological functions which they control it seems more desirable from an encyclopedic point of view to consider each group according to the customary classification of auxins, kinins,

Table 7.1. Applied hormones and their effect on ion transport processes

Tissue	Hormone	Effective concentrations (M)	Ion transport process affected	Ref.
Sunflower hypocotyl segments (*Helianthus annuus*)	IAA	10^{-5}	increase of H^+ secretion	Hager et al. (1971)
Sunflower hypocotyl segments (*Helianthus annuus*)	IAA	10^{-4}	increase of K^+ uptake	Ilan (1973)
Coleoptile segments (*Avena sativa*)	IAA	10^{-5}	increase of H^+ secretion	Rayle (1973)
Coleoptile segments (*Avena sativa*)	IAA	$10^{-6}-10^{-4}$	increase of Cl^- uptake	Rubinstein and Light (1973)
Callus culture cells (*Petroselinum sativum*)	IAA	10^{-5}	decrease of Cl^- uptake	Bentrup et al. (1973)
Bean stem slices (*Phaseolus vulgaris*)	BA	$5 \cdot 10^{-6}$	decrease of K^+ uptake in aged tissue	Rains (1969)
Bean leaf slices (*Phaseolus vulgaris*)	BA	$1.3 \cdot 10^{-4}$	decrease of Na^+ uptake in expanding leaves; no effect in mature leaves	Jacoby and Dagan (1970)
Sunflower leaf discs (*Helianthus annuus*)	BA/ Kinetin	$4.6 \cdot 10^{-5}$	increase of K^+ uptake; decrease of Na^+ uptake	Ilan (1971)
Sunflower cotyledons	Kinetin	$4.6 \cdot 10^{-5}$	increase of K^+ uptake; no effect on Na^+	Ilan et al. (1971)
Beetroot discs (*Beta vulgaris*)	BA/ Kinetin	$7 \cdot 10^{-5}$	decrease of K^+ and Na^+ uptake in aged tissue	van Steveninck (1972a)
Beetroot discs	ABA	$3.8 \cdot 10^{-5}$	increase of K^+, Na^+ and Cl^- uptake in aged tissue	van Steveninck (1972b)
Barley and maize roots (*Hordeum vulgare* and *Zea mays*)	ABA	$0.4-1.9 \cdot 10^{-5}$	decrease of K^+ and Cl^- in exudate from cut end	Cram and Pitman (1972)
Barley roots	ABA	$4 \cdot 10^{-6}$	decrease of ^{86}Rb in exudate from cut end	Pitman et al. (1974)
Broad bean leaf slices (*Vicia faba*)	ABA	$3.8 \cdot 10^{-5}$	decrease of K^+ uptake in expanding leaves; no effect in mature leaves	Horton and Bruce (1972)
Coleoptile segments (*Avena sativa*)	ABA	$5 \cdot 10^{-7}-5 \cdot 10^{-5}$	decrease of IAA induced H^+ secretion	Rayle and Johnson (1973)

gibberellins, abscisic acid (ABA), senescence factors (SF), ethylene etc. Several groups of substances are known which may mimic hormone effects because of structural similarities and interaction with endogenous hormones, or because they have evolved as molecules controlling interactions between hosts and parasites. These are discussed in 7.7. This is then followed by some concluding remarks regarding overall mechanism and a brief mention of those ion transport phenomena which may be related to hormonal control of plant morphogenesis (e.g. phytochrome action, cell elongation, tropisms, nastic movements, root adhesion, etc.). For a treatment of stomatal guard cell mechanism the reader is referred to Chap. *4.3.*

2. Auxins

Auxins in this context are defined as substances which stimulate cell-elongation or -expansion and are represented most commonly by IAA.

2.1 Effects on H$^+$ Ion Pumps

It was established some 40 years ago by BONNER (1934) that growth rates of *Avena* coleoptiles were stimulated in acid buffer solutions. This "acid growth" effect also occurred in *Helianthus* roots and hypocotyls (STRUGGER, 1932) and it was proposed that the non-dissociated form of auxin was more effective in stimulating cell elongation than the dissociated molecule (BONNER, 1934). Subsequently, in a comprehensive study of the effect of pH, concentration of buffer and buffering capacity of coleoptile cells, POHL (1948) developed a universal theory for elongation growth which was primarily based on the observation that added IAA mimicked the effect of high concentration of protons and that both protons and IAA stimulated cell elongation through an increase in permeability of the protoplasts to water.

More recently in a publication of major importance, HAGER et al. (1971) showed that in buffers of pH 4, auxin-starved sunflower hypocotyls could achieve growth rates which were equivalent to rates of auxin-stimulated growth while no growth occurred in buffers of pH 6 or higher. In contrast to auxin-induced growth, acid-induced growth did not require strict aerobic conditions, and in the case of auxin-induced growth, added nucleotides (GTP, ATP) could substitute for the aerobic requirement. It was established that growth of coleoptile cylinders could be switched on or off repeatedly by changing to an acid or alkaline medium. Similarly, auxin-induced growth at pH 5 could be immediately inhibited by alkaline buffers and this inhibition was reversed by lowering the pH value.

Using metabolic uncouplers and inhibitors such as Cu^{2+} ions, the authors proposed that auxin is an effector of a "proton pump" which transports or secretes protons into the cell-wall compartment and requires respiratory energy (ATP) for its operation. The accumulated protons in the cell wall would then trigger off an enzymatic process, causing a loosening of cell-wall material as a final step in the process of cell elongation.

These findings were recently confirmed by CLELAND (1973), who demonstrated that within 15–20 min of its application, auxin could induce excretion of H$^+$ from coleoptile sections from which the epidermis and cuticle had been removed (peeled) to facilitate the H^+ escape. This excretion of protons was blocked by CCCP, KCN and CHM. Rayle and Johnson (1973) showed that H^+ excretion could also be initiated by the synthetic auxin 2,4-D but not by 3,5-D, which lacks growth-promoting capacity. ABA partially ($5 \cdot 10^{-7}$ M) or completely ($5 \cdot 10^{-5}$ M) inhibited the auxin-induced excretion of protons, and at these concentrations had corresponding inhibitory effects on coleoptile elongation. Hence, it was concluded that auxin-induced extension of coleoptiles resulted from the excretion of H^+ ions (Rayle, 1973). However, Evans (1974) in a recent critical review on this topic has already cast some doubt on this conclusion.

2.2 Effects on Other Ions

It is of considerable importance to ascertain possible secondary effects of auxin on ion transport. Hager et al. (1971) suggested that transport of protons into the cell wall should be compensated by a flow of cations towards the interior of the cell (e.g. K^+ for H^+ exchange) or by a flow of anions to the cell periphery in order to maintain electrical neutrality. Haschke and Lüttge (1973) resolved this question by producing evidence for an IAA-controlled K^+-H^+ exchange pump with a 1:1 stoichiometry in *Avena* coleoptiles. A useful contribution in this respect was made by Lüttge et al. (1972) because their data included measurements of membrane potential (see 7.2.3). They found that auxin caused an acidification of the external medium and an enhancement of K^+ uptake by old *Mnium* leaves. Because the authors determined the electrochemical potential difference for K^+, it can be tentatively suggested that the auxin-stimulated secretion of H^+ ions was competing with an active K^+-efflux (see Fig. 7.1). Much earlier Higinbotham et al. (1953) recorded a stimulation of Rb^+ uptake by IAA within a period of 2 h

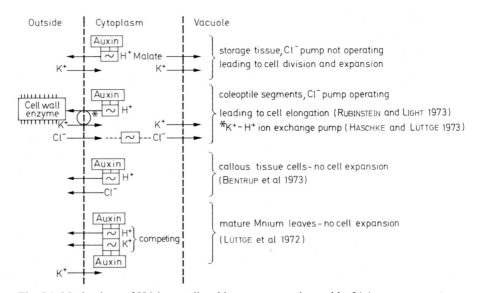

Fig. 7.1. Mechanisms of H^+ ion mediated ion transport triggered by IAA

for pea epicotyl segments but a similar response to IAA was much slower in slices of rutabaga (*Brassica napobrassica* Mill.) and potato tuber tissue. The slow response may have been due to induction phenomena (see Chap. *8*) since significant differences did not appear until the 3rd or 4th day and these were coupled to increases in dry and wet weight induced by auxin. Later, it was shown that Ca^{2+} played an essential role in the auxin stimulation of K^+ and Rb^+ uptake into the excised segments of etiolated pea epicotyls (HIGINBOTHAM et al., 1962), and also FISHER and ALBERSHEIM (1973) established that Ca^{2+} is essential for the operation of an auxin-induced H^+ ion efflux (up to 30 µmol g_{FW}^{-1} h^{-1}) in suspension-cultured bean or sycamore cells (see also 7.2.3).

ILAN (1962) showed that segments of sunflower hypocotyl, incubated in 10^{-2} M K^+-NH_4^+ phosphate buffer were capable of a much-enhanced rate of K^+ uptake in the presence of auxin (14.3 µmol g^{-1} and 3.4 µmol g^{-1} for IAA and control respectively); the uptake of NH_4^+ however was somewhat retarded by auxin treatment (10.5 µmol g^{-1} and 13.9 µmol g^{-1} for IAA and control respectively). Furthermore, the effect of IAA on K^+ uptake was much greater than on Na^+ and Li^+ uptake. Later, ILAN and REINHOLD (1963) confirmed that IAA promoted K^+ and Rb^+ uptake into the osmotic volume of segments of sunflower hypocotyl, and although the promotion of water uptake appeared to *precede* the effect of IAA on cation uptake, the authors did not reject the possibility that the promotion of K^+ or Rb^+ uptake was closely related to IAA-induced growth. Again IAA caused an inhibition of NH_4^+ uptake and ILAN and REINHOLD therefore proposed that K^+, Rb^+ and NH_4^+ compete for a carrier site which could be modified by IAA causing a change of affinities towards the three ions in question. Sucrose (0.05 M) counteracted the stimulatory effect of IAA on K^+ uptake if NH_4^+ was present but this counteraction was nullified if NH_4^+ in the medium was replaced by Na^+ (ILAN and REINHOLD, 1964).

ILAN (1973) observed an auxin-induced pH drop of the external solution within 2–3 h if the sunflower hypocotyl segments were floated in a K^+-containing solution, but with an NH_4^+-containing solution this auxin-induced pH drop was followed by a pH rise some hours later. ILAN (1973) therefore believed that IAA stimulation primarily concerned the influx of K^+, and this process would then be linked to a proton efflux or alternatively an auxin-induced NH_4 efflux (REINHOLD, 1958). Because of the apparent secondary nature of auxin stimulated proton efflux, ILAN rejected an hypothesis of auxin action which would primarily depend on an auxin-induced drop of pH in the cell-wall compartment.

It should be noted that root tissues generally have shown responses to auxin which are different from stem segments and coleoptile tissues, e.g. WEIGL (1969a) found that auxin had no effect whatever on fluxes of ^{86}Rb and ^{36}Cl in excised maize roots, and similarly no clear-cut effects on ^{86}Rb and ^{36}Cl fluxes were observed in beetroot slices (VAN STEVENINCK, 1974). NEIRINCKX (1968), however, did observe a 20% stimulation of net SO_4^{2-} uptake in aged beetroot slices.

It has already been pointed out that auxin-induced stimulation of ion uptake often depends on a long term induction of ion uptake capacity (see Chap. *8*), and in the absence of such an ion uptake capacity auxin-induced leakage of ions may become apparent. PICKLES and SUTCLIFFE (1955) observed that IAA caused pigment effusion from discs of beetroot and net Na^+ uptake was inhibited. VAN STEVENINCK (1965) noted that auxin, applied at quite low concentrations ($5.7 \cdot 10^{-9}$

to $5.7 \cdot 10^{-7}$ M) to freshly sliced discs of beetroot, that is, before a net ion uptake capacity for K^+ or Na^+ had developed, stimulated the release of K^+ ions to the external solution. However, no such stimulation of K^+ release occurred in aged discs unless IAA was applied at an unusually high concentration ($5.7 \cdot 10^{-4}$ M). From these results it is clear that the primary effect of auxin in inducing an increased permeability in freshly sliced storage tissue was apparent only while the tissue was in a receptive condition. Correspondingly, rates of K^+ uptake remained unaltered by auxin treatment in aged slices. PALMER's (1966) observations are essentially in agreement, because the complete prevention by IAA of invertase activity and development of a phosphate-uptake capacity was achieved by incubating fresh beetroot discs in solutions containing up to 10^{-3} M IAA (175 ppm).

Recently, BENTRUP et al. (1973) found that 10^{-5} M IAA did not significantly affect $^{86}Rb^+/K^+$ influx in a suspension of root callus-culture cells of *Petroselinum sativum*. The ^{36}Cl-influx however was substantially reduced within 30 min. These results were interpreted on the basis of HAGER et al. (1971) findings, suggesting that the primary event would be an IAA-induced H^+ ion extrusion so that less Cl^- need be taken up to balance the charge (see Fig. 7.1).

On the other hand, it is of considerable interest that auxin strongly stimulates ^{36}Cl uptake into coleoptile cells (RUBINSTEIN and LIGHT, 1973; RUBINSTEIN, 1973). This stimulation was observed within 15 min after the addition of IAA and occurred only with growth active auxins (e.g. with α-NAA but not β-NAA). The percentage enhancement by IAA was the same over a wide range of external chloride concentrations and did not occur at $0°$ C or in the presence of CCCP. It appears then that a wide range of secondary responses are possible depending on the tissue chosen and its physiological condition. These may be categorized into four basic mechanisms shown in Fig. 7.1 and from this it should be immediately apparent that auxin-induced stimulation of H^+ ion efflux may be regarded as the unifying principle, while for instance K^+ for H^+ exchange may be regarded as a special case. Apart from its role in triggering enzyme activity resulting in a cell-wall loosening and a change of turgor, the H^+ ion efflux seems to have an important regulatory function in controlling salt accumulation and thereby may provide the necessary osmotic adjustments for cell elongation.

The reader should consult Part A, Chap. *12* for a comprehensive discussion of this particular role of H^+ transport in controlling solute transport phenomena.

2.3 Effects on Transmembrane Electropotentials and Bioelectric Phenomena; Role of Ca^{2+}

The effect of auxin (IAA) on transmembrane potentials (PD) is variable. MACKLON and HIGINBOTHAM (1968) reported a depolarization in pea seedling tissue while an IAA-induced hyperpolarization was observed in oat coleoptiles by ETHERTON (1970). Furthermore, IAA treatment did not result in any change in PD in young or old *Mnium* leaves (LÜTTGE et al., 1972).

As suggested earlier, the response to auxin may depend on the type of tissue chosen and especially on the hormonal balance of the tissue at the time of the experiment (cf. VAN STEVENINCK, 1974). LÜTTGE et al. (1972) observed an auxin-induced acidification of the external solution leading to a pH of approximately 5 while the observed potential difference was -220 mV. It seems that auxin-enhanced proton transport can be expected nearly always to require expenditure of metabolic energy, a fact well substantiated by use of metabolic inhibitors or added ATP (HAGER et al., 1971; CLELAND, 1973).

The action of IAA on cell potentials could be related to its effect on proton transport (see Part A, Chap. 12). For instance, a striking example is the electrogenic potential of *Neurospora* (see Part A, Chap. 7), which depends on ion transport and internal ATP levels.

Curious rhythmic variations have been observed in the electric field generated by bean roots growing in water (SCOTT, 1957). These spontaneous electric oscillations with periods of about 5 min duration were ascribed to a feedback loop which would involve auxin supply, membrane permeability and membrane potentials as illustrated in Fig. 7.2 (JENKINSON and SCOTT, 1961; JENKINSON, 1962a). The effects were greatest at the elongating zone of the root (JENKINSON, 1962b) and further analysis showed that K^+ and Na^+ caused significantly different pathways in bioelectric current with Na^+ entering the root most strongly in a region 8 mm from the apex, and K^+ nearer the apex (SCOTT and MARTIN, 1962). These phenomena may influence ion selectivity in various regions of the growing root and may in turn be related to ontogenetic developments (ESHEL and WAISEL, 1973; see Chap. 8).

For many years Ca^{2+} has been regarded as an antagonist of auxin action generally because of its inhibitory effect in cell elongation (TAGAWA and BONNER, 1957). More specifically IAA was held to act as a chelator of Ca^{2+} (HEATH and CLARK, 1956, 1960). The work by CLELAND, 1960 produced such strong evidence against this hypothesis for the action of Ca^{2+}, that there has been a tendency to disregard evidence pointing towards an essential role of Ca^{2+} in auxin-mediated ion transport (HIGINBOTHAM et al., 1962; FISHER and ALBERSHEIM, 1973). RUBERY and SHELDRAKE (1973) believe that the plasmamembrane contains auxin-binding sites which may constitute primary sites of hormone action. Generally, a preponderance of negatively charged pectic polysaccharides will increase the concentration of cations including H^+ near the cell membrane. Hence the plasmalemma could be thought of as "transparent" to IAA and other acidic hormones ABA and GA_3 (RUBERY and SHELDRAKE, 1973; but see also Part A, 12.6). It is therefore of considerable interest that normal basipetal IAA transport in coleoptile tissue is practically prevented by washing the tissue in EDTA, while the transport is restored by subsequent application of Ca-solutions (DE LA FUENTE and LEOPOLD, 1973).

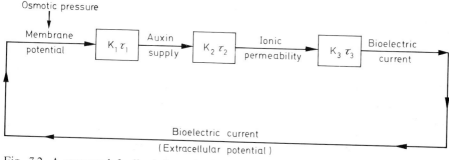

Fig. 7.2. A proposed feedback loop for bioelectric currents in plant roots featuring three exponential delays. (From JENKINSON and SCOTT, 1961)

Simple *ad hoc* postulates suggesting direct participation of Ca^{2+} in the mechanism of opening and closing of ionic "channels" produced by a shift in membrane potential were proposed by Fishman et al. (1971). These proposals of a molecular mechanism may be appropriate in cases of interaction between auxin and Ca^{2+} and some of the effects recorded which indicate a role of Ca^{2+} in ion selectivity (Epstein, 1961) and the frequently-encountered stimulation of monovalent cation uptake (Viets effect; Tanada, 1963; Waisel, 1962; Hourmant and Penot, 1971). Recently, Morré and Bracker (1974) reported that cycles of Ca^{2+} and IAA treatment can trigger a series of conformational responses of soybean plasma membranes which can be monitored by means of electron microscopy, but other evidence shows that the effectiveness of auxin in causing permeability changes bears no relation to Ca^{2+} content or cellular distribution of Ca^{2+} (Burström, 1963; van Steveninck, 1965).

3. Kinins

Apart from endogenous cytokinins such as zeatin, a number of compounds (kinetin, benzyladenine) have identical controlling influences on cell division, mobilization and retention of metabolites and retardation of senescence (Skoog and Armstrong, 1970).

Kinins have been reported to affect ion transport in a number of systems. The guard cell mechanism of stomatal opening and closing appears to depend on rapid changes in concentration of K^+ and an accompanying anion (Humble and Hsiao, 1970; Fischer, 1971, 1972). Kinins (and ABA) play a key role in regulation of stomatal opening (e.g. Livne and Vaadia, 1965) and hence appear to be acting on ion transport (see *4.3*.11). They also stimulate excretion by salt glands of *Statice* spp. (Bozcuk, 1972, see Table 7.2).

Table 7.2. The effect of kinetin (10^{-4} M) on ion excretion by leaf discs of *Statice sinuata* and *Statice latifolia*. (24 h incubation at 20° C in the light.) From Bozcuk (1972)

Treatment	Quantity of fluid excreted (mg)	Quantity of ions excreted (µg)				Ion concentration of excreted fluid (mM)			
		Na^+	K^+	Ca^{2+}	Cl^-	Na^+	K^+	Ca^{2+}	Cl^-
Statice sinuata									
H_2O	115	463	251	50	1,194	175	56	11	292
Kinetin	179	967	433	72	2,086	235	62	10	328
Statice latifolia									
H_2O	74	259	236	18	565	152	82	6	215
Kinetin	125	572	438	24	1,176	199	90	5	265

Kinins appear to affect the selectivity of ion transport processes. For instance, Jacoby and Dagan (1970) found that N-benzyl-adenine (BA) reduced Na^+ absorption rates in expanding primary leaves of intact bean plants. Ilan (1971) obtained clear-cut evidence of a change in K^+/Na^+ selectivity in favor of K^+ uptake in leaf discs and expanding cotyledons of *Helianthus* (Ilan et al., 1971). It was suggested that the increased "affinity" of the cell towards K^+ was related to a change in the DNA-RNA-protein system. Changes towards a greater selectivity for K^+ over Na^+ were also observed when fresh bean-stem slices were incubated in 0.5 mM $CaSO_4$ solutions for 20 h (RAINS, 1969). However, in this case BA reversed the trend by suppressing the development of a K^+-absorbing capacity in aged tissue.

Kinetin also prevented the development of invertase activity and the capacity for absorption of inorganic phosphate (PALMER, 1966) or SO_4^{2-} (NEIRINCKX, 1968) in discs of beetroot tissue. VAN STEVENINCK (1972a), who studied the short-term and long-term effects of kinetin and benzyladenine in discs of beetroot tissue, found that these compounds had a tendency to prevent net accumulation of K^+, Na^+ or Cl^- in aged discs. This kinetin-induced inhibition took several hours to establish and this period became more extended the longer the ion transport mechanisms had been in operation. In this respect the action of kinetin followed a pattern identical to that of CHM inhibition (VAN STEVENINCK and VAN STEVENINCK, 1972). When kinetin was added to freshly sliced discs, development of ion uptake capacities was completely prevented and thus the effects were identical in pattern to those obtained with actinomycin D. These results confirm that the effects of kinetin and BA mainly concern aspects of protein synthesis which play a role in the development and maintenance of ion transport mechanisms. Recent work by HOURMANT and PENOT (1973a, b) on the effect of benzyladenine on Rb^+ and on molybdate uptake by slices of potato tissue essentially support these conclusions.

The short term effects recorded (VAN STEVENINCK, 1972a), showed that kinetin or BA had no immediate inhibitory effect on the ion transport system as such. In fact, a distinct short-term promotion of cation uptake was recorded in freshly sliced discs treated with tris-buffer at pH 7.8–8. This observation seemed to be of some interest since tris buffer is known to stimulate glucose transport (VAN STEVENINCK, 1966), while BA is recorded to behave as a competitive inhibitor of hexokinase and pyruvic kinase (TULI et al., 1964). It seems pertinent that kinetin (and ABA, GA_3) has also a marked influence on the storage and hydrolysis of starch (TASSERON-DE JONG and VELDSTRA, 1971; MITTELHEUSER and VAN STEVENINCK, 1971a; JONES, 1973a; HADAČOVÁ et al., 1973; KUHL and UNGER, 1974) and hence the mobilization and storage of sugars (see Fig. 7.3). PITMAN et al. (1971) stressed the importance of sugar levels in determining rates of ion transport. A full understanding of the regulatory aspects of kinins on ion distributions in cells and tissues may therefore await a further clarification of their role in sugar metabolism of the cell.

Recently, it was shown that a 10 h kinetin treatment of Allium cepa cells may cause an increased permeability towards thiourea or urea (FENG and UNGER, 1972; FENG, 1973). Their suggestion that kinetin may affect the protein component of the membrane structure, thereby changing the properties of membrane pores found support recently with the isolation of a characteristic glycoprotein which is localized on the membrane surface of the unicellular coenocytic water-molds Achlya spp. and Blastocladiella emersonii (LE JOHN and CAMERON, 1973; LE JOHN and STEVENSON, 1973). This glycoprotein contained approximately 20 high- and low-affinity Ca^{2+} binding sites per mole, which appeared to be under allosteric control, specifically by N^6-substituted adenine derivatives and cytokinins. It was suggested that cytokinins produced in green plants function as allosteric regulators of metabolite and ion transport in fungal cells because they stimulated a massive release of Ca^{2+}

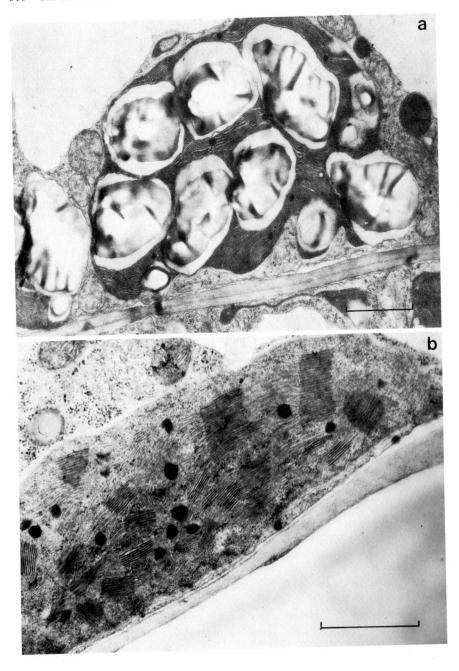

Fig. 7.3a and b. Chloroplasts in wheat leaf—parenchyma cells. (a) 3 days after treatment with $5 \cdot 10^{-5}$ M kinetin, (b) 3 days after treatment with $3.8 \cdot 10^{-6}$ M ABA. Bar represents 1 μm. Kinetin treatment causes starch grains to accumulate inside chloroplasts while ABA treatment causes an almost complete disappearance of starch grains. (From Mittelheuser and van Steveninck, 1971a)

localized on the membrane surface and simultaneously activated the transport of Ca^{2+} into the cells, thereby inhibiting their energy-linked transport of sugars, nucleosides and amino acids (LE JOHN and STEVENSON, 1973).

At present it is premature to ascribe a special role to proline but perhaps it is no mere coincidence that isolated potato lectin contains a massive amount of proline-OH (16% of all amino acid residues, see ALLEN and NEUBERGER, 1973), that incorporation of L-proline into acid-insoluble protein is immediately and completely inhibited by CHM (TIMBERLAKE and GRIFFIN, 1973), that massive proline accumulations may occur in barley seedlings during periods of water deficit which also induce high concentrations of ABA (SINGH et al., 1973; HIRON and WRIGHT, 1973), and finally the fact that patterns of ABA, kinetin and CHM-induced inhibitions are similar (LÄUCHLI et al., 1973; VAN STEVENINCK and VAN STEVENINCK, 1972). Lectins are proteins which interact with cell surfaces and cause cells to agglutinate. Potato lectin (a glycoprotein containing 50% carbohydrate of which 92% consists of arabinose with small amounts of glucose and glucosamine) is specifically inhibited in its agglutination reaction by oligosaccharides which contain N-acetyl glucosamine (ALLEN and NEUBERGER, 1973).

These observations no doubt will encourage further study on the role of membrane glycoproteins and possible hormone regulatory aspects of transport of ions, carbohydrates and other metabolites.

A complex situation such as the one illustrated above does not give much hope for a speedy solution. Nevertheless PITMAN's (1972) suggestion that kinins may play a key role in the regulation of ion transport from the root to the shoot provides a valuable starting point which ultimately may explain the controlling influences on growth and ion distribution in the whole plant (VAADIA and ITAI, 1969; PITMAN and CRAM, 1973). Recent detailed investigations (PITMAN et al., 1974) seem to indicate that kinin (BA) inhibits the export of ions from roots to shoots without affecting uptake from the external solution. These effects are mimicked by CHM, parafluoro-phenylalanine (FPA) and ABA (PITMAN et al., 1974; SCHAEFER et al., 1975; see Fig. 7.7), but also appear to be greatly dependent on environmental factors such as temperature and nutrient status. The latter should be kept in mind especially when conflicting results are obtained (TAL and IMBER, 1971; COLLINS and KERRIGAN, 1973, 1974).

4. Gibberellins

The physiological role of gibberellins (reviewed by JONES, 1973a) includes some interesting phenomena regarding electrolyte distribution and membrane permeability especially in germinating barley and wheat seeds. EASTWOOD and LAIDMAN (1971) found that within 12 h of imbibition a cytokinin-like hormone from the endosperm caused the cells to retain their reserves of K^+, Mg^{2+}, Ca^{2+} and inorganic phosphate (P_i), but later on during germination a gibberellin from the embryo induced the release of the retained ions. This induction process was inhibited by ABA and appeared to be independent of protein synthesis (EASTWOOD and LAIDMAN, 1971). Later work (JONES, 1973b), however, showed a dependence on protein synthesis. EVINS and VARNER (1971) showed that the incorporation of ^{14}C-choline into a semipurified, acid-insoluble ER fraction of barley aleurone cells was stimulated 4–8 fold by 1 µM GA_3 within 4 h of application of the hormone. This stimulation could be fully reversed by 0.25 µM ABA within 2 h of its addition. ^{14}C-choline incorporation was also depressed in cells stressed with 0.6 M polyethylene glycol solutions (Armstrong and Jones, 1973).

GA_3 caused an increase in the permeability of model membrane systems composed of various plant lipids, sterol and dicetyl phosphate (Wood and Paleg, 1972, 1974). Further studies on liposomes led to the hypothesis that hormonal regulation by GA_3 may consist of a direct effect on the physical properties of the phospholipid moiety of natural membranes (Paleg et al., 1973) and nuclear magnetic resonance work provided evidence of a GA_3-lecithin complex (Wood et al., 1974).

These effects may have a much more general application since it was found that GA_3 intensified the release of electrolytes from bean epicotyls placed in distilled water for 24 h (Artimonova and Artimonov, 1971), and from roots to shoots in pea seedlings (Lüttge et al., 1968). Pitman et al. (1974), however, showed that GA_3 had no effect on transport from roots to shoots in barley seedlings. In the case of pea seedlings a direct effect on membrane permeability was suggested since the response time was considered too short (4 h) to imply involvement of a transcription process, but in a later publication Köhler (1970) showed that complex interdependencies between growth and ion uptake may be involved.

Van Steveninck (1961) investigated the effect of GA_3 on the K^+ and Na^+ relations of freshly sliced discs of beetroot. At high concentrations (10^{-4} M) GA_3 enhanced leakage of K^+ and Na^+ from these discs. This is probably the result of a GA_3-induced increase of membrane permeability. Whereas the time course for the onset of K^+ uptake was not affected by GA_3 over a range of concentrations from 10^{-8} to 10^{-4} M, the time required for the onset of Na^+ uptake capacity was reduced from 22 to about 8 h by GA_3 over the range 10^{-8} to 10^{-5} M (see Table 7.3).

Table 7.3. The effects of GA_3 and IAA on rates of net leakage of K^+ and Na^+ from freshly sliced discs of beetroot tissue and on the time of commencement of net K^+- and net Na^+-uptake

	Rate of net leakage (nmol g^{-1} h^{-1})		Commencement of net uptake (h)	
	K^+ (over 0–14 h)	Na^+ (over 0–14 h)	K^+ uptake	Na^+ uptake
Control	310 ± 12	56 ± 2.7	38	20
GA_3, 10^{-8}M	290 ± 11	43 ± 2.5	38	17
GA_3, 10^{-7}M	310 ± 14	45 ± 1.2	38	14
GA_3, 10^{-6}M	280 ± 25	45 ± 3.3	38	14
GA_3, 10^{-5}M	370 ± 17	53 ± 4.3	38	14
GA_3, 10^{-4}M	500 ± 12	59 ± 1.2	38	32
	(over 0–8 h)	(over 0–8 h)		
Control	400 ± 17	126 ± 2.9	38	22
GA_3, 10^{-6}M	340 ± 23	117 ± 3.5	38	8
GA_3, 10^{-5}M	400 ± 21	126 ± 4.0	38	8
IAA, 10^{-5}M	470 ± 44	150 ± 8.8	38	40
IAA, 10^{-5}M + GA_3, 10^{-6}M	420 ± 23	131 ± 8.3	42	38
IAA, 10^{-5}M + GA_3, 10^{-5}M	380 ± 34	126 ± 5.0	40	20

5. Abscisic Acid (ABA)

Interest in abscisic acid as a regulator of ion transport arose with the discovery of its rapid action on the stomatal guard cell mechanism (MITTELHEUSER and VAN STEVENINCK, 1969, 1971 b; CUMMINS et al., 1971). IMBER and TAL especially (1970) and TAL and IMBER (1970) provided an elegant demonstration of the role of endogenous ABA and since then numerous publications have appeared (see also 4.3.11) most of which have indicated that ABA may participate in a wider regulatory system which controls both the salt and water balance of a plant.

The present impression is that the effects of ABA primarily concern the distribution of K^+ (MANSFIELD and JONES, 1971; HORTON, 1971; HORTON and MORAN, 1972). However, as was shown for IAA (see 7.2.2) the effect of ABA on K^+ fluxes may be ascribed to secondary effects such as ion exchange or the need for an accompanying ion. For instance the K^+ shuttle between guard cells and subsidiary cells in maize is dependent on Cl^- as the accompanying ion (RASCHKE and FELLOWS, 1971), while in *Vicia* or bean (*Vicia faba*) organic acids act as counter ions with H^+ ions being released in exchange for K^+ (PALLAS and WRIGHT, 1973; RASCHKE and HUMBLE, 1973). It seems possible that ABA may inhibit the H^+ ion release of bean guard cells since it was shown that ABA has this effect on H^+ ion secretion in *Avena* coleoptiles (RAYLE and JOHNSON, 1973).

Cyclic photophosphorylation appears to play an important role in producing the energy for K^+ mediated stomatal opening (DAS and RAGHAVENDRA, 1974) and the observation by TILLBERG (1968) that the inhibitor β-complex inhibits formation of ATP more than electron flow in isolated chloroplasts may be relevant. It indicates that the immediate effect of ABA in stomata is primarily on the energy supply for active ion transport in guard cells, thereby inhibiting proton extrusion (bean) or active Cl^--accumulation (maize). Apparently, ABA is ineffective as an uncoupler of photophosphorylation (KECK and BOYER, 1974), but it has been shown that extreme conditions of water or salt stress may have this effect (SANTARIUS and ERNST, 1967; SANTARIUS, 1967). MILLER et al. (1971), and MILLER et al. (1973) believe that mitochondria isolated from water-stressed seedlings are incapable of selective transport because of an apparent degradation of membrane structure.

Other membrane effects were recorded by GLINKA and REINHOLD (1971, 1972) who found that ABA caused an increase of hydraulic conductivity (L_p) in discs of carrot tissue within 30 min of treatment. Kinetin and CO_2 were found to have an opposing effect to ABA and the reflection coefficient σ apparently was not affected by ABA treatment. ABA in the root medium increased the exudate flow (J_v) in decapitated tomato plants and this also was taken as evidence that ABA affected the hydraulic conductivity (L_p) of the root systems rather than a change in driving force being responsible for the increased flow (TAL and IMBER, 1971; GLINKA, 1973). Considering that the main barriers to water movement in plant protoplasts appear to be at the plasmalemma (URL, 1971), one may presume that ABA primarily affects the conformation of structural elements of the plasmalemma. The observation that ABA may induce a shift in the degree of saturation of the fatty acid component of lipids in leaves and stalks of *Coleus* might be relevant (KUHL and UNGER, 1974).

A very different conclusion was reached by CRAM and PITMAN (1972) who found that ABA caused an 80–90% reduction in the rate of exudate volume flow (J_v) from barley and maize roots. The authors concluded on the basis of observed values for ΔJ_v, ΔJ_s and $[\Delta K]$ that the hydraulic conductivity L_p was unaffected by the ABA treatment (see also 3.4.3.5).

Results obtained from slices of leaf tissue contrast with those obtained from slices of root tissue. Horton and Bruce (1972) found that in slices of expanding leaf, K^+ uptake was greatly inhibited in the light or dark by $10\,\mathrm{g\,m^{-3}}$ ABA, while ABA had no effect on K^+ influx into tissue obtained from non-expanding mature leaves. ABA (0.03–$10\,\mathrm{g\,m^{-3}}$) also inhibited K^+-uptake and, to a lesser extent, Cl^- uptake in oat coleoptile segments (Reed and Bonner, 1974). There was no simple quantitative correspondence between ABA inhibition of coleoptile elongation and ABA inhibition of K^+ uptake (Reed and Bonner, 1974). Van Steveninck (1972b), on the other hand, found that ABA ($10\,\mathrm{g\,m^{-3}}$) strongly stimulated net uptake of Na^+, K^+ and Cl^- in discs of beetroot tissue once the cells had acquired a capacity for net uptake of these ions.

In beetroot ABA had a marked effect on ion selectivity causing the tissue cells to show a preference for Na^+ over K^+ if both ions were present in a solution (van Steveninck 1972b; see Fig. 7.4). In barley seedlings, discrimination between K^+ and Na^+ is affected by salt status of roots and by the degree of aeration of the root system (Pitman, 1969). Lack of aeration in *Nicotiana* roots will cause rapid wilting of the shoot which is prevented when approximately 100 mM NaCl

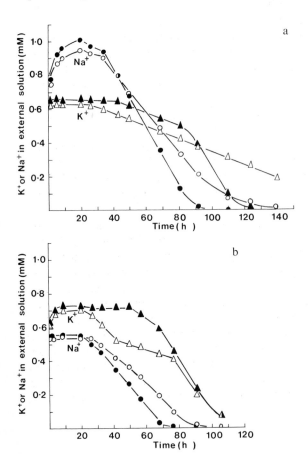

Fig. 7.4a and b. The effect of ABA ($3.8 \cdot 10^{-5}$M) on uptake and release of K^+ and Na^+ by freshly cut slices of beetroot placed in a solution of 0.5 mM KCl + 0.5 mM NaCl. $\triangle$, K^+ concentrations in the external solution; $\circ$, Na^+ concentrations in the external solution. Open symbols, control; closed symbols, +ABA. (a) Slices taken from a Na^+-selective beetroot ($[K_i^+]/[Na_i^+]=0.30$). (b) Slices taken from a K^+-selective beetroot ($[K_i^+]/[Na_i^+]=3.76$). In each case ABA causes a stimulation of Na^+ uptake while K^+ uptake is stimulated only when the Na^+ of the external solution has been depleted. (From van Steveninck, 1972b)

is added to the 1/2 Hoagland solution. This induced resistance to anaerobic conditions is associated with an increased concentration of ABA in the leaves, and also occurs when plants are pre-treated with ABA (MIZRAHI et al., 1972). LEGGET and STOLZY (1961) have previously claimed that anaerobic conditions favored Na-accumulation in barley seedlings irrespective of whether or not Na^+ is moved against an electrochemical gradient. ABA-controlled vacuolar accumulation of Na^+, especially in root tissue, seems to reflect an adaptation to osmotic stress under conditions of high salinity and/or anaerobiosis. This mechanism would allow for the release of K^+ in exchange for Na^+ and thus make an increased amount of K^+ available for transport to the shoot.

6. Senescence Factor (SF), C_2H_4 and CO_2

OSBORNE et al. (1972) have detected a new non-volatile, acidic growth substance which accelerates abscission and triggers a sequence of senescence processes especially those which depend on the rapid breakdown of permeability barriers (SACHER, 1957, 1962; EILAM, 1965; see Chap. 8). The release of this substance, called senescence factor (SF) may occur on wounding and it is suggested that either IAA, C_2H_4 or ABA in stimulating C_2H_4 production, may control the release of SF from a particular cell component. The interactions between growth hormones and C_2H_4 appear very complex and are incompletely understood (ABELES, 1972). Opposing effects have been observed between auxins, ABA, gibberellins and C_2H_4 (SCOTT and LEOPOLD, 1967; VALDOVINOS et al., 1967; PRATT and GOESCHL, 1969; GERTMAN and FUCH, 1972) and there is some doubt that C_2H_4 has any direct effects on membrane permeability (SACHER and SALMINEN, 1969; MEHARD et al., 1970). Recently, it has been suggested that C_2H_4 may cause a diversion of electrons from the respiratory chain to an alternate CN^- resistant path (SOLOMOS and LATIES, 1974). A similar phenomenon occurs during ageing of tissue slices (see 8.3.2.3) and may be relevant with respect to the development of ion transport capacities (VAN STEVENINCK, 1975).

CO_2 has come to be regarded an antagonist of ethylene action (PRATT and GOESCHL, 1969). SEARS and EISENBERG (1961) proposed a model which would involve an interaction between CO_2, HCO_3^- and the lipid components of the cell membranes, e.g. additional CO_2 would cause a decreased hydration, a more continuous lipid phase and hence create a less ion-permeable membrane. GLINKA and REINHOLD (1962) showed that in sunflower hypocotyl tissue, application of CO_2 markedly reduced H_2O influx and efflux. These effects took place in a few seconds (REINHOLD and GLINKA, 1966) and were shown to be rapidly reversible in carrot tissue (GLINKA and REINHOLD 1972). CO_2-induced stomatal closure was not mediated through ABA (LOVEYS, 1973), and PALLAGHY and RASCHKE (1972) showed that stomata did not respond to C_2H_4 treatment. Hence there seems to be little indication of any gaseous system of control of ion transport and permeability of membranes.

7. Phytochrome, Cyclic AMP, Sterols, Phytotoxins, Polylysins, Ionophores, etc.

A number of substances are grouped together here mainly because they are of special interest in providing a link between perception and control by hormone action (e.g. phytochrome), or between hormone action and cellular realization (2nd messenger, e.g. serotonin→cyclic AMP), or because they mimic hormone action due to characteristic peculiarities in molecular structure (e.g. sterols, polylysine etc.).

7.1 Phytochrome

The red far-red perception system finds its expression in a variety of aspects of morphogenesis (BRIGGS and RICE, 1972). Perhaps the most intensively studied from an ion transport point of view is the transition from etiolated to light-grown pea seedlings. KÖHLER et al. (1968) established that K^+/Rb^+ uptake of leaves and internodes parallelled developmental aspects as determined by light treatments, i.e. red light inhibited K^+-uptake into internodes and promoted uptake into the plumule. It was observed that the effect on K^+ transport preceded the red-light-induced inhibition of internode growth and the growth promotion of leaves and also the red light effect on K^+ transport was shown to be independent of the K^+ concentration supplied to the roots in the range between 0.2 and 125 mM (KÖHLER, 1969). DÖRFFLING (1973) subsequently showed that the red-light-induced growth inhibition was not correlated with an increase of ABA and hence it appears unlikely that the phytochrome control of K^+ transport is mediated by ABA.

Earlier suggestions that phytochrome action involved a change in membrane permeability in particular to K^+ were refuted by KENDRICK and HILLMAN (1972). They measured K^+ and Na^+ fluxes in pea epicotyl sections with different phytochrome contents and concluded that selective uptake of K^+ was not under phytochrome control. Obviously, further critical work is required to resolve this question. Phytochrome-mediated nyctinastic closure of leaflets in legumes was shown to depend on efflux of electrolyte from the pulvinule (JAFFE and GALSTON, 1967), electron probe microanalysis showing that primarily K^+ was involved (SATTER et al., 1970). Apparently, a shift of K^+ from ventral cells to dorsal cells occurred during closure while a reversal of flux occurred during light-induced opening. This shuttle also operated in circadian opening and closing under constant conditions (SATTER and GALSTON, 1971 a, b). It seems that both a phytochrome-induced alteration of membrane permeability and a direct effect on an active transport system may be involved.

In an elegant series of experiments using microbeams of polarized light it has been established that phytochrome in cells of filamentous algae is localized in the plasmalemma in a highly ordered state with dichroic orientations changing P_r to P_{fr} and vice versa (HAUPT, 1968, 1970a, b; HAUPT et al., 1969). It was shown (WEISENSEEL and SCHMEIBIDL, 1973) that immediate changes in water permeability of the plasmalemma of Mougeotia could be effected via the P_r, P_{fr} mechanism. Other evidence that phytochrome is particle bound comes from work by QUAIL (1974a, b), QUAIL and SCHÄFER (1974) and QUAIL et al. (1973).

Yet another phytochrome-controlled phenomenon concerns the surface adhesion properties of mung bean roots (TANADA, 1968; JAFFE, 1968; YUNGHANS and JAFFE, 1970, 1972). IAA and ABA seemed to have an effect on the light responses within minutes of their application (TANADA, 1973). RACUSEN (1973) and RACUSEN and ETHERTON (1973) consider that the reversal of electrical charge of the root cap cells is based on conformational changes of proteinaceous membrane components activated by changes of H^+ and Ca^{2+} densities.

7.2 Cyclic AMP

c-AMP has only recently been detected in plant tissues (POLLARD, 1970; AZHAR and KRISHNA MURTI, 1971; JANISTYN, 1972; JANISTYN and DRUMM, 1972; WELLBURN et al., 1973; RAYMOND et al., 1973; BREWIN and NORTHCOTE, 1973a; BROWN and NEWTON, 1973; NICOLAS and NIGON, 1973). Its role as a second messenger for a number of animal hormones (serotonin, epinephrine, vasopressin, catecholamines, glucagon, histamine, prostaglandins etc.) and its interaction with insulin and many other aspects of metabolism (for a review see ROBISON et al., 1971 or JOST and RICKENBERG, 1971) should justify exploration of a possible regulatory involvement between metabolism and ion distribution in plant tissues.

D.A. BAKER (personal communication) reported a considerable promotion of ^{86}Rb uptake by c-AMP in both stelar and cortical tissues of maize roots. The transport of label to the exudate compartment was also enhanced by about 50% after a lag of about 16 h. However, the volume flow of the exudate was unaffected. It was suggested that the effect of c-AMP might operate through a stimulation of ATPase activity. The applied concentrations of c-AMP were quite high (up to 1 mM), however it was argued that much of the exogenously supplied c-AMP might be broken down by c-AMP phosphodiesterase. A large proportion of c-AMP phosphodiesterase activity was shown to be located in the plasmalemma of soybean callus cells by BREWIN and NORTHCOTE (1973b).

c-AMP was shown to mimic GA_3 in barley aleurone tissue, exemplified by the induction of α-amylase (GALSKY and LIPPINCOTT, 1969; DUFFUS and DUFFUS, 1969; POLLARD, 1971), endosperm protease, acid phosphatase and ATPase (NICKELLS et al., 1971; EARLE and GALSKY, 1971; GILBERT and GALSKY, 1972), whereas ABA was shown to reverse the GA_3 and c-AMP-induced α-amylase synthesis (BARTON et al., 1973). Also, c-AMP and GA_3 or kinetin synergistically enhanced auxin-induced cell expansion in Jerusalem artichoke tuber tissue (KAMISAKA, 1972), while IAA was found to stimulate the synthesis of c-AMP in a number of different tissues (SOLOMON and MASCARENKAS, 1971, in oat coleoptile sections; AZHAR and KRISHNA MURTI, 1971, in bengal gram (*Cicer arietinum*) seedlings; JANISTYN 1972 in maize coleoptile sections).

GOLDTHWAITE (1974) observed that 1–5 mM c-AMP retards senescence in *Rumex* leaf disks incubated in the dark. Yet, GIANNATTASIO (1974) claimed that the level of c-AMP was very high in dormant tissues compared with those of sprouting tubers of Jerusalem artichoke, and also c-AMP contents in a wide range of monocots and dicots seemed to increase with ageing (KESSLER and LEVINSTEIN, 1974).

The validity of the above observations on effects of externally applied c-AMP is greatly reduced by recent surveys of the occurrence of c-AMP in the plant kingdom.

Although the green unicellular alga *Chlamydomonas reinhardtii* has endogenous c-AMP and a specific cyclic nucleotide phosphodiesterase (FISCHER and AMRHEIN, 1974; AMRHEIN, 1974a; AMRHEIN and FILNER, 1973), both c-AMP and the enzymes of its metabolism apparently do not occur in higher plant cells and tissues (AMRHEIN, 1974a, b; LIN, 1974). Hydrolysis of c-AMP observed with extracts of higher plant tissues appear to be due to nonspecific enzyme action (AMRHEIN, 1974a).

7.3 Sterols and Phytotoxins

CLARK (1938) established that the cholesterol derivative Na-glycocholate abolished polar transport of IAA in *Avena* coleoptiles, but had no effect on cell permeability or on Br^- accumulation in barley roots. VAN STEVENINCK (1961) observed an increased leakage of K^+ and Na^+ when 10^{-4} M Na-glycocholate was added to freshly sliced beet discs. Also the lag phase for development of net K^+ uptake was shortened from 38 to 22 h, while that for Na^+ was lengthened from 22 to 38 h. GRUNWALD (1968) tested a range of sterols and found that IAA did not reduce the betacyanin leakage from beet slices which was induced by β-sitosterol, stigmasterol, 16-dehydroxyprogesterone, astriole and progesterone. WEIGL (1969b, c), in model experiments, showed that auxin had a special affinity for lecithin, but not for cholesterol, oleic acid and stearic acid. It was implied that enhanced molecular incorporation of IAA in the membrane structure could have a regulatory effect on ion fluxes and on membrane potential.

Phytotoxins cause marked changes in salt balance even at remarkably low concentrations.

Victorin, a penta peptide linked to tricyclic secondary amine (PRINGLE and BRAUN, 1957) was shown to stimulate respiration at concentrations of $2 \cdot 10^{-4}$ µg of purified toxin per ml (SCHEFFER and PRINGLE, 1963), while a 50-fold lower concentration caused marked changes in permeability 5 min after application to susceptible tissues (WHEELER and BLACK, 1962, 1963).

Fusicoccin, a non-specific phytotoxin diterpene glucoside with a complex ring system (BALLIO et al., 1968a; Chain et al., 1971), which like victorin causes an increased cell permeability and increased respiration (BALLIO et al., 1968b), was also shown to have a pronounced effect on the stomatal guard cell mechanism in a wide range of plant species (TURNER and GRANITI, 1969). It increased stomatal opening through a stimulation of K^+ uptake in guard cells of bean epidermal strips (TURNER, 1972). The response time was of the order of 1 min (SQUIRE and MANSFIELD, 1974) while concentrations as low as $5 \cdot 10^{-6}$ M fusicoccin had the capacity to reverse a 10^{-4} M ABA-induced closure of stomata (SQUIRE and MANSFIELD, 1972).

The most exciting aspect is that fusicoccin-induced stimulation of cell enlargement in leaf slices, pea internode segments or squash cotyledons is accompanied by H^+ extrusion and a significant increase in negative transmembrane potential (MARRÉ et al., 1974a, b; see Fig. 7.5). Both the hyperpolarization and proton extrusion are greatly reduced by uncouplers and metabolic inhibitors (CCCP, DNP, CHM, puromycin etc.) indicating that the proton extrusion mechanism is dependent on metabolism. The proton extrusion is also stimulated by K^+ and to a lesser extent by Na^+, while Cs^+ and Li^+ have little effect. This order of activity corresponds

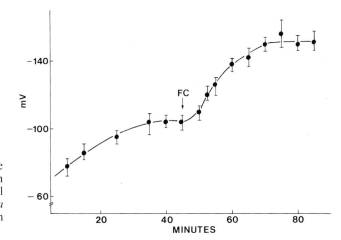

Fig. 7.5. Kinetics of the effect of fusioccin (FC) on transmembrane potential (mV) in squash (*Cucurbita maxima*) cotyledons. (From MARRÈ et al., 1974b)

to the capacity of these ions to activate K^+-dependent ATPase in "plasmalemma" preparations from pea internode segments. Dicyclohexylcarbodiimide (DCCD), an inhibitor of K^+-ATPase, completely blocked proton extrusion and cell enlargement in the presence or absence of fusicoccin or IAA (MARRÈ et al., 1974c).

The stimulative effects of fusicoccin were usually well in excess and clearly additive to auxin responses and did not show any inhibitory tendencies at supraoptimal concentrations (LADO et al., 1972, 1973). It will be an extremely useful tool in the study of auxin mediated H^+ pumps (CLELAND, 1974; DOHRMANN et al., 1974).

Filipin, an acylated steryl glucoside, isolated from *Streptomyces filipenensis,* induced leakage in beet and potato tissue, but not in apple discs or carrot (MUDD and KLEINSCHMIDT, 1970). This leakage could be reversed by addition of cholesterol to the medium. It was established in model experiments that artificial membranes are affected by filipin only when cholesterol had been incorporated into the membrane, hence it was suggested that filipin could be used as a diagnostic test for sterol-containing membranes. HENDRIX and HIGINBOTHAM (1973) established that filipin had no effect on cell electropotentials in etiolated pea stem segments and presumed that filipin interfered with the sterol stabilization of phospholipid layers of cell membranes.

Very recently, further interest has developed in a group of phytotoxins produced by *Helminthosporum maydis* race T which consist of steroid like compounds, but at least one of which is a glycoside (MERTZ and ARNTZEN, 1974). Some of the main features are a depolarization of membrane potential, inhibition of K^+ stimulated (microsomal) ATPase, inhibition of H^+ efflux, K^+ and Cl^- influx and K^+ transport in guard cells of maize leaves. Hence, the effects are similar to the mode of action of ABA on stomata (ARNTZEN et al., 1973). *Helminthosporum carbonum* toxin stimulated Na^+ and Cl^- uptake in maize roots (YODER and SCHEFFER, 1973). This again corresponds with the effects of ABA in root tissue (VAN STEVENINCK, 1972b; CRAM and PITMAN, 1972). Plant antifungal toxins (the phytoalexins pisatin, phaseolin, piletin and avenacin) received recent attention because this group of steroid-like compounds, some of which are also glycosides, similarly caused partial depolarization of membrane potential, inhibition of K^+ influx, H^+ efflux etc. (SLAYMAN and VAN ETTEN, 1974).

7.4 Polylysine, Histones

Polylysines have also been shown to have drastic effects on membrane permeability, especially on betacyanin efflux from beetroot (SIEGEL and DALY, 1966; SIEGEL, 1970; OSMOND and LATIES, 1970). The effect is Ca^{2+} reversible, and it has been suggested that polylysine displaces Ca^{2+} from membrane structures thereby increasing their permeability. These effects on membrane permeability appear to be universal and have also been observed on the hydropotes and epidermal cells of *Nymphaea* leaves (LÜTTGE et al., 1971), on barley roots (DREW and MCLAREN, 1970), on isolated corn mitochondria (HANSON, 1972), and photosystem I of isolated chloroplasts (BRAND et al., 1971). The latter effects were prevented by pre-incubation of chloroplasts in salt containing solutions, and also the uncoupling of the oxidative phosphorylation by basic proteins was reversed by K^+ (RIVENBARK and HANSON, 1962).

7.5 Ionophores and Related Compounds

Since the discovery of antibiotic-induced energy-linked mitochondrial ion transport by MOORE and PRESSMAN (1964), detailed information at the molecular level has become available which could assist in the study of natural active transport mechanisms and ultimately should assist in predicting some properties of carrier molecules.

The antibiotics, as well as their synthetic analogs, have complex molecular structures and are capable of forming a ligand inwardly with monovalent or divalent cations, often with a high degree of specificity. They are able to carry ions across lipid barriers on account of their outward lipophilic characteristics and hence the generic designation "ionophores" given by PRESSMAN et al. (1967). In spite of wide differences in molecular structure the classification of ionophores is simplified because of some basic mechanistic distinctions which are illustrated in Fig. 7.6.

Group 1, which includes valinomycin and its analogs, the gramicidins, the macro-tetralide actins and the crown polyethers have a symmetric ring structure, and are neutral, i.e. the cation complexes acquire the charge of the complexed cation. Valinomycin shows considerable specificity towards monovalent cations, the $K^+:Na^+$ preference being 10,000:1 (MOORE and PRESSMAN, 1964). Group 2, consists of the ionophores of the nigericin-type. They contain a carboxyl group and hence carry a negative charge when dissociated but are electrically neutral when they carry alkali ions or protons (Fig. 7.6). This explains their ability to catalyze an alkali ion-for-proton exchange across various lipid barriers without giving rise to the electrical phenomena which are observed with the charged valinomycin type complexes. Group 3 consists of molecules (e.g. alamethicin) which combine the properties of class one and two, i.e. carboxyl group and cyclic arrangement by covalent bonds, and this implies that they form both charged or neutral complexes (Fig. 7.6). Ionophores which bind divalent but not monovalent ions have recently been reported e.g. A23187 which is a carboxylic acid antibiotic (REED and LARDY, 1972; WONG et al., 1973). The complexation and decomplexation reactions have such favorable kinetics that valinomycin is able to catalyze K^+ transport in mitochondria with a turnover rate of 200 s^{-1}, while nigericin's turnover number for exchanging K^+ for H^+ across the mitochondrial membranes may reach 500 s^{-1} (PRESSMAN, 1973).

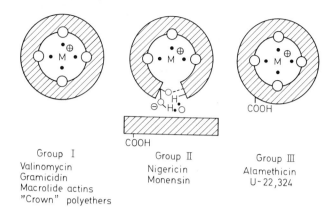

Fig. 7.6. A schematic representation of the three subclasses of ionophores. (From PRESSMAN, 1968)

Group I
Valinomycin
Gramicidin
Macrolide actins
"Crown" polyethers

Group II
Nigericin
Monensin

Group III
Alamethicin
U-22,324

By far the majority of work has been carried out on mitochondrial suspensions because of the interesting opposite effect of valinomycin which promotes movement of K^+ against a concentration gradient and nigericin which confers H^+ permeability and promotes downhill dissipation of K^+ gradients (PRESSMAN et al., 1967). However, DEGANI and SHAVIT (1972) recorded that nigericin-type ionophores stimulated K^+ and Na^+ uptake in isolated chloroplasts, while nigericin and valinomycin co-operated in uncoupling photophosphorylation. HODGES et al. (1971) found that gramicidin D and nigericin both stimulated K^+ influx 4–8 fold into cut roots over a 30 min absorption period. Valinomycin and nonactin on the other hand had little effect on K^+ influx. It was suggested that the stimulation which occurred mainly in the elongation zone and could be detected within one minute of the application of ionophores, would depend on the lipid composition of the particular membrane structure.

8. Hormonal Function and Ion Transport

It is aimed here to summarize those hormone-controlled phenomena which are mediated by ion transport or electrical events. It has always been believed that the first stages of cell differentiation, e.g. the formation of vacuoles are under hormonal control. While this phenomenon clearly involves basic aspects of cellular compartmentation, this fundamental question of morphogenesis is still wide open for speculation. It is also an aspect which sets plant cells apart from animal cells.

Some specialized functions which involve a localized response may offer some hope for a solution in the relatively near future. These are concerned mainly with tropisms and nastic movements. Perhaps the best known is the geoelectric effect (gravity related electrical polarization). The difference of potential between upper and lower side in an orthogeotropic organ placed horizontally is established within a few minutes of displacement (BRAUNER, 1927, 1928) with the lower side being up to 70 mV more positive than the upper side (HIGINBOTHAM, 1973). Since auxin at normal cell pH would be largely dissociated, it was originally believed that its relative enrichment on the lower side would depend on electrophoretic migration (BRAUNER, 1962). Although there is little doubt about a lateral movement of auxin induced by gravity in coleoptile segments (Goldsmith and Wilkins, 1964), it was conclusively shown that the potential difference is the result of auxin movement rather than its cause (Dedolph et al., 1965; Wilkins and Woodcock, 1965).

Recently, Etherton and Dedolph (1972) established that the lower ends of individual maize coleoptile cells are consistently more negative than the upper ends by about 2 mV. It was held that the geoelectric effect of a whole organ may be ascribed to a summation of electrically polarized cells which would act in series. Since the organ geoelectric effect is measured as a surface potential and the cell geoelectric effect as a transmembrane potential the sign of the potentials was consistent with the hypothesis (Etherton and Dedolph, 1972). So far it has been shown that IAA will cause a hyperpolarization of oat coleoptile cells (Etherton, 1970). It would be of considerable interest if it can be established that auxin gradients can cause the observed localized changes in membrane potential. Perhaps, the phenomenon depends on strictly localized stimulations of H^+ ion pumps within cells somewhat reminiscent of the acid and alkaline bands in *Nitella* (Spear et al., 1969).

In addition to the geoelectric effect, some consistent observations have been made on the redistribution of cations, e.g. there is an apparent movement of K^+ from the upper to the lower side (Bode, 1959), while Ca^{2+} has been shown to become relatively enriched in the upper side of *Helianthus* hypocotyls (Arslan-Çerim, 1966). The true significance of these ion distributions is difficult to ascertain at the present time, but it seems likely that localized H^+ ion secretion will prove to play an important role since a differential change of cell-wall pressure is probably the essential ingredient of the turgor mediated differential growth rates of the upper and lower sides (Brauner and Brauner, 1962; Reinhold and Ganot, 1972).

Reinhold et al. (1972) showed unequivocally that IAA can substitute for the contact stimulus in thigmotropism. In this case there is an innate difference between the two sides of the tendril which react differently to the same stimulus. The above authors showed that the tendrils of *Cucumis* and *Passiflora* possess interesting electrical properties. Bursts of spontaneous fluctuations in electric surface potential spikes (usually of 0.5 mV) at a rate of approximately 12 per second alternated with "quiet" periods of 5–10 min duration. The rhythmic nature of this phenomenon suggested that the fluctuations were related to the rhythmic movements of circumnutation.

9. Concluding Remarks

The available evidence of hormone effects on ion transport does not go much beyond the descriptive stage yet it has provided a definite notion that plant hormones have a decidedly regulatory function in matters of ion selectivity, distribution within tissues, long-distance transport and possibly also on cellular compartmentation (e.g. vacuolar *vs.* cytoplasmic). The present lack of knowledge about the actual mechanisms involved is due to the fact that as yet the means for a dissection of actual molecular steps incurred during transmembrane transport are lacking. This is largely due to the fact that plant tissues generally offer less scope for sophisticated achievements than one might find in more amenable biological systems showing greater specialization (Haydon and Hladky, 1972; Hille, 1972).

A clear-cut distinction needs to be made between the direct effect of hormones on existing ion pumps and membrane structure which are evident within minutes of application (e.g. auxin induced H^+ ion extrusion from coleoptile segments, K^+ in the stomatal guard cell mechanism), and those which take several hours to become established (e.g. induction of ion pumps in storage tissues, changes of ion selectivity, see Table 7.4). It seems that especially the first category of phenomena, will allow some real progress in elucidating the molecular steps involved. At this stage we can only speculate about these aspects, e.g. whether the hormone action involves molecular interactions within membrane "pores" (FENSOM and WANLESS, 1967; HILL, 1972; WANLESS et al., 1973; LÄUGER, 1973) and hence affect ion selectivity (DIAMOND and WRIGHT, 1969) in a manner similar to mechanisms which have been proposed for "ionophores" (PRESSMAN, 1973; HILLE, 1972; HAYDON and HLADKY, 1972). Alternatively, the mechanisms may involve structural aspects such as dynamics of membrane turnover in pinocytosis (LÜTTGE and KRAPF, 1968; HALL, 1970; POWER and COCKING, 1970; HEYN, 1971; YOUNG and SIMS, 1973), or movement from cell to cell through plasmodesmata (ARISZ, 1956; ZIEGLER and LÜTTGE, 1967; KAUFMAN et al., 1970; TYREE, 1970; PALLAS and MOLLENHAUER, 1972; VAN STEVENINCK and CHENOWETH, 1972; LÄUCHLI et al., 1974). In all the above aspects it is desirable to obtain a lot more information about the molecular interactions of the component lipids and proteins in the membranes, their charge distributions, influence of proton densities, hydration, CO_2, $H_2CO_3^-$, Ca^{2+} and possibly also of the almost ubiquitous presence of C_2H_4.

A further study of the second or slow category of hormone influences should be no less rewarding. Already some exciting aspects of hormone regulation of ion distribution in the whole plant have come to light (PITMAN and CRAM, 1973). However, interactions between endogenous factors and external stimuli have shown a bewildering complexity. ABA has been found to inhibit transport of ions to the xylem, but without inhibiting uptake to the root (CRAM and PITMAN, 1972). The response, however, depends on the conditions in which plants were grown and on the temperature; it may be an indirect interaction of ABA with endogenous hormones in the root (PITMAN et al., 1974). At present it is uncertain whether the divergent results are primarily due to species difference, the existing hormone balance of the experimental material, the degree of tissue differentiation (GOTÔ and ESASHI, 1974; WRIGHT, 1966), or to some environmental condition (e.g. light, temperature, humidity, etc.).

TAL and IMBER's (1971) proposal that cytokinins generally cause a reduction in plant turgor by decreasing stomatal resistance and increasing the resistance of the root to water is well matched by the observation that cytokinin concentration is decreased in stressed plants (ITAI and VAADIA, 1965). Added to this, an ABA-induced increase in plant turgor through an increase in stomatal resistance and a decrease in resistance of the root to water fits in with the observed increase of ABA concentration in the exudate of stressed plants. On the other hand, one could argue that the salt status of the root and the control of the exudate flow into the xylem vessels, possibly by cells with a special regulatory function, would have a more decisive effect on the water and salt balance than a change in the hydraulic conductivity of the root (see also *3.4.2.5*, *3.4.2.9* and *3.4.3.5*).

There have been differing views on the location of regulatory cells in the root. It has been shown by HIGINBOTHAM et al. (1973) that in maize roots the metaxylem

Table 7.4. Time lags between application and measurable responses following growth substance application

Type of response	Species	Substance	Time of response	Ref.
Stomatal opening	Commelina communis	10^{-5}M Fusicoccin (FC)	~1 min	Squire and Mansfield (1974)
Stomatal closure	Hordeum vulgare	10^{-7}M ABA	<5 min	Cummins et al. (1971)
Adhesion of root tips to a glass surface	Phaseolus vulgaris Hordeum vulgare	$\left.\begin{array}{l}10^{-9}\text{M ABA}\\10^{-8}\text{M IAA}\end{array}\right\}$	1–2 min	Tanada (1973)
Coleoptile segment elongation	Avena sativa	$\left\{\begin{array}{l}10^{-5}\text{M IAA}\\10^{-5}\text{M FC}\end{array}\right.$	1–2 min 1–2 min	Hager et al. (1971) Cleland (1974)
H$^+$-excretion by coleoptile segments	Avena sativa	10^{-5}M FC 10^{-5}M IAA	5 min 10–15 min	Cleland (1974) Dohrman et al. (1974)
Stimulation of Cl$^-$ uptake by coleoptile segments	Avena sativa	$10^{-6} - 10^{-4}$M IAA	15 min	Rubinstein and Light (1973)
Inhibition of Cl$^-$ uptake by callus culture cells	Petroselinum sativum	10^{-5}M IAA	<30 min	Bentrup et al. (1973)
Transport of Cl$^-$ or Rb$^+$ out of excised barley roots	Hordeum vulgare	0.4–$1.9 \cdot 10^{-5}$M ABA	40–60 min	Cram and Pitman (1972)
Stimulation of Cl$^-$ uptake in aged slices of beetroot	Beta vulgaris	$3.8 \cdot 10^{-5}$M ABA	<1h	Van Steveninck (1974)
Stimulation of K$^+$ (Rb$^+$) uptake in aged slices of beetroot	Beta vulgaris	$3.8 \cdot 10^{-5}$ ABA	2h	Van Steveninck (1974)
Inhibition of K$^+$ and Na$^+$ uptake in aged slices of beetroot	Beta vulgaris	$7 \cdot 10^{-5}$M kinetin or benzyladenine (BA)	1–6h depending on age of discs	Van Steveninck (1972a)
Stimulation of K$^+$ uptake in sunflower hypocotyl segments	Helianthus annuus	$4.6 \cdot 10^{-5}$M kinetin	>2h	Ilan (1971)
K$^+$/Na$^+$ selectivity in aged beetroot slices	Beta vulgaris	$3.8 \cdot 10^{-5}$M ABA	~2h	Van Steveninck (1972b)
K$^+$/Na$^+$ selectivity in bean stem slices	Phaseolus vulgaris	$5 \cdot 10^{-6}$M BA	~4h	Rains (1969)
K$^+$/Na$^+$ selectivity in sunflower leaf discs	Helianthus annuus	$4.6 \cdot 10^{-5}$M kinetin	<$11^{1}/_{2}$h	Ilan (1971)

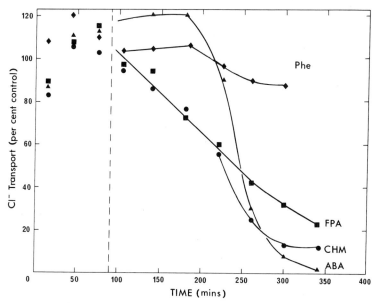

Fig. 7.7. Effect on transport of ^{36}Cl through barley roots of added 2 mM phenylalanine (♦); 2 mM FPA (■); 10 g m^{-3} CHM (●) and 10 μM ABA (▲). Transport is plotted relative to control (untreated) samples in each period. (From Schaeffer et al., 1975)

elements retained functional cytoplasm lining the cell walls as far as 10 cm from the apex. However, Läuchli et al. (1974) have found no living xylem vessels at 1 cm from the apex in barley roots. As transport to the xylem was operating in the mature vessels more than 1 cm from the apex, they suggested the xylem paren-chyma cells were the site of the hormone-sensitive pump. Wherever it is located, there seems to be evidence for a mechanism of secretion into the xylem that is sensitive to certain phytohormones and to inhibitors of protein synthesis (Fig. 7.7). Evidence from these kinds of studies favour the "two pump" hypothesis proposed at various times by different authors (see Chap. 3.4).

There is growing evidence too that hormone effects are particularly evident in cells and tissues with a specialized function. In most cases these regulatory aspects require several hours to become established and may resemble, or have been shown to mimic, the effects of inhibitors of protein synthesis, implying that these hormone effects are in some way related to protein synthesis. For instance, cycloheximide closely mimics the effect of ABA on transport of ions from the root to the shoot (Läuchli et al., 1973, Lüttge et al., 1974). This result is almost identical to the manner in which cycloheximide mimics the effect of kinetin in parenchyma root storage cells (van Steveninck, 1972a; van Steveninck and van Steveninck, 1972). Evidently, in both cases we are dealing with cytoplasmic control of ion transport perhaps involving membrane turnover and possible involvement of the symplasm, rather than with direct effects on membrane permeability. The second alternative most likely belongs to the category of rapidly induced phenomena (such as IAA or FC effects on H$^+$ release) which do not require the involvement of protein synthesis (see Table 7.4).

These are very interesting pointers towards hormonal regulation of ion distribution in the plant body. Especially in this field, best progress can be made by always making clear and logical distinctions between the location and the nature of the processes involved before attempting to correlate one piece of scientific evidence with another.

References

ABELES, F.B.: Biosynthesis and mechanism of action of ethylene. Ann. Rev. Plant Physiol. **23**, 259–292 (1972).

ALLEN, A.K., NEUBERGER, A.: The purification and properties of the lectin from potato tubers, a hydroxy proline containing glycoprotein. Biochem. J. **135**, 307–314 (1973).

AMRHEIN, N.: Cyclic nucleotide phosphodiesterase in plants. Z. Pflanzenphysiol. **72**, 249–261 (1974a).

AMRHEIN, N.: Evidence against the occurrence of adenosine—3′,5′-cyclic monophosphate in higher plants. Planta **118**, 241–258 (1974b).

AMRHEIN, N., FILNER, P.: Adenosine 3′:5′: cyclic monophosphate in *Chlamydomonas reinhardtii*: isolation and characterization. Proc. Natl. Acad. Sci. U.S. **70**, 1099–1103 (1973).

ARISZ, W.H.: Significance of the symplasm theory for transport across the root. Protoplasma (Wien) **46**, 5–62 (1956).

ARMSTRONG, J.E., JONES, R.L.: Osmotic regulation of α-amylase synthesis and polyribosome formation in aleurone cells of barley. J. Cell Biol. **59**, 444–455 (1973).

ARNTZEN, C.J., HAUGH, M.F., BOBACK, S.: Induction of stomatal closure by *Helminthosporum maidis* pathotoxin. Plant Physiol. **52**, 569–574 (1973).

ARSLAN-ÇERIM, N.: The redistribution of radioactivity in geotropically stimulated hypocotyls of *Helianthus annuus* pretreated with radioactive calcium. J. Exptl. Bot. **17**, 236–240 (1966).

ARTIMONOVA, G.M., ARTIMONOV, V.I.: Effect of gibberellin on electrolyte permeability of cells. Soviet Plant Physiol. **18**, 536–537 (1971).

AUDUS, L.J.: Plant growth substances, vol. 1: Chemistry and physiology. New York: Harper and Row 1972.

AZHAR, S., KRISHNA MURTI, C.R.: Effect of indole-3-acetic acid on the synthesis of cyclic 3′–5′ adenosine phosphate by bengal gram seeds. Biochem. Biophys. Res. Commun. **43**, 58–64 (1971).

BALLIO, A., BRUFANI, M., CASINOVI, C.G., CERRINI, S., FEDELI, W., PELLICCIARI, R., SANTURBANO, B., VACIAGO, A.: The structure of fusicoccin A. Experientia **24**, 631–635 (1968a).

BALLIO, A., GRANITI, A., POCCHIARI, F., SILANO, V.: Some effects of "fusicoccin A" on tomato leaf tissues. Life Sci. **7**, 751–760 (1968b).

BARTON, K.A., VERBEEK, R., ELLIS, R., KAHN, A.A.: Abscisic acid inhibition of gibberellic acid and cyclic 3′-5′-adenosine monophosphate induced α-amylase synthesis. Physiol. Plantarum **29**, 186–189 (1973).

BENTRUP, F.W., PFRÜNER, H., WAGNER, G.: Evidence for differential action of indole acetic acid upon ion fluxes in single cells of *Petroselinum sativum*. Planta **110**, 369–372 (1973).

BODE, H.R.: Über den Einfluß der Geoinduktion auf die Kationen-verteilung im Hypocotyl von *Helianthus annuus*. Planta **54**, 15–33 (1959).

BONNER, J.: The relation of hydrogen ions to the growth rate of the *Avena* coleoptile. Protoplasma **21**, 406–423 (1934).

BOZCUK, S.: Effect of growth regulators on the activity of salt glands of *Statice sinuata* and *Statice latifolia*. In: Hormonal regulation in plant growth and development (H. KALDEWEY, Y. VARDAR, eds.), Proc. Adv. Study Inst. Izmir 1971 p. 89–96. Weinheim: Verlag Chemie 1972.

BRAND, J., BASZYNSKI, T., CRANE, F.L., KROGMANN, D.W.: Photosystem I inhibition by polycations. Biochem. Biophys. Res. Commun. **45**, 538–543 (1971).

BRAUNER, L.: Untersuchungen über das geoelektrische Phänomen. Jahrb. Wiss. Botan. **66**, 381–428 (1927).

BRAUNER, L.: Untersuchungen über das geoelektrische Phänomen II. Membranstruktur und geoelektrischer Effekt. Jahrb. Wiss. Botan. **68**, 711–770 (1928).

BRAUNER, L.: Primäreffekte der Schwerkraft bei der geotropischen Reaktion. In: Encyclopedia of plant physiology (W. RUHLAND, ed.), vol. XVII, pt. 2, p. 74–102. Berlin-Heidelberg-New York: Springer 1962.

BRAUNER, L., BRAUNER, M.: Versuche zur Analyse der geotropischen Perzeption. III. Über den Einfluß des Schwerefeldes auf die Dehnbarkeit der Zellwand und den osmotischen Wert des Zellsaftes. Planta **58**, 301–325 (1962).

BREWIN, N.J., NORTHCOTE, D.H.: Variations in the amounts of 3′,5′-cyclic AMP in plant tissues. J. Exptl. Bot. **24**, 881–888 (1973a).

BREWIN, N.J., NORTHCOTE, D.H.: Partial purification of a cyclic AMP phosphodiesterase from soybean callus. Isolation of a non-dialysable inhibitor. Biochim. Biophys. Acta **320**, 104–122 (1973b).

BRIGGS, W.R., RICE, H.V.: Phytochrome: chemical and physical properties and mechanism of action. Ann. Rev. Plant Physiol. **23**, 293–334 (1972).

BROWN, E.G., NEWTON, R.P.: Occurrence of adenosine 3′:5′-cyclic monophosphate in plant tissues. Phytochemistry **12**, 2683–2685 (1973).

BURSTRÖM, H.: Growth regulation by metals and chelates. In: Advances in botanical research (R.D. PRESTON, ed.), vol. 1, p. 73–100. London-New York: Academic Press 1963.

CARR, D.J., ed.: Plant growth substances 1970, p. 1–837. Berlin-Heidelberg-New York: Springer 1972.

CHAIN, E.B., MANTLE, P.G., MILBORROW, B.V.: Further investigations of the toxicity of fusicoccins. Physiol. Plant Pathol. **1**, 495–514 (1971).

CLARK, W.G.: Electrical polarity and auxin transport. Plant Physiol. **13**, 529–552 (1938).

CLELAND, R.: Effect of auxin upon loss of calcium from cell walls. Plant Physiol. **35**, 581–584 (1960).

CLELAND, R.: An auxin-induced hydrogen ion pump in *Avena* coleoptiles. Plant Physiol., Suppl. **51**, 2 (1973).

CLELAND, R.: Fusicoccin as a tool for studying the mechanism of auxin action. Plant Physiol., Suppl. **53**, 43 (1974).

COLLINS, J.C., KERRIGAN, A.P.: Hormonal control of ion movements in the plant root. In: Ion transport in plants (W.P. ANDERSON, edd.), p. 589–593. London-New York: Academic Press 1973.

COLLINS, J.C., KERRIGAN, A.P.: The effect of kinetin and abscisic acid on water and ion transport in isolated maize roots. New Phytologist **73**, 309–314 (1974).

CRAM, W.J., PITMAN, M.G.: The action of abscisic acid on ion uptake and water flow in plant roots. Australian J. Biol. Sci. **25**, 1125–1132 (1972).

CUMMINS, W.R., KENDE, H., RASCHKE, K.: Specificity and reversibility of the rapid stomatal response to abscisic acid. Planta **99**, 347–351 (1971).

DAS, V.S.R., RAGHAVENDRA, A.S.: Role of cyclic photo-phosphorylation in the control of stomatal opening. In: Mechanisms of regulation of plant growth (R.L. BIELESKI, A.R. FERGUSON, M.M. CRESSWELL, eds.), Bulletin 12, p. 455–460. Wellington: Roy. Soc. of New Zealand 1974.

DEDOLPH, R.R., BREEN, J.J., GORDON, S.A.: Geoelectric effect and geotropic curvature. Science **148**, 1100–1101 (1965).

DEGANI, H., SHAVIT, N.: Ion movements in isolated chloroplasts. III. Ionophore-induced ion uptake and its effect on photophosphorylation. Arch. Biochem. Biophys. **152**, 339–346 (1972).

DE LA FUENTE, R.K., LEOPOLD, A.C.: A role for calcium in auxin transport. Plant Physiol. **51**, 845–847 (1973).

DIAMOND, J.M., WRIGHT, E.M.: Biological membranes: the physical basis of ion and non electrolyte selectivity. Ann. Rev. Physiol. **31**, 581–646 (1969).

DOHRMANN, U., JOHNSON, K.D., RAYLE, D.L.: Comparison of auxin and fusicoccin: growth, H^+ secretion, glycosidase activation, and binding to membrane vesicles in *Avena* coleoptiles. Plant Physiol., Suppl. **53**, 43 (1974).

DÖRFFLING, K.: Die Regulation des Internodienwachstums von Erbsenkeimlingen durch Licht und Abscisinsäure. Z. Pflanzenphysiol. **70**, 131–137 (1973).

DREW, M.C., McLAREN, A.D.: The effect of histones and other basic macro-molecules on cell permeability and elongation of barley roots. Physiol. Plantarum **23**, 544–560 (1970).

DUFFUS, C.M., DUFFUS, J.H.: A possible role for cyclic AMP in gibberellic acid triggered release of α-amylase in barley endosperm slices. Experientia **25**, 581 (1969).

Earle, K.M., Galsky, A.G.: The action of cyclic-AMP on GA3 controlled responses. II. Similarities in the induction of barley endosperm ATP-ase activity by gibberellic acid and cyclic-3′,5′-adenosine monophosphate. Plant Cell Physiol. **12**, 727–732 (1971).

Eastwood, D., Laidman, D.L.: The hormonal control of inorganic ion release from wheat aleurone tissue. Phytochemistry **10**, 1459–1467 (1971).

Eilam, Y.: Permeability changes in senescing tissue. J. Exptl. Bot. **16**, 614–627 (1965).

Epstein, E.: The essential role of calcium in selective transport of plant cells. Plant Physiol. **36**, 437–444 (1961).

Eshel, A., Waisel, Y.: Heterogeneity of ion uptake mechanisms along primary roots of maize seedlings. In: Ion transport in plants (W.P. Anderson, ed.), p. 531–537. London-New York: Academic Press 1973.

Etherton, B.: Effect of indole-3-acetic acid on membrane potentials of oat coleoptile cells. Plant Physiol. **45**, 527–528 (1970).

Etherton, B., Dedolph, R.R.: Gravity and intercellular differences in membrane potentials of plant cells. Plant Physiol. **49**, 1019–1020 (1972).

Evans, M.L.: Rapid responses to plant hormones. Ann. Rev. Plant Physiol. **25**, 195–223 (1974).

Evins, W.H., Varner, J.E.: Hormone-controlled synthesis of endoplasmic reticulum in barley aleurone cells. Proc. Natl. Acad. Sci. U.S. **68**, 1631–1633 (1971).

Feng, K.A.: Effects of kinetin on the permeability of *Allium cepa* cells. Plant Physiol. **51**, 868–870 (1973).

Feng, K.A., Unger, J.W.: Influence of kinetin on the membrane permeability of *Allium cepa* cells. Experientia **28**, 1310–1311 (1972).

Fensom, D.S., Wanless, I.R.: Further studies of electro-osmosis in *Nitella* in relation to pores in membranes. J. Exptl. Bot. **18**, 563–577 (1967).

Fischer, R.A.: Role of potassium in stomatal opening in the leaf of *Vicia faba*. Plant Physiol. **47**, 555–558 (1971).

Fischer, R.A.: Aspects of potassium accumulation by stomata of *Vicia faba*. Australian J. Biol. Sci. **25**, 1107–1123 (1972).

Fischer, U., Amrhein, N.: Cyclic nucleotide phosphodiesterase of *Chlamydomonas reinhardtii*. Biochim. Biophys. Acta **341**, 412–420 (1974).

Fisher, M.L., Albersheim, P.: A calcium dependent hydrogen pump in plant plasma membranes. Plant Physiol. Suppl. **51**, 2 (1973).

Fishman, S.N., Chodorov, B.I., Volkenstein, M.V.: Molecular mechanisms of membrane ionic permeability changes. Biochim. Biophys. Acta **225**, 1–10 (1971).

Galsky, A.G., Lippincott, J.A.: Promotion and inhibition of α-amylase production in barley endosperm by cyclic 3′,5′-adenosine monophosphate and adenosine di-phosphate. Plant Cell Physiol. **10**, 607–620 (1969).

Galston, A.W., Davies, P.J.: Hormonal regulation in plants. Science **163**, 1288–1297 (1969).

Gertman, E., Fuch, Y.: Effect of abscisic acid and its interactions with other plant hormones on ethylene productions in two plant systems. Plant Physiol. **50**, 194–195 (1972).

Giannattasio, M.: Content of 3′,5′ cyclic AMP and cyclic AMP phosphodiesterase in dormant and activated tissues of Jerusalem artichoke tubers. Biochem. Biophys. Res. Commun. **57**, 365–371 (1974).

Gilbert, M.L., Galsky, A.G.: The action of cyclic-AMP on GA3 controlled responses. III Characteristics of barley endosperm acid phosphatase induction by gibberellic acid and cyclic 3′,5′-adenosine monophosphate. Plant Cell Physiol. **13**, 867–873 (1972).

Glinka, Z.: Abscisic acid effect on root exudation related to increased permeability to water. Plant Physiol. **51**, 217–219 (1973).

Glinka, Z., Reinhold, L.: Rapid changes in permeability of cell membranes to water brought about by carbon dioxide and oxygen. Plant Physiol. **37**, 481–486 (1962).

Glinka, Z., Reinhold, L.: Abscisic acid raises the permeability of plant cells to water. Plant Physiol. **48**, 103–105 (1971).

Glinka, Z., Reinhold, L.: Induced changes in permeability of plant cell membranes to water. Plant Physiol. **49**, 602–606 (1972).

Goldsmith, M.H.M., Wilkins, M.B.: Movement of auxin in coleoptiles of *Zea mays* L during geotropic stimulation. Plant Physiol. **39**, 151–162 (1964).

Goldthwaite, J.J.: Activity of cyclic and noncyclic nucleotides as senescence inhibitors in *Rumex obtusifolius* leaf tissue. Plant Physiol., Suppl. **53**, 58 (1974).

GOTÔ, N., ESASHI, Y.: Differential hormone responses in different growing zones of the bean hypocotyl. Planta **116**, 225–241 (1974).

GRUNWALD, C.: Effect of sterols on the permeability of alcohol-treated red beet tissue. Plant Physiol. **43**, 484–488 (1968).

HADAČOVÁ, V., LUŠTINEC, J., KAMÍNEK, M.: Kinetin and naphthalene acetic acid and controlled starch formation in isolated roots of *Pisum sativum*. Biol. Plant. Acad. Sci. Bohemoslov. **15**, 427–429 (1973).

HAGER, A., MENZEL, H., KRAUSS, A.: Versuche und Hypothese zur Primärwirkung des Auxins beim Streckungswachstum. Planta **100**, 47–75 (1971).

HALL, J.L.: Pinocytotic vesicles and ion transport in plant cells. Nature **226**, 1253–1254 (1970).

HANSON, J.B.: Ion transport induced by polycations and its relationship to loose coupling of corn mitochondria. Plant Physiol. **49**, 707–715 (1972).

HASCHKE, H.-P., LÜTTGE, U.: β-Indolylessigsäure (-IES) — abhängiger K^+-H^+-Austauschmechanismus und Streckungswachstum bei *Avena* — Koleoptilen. Z. Naturforsch. **28C**, 555–558 (1973).

HAUPT, W.: Die Orientierung der Phytochrom-Moleküle in der *Mougeotia*zelle: Ein neues Modell zur Deutung der experimentellen Befunde. Z. Pflanzenphysiol. **58**, 331–346 (1968).

HAUPT, W.: Über den Diochroismus von Phytochrom $_{660}$ und Phytochrom$_{730}$ bei *Mougeotia*. Z. Pflanzenphysiol. **62**, 287–298 (1970a).

HAUPT, W.: Localization of phytochrom in the cell. Physiol. Vég. **8**, 551–563 (1970b).

HAUPT, W., MÖRTEL, G., WINKELNKEMPER, I.: Demonstration of different dichroic orientation of phytochrome P_R and P_{FR}. Planta **88**, 183–186 (1969).

HAYDON, D.A., HLADKY, S.B.: Ion transport across thin lipid membranes: a critical discussion of mechanisms in selected systems. Quart. Rev. Biophysics **5**, 187–282 (1972).

HEATH, O.V.S., CLARK, J.E.: Chelating agents as plant growth substances. A possible clue to the mode of action of auxin. Nature **177**, 1118–1121 (1956).

HEATH, O.V.S., CLARK, J.E.: Chelation in auxin action. I. A study of the interactions of 3-indolyl acetic acid and synthetic chelating agents as affecting the growth of wheat roots and coleoptile sections. J. Exptl. Bot. **11**, 167–187 (1960).

HENDRIX, D.L., HIGINBOTHAM, N.: Effects of filipin and cholesterol on K^+ movement in etiolated stem cells of *Pisum sativum* L. Plant Physiol. **52**, 93–97 (1973).

HEYN, A.N.J.: Observations on the exocytosis of secretory vesicles and their products in coleoptiles of *Avena*. J. Ultrastruct. Res. **37**, 69–81 (1971).

HIGINBOTHAM, N.: Electropotential of plant cells. Ann. Rev. Plant Physiol. **24**, 25–46 (1973).

HIGINBOTHAM, N., DAVIS, R.F., MERTZ, S.M., SHUMWAY, L.K.: Some evidence that radial transport in maize roots is into living vessels. In: Ion transport in plants (W.P. ANDERSON, ed.), p. 493–506. London-New York: Academic Press 1973.

HIGINBOTHAM, N., LATIMER, H., EPPLEY, R.: Stimulation of rubidium absorption by auxins. Science **118**, 243–245 (1953).

HIGINBOTHAM, N., PRATT, M.J., FOSTER, R.J.: Effects of calcium, indole-acetic acid, and distance from stem apex on potassium and rubidium absorption by excised segments of etiolated pea epicotyl. Plant Physiol. **37**, 203–214 (1962).

HILL, A.E.: A theory of hydro-osmotic hormone action. J. Theoret. Biol. **36**, 271–281 (1972).

HILLE, B.: The permeability of the sodium channel to metal cations in myelinated nerve. J. Gen. Physiol. **59**, 637–658 (1972).

HIRON, R.W.P., WRIGHT, S.T.C.: The role of endogenous abscisic acid in the response of plants to stress. J. Exptl. Bot. **24**, 769–781 (1973).

HODGES, T.K., DARDING, R.L., WEIDNER, T.: Gramicidin-D stimulated influx of monovalent cations into plant roots. Planta **97**, 245–256 (1971).

HORTON, R.F.: Stomatal opening: the role of abscisic acid. Canad. J. Bot. **49**, 583–585 (1971).

HORTON, R.F., BRUCE, K.R.: Inhibition by abscisic acid of the light and dark uptake of potassium by slices of *Vicia faba* leaves. Canad. J. Bot. **50**, 1915–1917 (1972).

HORTON, R.F., MORAN, L.: Abscisic acid inhibition of K^+ influx into stomatal guard cells. Z. Pflanzenphysiol. **66**, 193–196 (1972).

HOURMANT, A., PENOT, M.: Etude comparée de l'absorption du ^{86}Rb par différents organes végétaux en présence de calcium. Compt. Rend. **273^D**, 2518–2521 (1971).

HOURMANT, A., PENOT, M.: Action de la benzyladenine sur l'absorption du rubidium par des disques de tubercules de pomme de terre. Compt. Rend. **276^D**, 323–326 (1973a).

HOURMANT, A., PENOT, M.: Action de la benzyladenine sur l'absorption du molybdate par des disques de tubercules de pomme de terre. Compt. Rend. **277^D**, 297–300 (1973b).

Humble, G.D., Hsiao, T.C.: Light-dependent influx and efflux of potassium of guard cells during stomatal opening and closing. Plant Physiol. **46**, 483–487 (1970).

Ilan, I.: A specific stimulatory action of indolyl-3-acetic acid on potassium uptake by plant cells, with concomitant inhibition of ammonium uptake. Nature **194**, 203–204 (1962).

Ilan, I.: Evidence for hormonal regulation of the selectivity of ion uptake by plant cells. Physiol. Plantarum **25**, 230–233 (1971).

Ilan, I.: An auxin-induced pH drop and on the improbability of its involvement in the primary mechanism of auxin-induced growth promotion. Physiol. Plantarum **28**, 146–148 (1973).

Ilan, I., Gilad, T., Reinhold, L.: Specific effects of kinetin on the uptake of monovalent cations by sunflower cotyledons. Physiol. Plantarum **24**, 337–341 (1971).

Ilan, I., Reinhold, L.: Analysis of the effects of indole-3-acetic acid on the uptake of monovalent cations. Physiol. Plantarum **16**, 596–603 (1963).

Ilan, I., Reinhold, L.: Reversal by sucrose of the effects of indolyl-3-acetic acid on cation uptake by plant cells. Nature **201**, 726 (1964).

Imber, D., Tal, M.: Phenotypic reversion of flacca, a wilty mutant of tomato by abscisic acid. Science **169**, 592–593 (1970).

Itai, C., Vaadia, Y.: Kinetin-like activity in root exudate of water stressed sunflower plants Physiol. Plantarum **18**, 941–944 (1965).

Jacoby, B., Dagan, J.: Effects of ^{6}N-benzyladenine on primary leaves of intact bean plants and on their sodium absorption capacity. Physiol. Plantarum **23**, 397–403 (1970).

Jaffe, M.J.: Phytochrome-mediated bioelectric potentials in mung bean seedlings. Science **162**, 1061–1067 (1968).

Jaffe, M.J., Galston, A.W.: Phytochrome control of rapid nyctinastic movements and membrane permeability in *Albizzia julibrissin*. Planta **77**, 135–141 (1967).

Janistyn, B.: Indol-3-essigsäure-induzierte Nukleotid-abgabe bei gleichzeitig erhöhter Adenosin-3′:5′-monophosphorsäure (C-AMP)-Synthese in Maiskoleoptilzylindern. Z. Naturforsch. **27b**, 273–276 (1972).

Janistyn, B., Drumm, H.: Light-mediated changes in concentration of C-AMP in mustard seedlings. Naturwissenschaften **59**, 218 (1972).

Jenkinson, I.S.: Bioelectric oscillations of bean roots: further evidence for a feedback oscillator. II Intracellular plant root potentials. Australian J. Biol. Sci. **15**, 101–114 (1962a).

Jenkinson, I.S.: Bioelectric oscillations of bean roots: further evidence for a feedback oscillator. III Excitation and inhibition of oscillations by osmotic pressure, auxins, and antiauxins. Australian J. Biol. Sci. **15**, 115–125 (1962b).

Jenkinson, I.S., Scott, B.I.H.: Bioelectric oscillations of bean roots: further evidence for a feedback oscillator. Australian J. Biol. Sci. **14**, 231–243 (1961).

Jones, R.L.: Gibberellins: their physiological role. Ann. Rev. Plant Physiol. **24**, 571–598 (1973a).

Jones, R.L.: Gibberellic acid and ion release from barley aleurone tissue. Evidence for hormone-dependant ion transport capacity. Plant Physiol. **52**, 303–308 (1973b).

Jost, J.-P., Rickenberg, H.V.: Cyclic AMP. Ann. Rev. Biochem. **40**, 741–774 (1971).

Kaldewey, H., Vardar, Y., eds.: Hormonal regulation in plant growth and development. Proceedings of the Advanced Study Institute, Izmir 1971, p. 1–523. Weinheim: Verlag Chemie 1972.

Kamisaka, S.: Auxin-induced growth of tuber tissue of jerusalem artichokes VII Effect of cyclic 3′,5′-adenosine monophosphate on the auxin-induced cell expansion growth. In: Plant growth substances, 1970 (D.J. Carr, ed.), p. 654–660. Berlin-Heidelberg-New York: Springer 1972.

Kaufman, P.B., Petering, L.B., Yocum, C.S., Baic, D.: Ultrastructural studies on stomata development in internodes of *Avena sativa*. Amer. J. Bot. **57**, 33–49 (1970).

Keck, R.W., Boyer, J.S.: Chloroplast response to low leaf water potentials. III Differing inhibition of electron transport and phosphorylation. Plant Physiol. **53**, 474–479 (1974).

Kendrick, R.E., Hillman, W.S.: Ion relations, chlorophyll synthesis and the question of "bulk" phytochrome in *Pisum sativum*. Physiol. Plantarum **26**, 7–12 (1972).

Kessler, B., Levinstein, R.: Adenosine 3′,5′-cyclic monophosphate in higher plants: assay, distribution and age-dependency. Biochim. Biophys. Acta **343**, 156–166 (1974).

Köhler, D.: Phytochromabhängiger Ionentransport in Erbsensämlingen. Planta **84**, 158–165 (1969).

Köhler, D.: Über den Zusammenhang zwischen Achsen- und Blattwachstum und Ionenaufnahme bei Keimlingen von *Pisum sativum*. Z. Pflanzenphysiol. **63**, 185–193 (1970).

KÖHLER, D., WILLERT, K. VON, LÜTTGE, U.: Phytochromabhängige Veränderungen des Wachstums und der Ionenaufnahme etiolierter Erbsenkeimlinge. Planta **83**, 35–48 (1968).

KUHL, U., UNGER, M.: Wirkungen von Abscisinsäure auf den Kohlenhydrat- und Fettsäurehaushalt von *Coleus blumei*. Z. Pflanzenphysiol. **72**, 135–140 (1974).

LADO, P., PENNACHIONI, A., CADOGNO, F.R., RUSSI, S., SILANO, V.: Comparison between some effects of fusicoccin and indole-3-acetic acid on cell enlargement in various plant materials. Physiol. Plant Pathol. **2**, 75–85 (1972).

LADO, P., RASI-CALDOGNO, F., PENNAECHIONI, A., MARRÈ, E.: Mechanism of the growth promoting action of fusicoccin. Interaction with auxin, and effects of inhibitors or respiration and protein synthesis. Planta **110**, 311–320 (1973).

LÄUCHLI, A., KRAMER, D., PITMAN, M.G., LÜTTGE, U.: Ultrastructure of xylem parenchyma cells of barley roots in relation to ion transport to the xylem. Planta **119**, 85–99 (1974).

LÄUCHLI, A., LÜTTGE, U., PITMAN, M.G.: Ion uptake and transport through barley seedlings: differential effect of cycloheximide. Z. Naturforsch. **28 C**, 431–434 (1973).

LÄUGER, P.: Ion transport through pores: a rate-theory analysis. Biochim. Biophys. Acta **311**, 423–441 (1973).

LEGGETT, J.E., STOLZY, L.H.: Anaerobiosis and sodium accumulation. Nature **192**, 991–992 (1961).

LE JOHN, H.B., CAMERON, L.E.: Cytokinins regulate calcium binding to a glycoprotein from fungal cells. Biochem. Biophys. Res. Commun. **54**, 1053–1060 (1973).

LE JOHN, H.B., STEVENSON, R.M.: Cytokinins and magnesium ions may control the flow of metabolites and calcium ions through fungal cell membranes. Biochem. Biophys. Res. Commun. **54**, 1061–1066 (1973).

LIN, P.P.C.: Cyclic nucleotides in higher plants? In: Adv. Cycl. Nucleotide Res. (P. GREENGARD, G.A. ROBISON, eds.), vol. 4, p. 439–461. New York: Raven Press 1974.

LIVNÈ, A., VAADIA, Y.: Stimulation of transpiration rate in barley leaves by kinetin and gibberellic acid. Physiol. Plantarum **18**, 658–664 (1965).

LIVNÈ, A., VAADIA, Y.: Water deficits and hormone relations. In: Water deficits and plant growth (T.T. KOZLOWSKI, ed.), vol. 3, p. 255–275. New York-London: Academic Press 1972.

LOVEYS, B.R., KRIEDEMANN, P.E., TÖRÖKFALVY, E.: Is abscisic acid involved in stomatal response to carbon dioxide? Plant Sci. Letters **1**, 335–338 (1973).

LÜTTGE, U., BAUER, K., KÖHLER, D.: Frühwirkungen von Gibberellinsäure auf Membrantransporte in jungen Erbsenpflanzen. Biochim. Biophys. Acta **150**, 452–459 (1968).

LÜTTGE, U., HIGINBOTHAM, N., PALLAGHY, C.K.: Electrochemical evidence of specific action of indole acetic acid on membranes in *Mnium* leaves. Z. Naturforsch. **27** b, 1239–1242 (1972).

LÜTTGE, U., KRAPF, G.: Die Ultrastruktur der Blattzellen junger und alter *Mnium*-Sprosse und ihr Zusammenhang mit der Ionenaufnahme. Planta **81**, 132–139 (1968).

LÜTTGE, U., LÄUCHLI, A., BALL, E., PITMAN, M.G.: Cycloheximide: a specific inhibitor of proteinsynthesis and intercellular ion transport in plant roots. Experientia **30**, 470–471 (1974).

LÜTTGE, U., PALLAGHY, C.K., WILLERT, K. VON: Microautoradiographic investigations of sulfate uptake by glands and epidermal cells of water lily (*Nymphaea*) leaves with special reference to the effect of poly-1-lysine. J. Membrane Biol. **4**, 395–407 (1971).

MACKLON, A.E.S., HIGINBOTHAM, N.: Potassium and nitrate uptake and cell transmembrane electro-potential in excised pea epicotyls. Plant Physiol. **43**, 888–892 (1968).

MANSFIELD, T.A., JONES, R.J.: Effects of abscisic acid on potassium uptake and starch content of stomatal guard cells. Planta **101**, 147–158 (1971).

MARRÈ, E., COLOMBO, R., LADO, P., RASI-CALDOGNO, F.: Correlation between proton extrusion and stimulation of cell enlargement. Effects of fusicoccin and of cytokinins on leaf fragments and isolated cotyledons. Plant Sci. Letters **2**, 139–150 (1974a).

MARRÈ, E., LADO, P., FERRONI, A., BALLARIN DENTI, A.: Trans-membrane potential increase induced by auxin, benzyladenine and fusicoccin. Correlation with proton extrusion and cell enlargement. Plant Sci. Letters **2**, 257–265 (1974b).

MARRÈ, E., LADO, P., RASI-CALDOGNO, F., COLOMBO, R., DE MICHELIS, M.I.: Evidence for the coupling of proton extrusion to K$^+$ uptake in pea internode segments treated with fusicoccin or auxin. Plant Sci. Letters **3**, 365–379 (1974c).

MEHARD, C.W., LYONS, J.M., KUMAMOTO, J.: Utilization of model membranes in a test for the mechanism of ethylene action. J. Membrane Biol. **3**, 173–179 (1970).

Mertz, S.N., Arntzen, C.J.: Fungal antiplant toxins. Communicated at International Workshop on Membrane Transport in Plants and Plant Organelles, Jülich, February 4th–8th, 1974.

Milborrow, B.V.: The chemistry and physiology of abscisic acid. Ann. Rev. Plant Physiol. 25, 259–307 (1974).

Miller, R.J., Bell, D.T., Koeppe, D.E.: The effects of water stress on some membrane characteristics of corn mitochondria. Plant Physiol. 48, 229–231 (1971).

Miller, R.J., Koeppe, D.E., Comeau, G., Malone, C.P.: Water stress induced changes in mitochondrial membranes. Plant Physiol., Suppl. 51, 22 (1973).

Mittelheuser, C.J., van Steveninck, R.F.M.: Stomatal closure and inhibition of transpiration by (rs)-abscisic acid. Nature 221, 281–282 (1969).

Mittelheuser, C.J., van Steveninck, R.F.M.: The ultrastructure of wheat leaves. II The effects of kinetin and ABA on detached leaves incubated in the light. Protoplasma 73, 253–262 (1971a).

Mittelheuser, C.J., Steveninck, R.F.M. van: Rapid action of abscisic acid on photosynthesis and stomatal resistance. Planta 97, 83–86 (1971b).

Mizrahi, Y., Blumenfeld, A., Richmond, A.E.: The role of abscisic acid and salination in the adaptive response of plants to reduced root aeration. Plant Cell Physiol. 13, 15–21 (1972).

Moore, C., Pressman, B.C.: Mechanism of action of valinomycin. Biochem. Biophys. Res. Commun. 15, 562–567 (1964).

Morré, D.J., Bracker, C.E.: Conformational alteration of soybean plasma membranes induced by auxin and calcium ions. Plant Physiol., Suppl. 53, 442 (1974).

Mudd, J.B., Kleinschmidt, M.G.: Effect of filipin on the permeability of red beet and potato tuber discs. Plant Physiol. 45, 517–518 (1970).

Neirinckx, L.J.A.: Influence de quelques substances de croissance sur l'absorption du sulfate par le tissu radiculaire de beterave rouge (Beta vulgaris L. ssp vulgaris var rubra L.). Ann. Physiol. Vég. Bruxelles 13, 83–108 (1968).

Nickells, M.W., Schaefer, G.M., Galsky, A.G.: The action of cyclic AMP on GA3 controlled responses. I Induction of barley endosperm protease and acid phosphatase activity by cyclic-3',5'-adenosine monophosphate. Plant Cell Physiol. 12, 717–725 (1971).

Nicolas, P., Nigon, U.: L'adénosine -3',5'-monophosphate cyclique et sa proteine fixatrice chez Euglena gracilis. Compt. Rend. 277 D, 1641–1644 (1973).

Osborne, D.J., Jackson, M.B., Milborrow, B.V.: Physiological properties of abscission accelerator from senescent leaves. Nature New Biol. 240, 98–101 (1972).

Osmond, C.B., Laties, G.G.: Effect of poly-L-lysine on potassium fluxes in red beet tissue. J. Membrane Biol. 2, 85–94 (1970).

Paleg, L.G., Wood, A., Spotswood, T.M.: A possible mechanism of gibberellic acid action. Australian Soc. Plant Physiol., Abstracts 14th meeting, p. 52 (1973).

Pallaghy, C.K., Raschke, K.: No stomatal response to ethylene. Plant Physiol. 49, 275–276 (1972).

Pallas, J.E., Mollenhauer, H.H.: Electron microscopic evidence for plasmodesmata in dicotyledonous guard cells. Science 175, 1275–1276 (1972).

Pallas, J.E., Wright, B.G.: Organic acid changes in the epidermis of Vicia faba and their implication in stomatal movement. Plant Physiol. 51, 588–590 (1973).

Palmer, J.M.: The influence of growth regulating substances on the development of enhanced metabolic rates in thin slices of beetroot storage tissue. Plant Physiol. 41, 1173–1178 (1966).

Pickles, V.R., Sutcliffe, J.F.: The effects of 5-hydroxytryptamine, indole-3-acetic acid and some other substances on pigment effusion, sodium uptake, and potassium efflux by slices of red beetroot, in vitro. Biochim. Biophys. Acta 17, 244–251 (1955).

Pitman, M.G.: Adaptation of barley roots to low oxygen supply and its relation to potassium and sodium uptake. Plant Physiol. 44, 1233–1240 (1969).

Pitman, M.G.: Uptake and transport of ions in barley seedlings. III. Correlation between transport to the shoot and relative growth rate. Australian J. Biol. Sci. 25, 905–919 (1972).

Pitman, M.G., Cram, W.J.: Regulation of inorganic ion transport in plants In: Ion transport in plants (W.P. Anderson, ed.), p. 465–481. New York-London: Academic Press 1973.

Pitman, M.G., Lüttge, U., Läuchli, A., Ball, E.: Action of abscisic acid on ion transport as affected by root temperature and nutrient status. J. Exptl. Bot. 25, 147–155 (1974).

Pitman, M.G., Mowat, J., Nair, H.: Interactions of processes for accumulation of salt and sugar in barley plants. Australian J. Biol. Sci. 24, 619–631 (1971).

PITMAN, M.G., SCHAEFER, W., WILDES, R.A.: Effect of abscisic acid on fluxes of ions in barley roots. In: Membrane transport in plants (U. ZIMMERMANN, J. DAINTY, eds.), p. 391–396. Berlin-Heidelberg-New York: Springer 1974b.

POHL, R.: Ein Beitrag zur Analyse des Streckungswachstums der Pflanzen. Planta **36**, 230–261 (1948).

POHL, R.: Die Wirkung der Wuchsstoffe auf das Plasma. In: Encyclopedia of plant physiology (W. RUHLAND, ed.), vol. XIV, p. 729–742. Berlin-Heidelberg-New York: Springer 1961a.

POHL, R.: Wuchsstoffe und Wasseraufnahme. In: Encyclopedia of plant physiology (W. RUHLAND, ed.), vol. XIV, p. 743–753. Berlin-Heidelberg-New York: Springer 1961b.

POLLARD, C.J.: Influence of gibberellic acid on the incorporation of 8-^{14}C adenine into adenosine,-$3',5'$-cyclic phosphate in barley aleurone layers. Biochim. Biophys. Acta **201**, 511–512 (1970).

ROLLARD, C.J.: Rapid gibberellin responses and the action of adenosine $3',5'$-monophosphate in aleurone layers. Biochim. Biophys. Acta **252**, 553–560 (1971).

POWER, J.B., COCKING, E.C.: Isolation of leaf protoplasts: macromolecule uptake and growth substance response. J. Exptl. Bot. **21**, 64–70 (1970).

PRATT, H.K., GOESCHL, J.D.: Physiological roles of ethylene in plants. Ann. Rev. Plant Physiol. **20**, 541–584 (1969).

PRESSMAN, B.C.: Ionophorous antibiotics as models for biological transport. Federation Proc. **27**, 1283–1288 (1968).

PRESSMAN, B.C.: Properties of ionophores with broad range cation selectivity. Federation Proc. **32**, 1698–1703 (1973).

PRESSMAN, B.C., HARRIS, E.J., JAGGER, W.S., JOHNSON, J.H.: Antibiotic-mediated transport of alkali ions across lipid barriers. Proc. Natl. Acad. Sci. U.S. **58**, 1949–1956 (1967).

PRINGLE, R.B., BRAUN, A.C.: The isolation of the toxin of *Helminthosporium victoriae*. Phytopathology **47**, 369–371 (1957).

QUAIL, P.H.: Particle-bound phytochrome: spectral properties of bound and unbound fractions. Planta **118**, 345–355 (1974a).

QUAIL, P.H.: *In-vitro* binding of phytochrome to a particulate fraction: a function of light dose and steady-state P_{fr} level. Planta **118**, 357–360 (1974b).

QUAIL, P.H., MARMÉ, D., SCHÄFER, E.: Particle-bound phytochrome from maize and pumpkin. Nature New Biol. **245**, 189–191 (1973).

QUAIL, P.H., SCHÄFER, E.: Particle-bound phytochrome: a function of light dose and steady-state level of the far-red-absorbing form. J. Membrane Biol. **15**, 393–404 (1974).

RACUSEN, R.H.: Membrane protein conformational changes as a mechanism for the phytochrome-induced fixed charge reversal in root cap cells of mung bean. Plant Physiol., Suppl. **51**, 51 (1973).

RACUSEN, R.H., ETHERTON, B.: The role of root cap cell fixed charges in phytochrome mediated mung bean root tip adherence phenomena. Plant Physiol., Suppl. **51**, 51 (1973).

RAINS, D.W.: Sodium and potassium absorption by bean stem tissue. Plant Physiol. **44**, 547–554 (1969).

RASCHKE, K., FELLOWS, M.P.: Stomatal movement in *Zea Mays*: Shuttle of potassium and chloride between guard cells and subsidiary cells. Planta **101**, 296–316 (1971).

RASCHKE, K., HUMBLE, G.D.: No uptake of anions required by opening stomata of *Vicia faba*: guard cells release hydrogen ions. Planta **115**, 47–57 (1973).

RAYLE, D.L.: Auxin-induced hydrogen-ion secretion in *Avena* coleoptiles and its implications. Planta **114**, 63–73 (1973).

RAYLE, D.L., JOHNSON, K.D.: Direct evidence that auxin-induced growth is related to hydrogen ion secretion. Plant Physiol., Suppl. **51**, 2 (1973).

RAYMOND, P., NARAYANAN, A., PRADET, A.: Evidence for the presence of $3',5'$-cyclic AMP in plant tissues. Biochem. Biophys. Res. Commun. **53**, 1115–1121 (1973).

REED, N.R., BONNER, B.A.: The effect of abscisic acid on the uptake of potassium and chloride into *Avena* coleoptile sections. Planta **116**, 173–185 (1974).

REED, P.W., LARDY, H.A.: A23187: a divalent cation ionophore. J. Biol. Chem. **247**, 6970–6977 (1972).

REINHOLD, L.: Release of ammonia by plant tissues treated with indole-3-acetic acid. Nature **182**, 1022 (1958).

REINHOLD, L., GANOT, D.: Asymmetric "acid growth" response following gravistimulus. In: Plant growth substances, 1970 (D.J. CARR, ed.), p. 725–730. Berlin-Heidelberg-New York: Springer 1972.

Reinhold, L., Glinka, Z.: Reduction in turgor pressure as a result of extremely brief exposure to CO_2. Plant Physiol. **41**, 39–44 (1966).

Reinhold, L., Sachs, T., Vislovska, L.: The role of auxin in thigmotropism In: Plant growth substances, 1970 (D.J. Carr, ed.), p. 731–737. Berlin-Heidelberg-New York: Springer 1972.

Rivenbark, W.L., Hanson, J.B.: The uncoupling of oxidative phosphorylation by basic proteins and its reversal with potassium. Biochem. Biophys. Res. Commun. **7**, 318–321 (1962).

Robison, G.A., Butcher, R.W., Sutherland, E.W.: Cyclic AMP. New York-London: Academic Press 1971.

Rubery, P.H., Sheldrake, A.R.: Effect of pH and surface charge on cell uptake of auxin. Nature New Biol. **244**, 285–288 (1973).

Rubinstein, B.: Auxin stimulated ion uptake into coleoptile sections. Plant Physiol., Suppl. **51**, 3 (1973).

Rubinstein, B., Light, E.N.: Indole acetic acid-enhanced chloride uptake into coleoptile cells. Planta **110**, 43–56 (1973).

Sacher, J.A.: Relationship between auxin and membrane integrity in tissue senescence and abscission. Science **125**, 1199–1200 (1957).

Sacher, J.A.: Relations between changes in membrane permeability and the climacteric in banana and avocado. Nature **195**, 577–578 (1962).

Sacher, J.A., Salminen, S.O.: Comparative studies of effect of auxin and ethylene on permeability and synthesis of RNA and protein. Plant Physiol. **44**, 1371–1377 (1969).

Santarius, K.A.: Das Verhalten von CO_2-Assimilation, NADP- und PGS-Reduktion und ATP-Synthese intakter Blattzellen in Abhängigkeit vom Wassergehalt. Planta **73**, 228–242 (1967).

Santarius, K.A., Ernst, R.: Das Verhalten von Hill-Reaktion und Photophosphorylierung isolierter Chloroplasten in Abhängigkeit vom Wassergehalt. I. Wasserentzug mittels konzentrierter Lösungen. Planta **73**, 91–108 (1967).

Satter, R.L., Galston, A.W.: Potassium flux: a common feature of *Albizzia* leaflet movement controlled by phytochrome or endogenous rhythm. Science **174**, 518–520 (1971a).

Satter, R.L., Galston, A.W.: Phytochrome-controlled nyctinasty in *Albizzia julibrissin*. III Interaction between an endogenous rhythm and phytochrome in control of potassium flux and leaflet movement. Plant Physiol. **48**, 740–746 (1971b).

Satter, R.L., Marinoff, P., Galston, A.W.: Phytochrome controlled nyctonasty in *Albizzia julibrissin*. II Potassium fluxes as a basis for leaflet movement. Amer. J. Bot. **57**, 916–926 (1970).

Schaefer, N., Wildes, R.A., Pitman, M.G.: Inhibition by p-fluorophenylalanine of protein synthesis and transport of ions across the roots of barley seedlings. Australian J. Plant Physiol. **2**, 61–73 (1975).

Scheffer, R.P., Pringle, R.B.: Respiratory effects of the selective toxin of *Helminthosporium victoriae*. Phytopathology **53**, 465–468 (1963).

Scott, B.I.H.: Electric oscillations generated by plant roots and a possible feedback mechanism responsible for them. Australian J. Biol. Sci. **10**, 164–179 (1957).

Scott, B.I.H., Martin, D.W.: Bioelectric fields of bean roots and their relation to salt accumulation. Australian J. Biol. Sci. **15**, 83–100 (1962).

Scott, P.C., Leopold, A.C.: Opposing effects of gibberellin and ethylene. Plant Physiol. **42**, 1021–1022 (1967).

Sears, D.F., Eisenberg, R.M.: A model representing a physiological role of CO_2 at the cell membrane. J. Gen. Physiol. **44**, 869–887 (1961).

Siegel, S.M.: Further studies on regulation of betacyanin efflux from beetroot tissue: Ca-ion reversible effects of hydrochloric acid and ammonia water. Physiol. Plantarum **23**, 251–257 (1970).

Siegel, S.M., Daly, O.: Regulation of betacyanin efflux from beetroot by poly-L-lysine, Ca-ion and other substances. Plant Physiol. **41**, 1429–1438 (1966).

Singh, T.N., Paleg, L.G., Aspinall, D.: Stress metabolism. I Nitrogen metabolism and growth in the barley plant during water stress. Australian J. Biol. Sci. **26**, 45–56 (1973).

Skoog, F., Armstrong, D.J.: Cytokinins. Ann. Rev. Plant Physiol. **21**, 359–384 (1970).

Slayman, C.L., van Etten, H.: Are certain pterocarpanoid phytoalexins and steroid hormones inhibitors of membrane ATP-ase? Plant Physiol., Suppl. **53**, 24 (1974).

Solomon, D., Mascarenkas, J.P.: Auxin-induced synthesis of cyclic 3′,5′-adenosine monophosphate in *Avena* coleoptiles. Life Sci. **10**, 879–885 (1971).

SOLOMOS, T., LATIES, G.G.: Diversion of respiratory electrons by ethylene to the cyanide-resistant path. Plant Physiol., Suppl. **53**, 72 (1974).

SPEAR, D.G., BARR, J.K., BARR, C.E.: Localization of hydrogen ion and chloride ion fluxes in *Nitella*. J. Gen. Physiol. **54**, 397–414 (1969).

SQUIRE, G.R., MANSFIELD, T.A.: Studies of the mechanism of action of fusicoccin, the fungal toxin that induces wilting, and its interaction with abscisic acid. Planta **105**, 71–78 (1972).

SQUIRE, G.R., MANSFIELD, T.A.: The action of fusicoccin on stomatal guard cells and subsidiary cells. New Phytologist **73**, 433–440 (1974).

STEVENINCK, R.F.M. VAN: The lag phase in ion uptake by plant tissues. PhD Thesis, University of London (1961).

STEVENINCK, R.F.M. VAN: Effect of indolyl-3-acetic acid on the permeability of membranes in storage tissue. Nature **205**, 83–84 (1965).

STEVENINCK, R.F.M. VAN: Further aspects of the tris effect in beetroot tissue during its lag phase. Australian J. Biol. Sic. **19**, 283–290 (1966).

STEVENINCK, R.F.M. VAN: Inhibition of the development of a cation accumulating system and of tris-induced uptake in storage tissues by N^6-benzyladenine and kinetin. Physiol. Plantarum **27**, 43–47 (1972a).

STEVENINCK, R.F.M. VAN: Abscisic acid stimulation of ion transport and alteration in K^+/Na^+ selectivity. Z. Pflanzenphysiol. **67**, 282–286 (1972b).

STEVENINCK, R.F.M. VAN: Hormonal regulation of ion transport in parenchyma tissue. In: Membrane transport in plants (U. ZIMMERMANN, J. DAINTY, eds.), p. 450–456. Berlin-Heidelberg-New York: Springer 1974.

STEVENINCK, R.F.M. VAN: The "washing" or "aging" phenomenon in plant tissues. Ann. Rev. Plant Physiol. **26**, 237–258 (1975).

STEVENINCK, R.F.M. VAN, CHENOWETH, A.R.F.: Ultrastructural localization of ions. I. Effect of high external sodium chloride concentration on the apparent distribution of chloride in leaf parenchyma cells of barley seedlings. Australian J. Biol. Sci. **25**, 499–516 (1972).

STEVENINCK, R.F.M. VAN, STEVENINCK, M.E. VAN: Effect of inhibitors of protein and nucleic acid synthesis on the development of ion uptake mechanisms in beetroot slices (*Beta vulgaris*). Physiol. Plantarum **27**, 407–411 (1972).

STRUGGER, S.: Die Beeinflussung des Wachstums und des Geotropismus durch die Wasserstoffionen. Ber. Deut. Bot. Ges. (Anhang) **50**, 77–92 (1932).

TAGAWA, T., BONNER, J.: Mechanical properties of the *Avena* coleoptile as related to auxin and to ionic interactions. Plant Physiol. **32**, 207–212 (1957).

TAL, M., IMBER, D.: Abnormal stomatal behavior and hormonal imbalance in flacca, a wilty mutant of tomato. II Auxin and abscisic acid-like activity. Plant Physiol. **46**, 373–376 (1970).

TAL, M., IMBER, D.: Abnormal stomatal behavior and hormonal imbalance in flacca, a wilty mutant of tomato. III Hormonal effects on the water status in the plant. Plant Physiol. **47**, 849–850 (1971).

TANADA, T.: Localization and mechanism of calcium stimulation of rubidium absorption in the mung bean root. Amer. J. Bot. **49**, 1068–1072 (1963).

TANADA, T.: A rapid photo reversible response of barley root tips in the presence of 3-indole acetic acid. Proc. Natl. Acad. Sci. U.S. **59**, 376–380 (1968).

TANADA, T.: Indole-acetic acid and abscisic acid antagonism. I On the phytochrome-mediated attachment of mung bean root tips on glass. Plant Physiol. **51**, 150–153 (1973).

TASSERON-DE JONG, J.G., VELDSTRA, H.: Investigations on cytokinins. I. Effect of 6-benzylaminopurine on growth and starch content of *Lemna minor*. Physiol. Plantarum **24**, 235–238 (1971).

THIMANN, K.V.: The natural plant hormones. In: Plant physiology. A treatise (F.C. STEWARD, ed.), vol. VI B, p. 3–332. New York-London: Academic Press 1972.

TILLBERG, J.-E.: Effects of the inhibitor β-complex on photosynthetic activities in chloroplasts. Z. Pflanzenphysiol. **59**, 305–308 (1968).

TIMBERLAKE, W.E., GRIFFIN, D.H.: Direct inhibition of the uptake of proline by cycloheximide. Biochem. Biophys. Res. Commun. **54**, 216–221 (1973).

TULI, V., DILLEY, D.R., WITTWER, S.H.: N^6-benzyladenine: inhibitor of respiratory kinases. Science **146**, 1477–1479 (1964).

TURNER, N.C.: K^+ uptake of guard cells stimulated by fusicoccin. Nature **235**, 341–342 (1972).

TURNER, N.C., GRANITI, A.: Fusicoccin: a fungal toxin that opens stomata. Nature **223**, 1070–1071 (1969).

TYREE, M.T.: The symplast concept. A general theory of symplastic transport according to the thermodynamics of irreversible processes. J. Theoret. Biol. **26**, 181–214 (1970).

URL, W.G.: The site of penetration resistance to water in plant protoplasts Protoplasma **72**, 427–447 (1971).

VAADIA, Y., ITAI, C.: Interrelationships of growth with reference to the distribution of growth substances. In: Root growth (W.J. WHITTINGTON, ed.), p. 65–79. London: Butterworths 1969.

VALDOVINOS, J.G., ERNEST, L.C., HENRY, E.W.: Effect of ethylene and gibberellic acid on auxin synthesis in plant tissues. Plant Physiol. **42**, 1803–1806 (1967).

WAISEL, J.: The effect of Ca on the uptake of monovalent ions by excised barley roots. Physiol. Plantarum **15**, 709–724 (1962).

WANLESS, I.R., BRYNIAK, W., FENSOM, D.S.: The effect of some growth regulating compounds upon electro-osmotic measurements, transcellular water flow, and Na^+, K^+ and Cl^- influx in *Nitella flexilis*. Canad. J. Bot. **51**, 1055–1070 (1973).

WEIGL, J.: Efflux und Transport von Cl^- und Rb^+ in Maiswurzeln. Wirkung von Außenkonzentration, Ca^{++}, EDTA und IES. Planta **84**, 311–323 (1969a).

WEIGL, J.: Einbau von Auxin in gequollene Lecithin-Lamellen. Z. Naturforsch. **24b**, 365–366 (1969b).

WEIGL, J.: Specificität der Wechselwirkung zwischen Wuchsstoffen und Lecithin. Z. Naturforsch. **24b**, 367–368 (1969c).

WEISENSEEL, M., SCHMEIBIDL, E.: Phytochrome controls the water permeability in *Mougeotia*. Z. Pflanzenphysiol. **70**, 420–431 (1973).

WELLBURN, A.R., ASHBY, J.P., WELLBURN, F.A.M.: Occurrence and biosynthesis of adenosine 3′,5′-cyclic phosphate in isolated *Avena* etioplasts. Biochim. Biophys. Acta **320**, 363–371 (1973).

WHEELER, H., BLACK, H.S.: Changes in permeability induced by victorin. Science **137**, 983–984 (1962).

WHEELER, H., BLACK, H.S.: Effects of *Helminthosporium victoriae* and victorin upon permeability. Amer. J. Bot. **50**, 686–693 (1963).

WILKINS, M.B., WOODCOCK, A.E.R.: Origin of the geoelectric effect in plants. Nature **208**, 990–992 (1965).

WONG, D.T., WILKINSON, J.R., HAMILL, R.L., HORNG, J-S.: Effects of antibiotic ionphore, A23187, on oxidative phosphorylation and calcium transport of liver mitochondria. Arch. Biochem. Biophys. **156**, 578–585 (1973).

WOOD, A., PALEG, L.G.: The influence of gibberellic acid on the permeability of model membrane systems. Plant Physiol. **50**, 103–108 (1972).

WOOD, A., PALEG, L.G.: Alteration of liposomal membrane fluidity by gibberellic acid. Australian J. Plant Physiol. **1**, 31–40 (1974).

WOOD, A., PALEG, L.G., SPOTSWOOD, T.M.: Hormone-phospholipid interaction: a possible mechanism of action in the control of membrane permeability. Australian Plant Physiol. **1**, 167–169 (1974).

WRIGHT, S.T.C.: Growth and cellular differentiation in the wheat coleoptile (*Triticum vulgare*). II Factors influencing the growth response to gibberellic acid, kinetin and indolyl-3-acetic acid. J. Exptl. Bot. **17**, 165–176 (1966).

YODER, O.C., SCHEFFER, R.P.: Effects of *Helminthosporium carbonum* toxin on absorption of solutes by corn roots. Plant Physiol. **52**, 518–523 (1973).

YOUNG, M., SIMS, A.P.: The potassium relations of *Lemna minor* L. II. The mechanism of potassium uptake. J. Exptl. Bot. **24**, 317–327 (1973).

YUNGHANS, H., JAFFE, M.J.: Phytochrome controlled adhesion of mung bean root tips to glass: a detailed characterization of the phenomenon. Physiol. Plantarum **23**, 1004–1016 (1970).

YUNGHANS, H., JAFFE, M.J.: Rapid respiratory changes due to red light or acetyl choline during the early events of phytochrome-mediated photomorphogenesis. Plant Physiol. **49**, 1–7 (1972).

ZIEGLER, H., LÜTTGE, U.: Die Salzdrüsen von *Limonium vulgare*. II. Mitteilung. Die Lokalisierung des Chlorids. Planta **74**, 1–17 (1967).

8. Cellular Differentiation, Ageing and Ion Transport

R.F.M. VAN STEVENINCK

1. Introduction

It is axiomatic that in space and time a cell changes its character and functional capabilities from the moment it is generated until its death. The adequate maintenance of permeability barriers and the degree of complexity achieved in compartmentation are perhaps among the most important criteria in deciding the physiological competence of a cell. In this connection it should be realized that in plants, more so than in animals, the regulation and maintenance of ionic balance tends to become an important function of each individual cell. It is true that symplastic continuity makes sharing of a regulatory function possible and that sometimes special situations demand specialized excretory cells or tissues (glands), but the ubiquity of vacuolation as an early event in differentiation stresses the role individual cells have in controlling an important aspect of their secretory function. Another salient feature is that apart from these individualistic aspects plant cells have a remarkable capacity to change their physiological competence e.g., changes during the washing of excised tissues or the generation of complete plants from a single vegetative cell (STEWARD et al., 1964). It should be emphasized that the term ageing only refers to passage of time and is used purely operationally. Foremost consideration should be given to define in terms of physiological competence the *changes* which may occur *during the passage of time*. It is important to know the reasons for these changes and to be able to characterize them as accurately as possible.

Basically, nature provides two opposing tendencies:

Either, A. the process may lead to an ultimate, irreversible break-down of permeability barriers and hence loss of compartmentation and can be referred to as *senescent ageing* or *true senescence*.

Or, B. the process may lead to a further elaboration of physiological competence through a stimulation of protein synthesis generally involving fresh genetical information (de-repression). This process of acquisition of physiological and biochemical functions (e.g. the development of ion uptake capacities and induction of enzyme synthesis in washed slices of storage tissue) will be referred to as *adaptive ageing* or *adaptation*.

Apart from this basic distinction, plants seem to have a special ability temporarily to suspend the effects of time through a drastic reduction of all metabolic activities. This phenomenon is referred to as dormancy and is especially apparent in storage organs (seeds, roots, stolons etc.) and meristematic regions (buds).

Both adaptive and senescent ageing may either occur *in situ* or after the excision of the appropriate plant organ or tissue, hence the subject matter will be treated under four separate headings:

a) adaptation as part of the normal process of cellular differentiation *in situ* (e.g. change to mature cells of different general types in stem and root)

b) induced adaptation including release from dormancy due to excision and change of environmental circumstances (e.g. slices of storage, stem and leaf tissue, cell cultures)

c) true senescence of plant organs *in situ* (e.g. fruit, leaves)

d) induced senescence of excised plant parts (leaf disks, fruit slices, etc.).

Many excellent reviews have appeared which cover one or several aspects of cell differentiation and senescence, but they do not deal with the topic of transport, e.g. on de-repression (Kahl, 1973; Stewart and Letham; 1973), on enzyme synthesis (Filner et al., 1969; Glasziou, 1969; Marcus, 1971; Scandalios, 1974; Huffaker and Peterson, 1974) or on respiratory control (Hackett, 1963; Laties, 1963; Rowan, 1966; Kahl, 1974; Laties, 1975).

Cells have a vast potential to perform specific functions under certain conditions. This capacity is based on already existing components and also on the storage of information which can lead to the production of other new components. In the previous Chapter (7) it was shown that hormones may play an important role in adaptive changes which affect the overall ionic balance of whole plants or their individual cells. This Chapter aims at exploring how these adaptations occur in time and to what degree they depend on changes of compartmentation, membrane permeability or transport processes. For instance, compartmentation would be much affected by the production of new organelles, e.g. mitochondria, endoplasmic reticulum (ER) as part of the endomembrane system (Morré, 1975; van Steveninck, 1975). Changes of membrane permeability could readily result from the incorporation of different lipid or protein components (Branton, 1969; Kuiper, 1972; Mazliak, 1973). Changes of transport processes could depend either on the incorporation of new transporting agents in the membrane structures, e.g. ATPases (Hill and Hill, 1973; Leonard and Hanson, 1972a, b), or alternatively, to the activation of carriers which are already there.

At this stage it should be pointed out that nearly all present knowledge regarding the above changes is based on the study of excised tissues and very little on what happens during cellular differentiation *in situ*. This probably reflects the need for a relatively uniform system if any progress is to be made in this very complex problem of internal control of ionic balance and distribution. There is an obvious need to continue these studies in order to unravel the mechanistic aspects but it should be stressed that in matters of internal control these aspects need to be studied *in situ* before their significance can be assessed at the whole plant level.

2. Adaptation *in situ*

2.1 Roots

In a root, cells close to the tip represent a series of progressive differentiation which can be more or less accentuated depending on whether their position in the root pre-destines them to become epidermal, cortical, endodermal, parenchymatous or vascular elements of the stele. With this large variety of possibilities over a comparatively short distance (8–15 cm), it is to be expected that ion transport may show a great deal of variation depending on the functional aspects as well as the degree of differentiation of the cells in question. This important principle can be illustrated by differences of transport depending on temperature and the longitudinal position in the root (Fig. *8.1*a–d) (ESHEL and WAISEL, 1972, 1973), but so far it has been difficult to relate these phenomena to the differentiation of specific cell types. However, some valuable information is being generated which relates root transport function to the maturation of xylem vessels (ANDERSON and HOUSE, 1967; HIGINBOTHAM et al., 1973) or to the ultrastructure of xylem parenchyma cells (LÄUCHLI et al., 1974; see also Chaps. *3.1, 3.4*).

The electron probe analysis work of LÄUCHLI et al. (1971) shows promise for future research linking cell specialization and transport systems. Similarly, attempts by ROBARDS et al. (1973) to correlate successive states in the development of endodermal cells (formation of suberin lamellae) with movement of Ca^{2+} and K^+ (HARRISON-MURRAY and CLARKSON, 1973) show that a detailed evaluation of the effects of cell differentiation on ion transport can in fact be achieved.

It appears that in intact roots, cells close to the surface are more involved in the transport of ions from the external solution to the cytoplasm (ϕ_{oc}) than cells further below the surface (VAKHMISTROV, 1967; CHANEY et al., 1972; BANGE,

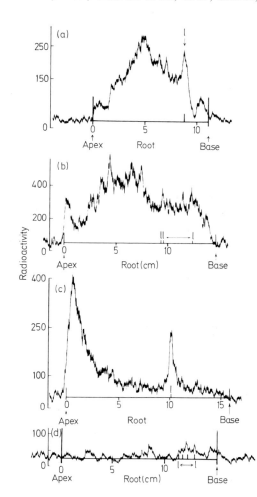

Fig. *8.1*a–d. Scans of $^{32}P_i$ uptake at 3 weeks along typical roots of *Pinus radiata* (a) grown at 11° C, uptake at 16° C; (b) grown at 14° C, uptake at 25° C; (c) grown at 25° C, uptake at 25° C; and (d) uptake in the presence of 10^{-3}M KCN. *l* position of lateral roots. Radioactivity in counts per min (from BOWEN, 1970)

1973). At low external concentrations this could be essentially due to the fact that diffusion rates to inner cells can be less than the intensities of uptake of cells at the surface (Ehwald et al., 1973). However, it has been suggested (Laties, 1967) that *in analogy* with the adaptations which occur at the surface of ageing slices of storage tissue (see 8.3.1.1), the transport capacity of the cell layers at the surface of the root may develop to a greater extent than that of the internal cells.

Leonard and Hanson (1972b) established an increase in $(Mg^{2+} + K^+)$-stimulated ATPase of the microsomal fraction of washed maize roots, which was closely paralleled by a stimulation of P_i-absorption. Although possibly a washing artifact, it would be of interest to establish whether this *in situ* stimulation is confined to the surface cell layers of the root or whether it is due to a developmental change in the specialized xylem parenchyma cells in the stele.

At present, there is little information on relevant metabolic aspects of *in situ* root cell differentiation. Fowler and Ap Rees (1970) established that carbohydrate oxidation in the relatively undifferentiated cells of the 3 mm apical region of pea roots mainly occurs *via* glycolysis with only a small contribution from the pentose phosphate pathway. Differentiation 4–6 mm from the apex was characterized by an increase of the pentose phosphate pathway relative to glycolysis. Both pathways contributed substantially in the 6–26 and 26–46 mm regions, but the maturation of the stele was accompanied by an increase in the capacity of the pentose phosphate pathway relative to glycolysis (Wong and Ap Rees, 1971). Again, this finding provides an interesting parallel with observed changes in the respiratory pathway of ageing slices of storage tissue (see 8.3.2.3). Cell differentiation in pea, tomato and maize roots also finds its expression in high activities of invertases in the apical region and a predominant activity in the cortex, while sucrose-phosphate-synthetase activity only occurs in the stele and in regions where root hairs occur (Vieweg, 1974).

2.2 Shoots

While there is no doubt that those factors which control the differentiation of meristematic cells in the shoot will profoundly affect the ion distribution of the resulting tissues, even less is known about the relation between cell differentiation and ion transport in shoots than in roots. Jacoby (1964) established by means of autoradiographs of intact bean plants that the basal part of the stem has a special capacity to retain $^{22}Na^+$, thereby preventing its transport towards the shoot apex. Na^+ distribution became uniform over the whole plant only when the stem tissue was saturated with Na^+. It must therefore be concluded that cell differentiation in root or shoot tissue may lead to a special capability for Na^+ transport into the vacuole (ϕ_{cv}) (see Waisel and Kuller, 1972 for bean hypocotyl and Rains and Epstein, 1967a, b for barley roots). This special capability for Na^+ transport is also a property of fresh slices of bean stem tissue, but changes with ageing (Rains, 1969; Rains and Floyd, 1970; see also 8.3).

Some aspects of differentiation which produce special features of ion transport are well documented, e.g. the appearance of the pulvinus mechanism (see 7.8), guard cells of stomata (see Chap. 4.3) and salt secretion from gland cells (see Chap. 5).

3. Induced Adaptation

3.1 De-Repression Phenomena and Change of Membrane Properties

Salt uptake studies using excised tissues date back to at least the turn of the century
(NATHANSON, 1904). The literature regarding the metabolism and ion transport
of excised tissues has subsequently shown an exponential rise which outstrips the
best of stimulations recorded for any of the excised tissues. The now classical reviews
by STEWARD and MILLAR (1954), ROBERTSON (1951, 1956, 1960), SUTCLIFFE (1954,
1959), and LATIES (1957, 1963) summarize the earlier work and therefore there
should be no need to elaborate on the invaluable contributions during this early
period.

For details see STEWARD and coworkers using mainly potato slices (STEWARD, 1932, 1937;
STEWARD and PRESTON, 1940, 1941 a, b; STEWARD et al., 1932), ROBERTSON and coworkers
using carrot and red beet slices (ROBERTSON and TURNER, 1945; ROBERTSON et al., 1947; ROBERT-
SON et al., 1951) and HOAGLAND using excised barley roots (HOAGLAND and BROYER, 1936,
1940, 1942).

3.1.1 Wound Effects Causing De-Repression

Obviously, slicing of tissue kills many cells and has a traumatic effect on neighboring
cells. The degree of damage influences the extent and rapidity of change as was
shown by experiments varying the surface to volume ratio of the excised tissue
(STEWARD et al., 1932), or the thickness of the slices cut (LATIES, 1962). These
experiments also established that the changes observed were mainly confined to
the surface of the excised tissue (STEWARD and HARRISON, 1939). Because potato
tissue responds by cell division at the cut surface, STEWARD et al. (1943) were
convinced that cell division and the ensuing protein synthesis provided the indispensable link with the observed stimulation of salt uptake. Much of STEWARD's subsequent work has been specifically devoted to this aspect, and comparisons were
made between non-dividing vacuolate cells and rapidly dividing callus tissues (STEW-
ARD and MILLAR, 1954; STEWARD and MOTT, 1970). The mere fact that slices of
beetroot and several other tissues can show a pronounced rise in salt uptake capacity
without a concomitant cell division (SUTCLIFFE, 1954) relegates the issue of stimulated
cell division to secondary importance. The same probably applies to the observed
induction of DNA synthesis caused by cutting potato tuber tissue (WATANABE and
IMASEKI, 1973; BORCHERT and MCCHESNEY, 1973).

Although the nature of the signal is unknown, de-repression (increased transcription) leading to a new state of biochemical and cytological differentiation must
be considered as the essential pre-requisite for the development of ion transport
capabilities. At the same time changes of membrane structure and permeability
will occur through turnover leading to possible replacement with new or different
physiological and biochemical entities, e.g. the rapid stimulation of phospholipid
synthesis during ageing of tissue slices is well documented (WILLEMOT and STUMPF,
1967 a; TANG and CASTELFRANCO, 1968 a; BEN ABDELKADER and MAZLIAK, 1968;
GALLIARD et al., 1968 a) and has been ascribed to a de-repression phenomenon
(WILLEMOT and STUMPF, 1967 b; TANG and CASTELFRANCO, 1968 b). KAHL (1973)
recently provided an outstanding critical discussion on the de-repression phenome-

non in cells of sliced storage tissue which has become a central issue following the now classical publication of CLICK and HACKETT (1963).

3.1.2 Time Course of Changes Related to RNA Metabolism and Net Ion Uptake Capacity

Certain manifestations perhaps rate a special mention as they could have a more direct bearing on the development of ion transport mechanisms. One of these is the time factor involved (see Fig. 8.2). The association of single ribosomes into polyribosomes is one of the first apparent cytological changes after the slicing of storage tissue. LEAVER and KEY (1967) observed a change in proportion of ribosomes present as polysomes from 10% to 65% during the ageing of carrot slices, but stressed that the greatest increase occurred during the first hour of incubation. However, the ribosomes from aged disks differed from those of fresh disks by a more effective binding with poly-U (LIN et al., 1973); apparently, ribosomes from aged disks contained at least two components not associated with ribosomes from fresh disks. These observations are readily explained by the burst of ribosomal-RNA synthesis shown to occur immediately after slicing of potato tubers (SAMPSON and LATIES, 1968; KAHL, 1971a). KAHL (1971a) also showed that polyribosome formation in potato is virtually complete 9 h after slicing and is dependent on formation of new m-RNA which is not available immediately after slicing of the tissue (KAHL, 1971b). On the other hand, BRYANT and AP REES (1971) detected a net synthesis of RNA in fresh carrot disks after 8 h which continued for at least 4 days. DUDA and CHERRY (1971) also detected a 7-fold rise in RNA polymerase activity in washed sugar beet tissue which reached a peak at 25 h. Considering the relative instability of m-RNA, e.g. the half time for peroxidase m-RNA decay

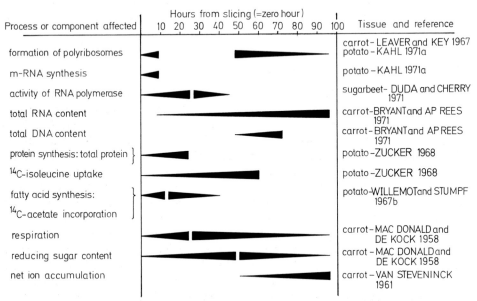

Fig. 8.2. Changes in processes or cellular components during adaptive ageing of tissue slices (◄, increase to a maximum value; ►, decline)

in washed sugar cane tissue is $1^1/_2$–2 h (GAYLER and GLASZIOU, 1968), continued synthesis of m-RNA would seem to be an essential requirement.

Traditionally, slices of storage tissue used for ion transport studies were washed for periods of up to one week. Storage tissues of various species exhibit appreciable differences in the duration of lag phase required for the development of a salt uptake capacity (VAN STEVENINCK, 1961), e.g. net K^+ uptake in parsnip (*Pastinaca sativa* L.) phloem tissue required a lag of more than 40 h while artichoke (*Helianthus tuberosus* L.) required less than 10 h to establish its net uptake capacity. The duration of the lag may also differ for different ions and may depend on the origin of the tissue. For instance, in beetroot slices derived from separate roots, net uptake of K^+, Na^+ and Cl^- may develop sequentially with a time difference of at least 10 h and with net Cl^- uptake usually commencing last and with net Na^+ uptake commencing before net K^+ uptake (VAN STEVENINCK and VAN STEVENINCK, 1972). However, such a sequence can be modified by hormone treatment (VAN STEVENINCK, 1972 a), addition of abscisic acid to the washing medium inducing a reversal in times of onset for net K^+ and Na^+ uptake. Hence, it appears that growing and/or storage conditions of the roots may determine the duration of the lag phase for net ion uptake. The effect of storage conditions on the subsequent accumulation of solutes was earlier given extensive consideration by STEWARD et al. (1943) for potato tubers, and more recently ROSENSTOCK et al. (1971) established that the time required for the attainment of a respiratory peak may vary between 0 and 48 h depending on the primary meristematic activity of the tuber tissue and on the duration of storage after harvesting of the tubers. Such studies indicate the necessity for using uniform genotypes (clones preferably, see Chap. 9) and the particular need to have the material raised (and stored if necessary) under welldefined and repeatable conditions. However, existing evidence shows that the times required for the development of ion uptake capacities are similar to those required for de-repression phenomena and also that de-repression may be specific for each characteristic transport mechanism or carrier.

3.1.3 The Role of the Endoplasmic Reticulum (ER)

Suggestions that the ER may play a role in salt uptake have come mainly from MACROBBIE's (1969, 1971) proposals based on kinetic considerations of ion transport in giant algal cells, and from ultrastructural observations (COSTERTON and MACROBBIE, 1970; JACKMAN and VAN STEVENINCK, 1967). Furthermore, the continuity of ER from cell to cell through plasmodesmata (KOLLMANN and SCHUMACHER, 1962, 1963; LÓPEZ-SÁEZ et al., 1966; ROBARDS, 1971; BURGESS, 1971; see also 2.2 and Fig. 2.1) together with the recent idea that the ER forms part of a much larger endomembrane complex consisting of a system of intricate ramifications which includes the Golgi apparatus (MORRÉ et al., 1971; CHARDARD, 1973; MARTY, 1973 b) adds weight to this proposition. The nature of this endomembrane system appears ill-defined at present, but the suggestion that the Golgi apparatus exhibits a certain polarity (NORTHCOTE, 1971; MARTY, 1973 a) with the proximal side being in contact with the ER and the distal side forming a network from which provacuoles may arise (MARTY, 1973 b), adds credibility to the idea of directional transport from external solution to vacuole *via* the ER, especially in rapidly differentiating cells. Ultrastructural evidence based on Cl^- precipitation with Ag^+ ions showed that the ER may be involved in transport of Cl⁻ (LÄUCHLI et al., 1974, for barley roots, VAN STEVENINCK et al., 1974, for *Nitella translucens*). Furthermore, it has been shown by means of ultrastructural cytochemistry that Mg^{2+} dependent ATPase activity in the meristematic cells of *Cucurbita* roots occurs in the plasmodesmata and in the ER, especially when the ER is in close vicinity or in actual connection with the plasmodesmata (COULOMB and COULOMB, 1972; see Fig. 8.3).

The concurrent development of ER and salt uptake capacity in beetroot slices following the disappearance of practically all ER at the time of slicing the tissue (JACKMAN and VAN STEVENINCK, 1967) may prove to be coincidental. Already, some facts on metabolism add interest to this finding. Significant changes in electron transport capabilities of microsomal fractions (derived from ER) of turnip (*Brassica rapa* L.), swede (*Brassica napobrassica* Mill.) and beetroot tissues were observed due to slicing and ageing of the tissue (RUNGIE and WISKICH, 1972a, b). Furthermore, microsomal fractions isolated from aged disks of beetroot and Jerusalem artichoke (*Helianthus tuberosus* L.) respectively have a much greater amino acid incorporating ability than those obtained from fresh disks (ELLIS and MACDONALD, 1967; CHAPMAN and EDELMAN, 1967).

3.1.4 The Effect of Inhibitors on the ER and Ion Transport

Inhibitors of protein and nucleic acid metabolism also affect ion transport processes (MACDONALD et al., 1966; POLYA, 1968; VAN STEVENINCK and VAN STEVENINCK, 1972), hence it is worthwhile to consider the separate contributions of the cytoplasmic and ER protein synthesizing systems in establishing the salt transport capacity.

A large range of inhibitors is available, each of these having a specific effect (POLYA, 1973; STEWART, 1973), but most frequently used are cycloheximide (CHM), puromycin, para-fluoro-phenylalanine (FPA) and actinomycin D (ACT-D). Cycloheximide is thought to inhibit protein synthesis mainly by interfering with the initiation and transfer reaction in peptide bond formation, while puromycin inhibits by prematurely terminating the unfinished polypeptide chain. FPA does not directly inhibit protein synthesis but interferes through the production of non-functional (nonsense) protein. Actinomycin D blocks RNA polymerase action, thus preventing the transcription process.

In observing the effect of puromycin, CHM, ACT-D and FPA on the development of ER in ageing disks of beetroot, VAN STEVENINCK and VAN STEVENINCK (1971) found that the re-assembly of the ER was not affected by ACT-D, but completely prevented by CHM and puromycin (see Table 8.1). Very little re-assembly occurred in the presence of FPA, and any ER present appeared very fragmented. The development of crystalloid protein bodies in the cisternae of the ER which normally appear during ageing was totally inhibited by CHM and puromycin but greatly stimulated

▷

Fig. 8.3a–d. Ultrastructural detail of ER and plasmodesmata connections between cells. (a) Adjoining phloem parenchyma cells in a tangential longitudinal section of *Metasequoia glyptostroboides* showing vesicles of ER and branched plasmodesmata in primary pit fields of the radial longitudinal wall. Original electron micrograph by courtesy of Prof. Dr. R. KOLLMANN (Kiel). (b) Meristematic cells of *Cucurbita* roots treated with modified WACHSTEIN and MEISEL medium to show the presence of Mg^{2+}-dependent ATPase activity. Dense reaction products occur in ER especially when in the vicinity of plasmodesmata (from COULOMB and COULOMB 1972). (c) and (d) Barley root cells treated with Ag^+ ions in order to precipitate Cl⁻. Dense precipitation products occur in plasmodesmata and also in vesicles in the vicinity of plasmodesmata (R.F.M. VAN STEVENINCK). Bar represents 1 μm

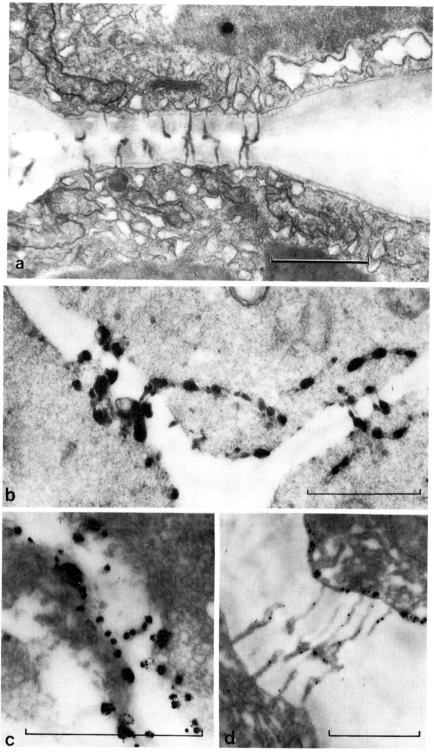

Fig. 8.3a–d. Legend see opposite page

by ACT-D and FPA. It was concluded that current protein synthesis is required for the re-assembly of the ER during ageing but that this protein synthesis does not require freshly synthesized m-RNA. It was also suggested that the protein synthesizing system of the rough ER is quite stable and can continue to function without synthesis of fresh m-RNA. In fact, it seemed that protein synthesis by this system became stimulated due to the inhibition of m-RNA synthesis required for the cytoplasmic protein synthesizing system. The greater stability towards RN-ase shown by membrane bound m-RNA compared with polyribosomal m-RNA for rat liver seems relevant (BLOBEL and VAN POTTER, 1967). In parallel experiments (VAN STEVENINCK and VAN STEVENINCK, 1971) it was shown that ACT-D prevented the development of a K^+, Na^+ or Cl^- uptake capacity when added immediately to fresh disks (see Table 8.1). If added later when the uptake capacity was established ACT-D had no effect on the system. Similarly, puromycin and CHM prevented the development of an uptake capacity when added to fresh disks. When added at a later stage during ageing, inhibition of ion uptake also occurred, but this inhibition in aged disks was not immediate and exhibited a lag which varied in duration for each individual ion (e.g. 3 h for Cl^-, 9 h for Na^+ and 24 h for K^+), but also progressively lengthened when the inhibitor was added later during ageing. The lag was not due to lack of penetration of the inhibitors since inhibition of protein synthesis by CHM in aged beetroot disks was almost immediate while inhibition of Cl^- uptake did not occur until $1^1/_2$ h after addition of the inhibitor (POLYA, 1968). This, according to POLYA (1968), contradicted STEWARD's hypothesis of a direct link between ion transport and protein synthesis. In yeast, CHM affected

Table 8.1. The effect of protein synthesis inhibitors on assembly of ER, the formation of crystalloid protein inclusions within the ER and on the rate of net accumulation of ions in beetroot tissue at various time intervals after slicing the tissue. (0=denotes not seen; + =occasional; + + =medium lamellae, several inclusions; + + + =long lamellae, many inclusions)

Treatment	ER lamellae at				Treatment	Protein inclusions at			
	0 h	24 h	48 h	64 h		0 h	24 h	48 h	64 h
Control	+	+ +	+ +	+ + +	Control	0	0	0	+
ACT-D	+	+ +	+ +	+ + +	ACT-D	0	+	+ +	+ + +
CHM	+	+	+	+	CHM	0	0	0	0
FPA	+	+	+	+	FPA	0	+	+ + +	+ + +

Treatment	Net ion accumulation (μmol g_{FW}^{-1} h^{-1}) at											
	0 h			24 h			48 h			64 h		
	K^+	Na^+	Cl^-	K^+	Na^+	Cl^-	K^+	Na^+	Cl^-	K^+	Na^+	Cl^-
Control	0	0	0	0.17	0.26	0.03	0.17	0.28	0.26	0.20	0.37	0.61
ACT-D	0	0	0	0	0	0	0	0.10	0.04	0.11	0.16	0.09
CHM	0	0	0	0	0	0	0	0.02	0	0	0	0
FPA	0	0	0	0.30	0.23	0	0	0.04	0	0	0	0

K^+ uptake and P_i-uptake quite differently, allowing a substantial K^+ uptake to proceed even at 20 mM CHM, a concentration which almost totally inhibited P_i-uptake and which must have completely blocked protein synthesis (REILLY et al., 1970). From these data it may be concluded that both transcription [1] and translation [2] processes are involved in the development of ion transport capacities. Specific proteins are required for each transport system or carrier and the turnover—or time constants for each system may be different. Finally, the ER-protein synthesizing system most likely is not primarily responsible for the development of the ion transport systems, at least *not until* specific new m-RNA is produced as part of the adaptation process (e.g. the ER protein synthesizing system remained active in the presence of FPA and ACT-D, while the development of transport systems was inhibited).

A differential effect of quite a different nature was reported by LÄUCHLI et al. (1973). CHM inhibited K^+ transport through excised barley roots but without inhibiting accumulation of K^+ in the cortical cells. It was suggested that membrane turnover and therefore protein synthesis may be an essential requirement of symplastic transport of ions. In a later publication LÜTTGE et al. (1974) placed particular emphasis on the fact that CHM had no effect on ion accumulation across plasmalemma and tonoplast in root cell. This, according to the authors, indicated that the role of CHM may primarily concern membrane turnover of compartments such as the ER (see also 3.4.2.5).

The induction of chloride pumping in salt glands of *Limonium* may be regarded as yet another adaptive ageing phenomenon (see Chap. 9). SHACHAR-HILL and HILL (1970), who made a comprehensive study employing a large variety of inhibitors of protein and nucleic acid synthesis, found that CHM (0.036 mM) caused an immediate inhibition of Cl^- pumping. They ascribed this to possible side effects of CHM on the respiratory pathway through an "uncoupling" effect somewhat similar to, but ten times more effective than those normally caused by DNP in beet disks (MacDONALD and ELLIS, 1969). The effects of CHM on O_2 uptake seemed to differ greatly in carrot-, beet-, potato disks and in wheat root and pea radicles. ELLIS and MacDONALD (1970) also found ion uptake by non-green tissues to be inhibited, while leaf tissue appeared to be insensitive to CHM treatment. Because the ion absorption mechanism in leaves may differ fundamentally from that in root tissue with respect to ATP supply or direct dependence on electron transport the authors suggested that CHM may affect ion uptake through disruption of the energy supply rather than protein synthesis. This view has failed to gain support (KIRK, 1970; AP REES and BRYANT, 1971; LÄUCHLI et al., 1973; LÜTTGE et al., 1974). Recently, COCUCCI and MARRÉ (1973) provided conclusive evidence that effects of CHM on respiration and energy supply are of a secondary nature only and may result from a greatly diminished utilization of high energy phosphate in protein synthesis.

3.1.5 The Effect of Hormones on the ER and Ion Transport

Kinetin and benzyladenine are frequently shown to delay senescence and maintain high rates of protein synthesis in excised tissues (see Chap. 7). In beetroot, however, kinetin and benzyladenine prevented the development of net K^+, Na^+ or Cl^- uptake when applied to freshly cut disks (VAN STEVENINCK, 1972b). On the whole, the pattern of inhibition was very similar to that of CHM because again the delays between application and inhibition became progressively longer with further ageing of the disks at the time of application (VAN STEVENINCK, 1972b). However, their inhibitory effect on the re-assembly of the ER during ageing of the disks was much

---

[1] Transcription is the transfer of genetic information from DNA to m-RNA.
[2] Translation is the synthesis of protein based on information provided by m-RNA.

less severe than with CHM. It may be significant that kinetin and benzyladenine had no effect on the Na^+ uptake in swede (*Brassica napobrassica* Mill.) disks during ageing, but this tissue differed markedly from beetroot in that the fresh disks were capable of net Na^+ uptake almost immediately after slicing (van Steveninck, 1972b).

After prolonged culture, crown gall cells may acquire autonomy towards auxin. Like cells of other habituated auxin-autonomous clones (*Vinca, Acacia*) they are richer in ER than their counterparts which require exogenous auxin for growth (Lipetz, 1969).

3.1.6 Other Membrane Changes during Ageing

Not all membrane changes during ageing are readily perceived at the ultrastructural level. Hanson et al. (1973) pointed out that because they were looking for more dramatic differences they had failed to notice that the electron density of the tonoplasts in fixed and stained tissue had increased as a result of washing of maize root segments. These authors suggested that the increases in solute absorption rate during the washing treatment may therefore relate to the tonoplast. These observations, although highly speculative at present, may prove to be valuable when they become reliable. Hitherto it has been difficult to pinpoint biochemical evidence of membrane transformation to a particular membrane system. However, rapid progress is being made in the separation and characterization of specific membrane systems. A notable attempt was made by Castelfranco et al. (1971), who monitored membrane transformations in ageing potato tuber slices by means of ^{14}C-choline incorporation studies and observed a marked progressive increase in the total lipoidal radioactivity with ageing of slices. However, more significantly, they were able to determine a shift in a predominant activity from microsomes to mitochondrial membranes after 24 h of ageing. At present, methods are becoming available which will make it possible to separate membrane fractions originating from the plasmalemma, endoplasmic reticulum and tonoplast respectively (Hodges et al., 1972; Ryan and Smith, 1972; Glaumann, 1973).

3.2 Changes of Salt Saturation, Ion Selectivity and Metabolism

3.2.1 Salt Saturation Levels and Ion Fluxes during Ageing

The development of new salt uptake capacities in excised tissues will result in a flux equilibrium at elevated internal ion concentrations, e.g. fresh beetroot slices with an internal K^+ concentration of say 60 mM and in equilibrium with a 1 mM KCl external solution may under suitable conditions attain an internal concentration (c_v) of 200 mM K^+ at a new equilibrium level when the slices are aged (Pitman, 1963). The new level of salt saturation should be an important parameter of cellular differentiation during the ageing of excised tissues. This compared with the original state of the tissue should provide information about various factors which may control the flux equilibrium at the respective saturation levels. However, surprisingly little systematic research has been carried out in this field. The new equilibrium levels seemed to arise from increasing rates of efflux counterbalancing a constant rate of influx when salt saturation is reached (Pitman, 1963; van Steveninck,

1964 working on beetroot tissue; JACKSON and ADAMS, 1963; JACKSON and STIEF, 1965; JACKSON and EDWARDS, 1966, barley roots), but more recently the accent has been on a relative decrease of influx to explain salt saturation (PITMAN et al., 1968; JOHANSEN et al., 1970; NEIRINCKX and BANGE, 1971, barley roots). In the latter case especially it would be important to establish the mechanism controlling the influx. CRAM (1972) showed that a change of hydrostatic pressure during the washing of carrot slices had no effect on influx during ageing. It was however established that rates of Cl^- and NO_3^- influx into carrot and barley root cells depended on the sum of Cl^- and NO_3^- concentrations in the vacuole, while Cl^- influx showed no correlation with the vacuolar concentrations of K^+, Na^+, $K^+ + Na^+$, malate, Cl^- or NO_3^- (CRAM, 1973; Part A, 11.3.2).

Extensive work by MacDONALD and DE KOCK (1958) using a range of sliced storage tissues (sugar beet, red beet, carrot, swede, potato) showed that concentrations of reducing sugar were positively correlated with rates of respiration, e.g. highest value for both being reached after 4 days in red beet (Fig. 8.4). The sugar concentrations would then drop again to low values at a time when the disks are capable of reaching high salt saturation values. This suggests that an internal control system may exist which regulates total solute content of the vacuoles by relating sugar and salt concentration, similar to that observed in barley roots by PITMAN et al. (1971). However, CRAM (1973) found that Cl^- influx was not related to sugar concentration during ageing drifts in excised carrot tissue.

Obviously, further work is required to define changes of salt saturation and ion selectivity and relate these to processes of cell differentiation which in turn may be controlled by hormones (see Chap. 7) and by carbohydrate metabolism. Kinetin, ABA and GA_3 are shown to have a marked effect on storage and hydrolysis of starch (TASSERON DE JONG and VELDSTRA, 1971; MITTELHEUSER and VAN STEVENINCK, 1971b; JONES, 1973; HADAČOVÁ et al., 1973; KULL and UNGER, 1974). Also a large body of evidence has accumulated concerning enzymes which involve the breakdown of starch to reducing sugars in ageing disks of storage tissue. Tissues with an active starch and sugar metabolism may show large increases in sucrose synthetase activity during the ageing of slices in potato (LAVINTMAN and CARDINI, 1968) or in invertase activity in sugar cane slices (GAYLER and GLASZIOU, 1969; GLASZIOU, 1969) in beetroot (BACON et al., 1965; VAUGHAN and MacDONALD, 1967a, b, c; STONE et al., 1970) and in carrot (RICARDO and AP REES, 1970).

3.2.2 Ion Selectivity

Marked changes in ion selectivity were observed in bean stem tissue slices during ageing (RAINS, 1969; FLOYD and RAINS, 1971). Fresh slices absorbed Na^+ but not K^+ at low external concentrations (0.02–0.5 mM) and in equal proportion at relatively high concentrations (0.5–50 mM) but in stem slices aged in 0.5 mM $CaSO_4$ the situation was reversed, i.e. the Na^+-absorbing capacity was lost and replaced by a K^+-absorbing capacity. This change in selectivity was not due to a significant change in rate of efflux of either ion, but seemed to be influenced by the presence of Ca^{2+} (RAINS and FLOYD, 1970) and could be modified by means of benzyladenine (see 7.3).

Rather the reverse was observed by POOLE (1971a, b). Beetroot slices washed for 1 day seemed to absorb K^+ readily but lacked a mechanism for Na^+ uptake.

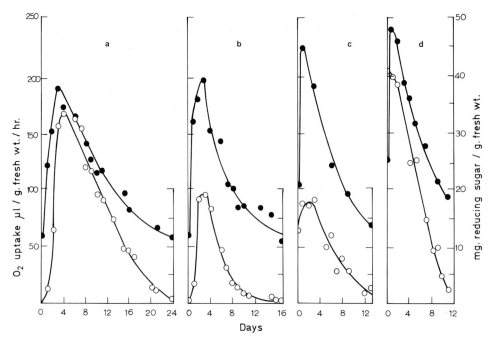

Fig. 8.4a–d. Respiration measured at 25° C (●) and reducing sugar values (○) in slices of storage tissue pretreated at 25° C. (a) sugar-beet; (b) red beet; (c) carrot; (d) swede (*Brassica napobrassica* Mill.). Insufficient information is available to relate reducing sugar levels directly with saturation levels of inorganic salts, however the data shown here suggest that the onset of net ion accumulation may coincide with the downward trend of reducing sugar levels (e.g. slices of swede tissue normally do not exhibit a lag phase in ion uptake, van Steveninck, 1972b). Also the loss in reducing sugars may be equivalent to the gain in inorganic ions when saturation levels are reached (from MacDonald and De Kock, 1958)

However after 5–6 days ageing, the situation was reversed and Na^+ was accumulated in preference to K^+ from solutions containing both ions. Unfortunately, no information is available to relate these changes in selectivity to changes in separate fluxes or to relative degrees of salt saturation. Possibly, changes of hormone balance may have a significant bearing on these K^+/Na^+ selectivity changes during ageing of tissue slices (van Steveninck, 1972a, see 7.5).

It appears that in certain situations the preference of particular tissues for Na^+ may constitute part of a salt-evading mechanism by which plant species can prevent Na^+ moving from the roots to the shoot (e.g. bean epicotyls, cf. Waisel and Kuller, 1972). It seems significant that the differentiation processes induced in excised tissue often involve changes in K^+/Na^+ selectivity, thereby altering the nature and salt balance of the tissue. Thus it is not surprising that plant hormones have been shown to modify these capacities for adaptive change.

3.2.3 Controlling Influences of Mitochondria and Ca^{2+}

A large amount has been written on changes in metabolism during adaptive ageing of storage tissues. It will only be possible here to sketch the briefest outline of metabolic factors which may have a bearing on ion transport processes during the ageing of tissues (see also Part A, Chap. 9). The 1940's and 1950's provided a comprehensive characterization of the respiratory rise in ageing storage tissue slices (see reviews by LATIES, 1957, 1963), and an understanding that this enhanced respiration and the observed changes in respiratory pathway were intimately related to the increases in salt transport observed during ageing (ROBERTSON, 1960, 1968). The changes include a much-increased participation of the pentose phosphate pathway (AP REES and BEEVERS, 1960; LATIES, 1964b; REED and KOLATTUKUDY, 1966) and an activation of the tricarboxylic acid cycle in which γ-hydroxy-α-ketoglutarate appears to play a key role (LATIES, 1967).

LATIES and HOELLE (1967) stressed the importance of the α-oxidation process as a respiratory component of storage slices and as an immediate means of degradation of massive amounts of long chain fatty acids in fresh slices, e.g. membranes of the ER in beetroot disappear on slicing (JACKMAN and VAN STEVENINCK, 1967), and mitochondria become very fragile on isolation from fresh disks (VAN STEVENINCK and JACKMAN, 1967) and may lose their capacity for oxidative phosphorylation through the presence of lipids (DALGARNO and BIRT, 1963). LATIES et al. (1972) recently confirmed by means of an elegant $^{13}C/^{12}C$ ratio technique (JACOBSON et al., 1970) that lipids constitute the predominant substrate for respiration in fresh tissues and carbohydrate in aged slices.

The changes in the respiratory process are typified by a change from malonate resistance to malonate sensitivity, and from CN^- sensitivity to CN^- resistance during ageing (LATIES and HOELLE, 1965). This latter effect may be due to the establishment after slicing of a CN^--resistant electron transport system by the incorporation of an auto-oxidizable CN^- resistant cytochrome b_7 (MARRÉ, 1961). However, the main point is that the increase in terminal respiratory chain activity reflects an increase of mitochondrial activity during ageing. For a review on the role of mitochondria providing the energy for active ion transport see ROBERTSON (1968).

An aspect of mitochondrial control which may be of particular relevance is the apparent capacity to regulate the concentration of Ca^{2+} in the cytoplasm (SPENCER and BYGRAVE, 1973). This particular role of controlling the ionic environment was suggested for rat liver mitochondria but could equally apply for plant mitochondria since they are capable of Ca^{2+} accumulation in vitro (GRUNWALD, 1966; HODGES and HANSON, 1965; TRUELOVE and HANSON, 1966; KENEFICK and HANSON, 1966; HODGES and ELZAM, 1967; EARNSHAW et al., 1973) as well as in vivo (MERTZ and LEVITT, 1961). Apparently relatively large amounts of Ca^{2+} may be accumulated (CARAFOLI, 1967) and deposited as relatively insoluble intramitochondrial granules which would allow some measure of ionic control in the limited space of the mitochondrion (PEACHEY, 1964). PASQUALI-RONCHETTI et al. (1969) however cast some doubt that the granules represent Ca- or Mg-phosphate deposits.

Over many years the major role of Ca^{2+} in maintaining the physical properties of membranes has been well recognized (EPSTEIN, 1973). Removal or deficiency of Ca^{2+} from the tissue causes a reversible leakage of vacuolar contents (VAN STEVENINCK, 1965) and structural disruption (MARINOS, 1962). Also the presence of Ca^{2+} in the external medium during the ageing of tissue has profound effects on the selectivity of ion transport (RAINS and EPSTEIN, 1967a; RAINS and FLOYD, 1970) and hence it may be assumed that Ca^{2+} influences the development of ion-absorption capacities during the adaptive ageing of tissue. Although most ion-stimulated ATPases isolated from plant tissues are activated by Mg^{2+} (Fisher and Hodges, 1969; Kylin and Gee, 1970; Hodges et al., 1972) Ca^{2+}-activation may also occur (Dodds and Ellis, 1966; Hall and Butt, 1969). It seems difficult to reconcile the above facts with the requirement for Ca^{2+} removal from homogenates (by means of EDTA) to obtain mitochondrial suspensions which are capable of oxidative phosphorylation with a satisfactory respiratory control and structural intactness (Lund et al., 1958; Chrispeels and Simon, 1964). It has been shown, however, that extramitochondrial lysophospholipase acitvity is greatly stimulated by Ca^{2+} (not Mg^{2+}), and greatly inhibited by EDTA (Björnstad, 1966). Free fatty acids are known to inhibit key glycolytic enzymes (Weber et al., 1966) and also oxidative phosphorylation, possibly by disrupting the mitochondrial membrane by detergent action (Strickland, 1961; Dalgarno and Birt, 1963; Baddeley and Hanson, 1967; Earnshaw and Truelove, 1970; Earnshaw et al., 1970).

The difficulty of isolating active mitochondria from freshly sliced tissue in contrast with the ease of isolation from aged disks may be relevant (van Steveninck and Jackman, 1967; Castelfranco et al., 1971). It seems most likely that a rapid reorganization allowing the membrane transformations necessary for adaptation can take place during the early stages of preparing tissue slices. For example, the presence of Ca^{2+} would help in limiting the amount of surplus membrane material, resulting from the disappearance of the endoplasmic reticulum (Jackman and van Steveninck, 1967) through a stimulation of phospholipase activity together with the removal of surplus fatty acids by α-oxidation (Laties et al., 1972). Subsequently, with the availability of new m-RNA, new membrane systems, mitochondria and other organelles may be synthesized which offer new permeability barriers and thus cells at the surface of the tissue may be better adapted to the new environment. This seems a logical extension of the de-repression process which is triggered during the slicing of tissue by the drastic destruction of permeability barriers causing leakage of a variety of metabolites and a change of gas conditions (O_2, CO_2, C_2H_4 etc.) (Wolley, 1962; Laties, 1962; MacDonald, 1968).

Apart from an increase in mitochondrial activity it has been reported that cytochrome oxidase and succinate dehydrogenase activities increased during the ageing of sliced sweet potato after a lag phase of 8–10 h, and more significantly the haem contents in isolated mitochondria increased after a lagphase of similar duration (Sakano and Asahi, 1971). These increases were completely suppressed by CHM. Verleur (1969) observed by electron microscopy that, especially in layers of cells with cambial activity just below the wound surface of potato slices, the number of mitochondria per cell had greatly increased. Mitochondrial fractions isolated from intact, freshly sliced and aged tissue displayed an increase in specific activities for cytochrome oxidase and succinate dehydrogenase during ageing. This heterogeneity of the mitochondrial population induced by slicing has been reported by a number of authors (Sakano et al., 1968; Ben Abdelkader et al., 1969; Verleur et al., 1970).

3.2.4 Salt-Stimulated Respiration

Salt-stimulated respiration (salt respiration) has often been ascribed to a direct involvement of mitochondria in the actual process of salt transport (Robertson, 1968, Packer et al., 1970). Although it is widely accepted that mitochondrial ATP is required for active transport, evidence has been presented that the high affinity system in aged beet and carrot disks may be directly dependent on the mitochondrial electron transport system (ATKINSON and POLYA, 1968; POLYA and ATKINSON, 1969). However, LÜTTGE et al. (1971) showed that the Cl^--stimulated respiration in carrot disks is related to transport at the tonoplast. According to ADAMS and ROWAN (1972) ADP requirement and phosphoglycerate kinase activity are the keys to the reaction mechanism of salt respiration. Other key reactions of the glycolytic pathway are similarly controlled by changes in the relative amounts of ADP and ATP during ageing (ADAMS and ROWAN, 1970), rather than by changes of enzyme activities *in vitro* which do not seem to reflect the metabolic changes during ageing (KAHL et al., 1969a, b). Recently, it was proposed that control of glycolysis in aged slices of carrot root tissue is by a negative feedback system involving PEP, phospho-glycerate and phosphoglycerate kinase (FAIZ-UR-RAHMAN et al., 1974).

Studies on the development of transport ATPase activity during the adaptive ageing of slices have given rather inconclusive results (JACOBY, 1965; ATKINSON and POLYA, 1967), but washing of excised maize roots has shown a close parallel between increases in ATPase activity and enhanced solute uptake (LEONARD and HANSON, 1972b). Ultrastructural localization of this enhanced activity (HALL, 1973) may soon provide further detail about membrane changes (Fig. 8.5) during adaptive ageing (HANSON et al., 1973).

3.2.5 Uptake of Organic Solutes

The process of adaptive ageing has also been shown to stimulate uptake of organic solutes, notably amino acids (DALGARNO and HIRD, 1960; SHTARKSHALL et al., 1970; ÖZER and BASTIN, 1972) and glucose (LATIES, 1964a; VAN STEVENINCK, 1966b; CARLIER, 1973). The principal effect of ageing appears to be on uptake in the lower concentration range (REINHOLD et al., 1970), but multiphasic systems in both uptake of amino acids and glucose have been detected in aged tissues (LINASK and LATIES, 1973; SHTARKSHALL and REINHOLD, 1974).

Tris buffer in the medium has been shown to stimulate glucose transport in slices of beetroot tissue (VAN STEVENINCK, 1966b). *Tris* also in certain tissues induced an immediate net cation uptake in fresh slices (VAN STEVENINCK, 1966a). D-glucos-amine completely inhibited the *tris*-stimulated cation uptake, an effect which could be specifically reversed by further additions of D-glucose. It was implied that the D-glucosamine blockage of cation uptake was mediated through an inhibi-tion of a glucose specific hexokinase (VAN STEVENINCK et al., 1973). The activity of this enzyme has been shown to increase up to four-fold during ageing of carrot disks for 24 h (RICARDO and AP REES, 1972).

4. Senescence *in situ*

Research in this field is mainly concerned with fruit ripening and senescent changes in foliage tissue, only a very small proportion being directed towards the investigation of salt relations.

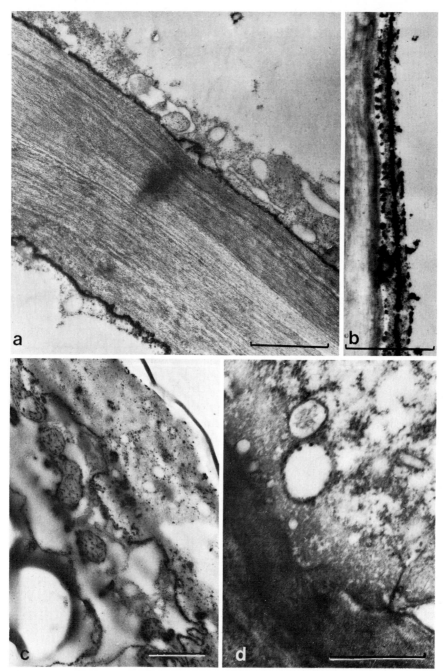

Fig. 8.5a–d. Ultrastructural evidence for ATPase activity. Aged beetroot disks (148 h) were incubated in modified Wachstein-Meisel medium for 15 min (a,c) or 1 h (b) after preliminary fixation with formaldehyde for 2 h; (d) control, 1 h in the medium in absence of ATP. Post fixation in OsO₄ for 1 h; a, stained in uranyl acetate; b–d, unstained. Lead deposits were particularly associated with the plasmalemma, tonoplast and ER (a, b), but also occurred in cytoplasmic vesicles and mitochondrial cristae (c). The control (d) showed some small lead deposits in vesicles but none associated with the mitochondria or plasmalemma. (From VAN STEVENINCK, 1970). Bar represents 1 μm

Changes due to senescence are mainly defined by a loss of physiological competence which is exemplified by a rapid decrease in synthesis (nucleic acids and total proteins), increased activity of lytic enzymes, respiratory rise (climacteric fruits), export of metabolites, loss of weight (and of chlorophyll in leaves), and above all a breakdown of permeability barriers (decrease in compartmentation). The latter leads to an increase in apparent free space (AFS) of the tissue (SACHER, 1962; SWARDT, DE and ROUSSEAU, 1973). Hence, in senescing tissue, ion accumulation becomes an irrelevant topic and concerns only loss of ions. Generally, there is little evidence of this export being an active process, requiring metabolic energy, however JACOBY and DAGAN (1969) reported some measure of control of Na^+ effluxes from ageing bean leaves. This was not related to an increase of permeability of leaf tissue and therefore should be regarded as preceding the senescence change (JACOBY et al., 1973). EILAM (1965) found that in bean cotyledons and in the *Arum* spadix permeability changes measured as K^+ leakage preceded changes in respiratory rates. FERGUSON and SIMON (1973) in a comprehensive study of permeability changes of tonoplast and plasmalemma membranes in ageing cotyledons found that the major membrane phospholipid, phosphatidylcholine, begins to disappear once cotyledons reach maximum fresh weight, and this is followed by disappearance of phosphatidylethanolamine and inositol when the tissue starts to lose water. The minor lipids phosphatidylglycerol and glycolipids already begin to disappear 2 weeks earlier at the same time as chlorophyll losses, and are therefore probably related to breakdown of the chloroplasts. Increased RN-ase activity, loss of chlorophyll, and breakdown of chloroplast structure seem to be senescence phenomena of leaf tissue which are under hormonal control and which can be delayed experimentally by external application of kinetin and auxin, or accelerated by application of abscisic acid (MITTELHEUSER and VAN STEVENINCK, 1971 a, b). Similarly, ethylene appears to play a major role in senescence changes of fruit tissues. Most of the research on hormonal control of senescence changes has been carried out on excised tissues, hence at present there is little evidence available to indicate that hormones have a direct effect on salt relations of senescing tissues *in situ* (see Chap. 7).

5. Induced Senescence

Again a breakdown of permeability barriers and the resulting increase of AFS may be regarded as chief criteria for senescence changes. Especially in excised tissue, these changes have been shown to be hormone-controlled (e.g. in avocado fruit slices, bean endocarp and *Rhoeo* leaf slices, SACHER, 1957, 1959, 1969; SACHER and GLASZIOU, 1959; BEN-YEHOSHUA, 1964; SACHER and SALMINEN, 1969). Ethylene appears to play a major role in the changes (GALLIARD et al., 1968b) but does not have a direct effect itself on permeability (SACHER and SALMINEN, 1969), and may operate through the release of another hormone which has not been fully characterized, i.e. the senescence factor (SF) of OSBORNE et al. (1972) (see 7.2.6).

6. Conclusion

It is evident from this survey that the processes of cell differentiation and development are of cardinal interest in problems of salt regulation in the whole plant. It appears that cells which are at the surface have an especially well-developed capacity for active accumulation or excretion of ions. Cells not so exposed may either have the capacity to acquire these characteristics when exposed by slicing because they still retain the ability to differentiate (storage organs, roots), or they may have reached a state where only a breakdown of permeability barriers may occur through a process of senescent ageing (climacteric fruit).

The primary stimulus which triggers the de-repression and which eventually results in the development of ion uptake capacities is unknown. It seems important to try to relate this response to the change in exposure of cells to a new environment which often consists of large volumes of a liquid medium. It may be significant that recent observations appear to suggest that only the surface cells (epidermis and outer cortical cells) of a root are actively engaged in the loading of the cytoplasmic continuum from the cortex to the stele (Vakhmistrov, 1967; Bange, 1973), and that the induced uptake of Rb^+ and Br^- in sliced storage tissue is limited to the surface layers (Steward and Harrison, 1939). Probably, in root transport one should consider the symplast more as a primary unit overriding to a degree the individuality of cortical cells and perceiving a gradient through certain external stimuli, i.e. a polarity created by aqueous medium outwards and plasma connections inwards.

Gradients in differentiation also occur from extreme apex (non-vacuolated cells) to basal regions of the root (vacuolated cells). The large differences which exist in solute regulation of these extremes in cellular differentiation have been elaborated by Steward (see Steward and Mott, 1970). It may serve to express briefly the importance of vacuolar differentiation in providing another dimension to the salt regulatory capacity of plant cells, especially with respect to ion selectivity. For instance the observations by Rains and Floyd (1970), and by Waisel and Kuller (1972), have illustrated that vacuolar accumulations in the basal parts of bean plants may be a means to regulate transport selectively from the root to the shoot. It is also evident that equilibrium saturation (Pitman, 1967; Cram, 1973) and hormonal control (van Steveninck, 1972a, b; Pitman and Cram, 1973) are profound aspects of salt regulation in differentiating cells.

References

Adams, P.B., Rowan, K.S.: Glycolytic control of respiration during ageing of carrot root tissue. Plant Physiol. **45**, 490–494 (1970).

Adams, P.B., Rowan, K.S.: Regulation of salt respiration in carrot slices. Plant Physiol. **50**, 682–686 (1972).

Anderson, W.P., House, C.R.: A correlation between structure and function in the root of *Zea mays*. J. Exptl. Bot. **18**, 544–555 (1967).

Ap Rees, T., Beevers, H.: Pentose phosphate pathway as a major component of induced respiration of carrot and potato slices. Plant Physiol. **35**, 839–847 (1960).

Ap Rees, T., Bryant, J.A.: Effects of cycloheximide on protein synthesis and respiration in disks of carrot storage tissue. Phytochemistry **10**, 1183–1190 (1971).

Atkinson, M.R., Polya, G.M.: Salt-stimulated adenosine triphosphatases from carrot, beet and *Chara australis*. Australian J. Biol. Sci. **20**, 1069–1086 (1967).

Atkinson, M.R., Polya, G.M.: Effects of L-ethionine on adenosine triphosphate levels, respiration, and salt accumulation in carrot xylem tissue. Australian J. Biol. Sci. **21**, 409–420 (1968).

BACON, J.S.D., MACDONALD, I.R., KNIGHT, A.H.: The development of invertase activity in slices of the root of *Beta vulgaris* L. washed under aseptic conditions. Biochem. J. **94**, 175–182 (1965).

BADDELEY, M.S., HANSON, J.B.: Uncoupling of energylinked functions of corn mitochondria by linoleic acid and monomethyldecenylsuccinic acid. Plant Physiol. **42**, 1702–1710 (1967).

BANGE, G.G.J.: Diffusion and absorption of ions in plant tissues. III. The role of the root cortex cells in ion absorption. Acta Botan. Neerl. **22**, 529–542 (1973).

BEN ABDELKADER, A., MAZLIAK, P.: Influence de la "survie" (ageing) sur la biosynthèse des phospholipides dans les cellules entières ou les mitochondries du tubercule de pomme de terre (*Solanum tuberosum* L.). Compt. Rend. **267**D, 609–612 (1968).

BEN ABDELKADER, A., MAZLIAK, P., CATESSON, A.-M.: Biogenèse des lipides mitochondriaux au cours de la "survie" (ageing) de disques de parenchyme de tubercule de pomme de terre. Phytochemistry **8**, 1121–1133 (1969).

BEN-YEHOSHUA, S.: Respiration and ripening of discs of the avocado fruit. Physiol. Plantarum **17**, 71–80 (1964).

BJÖRNSTAD, P.: Phospholipase activity of rat liver mitochondira studied by the use of exogenous substrates. J. Lipid Res. **7**, 612–620 (1966).

BLOBEL, G., POTTER, R. VAN: Studies on free and membranebound ribosomes in rat liver. II. Interaction of ribosomes and membranes. J. Mol. Biol. **26**, 279–292 (1967).

BORCHERT, R., McCHESNEY, J.D.: Time course and localization of DNA synthesis during wound healing of potato tuber tissue. Develop. Biol. **35**, 293–301 (1973).

BOWEN, G.D.: Effects of soil temperature on root growth and on phosphate uptake along *Pinus radiata* roots. Australian J. Soil Res. **8**, 31–42 (1970).

BRANTON, D.: Membrane structure. Ann. Rev. Plant Physiol. **20**, 209–238 (1969).

BRYANT, J.A., AP REES, T.: Nucleic acid synthesis and induced respiration by disks of carrot storage tissue. Phytochemistry **10**, 1191–1197 (1971).

BURGESS, J.: Observations on structure and differentiation in plasmodesmata. Protoplasma **73**, 83–95 (1971).

CARAFOLI, E.: *In vivo* effect of uncoupling agents on the incorporation of calcium and strontium into mitochondria and other subcellular fractions of rat liver. J. Gen. Physiol. **50**, 1849–1864 (1967).

CARLIER, G.: Influence de la survie sur la cinétique de l'absorption du glucose-^{14}C(U) par les disques foliaires de *Pelargonium zonale*, L. Compt. Rend. **276**D, 521–524 (1973).

CASTELFRANCO, P.A., TANG, W.-J., BOLAR, M.L.: Membrane transformations in ageing potato tuber slices. Plant Physiol. **48**, 795–800 (1971).

CHANEY, R.L., BROWN, J.C., TIFFIN, L.O.: Obligatory reduction of ferric chelates in iron uptake by soy beans. Plant Physiol. **50**, 208–213 (1972).

CHAPMAN, J.M., EDELMAN, J.: Relationship between protein synthesis in tuber discs and the protein synthetic activity of a cell-free preparation. Plant Physiol. **42**, 1140–1146 (1967).

CHARDARD, R.: Observations aux microscopes electroniques à bas et haut voltàges de coupes fines et épaisses de cellules méristématiques radiculaires de *Zea mays* L. impregnées par des sels métalliques. Compt. Rend. **276**D, 2155–2158 (1973).

CHRISPEELS, M.J., SIMON, E.W.: The isolation of mitochondria from plant tissues. J. Roy. Microscop. Soc. **83**, 271–276 (1964).

CLICK, R.E., HACKETT, D.P.: The role of protein and nucleic acid synthesis in the development of respiration in potato tuber slices. Proc. Natl. Acad. Sci. U.S. **50**, 243–250 (1963).

COCUCCI, M.C., MARRÉ, E.: The effect of cycloheximide on respiration, protein synthesis and adenosine nucleotide levels in *Rhodotorula gracilis*. Plant Sci. Letters **1**, 293–301 (1973).

COSTERTON, J.W.F., MACROBBIE, E.A.C.: Ultrastructure of *Nitella translucens* in relation to ion transport. J. Exptl. Bot. **21**, 535–542 (1970).

COULOMB, P., COULOMB, C.: Localisation cytochimique ultrastructurale d'une adénosine triphosphatase—Mg^{++} dépendante dans des cellules de méristèmes radiculaires de la courge (*Cucurbita pepo* L. Cucurbitacée). Compt. Rend. **275**D, 1035–1038 (1972).

CRAM, W.J.: The initiation of developmental drifts in excised plant tissues. Australian J. Biol. Sci. **25**, 855–859 (1972).

CRAM, W.J.: Internal factors regulating nitrate and chloride influx in plant cells. J. Exptl. Bot. **24**, 328–341 (1973).

DALGARNO, L., BIRT, L.M.: Free fatty acids in carrot tissue preparations and their effect on isolated mitochondria. Biochem. J. **87**, 586–596 (1963).

Dalgarno, L., Hird, F.J.R.: Increase in the process of accumulation of amino acids in carrot slices with prolonged aerobic washing. Biochem. J. **76**, 209–215 (1960).

Dodds, J.J.A., Ellis, R.J.: Cation-stimulated adenosine triphosphatase activity in plant cell walls. Biochem. J. **101**, 31P–32P (1966).

Duda, C.T., Cherry, J.H.: Chromatin- and nuclei-directed ribonucleic acid synthesis in sugar beet root. Plant Physiol. **47**, 262–268 (1971).

Earnshaw, M.J., Madden, D.M., Hanson, J.B.: Calcium accumulation by corn mitochondria. J. Exptl. Bot. **24**, 828–840 (1973).

Earnshaw, M.J., Truelove, B.: Swelling of *Phaseolus* mitochondria induced by the action of phospholipase A. Plant Physiol. **45**, 322–326 (1970).

Earnshaw, M.J., Truelove, B., Butler, R.D.: Swelling of *Phaseolus* mitochondria in relation to free fatty acid levels. Plant Physiol. **45**, 318–321 (1970).

Ehwald, R., Sammler, P., Göring, H.: Die Bedeutung der Diffusion im „Freien Raum" für die Konzentrationsabhängigkeit der Aufnahme von Zuckern und Ionen durch pflanzliche Gewebe. Biochem. Physiol. Pflanzen **164**, 596–613 (1973).

Eilam, Y.: Permeability changes in senescing tissue. J. Exptl. Bot. **16**, 614–627 (1965).

Ellis, R.J., MacDonald, I.R.: Activation of protein synthesis by microsomes from ageing beet disks. Plant Physiol. **42**, 1297–1302 (1967).

Ellis, R.J., MacDonald, I.R.: Specificity of cycloheximide in higher plant systems. Plant Physiol. **46**, 227–232 (1970).

Epstein, E.: Mechanisms of ion transport through plant cell membranes. Intern. Rev. Cytol. **34**, 123–168 (1973).

Eshel, A., Waisel, Y.: Variations in sodium uptake along primary roots of corn seedlings. Plant Physiol. **49**, 585–589 (1972).

Eshel, A., Waisel, Y.: Variations in uptake of sodium and rubidium along barley roots. Physiol. Plantarum **28**, 557–560 (1973).

Faiz-Ur-Rahman, A.T.M., Davies, D.D., Trewavas, A.J.: The control of glycolysis in aged slices of carrot root tissue. Planta **118**, 211–224 (1974).

Ferguson, C.H.R., Simon, E.W.: Membrane lipids in senescing green tissues. J. Exptl. Bot. **24**, 307–316 (1973).

Filner, P., Wray, J.L., Varner, J.E.: Enzyme induction in higher plants. Science **165**, 358–367 (1969).

Fisher, J., Hodges, T.K.: Monovalent ion stimulated adenosine triphosphatase from oat root. Plant Physiol. **44**, 385–395 (1969).

Floyd, R.A., Rains, D.W.: Investigation of respiratory and ion transport properties of ageing bean stem slices. Plant Physiol. **47**, 663–667 (1971).

Fowler, M.W., Ap Rees, T.: Carbohydrate oxidation during differentiation in roots of *Pisum sativum*. Biochim. Biophys. Acta **201**, 33–44 (1970).

Galliard, T., Rhodes, M.J.C., Wooltorton, L.S.C., Hulme, A.C.: Metabolic changes in excised fruit tissue. II. The development of a lipid synthesis system during the ageing of peel disks from pre-climacteric apples. Phytochemistry **7**, 1453–1463 (1968a).

Galliard, T., Rhodes, M.J.C., Wooltorton, L.S.C., Hulme, A.C.: Metabolic changes in excised fruit tissue. III. The development of ethylene biosynthesis during the ageing of disks of apple peel. Phytochemistry **7**, 1465–1470 (1968b).

Gayler, K.R., Glasziou, K.T.: Plant enzyme synthesis: decay of messenger RNA for peroxidase in sugarcane stem tissue. Phytochemistry **7**, 1247–1251 (1968).

Gayler, K.R., Glasziou, K.T.: Plant enzyme synthesis: hormonal regulation of invertase and peroxidase synthesis in sugar cane. Planta **84**, 185–194 (1969).

Glasziou, K.T.: Control of enzyme formation and inactivation in plants. Ann. Rev. Plant Physiol. **20**, 63–88 (1969).

Glaumann, H.: Techniques for fractionating intracellular membranes with particular reference to the synthesis and transport of protons. In: Techniques in protein biosynthesis (P.N. Campbell, J.R. Sargent, eds.), p. 191–248. London-New York: Academic Press 1973.

Grunwald, C.: Calcium uptake by potato tuber mitochondria. I. Isolation and identification of calcium-45 complexes. Physiol. Plantarum **19**, 335–347 (1966).

Hackett, D.P.: Respiratory mechanisms and control in higher plant tissues. In: Control mechanisms in respiration and fermentation (B. Wright, ed.), p. 105–127. New York: Ronald Press 1963.

HADAČOVÁ, V., LUŠTINEC, J., KAMIŃEK, M.: Kinetin and naphthalene acetic acid controlled starch formation in isolated roots of *Pisum sativum*. Biol. Plant. Acad. Sci. Bohemoslov. **15**, 427–429 (1973).

HALL, J.L.: Enzyme localization and ion transport. In: Ion transport in plants (W.P. ANDERSON, ed.), p. 11–24. London-New York: Academic Press 1973.

HALL, J.L., BUTT, V.S.: Adenosine triphosphatase activity in cell wall preparations and excised roots of barley. J. Exptl. Bot. **20**, 751–762 (1969).

HANSON, J.B., LEONARD, R.T., MOLLENHAUER, H.H.: Increased electron density of tonoplast membranes in washed corn root tissue. Plant Physiol. **52**, 298–300 (1973).

HARRISON-MURRAY, R.S., CLARKSON, D.T.: Relationships between structural development and the absorption of ions by the root system of *Cucurbita pepo*. Planta **114**, 1–16 (1973).

HIGINBOTHAM, N., DAVIS, R.F., MERTZ, S.M., SHUMWAY, L.K.: Some evidence that radial transport in maize roots is into living vessels. In: Ion transport in plants (W.P. ANDERSON, ed.), p. 493–506. London-New York: Academic Press 1973.

HILL, B.S., HILL, A.E.: Enzymatic approaches to chloride transport in the *Limonium* salt gland. In: Ion transport in plants (W.P. ANDERSON, ed.), p. 379–384. London-New York: Academic Press 1973.

HOAGLAND, D.R., BROYER, T.C.: General nature of the process of salt accumulation by roots with description of experimental methods. Plant Physiol. **11**, 471–507 (1936).

HOAGLAND, D.R., BROYER, T.C.: Hydrogen-ion effects and the accumulation of salt by barley roots as influenced by metabolism. Amer. J. Bot. **27**, 173–185 (1940).

HOAGLAND, D.R., BROYER, T.C.: Accumulation of salt and permeability in plant cells. J. Gen. Physiol. **25**, 865–880 (1942).

HODGES, T.K., ELZAM, O.E.: Effect of azide and oligomycin on the transport of calcium ions in corn mitochondria. Nature **215**, 970–972 (1967).

HODGES, T.K., HANSON, J.B.: Calcium accumulation by maize mitochondria. Plant Physiol. **40**, 101–109 (1965).

HODGES, T.K., LEONARD, R.T., BRACKER, C.E., KEENAN, T.W.: Purification of ion-stimulated adenosine triphosphatase from plant roots: association with plasma membranes. Proc. Natl. Acad. Sci. U.S. **69**, 3307–3311 (1972).

HUFFAKER, R.C., PETERSON, L.W.: Protein turnover in plants and possible means of its regulation. Ann. Rev. Plant Physiol. **25**, 363–392 (1974).

JACKMAN, M.E., STEVENINCK, R.F.M. VAN: Changes in the endoplasmic reticulum of beetroot slices during ageing. Australian J. Biol. Sci. **20**, 1063–1068 (1967).

JACKSON, P.C., ADAMS, H.R.: Ion exchange features of K$^+$ absorption by barley roots. Plant Physiol. **38**, XXVII (1963).

JACKSON, P.C., EDWARDS, D.G.: Cation effects on chloride fluxes and accumulation levels in barley roots. J. Gen. Physiol. **50**, 225–241 (1966).

JACKSON, P.C., STIEF, K.J.: Equilibrium and ion exchange characteristics of potassium and sodium accumulation by barley roots. J. Gen. Physiol. **48**, 601–616 (1965).

JACOBSON, B.S., SMITH, B.N., EPSTEIN, S., LATIES, G.G.: The prevalence of carbon-13 in respiratory carbon dioxide as an indicator of the type of endogenous substrate. J. Gen. Physiol. **55**, 1–17 (1970).

JACOBY, B.: Function of bean roots and stems in sodium retention. Plant Physiol. **39**, 445–449 (1964).

JACOBY, B.: The effect of ATP on uptake of monovalent cations and anions by red beet slices. J. Exptl. Bot. **16**, 243–248 (1965).

JACOBY, B., DAGAN, J.: Effect of age on sodium fluxes in primary bean leaves. Physiol. Plantarum **22**, 29–36 (1969).

JACOBY, B., TIROSH, T., PLESSNER, O.E.: Relationship between age of bean leaves, sodium export, and permeability of leaf tissue. Botan. Gaz. **134**, 46–49 (1973).

JOHANSEN, C., EDWARDS, D.G., LONERAGAN, J.F.: Potassium fluxes during potassium absorption by intact barley plants of increasing potassium content. Plant Physiol. **45**, 601–603 (1970).

JONES, R.L.: Gibberellins: their physiological role. Ann. Rev. Plant Physiol. **24**, 571–598 (1973).

KAHL, G.: Synthesis of r-RNA, t-RNA and other RNA-species concomitant with polyribosome formation in ageing potato tuber slices. Z. Naturforsch. **26B**, 1058–1064 (1971a).

KAHL, G.: Activation of protein synthesis in ageing potato tuber slices. Z. Naturforsch. **26B**, 1064–1067 (1971b).

KAHL, G.: Genetic and metabolic regulation in differentiating plant storage tissue cells. Bot. Rev. **39**, 274–299 (1973).

KAHL, G.: Metabolism in plant storage slices. Bot. Rev. **40**, 263–314 (1974).

KAHL, G., LANGE, H., ROSENSTOCK, G.: Substratspiegel, Enzymaktivitäten und genetische Regulation nach Derepression in pflanzlichen Speichergeweben. Z. Naturforsch. **24b**, 911–918 (1969a).

KAHL, G., LANGE, H., ROSENSTOCK, G.: Regulation glykolytischen Umsatzes durch Synthese und Abbau von Enzymen. Z. Naturforsch. **24b**, 1544–1549 (1969b).

KENEFICK, D.G., HANSON, J.B.: Contracted state as an energy source for Ca-binding and Ca + organic phosphate accumulation by corn mitochondria. Plant Physiol. **41**, 1601–1609 (1966).

KIRK, J.T.O.: Failure to detect effects of cycloheximide on energy metabolism in *Euglena gracilis*. Nature **226**, 182 (1970).

KOLLMANN, R., SCHUMACHER, W.: Über die Feinstruktur des Phloems von *Metasequoia glyptostroboides* und seine jahreszeitlichen Veränderungen. II. Mitteilung. Vergleichende Untersuchungen der plasmatischen Verbindungsbrücken in Phloemparenchymzellen und Siebzellen. Planta **58**, 366–386 (1962).

KOLLMANN, R., SCHUMACHER, W.: Über die Feinstruktur des Phloems von *Metasequoia glyptostroboides* und seine jahreszeitlichen Veränderungen. IV. Mitteilung. Weitere Beobachtungen zum Feinbau der Plasmabrücken in den Siebzellen. Planta **60**, 360–389 (1963).

KUIPER, P.J.C.: Water transport across membranes. Ann. Rev. Plant Physiol. **23**, 157–172 (1972).

KULL, U., UNGER, M.: Wirkungen von Abscisinsäure auf den Kohlenhydrat- und Fettsäurehaushalt von *Coleus blumei*. Z. Pflanzenphysiol. **72**, 135–140 (1974).

KYLIN, A., GEE, R.: Adenosine triphosphatase activities in leaves of the mangrove *Avicennia nitida* Jacq. Plant Physiol. **45**, 169–172 (1970).

LÄUCHLI, A., KRAMER, D., PITMAN, M.G., LÜTTGE, U.: Ultrastructure of xylem parenchyma cells of barley roots in relation to ion transport to the xylem. Planta **119**, 85–99 (1974).

LÄUCHLI, A., KRAMER, D., STELZER, R.: Ultrastructure and ion localization in xylem parenchyma cells of roots. In: Membrane transport in plants (U. ZIMMERMANN, J. DAINTY, eds.), p. 363–371. Berlin-Heidelberg-New York: Springer 1974.

LÄUCHLI, A., LÜTTGE, U., PITMAN, M.G.: Ion uptake and transport through barley seedlings: differential effect of cycloheximide. Z. Naturforsch. **28c**, 431–434 (1973).

LÄUCHLI, A., SPURR, A.R., EPSTEIN, E.: Lateral transport of ions into the xylem of corn roots. II. Evaluation of a stelar pump. Plant Physiol. **48**, 118–124 (1971).

LATIES, G.G.: Respiration and cellular work and the regulation of the respiration rate in plants. Surv. Biol. Progr. **3**, 215–299 (1957).

LATIES, G.G.: Controlling influence of thickness in development and type of respiratory activity in potato slices. Plant Physiol. **37**, 679–690 (1962).

LATIES, G.G.: Control of respiratory quality and magnitude during development. In: Control mechanisms in respiration and fermentation (B. WRIGHT, ed.), p. 129–155. New York: Ronald Press 1963.

LATIES, G.G.: The relation of glucose absorption to respiration in potato slices. Plant Physiol. **39**, 391–397 (1964a).

LATIES, G.G.: The onset of tricarboxylic acid cycle activity with ageing in potato slices. Plant Physiol. **39**, 654–663 (1964b).

LATIES, G.G.: Metabolic and physiological development in plant tissues. Australian J. Sci. **30**, 193–203 (1967).

LATIES, G.G.: Solute transport in relation to metabolism and membrane permeability in plant tissues. In: Historical and current aspects of plant physiology: A symposium honoring Professor F.C. STEWARD (P.J. DAVIES, ed.) p. 98–151. Ithaca: New York State College of Agriculture and Life Sciences 1975.

LATIES, G.G., HOELLE, C.: Malonate and cyanide insensitivity in relation to respiratory compensation in potato slices. Plant Physiol. **40**, 757–764 (1965).

LATIES, G.G., HOELLE, C.: The α-oxidation of long-chain fatty acids as a possible component of the basal respiration of potato slices. Phytochemistry **6**, 49–57 (1967).

LATIES, G.G., HOELLE, C., JACOBSON, B.S.: α-Oxidation of endogenous fatty acids in fresh potato slices. Phytochemistry **11**, 3403–3411 (1972).

LAVINTMAN, N., CARDINI, C.E.: Changes in sucrose synthetase activities in ageing potato tuber slices. Plant Physiol. **43**, 434–436 (1968).

LEAVER, C.J., KEY, J.L.: Polyribosome formation and RNA synthesis during ageing of carrot root tissue. Proc. Natl. Acad. Sci. U.S. **57**, 1338–1344 (1967).

LEONARD, R.T., HANSON, J.B.: Induction and development of increased ion absorption in corn root tissue. Plant Physiol. **49**, 430–435 (1972a).

LEONARD, R.T., HANSON, J.B.: Increased membrane bound adenosine triphosphatase activity accompanying development of enhanced solute uptake in washed corn root tissue. Plant Physiol. **49**, 436–440 (1972b).

LIN, C.-Y., TRAVIS, R.L., CHIA, S.Y., KEY, J.L.: Changes in ribosomal activity during incubation of *Daucus carota* root disks. Phytochemistry **12**, 2801–2807 (1973).

LINASK, J., LATIES, G.G.: Multiphasic absorption of glucose and 3-0-methylglucose by aged potato slices. Plant Physiol. **51**, 289–294 (1973).

LIPETZ, J.: Abundance of endoplasmic reticulum in auxin autonomous plant cells. J. Cell Biol. **43**, 80[a] (1969).

LÓPEZ-SÁEZ, J.F., GIMÉNEZ-MARTÍN, G., RISUEÑO, M.C.: Fine structure of the plasmodesm. Protoplasma **61**, 81–84 (1966).

LÜTTGE, U., CRAM, W.J., LATIES, G.G.: The relationship of salt stimulated respiration to localised ion transport in carrot tissue. Z. Pflanzenphysiol. **64**, 418–426 (1971).

LÜTTGE, U., LÄUCHLI, A., BALL, E., PITMAN, M.G.: Cycloheximide: a specific inhibitor of protein synthesis and intercellular ion transport in plant roots. Experientia **30**, 470–471 (1974).

LUND, H.A., VATTER, A.E., HANSON, J.B.: Biochemical and cytological changes accompanying growth and differentiation in the roots of *Zea mays*. J. Biophys. Biochem. Cytol. **4**, 87–98 (1958).

MACDONALD, I.R.: Further evidence of oxygen diffusion as the determining factor in the relation between disk thickness and respiration of potato tissue. Plant Physiol. **43**, 274–280 (1968).

MACDONALD, I.R., BACON, J.S.D., VAUGHAN, D., ELLIS, R.J.: The relation between ion absorption and protein synthesis in beet disks. J. Exptl. Bot. **17**, 822–837 (1966).

MACDONALD, I.R., ELLIS, R.J.: Does cycloheximide inhibit protein synthesis specifically in plant tissues. Nature **222**, 791–792 (1969).

MACDONALD, I.R., KOCK, D.C. DE: Temperature control and metabolic drifts in ageing disks of storage tissue. Ann. Bot. **22**, 429–448 (1958).

MACROBBIE, E.A.C.: Ion fluxes to the vacuole of *Nitella translucens*. J. Exptl. Bot. **20**, 236–256 (1969).

MACROBBIE, E.A.C.: Vacuolar fluxes of chloride and bromide in *Nitella translucens*. J. Exptl. Bot. **22**, 487–502 (1971).

MARCUS, A.: Enzyme induction in plants. Ann. Rev. Plant Physiol. **22**, 313–336 (1971).

MARINOS, N.G.: Studies on submicroscopic aspects of mineral deficiencies. I. Calcium deficiency in the shoot apex of barley. Amer. J. Bot. **49**, 834–841 (1962).

MARRÉ, E.: Phosphorylation in higher plants. Ann. Rev. Plant Physiol. **12**, 195–218 (1961).

MARTY, F.: Sites réactifs à l'iodure de zinc-tetroxyde d'osmium dans les cellules de la racine d'*Euphorbia characias*. Compt. Rend. **277 D**, 1317–1320 (1973a).

MARTY, F.: Dissemblance des faces golgiennes et activité des dictyosomes dans les cellules en cours de vacuolisation de la racine d'*Euphorbia characias* L. Compt. Rend. **277 D**, 1749–1752 (1973b).

MAZLIAK, P.: Lipid metabolism in plants. Ann. Rev. Plant Physiol. **24**, 287–310 (1973).

MERTZ, D., LEVITT, J.: The relation between ion absorption on the cell wall and active uptake. Physiol. Plantarum **14**, 57–61 (1961).

MITTELHEUSER, C.J., STEVENINCK, R.F.M. VAN: The ultrastructure of wheat leaves. I. Changes due to natural senescence and the effects of kinetin and ABA on detached leaves incubated in the dark. Protoplasma **73**, 239–252 (1971a).

MITTELHEUSER, C.J., STEVENINCK, R.F.M. VAN: The ultrastructure of wheat leaves. II. The effects of kinetin and ABA on detached leaves incubated in the dark. Protoplasma **73**, 253–262 (1971b).

MORRÉ, D.J.: Membrane biogenesis. Ann. Rev. Plant Physiol. **26**, 441–481 (1975).

MORRÉ, D.J., MOLLENHAUER, H.H., BRACKER, C.E.: Origin and continuity of Golgi apparatus. In: Results and problems in cell differentiation. II. Origin and continuity of cell organelles (T. REINERT, H. URSPRUNG, eds.), p. 82–126. Berlin-Heidelberg-New York: Springer 1971.

NATHANSON, A.: Über die Regulation der Aufnahme anorganischer Salze durch die Knollen von *Dahlia*. Jahrb. Wiss. Bot. **39**, 607–644 (1904).

NEIRINCKX, L.J.A., BANGE, G.G.J.: Irreversible equilibration of barley roots with Na⁺ ions of different external Na⁺ concentrations. Acta Botan. Néerl. **20**, 481–488 (1971).

NORTHCOTE, D.H.: The Golgi apparatus. Endeavour **30**, 26–33 (1971).

ÖZER, N., BASTIN, M.: Some recent and original results on the role of some regulators in the active uptake and transport metabolism with special reference to amino acid metabolism in Jerusalem artichoke. In: Hormonal regulation in plant growth and development (H. KALDEWEY, Y. VARDAR, eds.), p. 71–82. Weinheim: Verlag Chemie 1972.

OSBORNE, D.J., JACKSON, M.B., MILBORROW, B.V.: Physiological properties of abscission accelerator from senescent leaves. Nature New Biol. **240**, 98–101 (1972).

PACKER, L., MURAKAMI, S., MEHARD, C.W.: Ion transport in chloroplasts and plant mitochondria. Ann. Rev. Plant Physiol. **21**, 271–304 (1970).

PASQUALI-RONCHETTI, I., GREENWALT, J.W., CARAFOLI, E.: On the nature of the dense matrix granules of normal mitochondria. J. Cell Biol. **40**, 565–568 (1969).

PEACHEY, L.D.: Electron microscopic observations on the accumulation of divalent cations in intramitochondrial granules. J. Cell Biol. **20**, 95–111 (1964).

PITMAN, M.G.: The determination of the salt relations of the cytoplasmic phase in cells of beetroot tissue. Australian J. Biol. Sci. **16**, 647–668 (1963).

PITMAN, M.G.: Conflicting measurements of sodium and potassium uptake by barley roots. Nature **216**, 1343–1344 (1967).

PITMAN, M.G., COURTICE, A.C., LEE, B.: Comparison of potassium and sodium uptake by barley roots at high and low salt status. Australian J. Biol. Sci. **21**, 871–881 (1968).

PITMAN, M.G., CRAM, W.J.: Regulation of inorganic ion transport in plants. In: Ion transport in plants (W.P. ANDERSON, ed.), p. 465–481. London-New York: Academic Press 1973.

PITMAN, M.G., MOWAT, J., NAIR, H.: Interaction of processes for accumulation of salt and sugar in barley roots. Australian J. Biol. Sci. **24**, 619–631 (1971).

POLYA, G.M.: Inhibition of protein synthesis and cation uptake in beetroot tissue by cycloheximide and cryptopleurine. Australian J. Biol. Sci. **21**, 1107–1118 (1968).

POLYA, G.M.: Transcription. In: The ribonucleic acids (P.R. STEWART, D.S. LETHAM, eds.), p. 7–36. Berlin-Heidelberg-New York: Springer 1973.

POLYA, G.M., ATKINSON, M.R.: Evidence for a direct involvement of electron transport in the high affinity ion accumulation system of aged beet parenchyma. Australian J. Biol. Sci. **22**, 573–584 (1969).

POOLE, R.J.: Effect of sodium on potassium fluxes at the cell membrane and vacuole membrane of red beet. Plant Physiol. **47**, 731–734 (1971a).

POOLE, R.J.: Development and characteristics of sodium selective transport in red beet. Plant Physiol. **47**, 735–739 (1971b).

RAINS, D.W.: Sodium and potassium absorption by bean stem tissue. Plant Physiol. **44**, 547–554 (1969).

RAINS, D.W., EPSTEIN, E.: Sodium absorption by barley roots: role of the dual mechanisms of alkali cation transport. Plant Physiol. **42**, 314–318 (1967a).

RAINS, D.W., EPSTEIN, E.: Sodium absorption by barley roots: its mediation by mechanism 2 of alkali cation transport. Plant Physiol. **42**, 319–323 (1967b).

RAINS, D.W., FLOYD, R.A.: Influence of calcium on sodium and potassium absorption by fresh and aged bean slices. Plant Physiol. **46**, 93–98 (1970).

REED, D.J., KOLATTUKUDY, P.E.: Metabolism of red beet slices. I. Effects of washing. Plant Physiol. **41**, 653–660 (1966).

REILLY, C., FUHRMANN, G.-F., ROTHSTEIN, A.: The inhibition of K⁺ and phosphate uptake in yeast by cycloheximide. Biochim. Biophys. Acta **203**, 583–585 (1970).

REINHOLD, L., SHTARKSHALL, R.A., GANOT, D.: Transport of amino acids in barley leaf tissue. II. The kinetics of uptake of an unnatural analogue. J. Exptl. Bot. **21**, 926–932 (1970).

RICARDO, C.P.P., AP REES, T.: Invertase activity during the development of carrot roots. Phytochemistry **9**, 239–247 (1970).

RICARDO, C.P.P., AP REES, T.: Activities of key enzymes of carbohydrate oxidation in disks of carrot storage tissue. Phytochemistry **11**, 623–626 (1972).

ROBARDS, A.W.: The ultrastructure of plasmodesmata. Protoplasma **72**, 315–323 (1971).

ROBARDS, A.W., JACKSON, S.M., CLARKSON, D.T., SANDERSON, J.: The structure of barley roots in relation to the transport of ions into the stele. Protoplasma **77**, 291–311 (1973).

ROBERTSON, R.N.: Mechanism of absorption and transport of inorganic nutrients in plants. Ann. Rev. Plant Physiol. **2**, 1–24 (1951).

ROBERTSON, R.N.: The mechanism of absorption. In: Encyclopedia of plant physiology (W. RUHLAND, ed.), vol. II, p. 449–467. Berlin-Heidelberg-New York: Springer 1956.

ROBERTSON, R.N.: Ion transport and respiration. Biol. Rev. **35**, 231–264 (1960).

ROBERTSON, R.N.: Protons, electrons, phosphorylation and active transport. London-New York: Cambridge University Press 1968.

ROBERTSON, R.N., TURNER, J.S.: Studies in the metabolism of plant cells. III. The effects of cyanide on the accumulation of potassium chloride and on respiration; the nature of salt respiration. Australian J. Exptl. Biol. Med. Sci. **23**, 63–73 (1945).

ROBERTSON, R.N., TURNER, J.S., WILKINS, M.J.: Studies in the metabolism of plant cells. V. Salt respiration and accumulation in red beet tissue. Australian J. Exptl. Biol. Med. Sci. **25**, 1–8 (1947).

ROBERTSON, R.N., WILKINS, M.J., WEEKS, D.C.: Studies in the metabolism of plant cells. IX. The effects of 2, 4-dinitrophenol on salt accumulation and salt respiration. Australian J. Sci., Res. Ser. **B4**, 248–264 (1951).

ROSENSTOCK, G., KAHL, G., LANGE, H.: Beziehungen zwischen Entwicklungszustand und Wund-atmung beim Speicherparenchym von *Solanum tuberosum*. Z. Pflanzenphysiol. **64**, 130–138 (1971).

ROWAN, K.S.: Phosphorus metabolism in plants. Intern. Rev. Cytol. **19**, 301–391 (1966).

RUNGIE, J.M., WISKICH, J.T.: Changes in microsomal electron transport of plant storage tissues induced by slicing and ageing. Australian J. Biol. Sci. **25**, 103–113 (1972a).

RUNGIE, J.M., WISKICH, J.T.: Soluble electron-transport activities in fresh and aged turnip tissue. Planta **102**, 190–205 (1972b).

RYAN, J.W., SMITH, U.: A rapid, simple method for isolating pinocytotic vesicles and plasma membrane of lung. Biochim. Biophys. Acta **249**, 177–180 (1972).

SACHER, J.A.: Relationship between auxin and membrane integrity in tissue senescence and abscission. Science **125**, 1199–1200 (1957).

SACHER, J.A.: Studies on auxin-membrane permeability relations in fruit and leaf tissues. Plant Physiol. **34**, 365–372 (1959).

SACHER, J.A.: Relations between changes in membrane permeability and the climacteric in banana and avocado. Nature **195**, 577–578 (1962).

SACHER, J.A.: Hormonal control of senescence of bean endocarp: auxin-suppression of RN-ase. Plant Physiol. **44**, 313–314 (1969).

SACHER, J.A., GLASZIOU, K.T.: Effects of auxins on membrane permeability and pectic substances in bean endocarp. Nature **183**, 757–758 (1959).

SACHER, J.A., SALMINEN, S.O.: Comparative studies of effect of auxin and ethylene on permeabilty and synthesis of RNA and protein. Plant Physiol. **44**, 1371–1377 (1969).

SAKANO, K., ASAHI, T.: Biochemical studies on biogenesis of mitochondria in sweet potato root tissue. I. Time course analysis of increase in mitochondrial enzymes. Plant Cell Physiol. **12**, 417–426 (1971).

SAKANO, K., ASAHI, T., URITANI, I.: Heterogeneity of mitochondrial particles in fresh and wounded tissue of sweet potato roots. Plant Cell Physiol. **9**, 49–60 (1968).

SAMPSON, M.J., LATIES, G.G.: Ribosomal RNA synthesis in newly sliced discs of potato tuber. Plant Physiol. **43**, 1011–1016 (1968).

SCANDALIOS, J.G.: Isozymes in development and differentiation. Ann. Rev. Plant Physiol. **25**, 225–258 (1974).

SHACHAR-HILL, B., HILL, A.E.: Ion and water transport in *Limonium*. VI. The induction of chloride pumping. Biochim. Biophys. Acta **211**, 313–317 (1970).

SHTARKSHALL, R., REINHOLD, L.: Multiphasic amino acid transport in leaf cells. In: Membrane transport in plants (U. ZIMMERMANN, J. DAINTY, eds.), p. 338–342. Berlin-Heidelberg-New York: Springer 1974.

SHTARKSHALL, R., REINHOLD, L., HAREL, H.: Transport of amino acids in barley leaf tissue. I. Evidence for a specific uptake mechanism and the influence of "ageing" on accumulatory capacity. J. Exptl. Bot. **21**, 915–925 (1970).

SPENCER, T., BYGRAVE, F.L.: The role of mitochondria in modifying the cellular ionic environ-ment: Studies of the kinetic accumulation of calcium by rat liver mitochondria. Bioenergetics **4**, 347–362 (1973).

STEVENINCK, M.E. van: Fine structure of plant cells in relation to salt accumulation. M.Sc. Thesis, University of Adelaide (1970).

STEVENINCK, M.E. van, STEVENINCK, R.F.M. van: Effect of protein synthesis inhibitors on the formation of crystalloid inclusions in the endoplasmic reticulum of beetroot cells. Protoplasma **73**, 107–119 (1971).

Steveninck, R.F.M. van: The "lag-phase" in salt uptake of storage tissue. Nature **190**, 1072–1075 (1961).

Steveninck, R.F.M. van: A comparison of chloride and potassium fluxes in red beet tissue. Physiol. Plantarum **17**, 757–770 (1964).

Steveninck, R.F.M. van: The effects of calcium and tris (hydroxymethyl) aminomethane on potassium uptake during and after the lag phase in red beet tissue. Australian J. Biol. Sci. **18**, 227–233 (1965).

Steveninck, R.F.M. van: Some metabolic implications of the tris effect in beetroot tissue. Australian J. Biol. Sci. **19**, 271–281 (1966a).

Steveninck, R.F.M. van: Further aspects of the tris effect in beetroot tissue during its lag phase. Australian J. Biol. Sci. **19**, 283–290 (1966b).

Steveninck, R.F.M. van: Abscissic acid stimulation of ion transport and alteration in K^+/Na^+ selectivity. Z. Pflanzenphysiol. **67**, 282–286 (1972a).

Steveninck, R.F.M. van: Inhibition of the development of a cation accumulatory system and of tris-induced uptake in storage tissue by N^6-benzyl adenine and kinetin. Physiol. Plantarum **27**, 43–47 (1972b).

Steveninck, R.F.M. van: The "washing" or "aging" phenomenon in plant tissues. Ann. Rev. Plant Physiol. **26**, 237–258 (1975).

Steveninck, R.F.M. van, Jackman, M.E.: Respiratory activity and morphology of mitochondria isolated from whole and sliced storage tissue. Australian J. Biol. Sci. **20**, 749–760 (1967).

Steveninck, R.F.M. van, Mittelheuser, C.J., Steveninck, M.E. van: Effects of tris-buffer on ion uptake and cellular ultrastructure. In: Ion transport in plant cells (W.P. Anderson, ed.), p. 251–269. London-New York: Academic Press 1973.

Steveninck, R.F.M. van, Steveninck, M.E. van: Effect of inhibitors of protein and nucleic acid synthesis on the development of ion uptake mechanisms in beetroot slices (*Beta vulgaris*). Physiol. Plantarum **27**, 407–411 (1972).

Steveninck, R.F.M. van, Steveninck, M.E. van, Hall, T.A., Peters, P.D.: X-ray Microanalysis and distribution of halides in *Nitella translucens*. In: Electron microscopy 1974 (J.V. Sandars, D.J. Goodchild, eds.), vol. II, p. 602–603. Canberra, A.C.T.: The Australian Academy of Science 1974.

Steward, F.C.: The absorption and accumulation of solutes by living plant cells. I. Experimental conditions which determine salt absorption by storage tissue. Protoplasma **15**, 29–58 (1932).

Steward, F.C.: Salt accumulation by plants—role of growth and metabolism. Trans. Faraday Soc. **33**, 1006–1016 (1937).

Steward, F.C., Berry, W.E., Preston, C., Ramamurti, T.K.: The absorption and accumulation of solutes by living plant cells. X. Time and temperature effects on salt uptake by potato discs and the influence of the storage conditions of the tubers on metabolism and other properties. Ann. Bot. (London), N.S. **7**, 221–260 (1943).

Steward, F.C., Harrison, J.A.: The absorption and accumulation of salts by living plant cells. IX. The absorption or rubidium bromide by potato discs. Ann. Bot. (London), N.S. **3**, 427–453 (1939).

Steward, F.C., Mapes, M.O., Kent, A.E., Holsten, R.D.: Growth and development of cultured plant cell. Biochemical and morphogenetic studies with cells yield new evidence on their metabolism and totipotency. Science **143**, 20–27 (1964).

Steward, F.C., Millar, F.K.: Salt accumulation in plants: a reconsideration of the role of growth and metabolism. Symp. Soc. Exptl. Biol. **8**, 367–406 (1954).

Steward, F.C., Mott, R.L.: Cells, solutes and growth: salt accumulation in plants re-examined. Intern. Rev. Cytol. **28**, 275–370 (1970).

Steward, F.C., Preston, C.: Metabolic processes of potato discs under conditions conductive to salt accumulation. Plant Physiol. **15**, 23–61 (1940).

Steward, F.C., Preston, C.: The effect of salt concentration upon the metabolism of potato discs and the contrasted effect of potassium and calcium salts which have a common ion. Plant Physiol. **16**, 85–116 (1941a).

Steward, F.C., Preston, C.: Effects of pH and the components of bicarbonate and phosphate buffered solutions on the metabolism of potato discs and their ability to absorb ions. Plant Physiol. **16**, 481–519 (1941b).

STEWARD, F.C., WRIGHT, R., BERRY, W.E.: The absorption and accumulation of solutes by living plant cells. III. The respiration of cut discs of potato tuber in air and immersed in water, with observations upon surface: volume effects and salt accumulation. Protoplasma **16**, 576–611 (1932).

STEWART, P.R.: Inhibitors of translation: In: The ribo-nucleic acids (P.R. STEWART, D.S. LETHAM, eds.), p. 151–158. Berlin-Heidelberg-New York: Springer 1973.

STEWART, P.R., LETHAM, D.S.: The ribonucleic acids. Berlin-Heidelberg-New York: Springer 1973.

STONE, B.P., WHITTY, C.D., CHERRY, J.H.: Effect of ethionine on development of invertase in slices of sugar beet tissue (*Beta vulgaris* L.). Plant Physiol., Suppl. **44**, 36 (1970).

STRICKLAND, R.G.: Effect of protein and fatty acids on the oxidation of succinate and α-oxoglutarate by pea-root mitochondria. Biochem. J. **81**, 286–291 (1961).

SUTCLIFFE, J.F.: Cation absorption by non-growing plant cells. Symp. Soc. Exptl. Biol. **8**, 325–342 (1954).

SUTCLIFFE, J.F.: Salt uptake in plants. Biol. Rev. **34**, 159–220 (1959).

SWARDT, G.H. DE, ROUSSEAU, G.G.: Relationship between changes in membrane permeability and the respiration climacteric in pericarp tissue of tomatoes. Planta **112**, 83–86 (1973).

TANG, W.-J., CASTELFRANCO, P.A.: Phospholipid synthesis in ageing potato slices. Plant Physiol. **43**, S-41 (1968a).

TANG, W.J., CASTELFRANCO, P.A.: Phospholipid synthesis in ageing potato tuber tissue. Plant Physiol. **43**, 1232–1238 (1968b).

TASSERON-DE JONG, J.G., VELDSTRA, H.: Investigations on cytokinins. I. Effect of 6-benzylamino purine on growth and starch content of *Lemna minor*. Physiol. Plantarum **24**, 235–238 (1971).

TRUELOVE, B., HANSON, J.B.: Calcium-activated phosphate uptake in conducting corn mitochondria. Plant Physiol. **41**, 1004–1013 (1966).

VAKHMISTROV, D.B.: On the function of apparent free space in plant roots. A study of the absorbing power of epidermal and cortical cells in barley roots. Soviet Plant Physiol. **14**, 103–107 (1967).

VAUGHAN, D., MacDONALD, I.R.: Invertase development in storage tissue disks of *Beta vulgaris*; its nature, extent and location. J. Exptl. Bot. **18**, 578–586 (1967a).

VAUGHAN, D., MacDONALD, I.R.: The effect of inhibitors on the increase in invertase activity and RNA content of beet disks during ageing. J. Exptl. Bot. **18**, 587–593 (1967b).

VAUGHAN, D., MacDONALD, I.R.: Development of soluble and insoluble invertase activity in washed storage tissue. Plant Physiol. **42**, 456–458 (1967c).

VERLEUR, J.D.: Observations on the induction of mitochondrial particles in potato tuber tissue after wounding. Z. Pflanzenphysiol. **61**, 299–309 (1969).

VERLEUR, J.D., VAN DER VELDE, H.H., SMINIA, T.: The heterogeneity of the mitochondrial fraction in relation to the increase of mitochondrial content in wounded and diseased potato tuber tissue. Z. Pflanzenphysiol. **62**, 352–361 (1970).

VIEWEG, G.-H.: Enzyme des Saccharosestoffwechsels in Wurzeln. Planta **116**, 347–359 (1974).

WAISEL, Y., KULLER, Z.: Control of selectivity and ion fluxes in bean hypocotyls. Experientia **28**, 1377–1378 (1972).

WATANABE, A., IMASEKI, H.: Induction of deoxyribonucleic acid synthesis in potato tuber tissue by cutting. Plant Physiol. **51**, 772–776 (1973).

WEBER, G., CONVERY, H.J.H., LEA, M.A., STAMM, N.B.: Feedback inhibition of key glycolytic enzymes in liver: action of free fatty acids. Science **154**, 1357–1366 (1966).

WILLEMOT, C., STUMPF, P.K.: Fat metabolism in higher plants. Development of fatty acid synthetase during the "ageing" of storage tissue. Canad. J. Bot. **45**, 579–584 (1967a).

WILLEMOT, C., STUMPF, P.K.: Fat metabolism in higher plants. XXXIV. Development of fatty acid synthetase as a function of protein synthesis in ageing potato tuber slices. Plant Physiol. **42**, 391–397 (1967b).

WOLLEY, J.T.: Potato tuber tissue respiration and ventilation. Plant Physiol. **37**, 793–798 (1962).

WONG, W.L., AP REES, T.: Carbohydrate oxidation in stele and cortex isolated from roots of *Pisum sativum*. Biochim. Biophys. Acta **252**, 296–304 (1971).

ZUCKER, M.: Sequential induction of phenylalanine ammonia-lyase and a lyase-inactivating system in potato tuber disks. Plant Physiol. **43**, 365–374 (1968).

9. Genotypic Variation in Transport

A. LÄUCHLI

1. Introduction

Ecologists have recognized for a long time that nature has confronted them with a great number of natural examples where plants are adapted to different nutrient regimes. For instance, mangroves growing in sea water can be contrasted with trees on the non-saline land next to them. The question arises as to why mangroves grow in sea water, and why do they not live further inland in the community of the less salt-tolerant tree species? A second example is given by the strikingly different floras on calcareous and acid soils, respectively. Why are there species which thrive on soils rich in lime and high in pH, whereas others occur only on acid soils low in Ca^{2+} and pH? Thirdly, serpentine soils bear a sparse yet characteristic flora. These soils are high in Mg^{2+} and certain heavy metals but contain little Ca^{2+}. Why, on the one hand, is there such a characteristic serpentine flora, and why, on the other hand, are most species not able to exist on serpentine soils? Many more such examples could be added. All of them have in common that their existence follows from variation in the genotype of plants and from adaptation during the course of evolution.

The subject of adaptation of plants to different nutrient regimes will be dealt with in 9.2. It will be emphasized here that adaptation to nutritional factors may be correlated with genotypic variation in ion transport processes. From this circumstantial correlation the idea is developed that membrane transport *per se* is genetically controlled. Genetic control of transport in plants will be discussed in 9.3.

2. Ecotypes—Their Adaptation to Nutritional Factors

Species may have phenotypic plasticity in order to adjust their metabolic processes to take maximum advantage of changes in the environment. The existence of such systems is known for morphological and also for some physiological characters. However, the latitude of phenotypic responses to changes in the environment is limited within a genotype (BRADSHAW, 1965). What this simply means in terms of nutritional adaptation of roots is that particular roots do not cope with all mineral substrates. Edaphic factors have, however, given rise to adaptation to the mineral habitats of the earth. Such adaptation has become evident on two levels. Differences in nutritional adaptation are well-known among species and may be also recognized among families. This type of adaptation led KINZEL (1969) to develop the concept of the comparative physiology of mineral metabolism. It is beyond the scope of this chapter to present an extensive survey of the comparative physiology on the species, genus and family levels, though a brief introduction will be presented as this is deemed necessary for the understanding of the other type of adaptation, i.e. genotypic adaptation to nutritional factors. This second type is a consequence of evolutionary processes leading to ranges of genotypes known as ecological races or ecotypes. For a recent introduction to nutritional adaptation the reader is referred to EPSTEIN (1972).

Possibilities for a comparative physiology of mineral metabolism as suggested by KINZEL (1969) are given with respect to three nutritional factors, i.e. Ca^{2+} heavy metals, and salt.

2.1 Calcium

It is well known that plants common on calcareous soils (*calcicole* plants) show high contents of soluble Ca^{2+} and malate while those confined to acid, low-calcium soils (*calcifuge* plants) contain little soluble Ca^{2+} because of the presence of oxalate precipitating Ca^{2+} as oxalate salt (KINZEL, 1963). One way of distinction is by comparing the ratios of the soluble fractions of K^+/Ca^{2+} in the plants. Using such an approach, all examined species of the *Crassulaceae* showed molar K^+/Ca^{2+} ratios below 1 and belong to the calcicole plants, but many species of *Caryophyllaceae* and *Labiatae* had K^+/Ca^{2+} ratios up to 100 and are considered to be calcifuge (KINZEL, 1969). That the extreme differences in K^+/Ca^{2+} ratios between calcicole and calcifuge species are under genetic control is indicated by the following example. *Anthyllis vulneraria*, a calcicole species which is also able to grow on acid soil, was collected from a calcareous and an acid soil and the K^+/Ca^{2+} ratio in the plant determined. Plants from both sites contained more soluble Ca^{2+} than K^+ emphasizing the inherent calcicole character (HORAK and KINZEL, 1971).

CLARKSON (1965) presented evidence that calcicole species may be discernible from calcifuge species through differences in mechanisms of Ca^{2+} uptake. *Agrostis setacea*, common on acid soil, appeared to possess an uptake system with a high affinity for Ca^{2+} approaching its maximal rate at 0.25 mM Ca^{2+}. The calcicole species *Agrostis stolonifera*, however, accumulated progressively more Ca^{2+} in the shoots with increasing Ca^{2+} concentration in the medium. Since uptake by whole plants was measured, transpiration-dependent Ca^{2+} uptake may have been important in CLARKSON's experiments. (Uptake of Ca^{2+} by roots is discussed in 3.3.5.) In a comparative study involving *Vicia faba* (calcicole) and *Lupinus luteus* (calcifuge), SALSAC (1973) found that a much greater fraction of Ca^{2+} was exchangeable in the root of the calcicole than in that of the calcifuge species.

ATPases associated with membranes appear to be involved in ion transport (Part A, Chap. *10*). In KYLIN's laboratory it has been shown that the activity of membrane-associated ATPases from roots of wheat, which has a high demand for Ca^{2+}, was stimulated more by Ca^{2+} than by Mg^{2+}. On the other hand, ATPase activity from roots of oat having a high need for Mg^{2+} was stimulated more by Mg^{2+} than by Ca^{2+} (KYLIN and KÄHR, 1973; KÄHR and KYLIN, 1974). Important physiological characteristics of membranes (i.e. mechanism of Ca^{2+} uptake, ATPases) may thus be fundamentally different in calcicole and calcifuge species, enabling the respective plants to be fit for their natural substrate. Species may differ too in their selectivity to Ca^{2+} and Mg^{2+} taken up to the shoot by largely transpiration-dependent processes. Transport to the shoot may be 10 or more times greater than uptake or accumulation in the roots (3.3.5.1; see also LEGGETT and GILBERT, 1969).

Genotypic adaptation to Ca^{2+} was found in *Trifolium repens*. Populations from acid and calcareous soils differed in Ca^{2+} contents of the shoot (SNAYDON and BRADSHAW, 1969). The differences were suggested to be largely due to variations in selective Ca^{2+} uptake.

2.2 Heavy Metals

Certain plants are able to grow and develop on habitats contaminated with toxic levels of heavy metals. The subject of heavy metal tolerance in plants has been reviewed thoroughly by ANTONOVICS et al. (1971). Survival of heavy metal tolerant plants is not due to metal exclusion from the plant (e.g. LANGE and ZIEGLER, 1963; TURNER, 1969; SEVERNE and BROOKS, 1972; JAFFRÉ and SCHMID, 1974). Rather, heavy metal tolerant plants may be capable of synthesizing chelating compounds which form non-toxic complexes with the heavy metals (JOWETT, 1958). An extreme case is represented by certain encrusting lichens which are able to colonize the bare rock on slags contaminated with mining debris. Heavy metal tolerance in these lichens may be due, in part, to a predominant deposition of insoluble metal compounds on the surface of a lichen (NOESKE et al., 1970). The specialized habitat of a metal-contaminated area has proved extremely valuable to studies on the evolution of ecotypes.

Genotypic adaptation to high levels of heavy metals (e.g. Zn, Cu, Ni, Pb) mainly studied by BRADSHAW and his group (cf. ANTONOVICS et al., 1971), does not seem to be correlated with variations in rates of uptake for heavy metals. In populations of *Agrostis tenuis* differing in tolerance to Zn, the subcellular distribution of bound Zn in roots varied with the degree of Zn tolerance (MATHYS, 1973). With increasing Zn^{2+} concentration in the medium, the percentage fraction of Zn bound to the cell walls increased in the roots of the tolerant population but decreased in those of the sensitive ones. The reverse was evident for Zn bound to a cytoplasmic fraction. Nonetheless, the study by MATHYS (1973) does not allow us to assume a clear relation between Zn tolerance and the mechanism of Zn^{2+} uptake. The nature of tolerance to heavy metals is probably a matter of degree; there is not an all-or-nothing effect as demonstrated for root growth of Cu tolerant and non-tolerant populations of *Agrostis tenuis* (Fig. 9.1). Tolerance to Cu in these populations is inherited in a dominant rather than a recessive manner (Fig. 9.1). From a genetic point of view, particularly important is the finding by GREGORY and BRADSHAW (1965) that the tolerance of plants to a given heavy metal is metal-specific. Recent evidence suggests that selection for tolerance can occur easily. WU and BRADSHAW (1972) found populations of *Agrostis stolonifera* near a Cu-refining industry in England, which obviously developed from plants that were tolerant to aerial copper pollution and had formed by mutation or segregation. This demonstrates evolution related to aerial pollution. Of particular interest is that this evolution must be of very recent origin and may still be in the process of occurring.

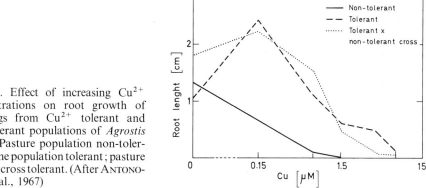

Fig. 9.1. Effect of increasing Cu^{2+} concentrations on root growth of seedlings from Cu^{2+} tolerant and non-tolerant populations of *Agrostis tenuis*. Pasture population non-tolerant; mine population tolerant; pasture × mine cross tolerant. (After ANTONOVICS et al., 1967)

2.3 Salinity

The third nutritional factor, i.e. salt (NaCl), can only be touched upon here so far as comparative physiology is concerned. The biology of halophytes has found extensive coverage in the monograph by WAISEL (1972), and various aspects of salt tolerance in halophytes as compared to glycophytes were reviewed recently (EPSTEIN, 1969, 1972; CAMPBELL and PITMAN, 1971; RAINS, 1972; GREENWAY, 1973, see also Chap. 5.2 and Part A, 13.4.3). As with calcicole and calcifuge species, halophytes are not uniformly distributed over the plant kingdom but are more common in certain families such as *Chenopodiaceae* than in others. Halophytes can also be grouped into different physiotypes according to physiological characters such as a certain pattern of ion content as a result of specific ion uptake processes and differential synthesis of organic acids (ALBERT and KINZEL, 1973).

    Attempts have been made to compare the salt relations of related species which differ in salt tolerance. ELZAM and EPSTEIN (1969a, b) investigated such a case by comparing *Agropyron intermedium*, a salt-sensitive species, with *Agropyron elongatum*, a salt-tolerant one of the same genus. At high salt concentrations, roots of the salt-tolerant species absorbed K$^+$ and Na$^+$ at high rates, whereas roots of the salt-sensitive species absorbed K$^+$ preferentially and at a low rate (ELZAM and EPSTEIN, 1969b).

    What about genotypic adaptation in saline habitats? Such ecotypes are best known in populations of salt-marsh species (cf. JEFFERIES, 1972). Ecotypes of *Festuca rubra* were demonstrated on salt marshes which are tolerant to salt concentrations as high as one-half of that in sea water; such high salt levels are lethal to normal populations (LANE and LYON, 1966; ANTONOVICS, LOVETT and BRADSHAW, 1967). The existence of a high degree of biological variation over very short distances indicates that selection is sufficiently strong to maintain this variation. Most of the variation recognized in salt-marsh populations concerns morphological characters but little direct evidence is known of variation in physiological processes (JEFFERIES, 1972). That differences in physiological processes do occur between salt-sensitive and salt-resistant populations has become apparent from studies of such ecotypes in artificial culture media under defined conditions. In populations of *Agrostis stolonifera* (TIKU and SNAYDON, 1971), collected from maritime and inland habitats, a correlation was found between the reduction in dry weight yield and the Na^+ content of the native soil. Furthermore, at low as well as high NaCl concentrations in the culture medium, populations from soils of low Na^+ content contained more Na^+ in the shoot and had higher Na^+/K^+ ratios than populations from high Na^+ soils (TIKU and SNAYDON, 1971). Obviously, salt tolerance in ecotypes of *Agrostis stolonifera* depends on the degree of Na^+ exclusion from the leaves, but the mechanism of this Na^+ exclusion is unknown. Populations of *Festuca rubra* collected from a salt marsh and from a pasture, respectively, also showed different salt relations in solution culture (KHAN, 1972). Salt tolerance in these ecotypes was achieved by restricting the uptake of Na^+ and Cl^- and by accumulating K^+ in preference to Na^+. (Preferential accumulation of K^+ is an outstanding feature of the halophyte *Avicennia marina* growing in sea water; see RAINS and EPSTEIN, 1967.) In addition, salt reduced transpiration, photosynthesis and respiration of the non-tolerant populations to a considerable extent, the magnitude of inhibition of photosynthesis being of the same order as the reduction in growth.

The few experimentally investigated examples described in 9.2.1–3 emphasize the point that genotypic adaptation to some edaphic factors may be brought about by genetic selection of mechanisms of ion uptake or transport which allow certain ecotypes to cope with a harsh mineral habitat. The conclusion can be drawn that plants different genetically (ecotypes) have different mechanisms of uptake of minerals or vary in degree of uptake. One may then also say that ion transport across membranes is under genetic control. Such an implication should prove extremely useful for breeding strains with more desirable transport properties (cf. EPSTEIN, 1972). Perhaps one could proceed, as geneticists have done for fungi, to isolate specific transport mutants. It is more likely that breeders will utilize crop species, whose gene pools are known and well controlled, and possibly related wild species. For example, breeding for improved salt tolerance (cf. DEWEY, 1962; EPSTEIN, 1972; RAINS, 1972; GREENWAY, 1973) involves, among other features, the genetic manipulation of salt-transport processes. This point is emphasized in the following examples. TAL (1971) and TAL and GAVISH (1973) studied various aspects of salt tolerance in the cultivated tomato (*Lycopersicon esculentum*) and more salt-tolerant wild relatives which originated in an arid habitat in Chile. The Na^+ content in the leaves of salt-treated plants was much higher in the salt-tolerant wild species and hybrid plants than in the cultivated salt-sensitive tomato (Table 9.1). Similar results were

Table 9.1. Content of Na^+ in leaves of 4-week-old control and salt-treated plants: comparison between the cultivated tomato (*Lycopersicon esculentum*) and more salt-tolerant wild relatives (*L. peruvianum*, and hybrid salt-tolerant plants selected from salt-sensitive *L. esculentum minor*). Salt treatment: NaCl, 196 mM. (From TAL, 1971)

Plant Type	Na^+ content (μmol g_{DW}^{-1})	
	Control	Salt-treated
L. esculentum	51	447
L. peruvianum	71	1,538
Hybrid plants	33	1,034

obtained for Cl^-. Furthermore, the increase in water deficit by the salt treatment, as indicated by the relative water content and succulence of the leaf, was smaller in the wild plants. Better osmotic adjustment, brought about by salt accumulation in the leaves, was suggested to be one of the main factors in the superior adaptation of the wild species to salinity. Ecotypes of the wild tomato species *Lycopersicon cheesmanii* from the Galapagos Islands which are far more salt tolerant than is the cultivated tomato also accumulate large concentrations of Na^+ in the leaf tissue (up to 7% of dry weight), a probable mechanism of compensating for the low water potential (RUSH and EPSTEIN, 1976).

3. Genetic Control of Transport

Ion uptake by plant cells probably involves some kind of carriers (proteins or enzymes, Chap. *3.2*). The synthesis of a particular enzyme or protein is controlled by a single gene pair ("one-gene—one-enzyme hypothesis", which is discussed in most genetics textbooks, e.g. BRESCH and HAUSMANN, 1970). The one-gene-one-enzyme hypothesis appears applicable to the understanding of the synthesis of specific ion transport systems in plants. Even though few instances are known from higher plants in which the existence of specific ion transport systems follows from single gene control of an ion transport process, it must be emphasized that there is convincing evidence for specific ion transport proteins in bacteria (PARDEE, 1968). In addition, several ion transport mutants have been described in microorganisms, i.e. in *E. coli, Salmonella typhimurium, Neurospora crassa* (review: SLAYMAN, 1973). With *Neurospora* it appears furthermore feasible to study the energetics of ion transport by using electron transport mutants such as *poky*, a cytochrome-deficient mutant (LAMBOWITZ et al., 1972a, b, c). The energetics of Cl^- uptake by *Scenedesmus obliquus* have been investigated by HOPE et al. (1974) using a wild-type strain and mutants deficient in photosystem I, photosystem II and cytochrome f, respectively. Our scanty knowledge of genetic control of ion transport in higher plants is summarized in reviews by EPSTEIN and JEFFERIES (1964) and EPSTEIN (1972).

3.1 Transport Mutants in Higher Plants

Discovery of transport mutants in higher plants dates back to 1943, when WEISS (1943) reported on the inheritance and physiology of efficiency in Fe utilization in soybean, *Glycine max*. The varieties PI-54619-5-1 (PI) and Hawkeye (HA) are inefficient and efficient respectively in their capacity to absorb and translocate Fe, the difference in Fe utilization being conditioned by a single gene (WEISS, 1943). Thus, the varieties PI and HA represent single-gene mutants in which some transport process is under genetic control.

Transport mutants, in which the physiological expression of the mutation is mediated by some ion-transport process, are listed in Table 9.2. Most of the mutants described were recognized because they exhibited micronutrient deficiencies and were thus easy to select. Much less is known about mutants where transport of a macronutrient ion is involved. Of great practical importance is the fact that varieties of plants exist which differ in salt tolerance, the differential response to salt being heritable and correlated with transport processes.

Attempts to elucidate the site and mechanism of genetically controlled ion transport in transport mutants showed that control may be exerted during uptake by the root or transport to the shoot. Regulation of ion content in shoots appears to be mainly due to regulation of ion input from root to shoot (PITMAN, 1972; PITMAN et al., 1974a). The situation of Fe utilization in the soybean mutants PI and HA is very complex (Table 9.2). Through grafting scions onto rootstocks in all possible combinations, it was ascertained that the genetic control is located in the roots (BROWN et al., 1958). The controlling biochemical factor in Fe utilization may have to do with the rate of reduction of Fe^{3+} to Fe^{2+} at the surface of the root which in turn depends upon the capacity of the root to release reducing compounds and H^+ (BROWN et al., 1972). Thus, the process of Fe uptake itself may be genetically controlled. In addition, transport of Fe to the shoot was also considered. BROWN (1963) proposed that two steps are implicated in the uptake-

Table 9.2. Ion-transport mutants in higher plants (single-gene mutants)[a]

Species	Mutant	Physiological expression of mutation	In-heritance	Transport process possibly involved	Ref.
Glycine max	PI-54619-5-1 (PI)	Inefficient in Fe utilization	Recessive	Fe uptake after reduction of Fe^{3+} to Fe^{2+}; Fe transport to shoot in PI	BROWN et al. (1958), BROWN et al. (1972); BROWN (1963)
	Hawkeye (HA)	Efficient in Fe utilization	Dominant		
Lycopersicon esculentum	T3238fe	Inefficient in Fe utilization	Recessive	Low uptake and transport in Fe	BROWN et al. (1971); WANN and HILLS (1973)
	T3238	Inefficient in B utilization	Recessive	B transport to shoot, controlled in the root	WALL and ANDRUS (1962), BROWN and JONES (1971)
	Rutgers	Efficient in B utilization	Dominant		
Apium graveolens	Utah 10B	Mg deficiency chlorosis	Recessive	Mg^{2+} uptake or transport to shoot?	POPE and MUNGER (1953)
Glycine max	Jackson	Salt sensitive	Recessive	Cl^- transport to shoot; Regulation of salt transport to shoot by root xylem parenchyma	ABEL and MACKENZIE (1964), ABEL (1969), LÄUCHLI et al. (1974b)
	Lee	Salt tolerant	Dominant		

[a] Other instances have been cited by EPSTEIN (1972, pp. 327–331).

transport process of Fe in the efficient HA-mutant, while the transport step is missing in the inefficient PI-mutant. The linkage between transport step and Fe^{3+}-reduction at the root surface remains uncertain. (Uptake of $Fe^{3+/2+}$ is also discussed in 3.3.4.1.5.)

ABEL and MACKENZIE (1964) studied uptake and transport of Cl^- in other soybean mutants differing in salt tolerance, under saline field conditions (Table 9.2). While there was not much difference between the mutants in Cl^- content of the roots, transport of Cl^- to the shoot of the salt-tolerant "Lee" was reduced to an exceedingly low level. On the other hand, the salt-sensitive mutants such as "Jackson" accumulated large quantities of Cl^- in the leaves, leading to salt damage. Thus, the moderately salt-tolerant mutant "Lee" is a Cl^- excluder, whereas in the genus *Lycopersicon* relative salt tolerance is correlated with the extent of salt accumulation in the leaves (see 9.2.3). The Cl^- transport condition might be controlled at the level of transport through the root to the xylem or, more probably, during secretion into or reabsorption from the xylem sap through the activity of the xylem parenchyma cells in the root (LÄUCHLI et al., 1974b). Ultrastructural investigations established the xylem parenchyma cells as transfer cells with wall ingrowths, particularly in "Lee" under mild salt stress, whereas there was heavy salt damage in "Jackson" at comparable conditions (LÄUCHLI et al., 1974b). This emphasizes the role of the xylem parenchyma cells of the root in controlling the Cl^- transport condition in these soybean mutants.

3.2 Varietal Differences

Frequently, one finds large differences in nutrient contents between different varieties of a species. Such variation may reflect genetically controlled differences in mechanisms of uptake or transport of an ion (VOSE, 1963; EPSTEIN, 1972). Pertinent examples are given in Table 9.3.

The varietal differences in toxicity of Mn and Al, studied extensively by FOY and coworkers, are not due to variation in a single, uniform transport mechanism. Nutritional differences in respect to several ions between varieties is also caused by a variety of transport processes such as uptake by the root, transport through the root, accumulation in stem tissues and uptake by leaf cells.

Occasionally, studies on mechanisms of ion uptake by excised roots of different varieties were conducted (EPSTEIN and JEFFERIES, 1964). However, the results were, in general, not conclusive. The carrier-mediated uptake of Fe by roots of three varieties of rice was found to vary greatly; it was suggested that these varietal differences may be due to differences in carrier concentration (SHIM and VOSE, 1965). Some differences were found in NO_3^- uptake by roots of two varieties of barley (SMITH, 1973). They were considered in relation to internal control of uptake, i.e. levels of NO_3^-, Cl^- and organic anions (cf. CRAM, 1973, Part A, 11.3.2.1), but may also be connected to variation in nitrate reductase activity, since synthesis of this enzyme (SHAFER et al., 1961) as well as its activity (WARNER et al., 1969) are genetically controlled (see 9.3.3.3).

An interesting example of varietal differences in Mn^{2+} transport to the shoot was described by MUNNS et al. (1963a, b, c). Cell compartmentation with respect to Mn was apparent in the root with the mobile fraction (cytoplasmic pool?) presum-

Table 9.3. Selected examples of varietal differences in mineral nutrition (phenomenon) of higher plants, possibly associated with transport processes (transport)[a,b]

Species	Phenomenon	Transport	Ref.
Gossypium hirsutum, G. barbadense	Mn toxicity	Differential compartmentation of Mn in leaves?	FOY et al. (1969)
Triticum aestivum	Al, Mn toxicity	Differential compartmentation of Mn in leaves?	FOY et al. (1973a)
Triticum vulgare, Hordeum vulgare	Al toxicity	Uptake of Al^{3+} related to cation exchange capacity of root	FOY et al. (1967)
Phaseolus vulgaris	Al toxicity	Effect of Al on Ca^{2+} uptake	FOY et al. (1972)
Lycopersicon esculentum	Al toxicity	Uptake of Al^{3+} and interference with P uptake	FOY et al. (1973b)
Oryza sativa	Fe nutrition	Fe uptake, related to carrier concentration?	SHIM and VOSE (1965)
Zea mays	Fe nutrition	Fe transport into xylem of root	CLARK et al. (1973)
Avena sativa A. byzantina	Mn nutrition	Mn^{2+} transport to shoot, related to mobile Mn^{2+} in root (cytoplasmic pool?)	MUNNS et al. (1963a, b, c);
Glycine max	Inorganic nutrition of seeds	Accumulation of inorganic nutrients in stem tissues	KLEESE (1967, 1968); KLEESE and SMITH (1970)
Zea mays	P toxicity	P uptake	FOOTE and HOWELL (1964)
	Mg^{2+} nutrition of leaves	Mg^{2+} accumulation in stem tissues	FOY and BARBER (1958)
	P nutrition of leaves	P uptake by excised leaves, related to strength of P-carrier bond	PHILLIPS et al. (1971)
Hordeum vulgare	N nutrition	NO_3^- uptake, related to internal control of uptake?	SMITH (1973)
	Salt tolerance	Cl^- uptake	GREENWAY (1962, 1965)
Lycopersicon esculentum	Salt tolerance	Na^+ uptake	PICCIURRO and BRUNETTI (1969)
Vitis vinifera	Salt tolerance	Cl^- transport to shoot	EHLIG (1960), BERNSTEIN et al. (1969)
		Cl^- transport to shoot related to monogalactose diglyceride concentration in root	KUIPER (1968a)
Beta vulgaris	Salt tolerance	Differential activation of root ATPases by Na^+ and K^+	KYLIN and HANSSON (1971); HANSSON (1975)

[a] Further examples in EPSTEIN (1972, Table 12.1).
[b] Effects of rootstocks on mineral composition of scions compiled in EPSTEIN (1972, Table 12.2).

ably being transportable to the shoot. The varietal differences in Mn content of the shoot were thought to be attributed to variation in size and rate of turnover of the mobile fraction.

Salt tolerance in agronomic species is in many cases associated with salt exclusion from the leaves (GREENWAY, 1973). This was also found to be true for barley varieties differing in salt tolerance (GREENWAY, 1962). The shoots of a salt-sensitive variety had higher Cl^- and Na^+ contents than those of tolerant varieties. Experiments on uptake of Cl^- by the same varieties revealed that the high salt content in the shoot of the sensitive variety was correlated with a high rate of Cl^- uptake by the root (GREENWAY, 1965). In tomato, however, salt-tolerant varieties show a high rate of Na^+ uptake (PICCIURRO and BRUNETTI, 1969; cf. Table 9.1). The degree of salt tolerance in varieties of grapes is inversely related to the extent of Cl^- accumulation in the leaves, similar to the situation in barley. The site of control lies in the root since scions grafted onto rootstocks of relatively salt-tolerant varieties accumulated much less Cl^- than did scions on sensitive roots (BERNSTEIN et al., 1969). These authors suggested that differential regulation of Cl^- transport from the root to the shoot may be involved, as the kinetics of Cl^- uptake by the rootstocks did not differ significantly among the varieties. In particular, the extent of Cl^- transport to the shoot of these grape varieties appears to be directly related to the concentration of the lipid component monogalactose diglyceride in the root (KUIPER, 1968a). When the liquid-membrane permeability of isolated grape-root lipids was measured, monogalactose diglyceride was indeed the most Cl^- permeable lipid (KUIPER, 1968b).

Nature has provided the crop species with various modes of coping with moderate levels of salinity. In general, salt tolerance is associated with exclusion of salt from the shoot, as for instance found in barley, soybeans and grapes (salt excluders); salt-tolerant tomato varieties are, on the other hand, salt includers. It turns out that this is beneficial to the consumer, as tomato juice is usually consumed salted, but just imagine how "salty wine" would taste!

Inbred lines of sugar beet, varying in salt tolerance and ranging from low-salt status, high proportion of K^+ (ADA) to high-salt status, high proportion of Na^+ (FIA) were analyzed for their monovalent cation activated root ATPases (KYLIN and HANSSON, 1971). Fig. 9.2 shows that the activity of the ATPase from the salt-sensitive variety ADA was stimulated more by K^+ than by Na^+; the ATPase activity from the salt tolerant variety FIA, however, was stimulated by Na^+ and not influenced or inhibited by K^+. In view of the fact that membrane ATPases are involved in cellular ion transport (Part A, Chap. 10), the differential activation of the root ATPases in these varieties may be correlated with selective uptake of K^+ by the salt-sensitive variety and with a high rate of Na^+ uptake by the salt-tolerant one. More extensive data on the ionic relations and ATPases of the inbred lines of sugar beet were compiled by HANSSON (1975).

In addition to varietal differences, noticeable variability in mineral nutrition of individual plants within a species was also observed. C.J. ASHER and P.G. OZANNE (unpublished results) found that in *Cryptostemma calendula*, grown in nutrient solution under controlled conditions, the K^+ content of the shoots of individual plants ranged from 0.29–4.82% of dry weight. In those plants with low K^+ contents, variation in shoot growth was correlated with levels of K^+. ASHER and OZANNE's observations represent a striking instance of genotypic control of mineral nutrient

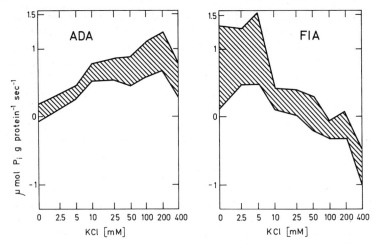

Fig. 9.2. Differential activation by Na^+ and K^+ of root ATPases from inbred lines of *Beta vulgaris* differing in salt tolerance. ADA=low salt, high proportion of K^+ variety. FIA=high salt, high proportion of Na^+ variety. Abscissa: KCl concentration in reaction medium. Ordinate: Specific activity (μmol $P_i\, g_{protein}^{-1}\, s^{-1}$) (the activity obtained in the basic medium without Na^+ and K^+ is taken as zero value). The upper and lower line denote the highest and lowest activity obtained with NaCl concentrations between 0 and 100 mM added to the reaction medium with KCl. (After KYLIN and HANSSON, 1971)

content in plants, revealing a range of variation which unfortunately most plant physiologists ignore.

3.3 Induction of Transport Processes

The concept of genetic control of ion transport thus far has been based exclusively on our knowledge of transport mutants and of varieties differing in some ion transport process. The implication was that any genetically controlled transport process must involve membrane-associated proteins or enzymes. One can then argue that ion transport may be inducible, as there is increasing evidence that enzymes in plants can be induced (MARCUS, 1971). In 9.3.3.1–5, enzyme induction in plants is reviewed briefly, and it is then examined whether transport processes in plants too are under inductive control. Only pertinent examples dealing with eucaryotic plant cells will be discussed.

3.3.1 Enzyme Induction in Plants

Not many examples are known of *genuine enzyme induction*—an increase in rate of *synthesis of an enzyme* in response to a change in a specific environmental parameter. Far more abundant is *apparent enzyme induction*—an increase in *enzyme activity* in response to a change in a specific environmental parameter (FILNER et al., 1969). The most important controlling environmental parameters encompass substrates, hormones and light. Nitrate reductase is the most thoroughly studied enzyme whose synthesis is substrate induced (see 9.3.3.3 and Part A, 13.2.3). Many cases of hormonal control of enzyme synthesis or enzyme activity are known, such as the control of α-amylase by gibberellic acid and abscisic acid in barley aleuron layers (CHRISPEELS and VARNER, 1967) and other examples (cf. Table 1 in FILNER et al., 1969). Light control of enzyme activities may be mediated through the phytochrome system, e.g. the activity of phenylalanine deaminase in mustard seedlings (DURST and MOHR, 1966), but phytochrome is not always involved and responses to light other than red-far red are also known.

Control of enzyme induction (MARCUS, 1971) could be at the level of transcription of DNA to mRNA or at the level of translation (attachment of messenger to ribosomes or tRNA). How does one go about testing experimentally these possibilities relative to enzyme induction? The application of specific inhibitors of steps involved in enzyme synthesis is a useful tool. The inhibitors of RNA synthesis, actinomycin D and 6-methyl purine, are used to test control at the level of transcription and the protein-synthesis inhibitors, cycloheximide and puromycin, to test control at the level of translation (*8.3.1.4*). A word of caution ought to be added when such inhibitors are used: the specificity of the inhibitor must be checked on the particular tissue before any conclusions are to be drawn, for secondary effects on respiratory metabolism or membrane properties have been detected in certain tissues (GLASZIOU, 1969; ELLIS and MacDONALD, 1970; VAN STEVENINCK and VAN STEVENINCK, 1972; LÄUCHLI et al., 1973; LÜTTGE et al., 1974).

3.3.2 Induction of Hexose Uptake in *Chlorella*

TANNER and KANDLER (1967) discovered that uptake of glucose by *Chlorella vulgaris* (see also *3.3.1, 5.3.1.1.2.2, 5.3.1.1.2.3*, and Part A, *6.7.2.3*) proceeds with a lag phase of 40 to 60 min, after which the time course of uptake is linear. Hence, it was suggested that *Chlorella* cells possess an inducible hexose-uptake system. TANNER (1969) was then able to show that glucose-adapted *Chlorella* cells take up the non-metabolized hexose analog 3-0-methyl-glucose linearly with time, but cells not adapted show a lag phase of 60 min. The induction of hexose uptake is energy-dependent and probably involves the synthesis of one or several proteins (TANNER and KANDLER, 1967). An important feature of the inducible uptake system is that it shows a *turnover* (Fig. 9.3). It can be seen that the rate of glucose uptake remains constant up to about 10 h following induction. After 13 h, the hexose uptake system has become inactive and must be induced again. The induced uptake system turns over with a half-life of 4 to 6 h (HAASS and TANNER, 1974). The turnover of the uptake system is in accord with the participation of protein(s).

Induction of hexose uptake was investigated further by HAASS and TANNER (1974). They found that only *Chlorella vulgaris* exhibited a large increase in the rate of uptake due to induction. In other species of *Chlorella,* and also in species of *Scenedesmus* and *Ankistrodesmus,* the transport to a large extent appeared constitutive. The experiments by HAASS and TANNER (1974) indicated furthermore that the inducing sugars have to penetrate the cells to be effective, and that regulation of induction occurs at the level of transcription or, in other words, that induction of the synthesis of protein(s) for hexose uptake is probably preceeded by RNA synthesis.

Induction of hexose uptake in *Chlorella* involves a membrane component of the cells. TANNER et al. (1974) found that a membrane fraction (protein?) became

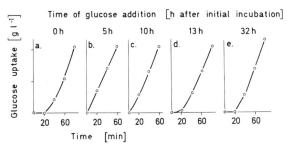

Fig. 9.3. Induction of hexose uptake in *Chlorella vulgaris* by glucose and turnover of the hexose uptake system. At time 0 h, glucose was added to 5 parallel samples, and the time course of glucose uptake was measured immediately in sample *a*. The other samples completely consumed the glucose, added at time 0 h, within 1.5 h; they received additional glucose after 5, 10, 13 and 32 h, respectively. The time course of glucose uptake in samples *b* to *e* was measured after the second addition of glucose, and the lag phase becomes discernible again at 13 h. (After Tanner et al., 1970)

labeled when ^{14}C-phenylalanine was fed to induced cells; labeling of the membrane fraction did not occur in non-induced cells. The findings of Tanner and coworkers clearly demonstrate the occurrence in a *eucaryotic cell* of an inducible uptake system for a class of organic solutes and suggest the participation of membrane protein(s) in a transport process.

3.3.3 Induction of Ion Uptake and Transport in Roots

The synthesis of NO_3^- reductase in higher plants is substrate induced; NO_3^- is required for induction of the enzyme, and the increase in enzyme activity is accompanied by *de novo* protein synthesis (Beevers and Hageman, 1969; Part A, *13*.2.3). Recent evidence suggests that not only NO_3^- reductase is substrate-induced but also uptake of NO_3^- *per se*. This would represent a case of substrate-induced synthesis of a transport protein in higher plants. In NO_3^--depleted seedlings of wheat, corn, cotton and tobacco, the pattern of NO_3^- uptake exhibited an initial lag phase upon exposure to NO_3^-; subsequently, the rate of NO_3^- uptake increased (Minotti et al., 1968; Jackson et al., 1972). Experiments with excised roots of corn seedlings (Jackson et al., 1973), grown in the absence of NO_3^- and Cl^-, revealed that the initial lag phase was observed only for uptake of NO_3^-, whereas the rate of Cl^- uptake did not change with time (Fig. 9.4). These authors proposed that in roots, NO_3^- specifically induced the development of a NO_3^- uptake system. Inhibitors of RNA and protein synthesis restricted maximum development of the accelerated NO_3^- uptake rate, but induction of NO_3^- reductase was inhibited too. One may argue that NO_3^- reduction in the root, providing a sink for NO_3^-, could have caused the increase in rate of NO_3^- uptake. Therefore, the possibility remains that the induction of NO_3^- reductase itself promoted NO_3^- uptake. In this context, however, studies by Filner and coworkers with cultured tobacco pith cells are worth mentioning. They were successful in measuring NO_3^- uptake separately from NO_3^- reductase activity by using tungstate, whose presence allows only the formation of non-functional NO_3^- reductase without impairing NO_3^- accumulation (Heimer et al., 1969). Thus, NO_3^- appeared to induce specifically the development of an

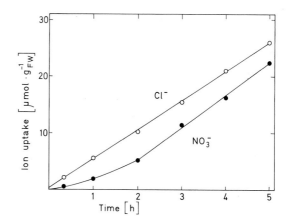

Fig. 9.4. Time course of NO$_3^-$ and Cl$^-$ uptake by excised corn roots exposed to 0.5 mM solutions of Ca(NO$_3$)$_2$ and CaCl$_2$, respectively. (After JACKSON et al., 1973)

active NO$_3^-$ uptake system (HEIMER and FILNER, 1971). It seems pertinent to refer to VAN STEVENINCK and VAN STEVENINCK (1972) who showed that the ion transport systems in cells of beetroot slices depend on the synthesis and decay of specific proteins (see 8.3).

Roots of low-salt seedlings need several hours for maximum development of K$^+$ and Cl$^-$ transport to the xylem (LÄUCHLI and EPSTEIN, 1970, 1971; PITMAN, 1971). The time lag of transport to the xylem is, however, no proof for an induced process, since ions taken up by the cortex cells could initially be diverted to the vacuoles before reaching the vessels. Yet studies on regulation of ion transport through roots indicate possible induction. Hormones as well as the presence of substrate ions in the symplast could be the controlling factors. The phytohormones cytokinin and abscisic acid were suggested as regulators of ion transport through the root (PITMAN et al., 1974b). Both hormones are considered to exert their action at the level of RNA activity or synthesis, but specific membrane action is an alternative possibility. Such a conclusion is consistent with the observation that both cyclo-heximide and p-fluorophenylalanine inhibit transport of K$^+$ through barley roots without having much effect on uptake (LÄUCHLI et al., 1973; SCHAEFER et al., 1975). Since p-fluorophenylalanine appears to act on the final product of protein synthesis, SCHAEFER et al. (1975) concluded that ion transport into xylem vessels involves some kind of protein which is used in transport either in a pinocytotic process or as a permease, and whose site of action is probably at the plasmalemma of the xylem parenchyma cells facing the vessel wall. If this protein would indeed act as a permease, it could adapt readily to fluctuations in ion flow through the symplast, which is considered to proceed in the rough endoplasmic reticulum (STELZER et al., 1975).

An example of the dependence of ion transport into the xylem on a specific protein was described by CLARK et al. (1973) who showed that a maize genotype lacked a mechanism for transport of iron from the root tissue to the xylem. From a structural point of view the xylem parenchyma cells with abundant rough endoplasmic reticulum appear fit to be the sites of regulation of ion transport into the xylem vessels (LÄUCHLI et al., 1974a).

3.3.4 Induction of a Chloride Pump in the Salt Gland of *Limonium*

SHACHAR-HILL and HILL (1970) made the discovery that pumping of Cl^- in low-salt-adapted *Limonium* glands develops in response to a salt load with a time lag of more than one hour, followed by a sigmoid rise in excretory activity (Fig. 9.5). When low-salt leaf discs were pretreated for 6–10 h with actinomycin D or puromycin and then salt-loaded, no activity developed. These inhibitors, however, did not have any effect on the sigmoid phase nor on the steady-state pumping (Fig. 9.5). It thus seems that both transcription and translation are involved in this induction process (phase A) and that synthesis of the Cl^- transport system takes place during induction (HILL and HILL, 1973a). The involvement of RNA synthesis in the production of a membrane protein for transport of Cl^- is consistent with ultrastructural observations showing that the cytoplasm of the gland cell is rich in polyribosomes with RNA threads and in rough endoplasmic reticulum (SHACHAR-HILL and HILL, 1970). Membrane preparations (microsomes) from salt-loaded tissue contain Cl^--stimulated ATPase activity (Fig. 9.5). The ATPase activity is induced by salt-loading, and the induction can be blocked by puromycin. The parallels between inducible ATPase and development of Cl^- pumping are such as to assume that the Cl^--stimulated ATPase activity is that of the pump itself. (The elimination of ions by glands is described fully in Chap. 5.2.)

Plants have apparently developed inducible enzyme or protein systems to cope with fluctuations in mineral-ion concentration in the environment. (i) The NO_3^--induced NO_3^- reductase appears common to all higher plants examined and facilitates the utilization of available NO_3^- (9.3.3.3). (ii) The possible involvement of specific

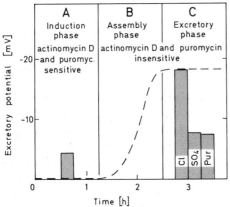

Fig. 9.5. Development of excretory activity in low-salt-adapted leaf discs of *Limonium* in response to a load of 100 mM NaCl at 25° C. Broken line: gland activity (excretory potential ψ [mV]) (from Fig. 9 of HILL and HILL, 1973a). Bars represent relative Cl^--ATPase activities compiled from Figs. 4, 5, 7 and 8 of HILL and HILL (1973b). These ATPase data were normalized for the ATPase activity of membrane preparations (microsomes) of induced leaves tested in the presence of Cl^- (= large bar in phase C:0.014 to 0.14 μmol P_i $g^{-1}_{protein}$ s^{-1}, depending on the leaf sample and preparation). Bar in phase A: ATPase activity of non-induced leaves; bar-SO_4 in phase C: ATPase activity of induced leaves tested in the presence of SO_4^{2-} instead of Cl^-; bar-Pur in phase C: ATPase activity of leaves treated with NaCl in the presence of puromycin. (From LÜTTGE, 1975)

proteins in transport of ions from the root symplast into the xylem vessels could be the key for regulation of ion transport to the shoot in response to fluctuations in the ionic composition of the substrate (9.3.3.3). (iii) The Cl^- induced salt transporting system in *Limonium* glands responds to changes in the salt load of the leaf tissue and presumably accommodates fluctuations in salinity levels of the substrate. It is not clear whether induction of excretory capacity by salt is common to all salt glands; nonetheless, the light-dependent increased rate of salt secretion to epidermal bladders of *Atriplex* (LÜTTGE and OSMOND, 1970) may reflect a similar phenomenon. In addition, it seems relevant here that induction of NO_3^- reductase in leaves is controlled by NO_3^- and by light (HAGEMAN and FLESHER, 1960).

3.3.5 Induction of Choline Sulfate Permease

Barley roots possess constitutive mechanisms for uptake of the organic zwitterion choline sulfate (NISSEN and BENSON, 1964). The uptake of choline sulfate is linear for short periods, but increases sharply after several hours (NISSEN, 1968). The increase in uptake requires choline sulfate as an inducer and is not observed when the roots take up the zwitterion under sterile conditions. NISSEN (1968) isolated several species of bacteria from barley roots, among which *Pseudomonas tolaasii* was found to be effective in inducing uptake of choline sulfate by barley roots. In addition, NISSEN's experiments revealed that: (i) the time courses of uptake in bacteria and roots do not coincide; (ii) the inducible and constitutive plant-uptake mechanisms are dependent on Ca^{2+}, but not the bacterial uptake of choline sulfate; (iii) cycloheximide inhibits induced uptake by barley roots without having an effect on bacterial uptake, indicating the requirement of protein synthesis in the root (cf. LÄUCHLI et al., 1973). These findings strongly suggest that bacteria are required for induction of uptake in the higher plant and indicate that the induced uptake observed in the plant is indeed carried out by the plant and is not due to bacterial contamination. NISSEN (1971) proposed a sequence of three induction processes to be involved in the uptake of choline sulfate by plant roots mediated by certain gram-negative bacteria (Fig. 9.6): firstly, substrate induction of a bacterial permease highly specific for choline sulfate (phase I). The bacteria are not yet able in phase I to mediate plant uptake; in other words, they are not yet effective. Bacterial effectiveness does not start before phase I is completed and is induced by either choline sulfate or choline (phase II). When the bacteria have become effective and are in contact with the root, an uptake mechanism is induced in the root with a specificity resembling that of the bacterial permease and different from the constitutive uptake mechanism (phase III). After completion of the induction in the root (2–3 h), the bacteria can be removed with no loss in rate of uptake (Fig. 9.6). A complete account of these experiments has been published (NISSEN, 1973).

These puzzling results cannot as yet be explained. A transfer of some form of information from the bacteria to the root is a feasible explanation. Whether contact alone provides the information, or whether a transfer of molecules occurs is not known. Equally unknown is the genetic origin of the inducible uptake mechanism in the higher plant. Of significance, however, may be the fact that the effective bacteria include genera prevalent in the rhizosphere. Undoubtedly, investigations of the interactions between microorganisms and roots in the rhizosphere with emphasis on uptake of substances by roots could be rewarding.

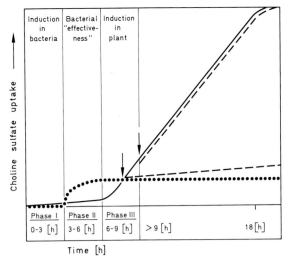

Fig. 9.6. Induction processes and time course of choline sulfate uptake by bacteria and higher plant. Bacteria: gram-negative bacteria, e.g. *Agrobacterium, Pseudomonas, Rhizobium;* higher plant: roots and leaf slices of barley. Solid line: plant with bacteria. Broken line: plant without bacteria (bacteria removed at time indicated by arrow). Dots: bacteria. Phase I: Induction of a specific permease in bacteria. Phase II: Bacterial "effectiveness" and contact between effective bacteria and plant tissue. Phase III: Induction of a specific permease in plant tissue. (After NISSEN, 1971)

4. Conclusions

The concept was advanced in this chapter that membrane transport *per se* is under genetic control. The evidence summarized here in support of this concept comes from two lines of experimental data: (i) there is good evidence for genotypic variation in mineral nutrition as related to transport processes (transport mutants, varietal differences); (ii) specific transport systems are known that are inducible, just like inducible enzymes. And indeed, induction of a transport process must involve new synthesis or activation of proteins or enzymes. Thus firstly, genotypic variation is expressed as variation in synthesis or activity of a particular protein(s) or enzyme(s). Secondly, induction involves proteins or enzymes too. Hence, the two lines of evidence are clearly connected. What is needed now is the unequivocal demonstration in higher plants of specific ion transport proteins as well as their isolation and characterization, following the type of work on bacteria by PARDEE (1968). This aim could possibly be achieved best with plants having inducible transport systems (cf. TANNER et al., 1974). But let us not forget that nature has already provided the definitive proof of genotypic variation in transport through the evolution of a great variety of ecotypes adapted to nutritional factors.

References

ABEL, G.H.: Inheritance of the capacity for chloride inclusion and chloride exclusion by soybeans. Crop Sci. **9**, 697–698 (1969).

ABEL, G.H., MACKENZIE, A.J.: Salt tolerance of soybean varieties (*Glycine max* L. Merrill) during germination and later growth. Crop Sci. **4**, 157–161 (1964).

ALBERT, R., KINZEL, H.: Unterscheidung von Physiotypen bei Halophyten des Neusiedlerseegebietes (Österreich). Z. Pflanzenphysiol. **70**, 138–157 (1973).

ANTONOVICS, J., BRADSHAW, A.D., TURNER, R.G.: Heavy metal tolerance in plants. Advan. Ecol. Res. **7**, 1–85 (1971).

ANTONOVICS, J., LOVETT, J., BRADSHAW, A.D.: The evolution of adaptation to nutritional factors in populations of herbage plants. In: Isotopes in plant nutrition and physiology, p. 549–567. Vienna: IAEA 1967.

ASHER, C.J., OZANNE, P.G.: Observations on nutritional variation in capeweed (in preparation).

BEEVERS, L., HAGEMAN, R.H.: Nitrate reduction in higher plants. Ann. Rev. Plant Physiol. **20**, 495–522 (1969).

BERNSTEIN, L., EHLIG, C.F., CLARK, R.A.: Effect of grape rootstocks on chloride accumulation in leaves. J. Amer. Soc. Hort. Sci. **94**, 584–590 (1969).

BRADSHAW, A.D.: Evolutionary significance of phenotypic plasticity in plants. Advan. Genetics **13**, 115–155 (1965).

BRESCH, C., HAUSMANN, R.: Klassische und molekulare Genetik. 2. Auflage. Berlin-Heidelberg-New York: Springer 1970.

BROWN, J.C.: Iron chlorosis in soybeans as related to the genotype of rootstock. Soil Sci. **96**, 387–394 (1963).

BROWN, J.C., AMBLER, J.E., CHANEY, R.L., FOY, C.D.: Differential responses of plant genotypes to micronutrients. In: Micronutrients in agriculture (J.J. MORTVEDT, P.M. GIORDANO, W.L. LINDSAY, eds.), p. 389–418. Madison, Wisconsin: Soil Sci. Soc. Amer. 1972.

BROWN, J.C., CHANEY, R.L., AMBLER, J.E.: A new tomato mutant inefficient in the transport of iron. Physiol. Plantarum **25**, 48–53 (1971).

BROWN, J.C., HOLMES, R.S., TIFFIN, L.O.: Iron chlorosis in soybeans as related to the genotype of rootstalk. Soil Sci. **86**, 75–82 (1958).

BROWN, J.C., JONES, W.E.: Differential transport of boron in tomato (*Lycopersicon esculentum* Mill.). Physiol. Plantarum **25**, 279–282 (1971).

CAMPBELL, L.C., PITMAN, M.G.: Salinity and plant cells. In: Salinity and water use (T. TALSMA, J.R. PHILIP, eds.), p. 207–224. London: Macmillan Press 1971.

CHRISPEELS, M.J., VARNER, J.E.: Hormonal control of enzyme synthesis: on the mode of action of gibberellic acid and abscisin in aleurone layers of barley. Plant Physiol. **42**, 1008–1016 (1967).

CLARK, R.B., TIFFIN, L.O., BROWN, J.C.: Organic acids and iron translocation in maize genotypes. Plant Physiol. **52**, 147–150 (1973).

CLARKSON, D.T.: Calcium uptake by calcicole and calcifuge species in the genus *Agrostis* L. J. Ecol. **53**, 427–435 (1965).

CRAM, W.J.: Internal factors regulating nitrate and chloride influx in plant cells. J. Exptl. Bot. **24**, 328–341 (1973).

DEWEY, D.R.: Breeding crested wheatgrass for salt tolerance. Crop Sci. **2**, 403–407 (1962).

DURST, F., MOHR, H.: Phytochrome-mediated induction of enzyme synthesis in mustard seedlings (*Sinapis alba* L.). Naturwissenschaften **53**, 531–532 (1966).

EHLIG, C.F.: Effects of salinity on four varieties of table grapes grown in sand culture. Proc. Amer. Soc. Hort. Sci. **76**, 323–331 (1960).

ELLIS, R.J., MACDONALD, I.R.: Specificity of cycloheximide in higher plant systems. Plant Physiol. **46**, 227–232 (1970).

ELZAM, O.E., EPSTEIN, E.: Salt relations of two grass species differing in salt tolerance. I. Growth and salt content at different salt concentrations. Agrochimica **13**, 187–195 (1969a).

ELZAM, O.E., EPSTEIN, E.: Salt relations of two grass species differing in salt tolerance. II. Kinetics of the absorption of K, Na and Cl by their excised roots. Agrochimica **13**, 196–206 (1969b).

EPSTEIN, E.: Mineral metabolism of halophytes. In: Ecological aspects of the mineral nutrition of plants (I.H. RORISON, ed.), p. 345–355. Oxford and Edinburgh: Blackwell Sci. Publ. 1969.

EPSTEIN, E.: Mineral nutrition of plants: principles and perspectives. New York-London-Sydney-Toronto: John Wiley and Sons 1972.

EPSTEIN, E., JEFFERIES, R.L.: The genetic basis of selective ion transport in plants. Ann. Rev. Plant Physiol. **15**, 169–184 (1964).

FILNER, P., WRAY, J.L., VARNER, J.E.: Enzyme induction in higher plants. Science **165**, 358–367 (1969).

FOOTE, B.D., HOWELL, R.W.: Phosphorus tolerance and sensitivity of soybeans as related to uptake and translocation. Plant Physiol. **39**, 610–613 (1964).

FOY, C.D., BARBER, S.A.: Magnesium absorption and utilization by two inbred lines of corn. Soil Sci. Soc. Amer. Proc. **22**, 57–62 (1958).

Foy, C.D., Fleming, A.L., Armiger, W.H.: Differential tolerance of cotton varieties to excess manganese. Agron. J. **61**, 690–694 (1969).

Foy, C.D., Fleming, A.L., Burns, G.R., Armiger, W.H.: Characterization of differential aluminium tolerance among varieties of wheat and barley. Soil Sci. Soc. Amer. Proc. **31**, 513–521 (1967).

Foy, C.D., Fleming, A.L., Gerloff, G.C.: Differential aluminium tolerance in two snapbean varieties. Agron. J. **64**, 815–818 (1972).

Foy, C.D., Fleming, A.L., Schwartz, J.W.: Opposite aluminium and manganese tolerances of two wheat varieties. Agron. J. **65**, 123–126 (1973a).

Foy, C.D., Gerloff, G.C., Gabelman, W.H.: Differential effects of aluminium on the vegetative growth of tomato cultivars in acid soil and nutrient solution. J. Amer. Soc. Hort. Sci. **98**, 427–432 (1973b).

Glasziou, K.T.: Control of enzyme formation and inactivation in plants. Ann. Rev. Plant Physiol. **20**, 63–88 (1969).

Greenway, H.: Plant response to saline substrates. I. Growth and ion uptake of several varieties of *Hordeum* during and after sodium chloride treatment. Australian J. Biol. Sci. **15**, 16–38 (1962).

Greenway, H.: Plant response to saline substrates. VII. Growth and ion uptake throughout plant development in two varieties of *Hordeum vulgare*. Australian J. Biol. Sci. **18**, 763–779 (1965).

Greenway, H.: Salinity, plant growth, and metabolism. J. Australian Inst. Agr. Sci. **39**, 24–34 (1973).

Gregory, R.P.G., Bradshaw, A.D.: Heavy metal tolerance in populations of *Agrostis tenuis* and other grasses. New Phytologist **64**, 131–143 (1965).

Haass, D., Tanner, W.: Regulation of hexose transport in *Chlorella vulgaris*. Characteristics of induction and turnover. Plant Physiol. **53**, 14–20 (1974).

Hageman, R.H., Flesher, D.: Nitrate reductase activity in corn seedlings as affected by light and nitrate content of nutrient media. Plant Physiol. **35**, 700–708 (1960).

Hansson, G.: Patterns of ionic influences on sugar beet ATPases. Dissertation, University of Stockholm (1975).

Heimer, Y.M., Filner, P.: Regulation of the nitrate assimilation pathway in cultured tobacco cells. III. The nitrate uptake system. Biochim. Biophys. Acta **230**, 362–372 (1971).

Heimer, Y.M., Wray, J.L., Filner, P.: The effect of tungstate on nitrate assimilation in higher plant tissues. Plant Physiol. **44**, 1197–1199 (1969).

Hill, A.E., Hill, B.S.: The *Limonium* salt gland: a biophysical and structural study. Intern. Rev. Cytol. **35**, 299–319 (1973a).

Hill, B.S., Hill, A.E.: ATP-driven chloride pumping and ATPase activity in the *Limonium* salt gland. J. Membrane Biol. **12**, 145–158 (1973b).

Hope, A.B., Lüttge, U., Ball, E.: Chloride uptake in strains of *Scenedesmus obliquus*. Z. Pflanzenphysiol. **72**, 1–10 (1974).

Horak, O., Kinzel, H.: Typen des Mineralstoffwechsels bei den höheren Pflanzen. Österr. Bot. Z. **119**, 475–495 (1971).

Jackson, W.A., Flesher, D., Hageman, R.H.: Nitrate uptake by dark-grown corn seedlings. Some characteristics of apparent induction. Plant Physiol. **51**, 120–127 (1973).

Jackson, W.A., Volk, R.J., Tucker, T.C.: Apparent induction of nitrate uptake in nitrate-depleted plants. Agron. J. **64**, 518–521 (1972).

Jaffré, T., Schmid, M.: Accumulation du nickel par une Rubiacée de Nouvelle-Calédonie. Compt. Rend. **278**, 1727–1730 (1974).

Jefferies, R.L.: Aspects of salt-marsh ecology with particular reference to inorganic plant nutrition. In: The estuarine environment, p. 61–85. London: Applied Science Publ. 1972.

Jowett, D.: Populations of *Agrostis* spp. tolerant of heavy metals. Nature **182**, 816–817 (1958).

Kähr, M., Kylin, A.: Effects of divalent cations and oligomycin on membrane ATPases from roots of wheat and oat in relation to salt status and cultivation. In: Membrane transport in plants (U. Zimmermann, J. Dainty, eds.), p. 321–325. Berlin-Heidelberg-New York: Springer 1974.

Khan, A.M.: The physiology of salt tolerance in *Festuca rubra* L. Ph. D. thesis, University of Wales (1972).

Kinzel, H.: Zellsaft-Analysen zum pflanzlichen Calcium- und Säurestoffwechsel und zum Problem der Kalk- und Silikatpflanzen. Protoplasma **57**, 522–555 (1963).

KINZEL, H.: Ansätze zu einer vergleichenden Physiologie des Mineralstoffwechsels und ihre ökologischen Konsequenzen. Ber. Deut. Bot. Ges. **82**, 143–158 (1969).

KLEESE, R.A.: Relative importance of stem and root in determining genotypic differences in Sr-89 and Ca-45 accumulation in soybeans (*Glycine max* L.). Crop Sci. **7**, 53–55 (1967).

KLEESE, R.A.: Scion control of genotypic differences in Sr and Ca accumulation in soybeans under field conditions. Crop Sci. **8**, 128–129 (1968).

KLEESE, R.A., SMITH, L.J.: Scion control of genotypic differences in mineral salts accumulation in soyabean (*Glycine max* L. Merr.) seeds. Ann. Bot. (London), N.S. **34**, 183–188 (1970).

KUIPER, P.J.C.: Lipids in grape roots in relation to chloride transport. Plant Physiol. **43**, 1367–1371 (1968a).

KUIPER, P.J.C.: Ion transport characteristics of grape root lipids in relation to chloride transport. Plant Physiol. **43**, 1372–1374 (1968b).

KYLIN, A., HANSSON, G.: Transport of sodium and potassium, and properties of (sodium+potassium)-activated adenosine triphosphatases: possible connection with salt tolerance in plants. In: Proc. 8th Colloq. Intern. Potash Inst., p. 64–68. Berne: Intern. Potash Inst. 1971.

KYLIN, A., KÄHR, M.: The effect of magnesium and calcium ions on adenosine triphosphatases from wheat and oat roots at different pH. Physiol. Plantarum **28**, 452–457 (1973).

LÄUCHLI, A., EPSTEIN, E.: Transport of potassium and rubidium in plant roots. The significance of calcium. Plant Physiol. **45**, 639–641 (1970).

LÄUCHLI, A., EPSTEIN, E.: Lateral transport of ions into the xylem of corn roots. I. Kinetics and energetics. Plant Physiol. **48**, 111–117 (1971).

LÄUCHLI, A., KRAMER, D., PITMAN, M.G., LÜTTGE, U.: Ultrastructure of xylem parenchyma cells of barley roots in relation to ion transport to the xylem. Planta **119**, 85–99 (1974a).

LÄUCHLI, A., KRAMER, D., STELZER, R.: Ultrastructure and ion localization in xylem parenchyma cells of roots. In: Membrane transport in plants (U. ZIMMERMANN, J. DAINTY, eds.), p. 363–371. Berlin-Heidelberg-New York: Springer 1974b.

LÄUCHLI, A., LÜTTGE, U., PITMAN, M.G.: Ion uptake and transport through barley seedlings: differential effect of cycloheximide. Z. Naturforsch. **28c**, 431–434 (1973).

LAMBOWITZ, A.M., SLAYMAN, C.W., SLAYMAN, C.L., BONNER, W.D. JR.: The electron transport components of wild type and *poky* strains of *Neurospora crassa*. J. Biol. Chem. **247**, 1536–1545 (1972a).

LAMBOWITZ, A.M., SMITH, E.W., SLAYMAN, C.W.: Electron transport in *Neurospora* mitochondria. Studies on wild type and *poky*. J. Biol. Chem. **247**, 4850–4858 (1972b).

LAMBOWITZ, A.M., SMITH, E.W., SLAYMAN, C.W.: Oxidative phosphorylation in *Neurospora* mitochondria. Studies on wild type, *poky* and chloramphenicol-induced wild type. J. Biol. Chem. **247**, 4859–4865 (1972c).

LANE, I.R., LYON, G.D.: Differential response to salt within the species *Festuca rubra*. B. Sc. Thesis, University of Wales (1966).

LANGE, O.L., ZIEGLER, H.: Der Schwermetallgehalt von Flechten aus dem *Acarosporeteum sinopicae* auf Erzschlackenhalden des Harzes. I. Eisen und Kupfer. Mitt. Flor.-Soz. Arbeitsgem., N.F. **10**, 156–183 (1963).

LEGGETT, J.E., GILBERT, W.A.: Magnesium uptake by soybeans. Plant Physiol. **44**, 1182–1186 (1969).

LÜTTGE, U.: Salt glands. In: Ion transport in plant cells and tissues (D.A. BAKER, J.L. HALL, eds.), p. 335–376. Amsterdam-Oxford-New York: North-Holland Publishing Company 1975.

LÜTTGE, U., LÄUCHLI, A., BALL, E., PITMAN, M.G.: Cycloheximide: a specific inhibitor of protein synthesis and intercellular ion transport in plant roots. Experientia **30**, 470–471 (1974).

LÜTTGE, U., OSMOND, C.B.: Ion absorption in *Atriplex* leaf tissue. III. Site of metabolic control of light-dependent chloride secretion to epidermal bladders. Australian J. Biol. Sci. **23**, 17–25 (1970).

MARCUS, A.: Enzyme induction in plants. Ann. Rev. Plant Physiol. **22**, 313–336 (1971).

MATHYS, W.: Vergleichende Untersuchungen der Zinkaufnahme von resistenten und sensitiven Populationen von *Agrostis tenuis* Sibth. Flora (Jena) **162**, 492–499 (1973).

MINOTTI, P.L., WILLIAMS, D.G., JACKSON, W.A.: Nitrate uptake and reduction as affected by calcium and potassium. Soil Sci. Soc. Amer. Proc. **32**, 692–698 (1968).

MUNNS, D.N., JACOBSON, L., JOHNSON, C.M.: Uptake and distribution of manganese in oat plants. II. A kinetic model. Plant Soil **19**, 193–204 (1963a).

MUNNS, D.N., JOHNSON, C.M., JACOBSON, L.: Uptake and distribution of manganese in oat plants. I. Varietal variation. Plant Soil **19**, 115–126 (1963b).

MUNNS, D.N., JOHNSON, C.M., JACOBSON, L.: Uptake and distribution of manganese in oat plants. III. An analysis of biotic and environmental effects. Plant Soil **19**, 285–295 (1963c).

NISSEN, P.: Choline sulfate permease: transfer of information from bacteria to higher plants? Biochem. Biophys. Res. Commun. **32**, 696–703 (1968).

NISSEN, P.: Choline sulfate permease: transfer of information from bacteria to higher plants? II. Induction processes. In: Informative molecules in biological systems (L.G.H. LEDOUX, ed.), p. 201–212. Amsterdam: North-Holland Publ. Co. 1971.

NISSEN, P.: Bacteria-mediated uptake of choline sulfate by plants. Sci. Rep. Agr. Univ. Norway **52**, 1–53 (1973).

NISSEN, P., BENSON, A.A.: Active transport of choline sulfate by barley roots. Plant Physiol. **39**, 586–589 (1964).

NOESKE, O., LÄUCHLI, A., LANGE, O.L., VIEWEG, G.H., ZIEGLER, H.: Konzentration und Lokalisierung von Schwermetallen in Flechten der Erzschlackenhalden des Harzes. Deut. Bot. Ges., N.F. **4**, 67–79 (1970).

PARDEE, A.B.: Membrane transport proteins. Science **162**, 632–637 (1968).

PHILLIPS, J.W., BAKER, D.E., CLAGETT, C.O.: Kinetics of P absorption by excised roots and leaves of corn hybrids. Agron. J. **63**, 517–520 (1971).

PICCIURRO, G., BRUNETTI, N.: Assorbimento del sodio (Na^{22}) in radici escisse di alcune varietà di *Lycopersicum esculentum*. Agrochimica **13**, 347–357 (1969).

PITMAN, M.G.: Uptake and transport of ions in barley seedlings. I. Estimation of chloride fluxes in cells of excised roots. Australian J. Biol. Sci. **24**, 407–421 (1971).

PITMAN, M.G.: Uptake and transport of ions in barley seedlings. III. Correlation between transport to the shoot and relative growth rate. Australian J. Biol. Sci. **25**, 905–919 (1972).

PITMAN, M.G., LÜTTGE, U., LÄUCHLI, A., BALL, E.: Ion uptake to slices of barley leaves, and regulation of K content in cells of the leaves. Z. Pflanzenphysiol. **72**, 75–88 (1974a).

PITMAN, M.G., LÜTTGE, U., LÄUCHLI, A., BALL, E.: Action of abscisic acid on ion transport as affected by root temperature and nutrient status. J. Exptl. Bot. **25**, 147–155 (1974b).

POPE, D.T., MUNGER, H.M.: Heredity and nutrition in relation to magnesium deficiency chlorosis in celery. Proc. Amer. Soc. Hort. Sci. **61**, 472–480 (1953).

RAINS, D.W.: Salt transport by plants in relation to salinity. Ann. Rev. Plant Physiol. **23**, 367–388 (1972).

RAINS, D.W., EPSTEIN, E.: Preferential absorption of potassium by leaf tissue of the mangrove *Avicennia marina*: an aspect of halophytic competence in coping with salt. Australian J. Biol. Sci. **20**, 847–857 (1967).

RUSH, D.W., EPSTEIN, E.: Genotypic responses to salinity: differences between salt sensitive and salt tolerant genotypes of the tomato. Plant Physiol. **57** (1976, in press).

SALSAC, L.: Absorption du calcium par les racines de féverole (calcicole) et de lupin jaune (calcifuge). Physiol. Vég. **11**, 95–119 (1973).

SCHAEFER, N., WILDES, R.A., PITMAN, M.G.: Inhibition by p-fluorophenylalanine of protein synthesis and of ion transport across the root in barley seedlings. Australian J. Plant Physiol. **2**, 61–73 (1975).

SEVERNE, B.C., BROOKS, R.R.: A nickel-accumulating plant from Western Australia. Planta **103**, 91–94 (1972).

SHACHAR-HILL, B., HILL, A.E.: Ion and water transport in *Limonium*. VI. The induction of chloride pumping. Biochim. Biophys. Acta **211**, 313–317 (1970).

SHAFER, J., JR., BAKER, J.E., THOMPSON, J.F.: A *Chlorella* mutant lacking nitrate reductase. Amer. J. Bot. **48**, 896–899 (1961).

SHIM, S.C., VOSE, P.B.: Varietal differences in the kinetics of iron uptake by excised rice roots. J. Exptl. Bot. **16**, 216–232 (1965).

SLAYMAN, C.W.: The genetic control of membrane transport. In: Current topics in membranes and transport (F. BRONNER, A. KLEINZELLER, eds.), vol. 4, p. 1–174. New York and London: Academic Press 1973.

SMITH, F.A.: The internal control of nitrate uptake into excised barley roots with differing salt contents. New Phytologist **72**, 769–782 (1973).

SNAYDON, R.W., BRADSHAW, A.D.: Differences between natural populations of *Trifolium repens* L. in response to mineral nutrients. II. Calcium, magnesium and potassium. J. Appl. Ecol. **6**, 185–202 (1969).

STELZER, R., LÄUCHLI, A., KRAMER, D.: Interzelluläre Transportwege des Chlorids in Wurzeln intakter Gerstepflanzen. Cytobiologie **10**, 449–457 (1975).

STEVENINCK, R.F.M. VAN, STEVENINCK, M.E. VAN: Effects of inhibitors of protein and nucleic acid synthesis on the development of ion uptake mechanisms in beetroot slices (*Beta vulgaris*). Physiol. Plantarum **27**, 407–411 (1972).

TAL, M.: Salt tolerance in the wild relatives of the cultivated tomato: responses of *Lycopersicon esculentum*, *L. peruvianum*, and *L. esculentum minor* to sodium chloride solution. Australian J. Agr. Res. **22**, 631–638 (1971).

TAL, M., GAVISH, U.: Salt tolerance in the wild relatives of the cultivated tomato: water balance and abscisic acid in *Lycopersicon esculentum* and *L. peruvianum* under low and high salinity. Australian J. Agr. Res. **24**, 353–361 (1973).

TANNER, W.: Light-driven active uptake of 3-O-methylglucose via an inducible hexose uptake system of *Chlorella*. Biochem. Biophys. Res. Commun. **36**, 278–283 (1969).

TANNER, W., GRÜNES, R., KANDLER, O.: Spezifität und Turnover des induzierbaren Hexose-Aufnahmesystems von *Chlorella*. Z. Pflanzenphysiol. **62**, 376–386 (1970).

TANNER, W., HAASS, D., DECKER, M., LOOS, E., KOMOR, B., KOMOR, E.: Active hexose transport in *Chlorella vulgaris*. In: Membrane transport in plants (U. ZIMMERMANN, J. DAINTY, eds.), p. 202–208. Berlin-Heidelberg-New York: Springer 1974.

TANNER, W., KANDLER, O.: Die Abhängigkeit der Adaptation der Glucose-Aufnahme von der oxydativen und der photosynthetischen Phosphorylierung bei *Chlorella vulgaris*. Z. Pflanzenphysiol. **58**, 24–32 (1967).

TIKU, B.L., SNAYDON, R.W.: Salinity tolerance within the grass species *Agrostis stolonifera* L. Plant Soil **35**, 421–431 (1971).

TURNER, R.G.: Heavy metal tolerance in plants. In: Ecological aspects of the mineral nutrition of plants (I.H. RORISON, ed.), p. 399–410. Oxford and Edinburgh: Blackwell Sci. Publ. 1969.

VOSE, P.B.: Varietal differences in plant nutrition. Herbage Abstr. **33**, 1–13 (1963).

WAISEL, Y.: Biology of halophytes. New York and London: Academic Press 1972.

WALL, J.R., ANDRUS, C.F.: The inheritance and physiology of boron response in the tomato. Amer. J. Bot. **49**, 758–762 (1962).

WANN, E.V., HILLS, W.A.: The genetics of boron and iron transport in the tomato. J. Heredity **64**, 370–371 (1973).

WARNER, R.L., HAGEMAN, R.H., DUDLEY, J.W., LAMBERT, R.J.: Inheritance of nitrate reductase activity in *Zea mays* L. Proc. Natl. Acad. Sci. U.S. **62**, 785–792 (1969).

WEISS, M.G.: Inheritance and physiology of efficiency in iron utilization in soybeans. Genetics **28**, 253–268 (1943).

WU, L., BRADSHAW, A.D.: Aerial pollution and the rapid evolution of copper tolerance. Nature **238**, 167–169 (1972).

10. Regulation in the Whole Plant

J.F. SUTCLIFFE

1. Introduction

Most vascular plants absorb the bulk of their inorganic nutrients from the soil solution through the roots. From the site of absorption, ions are distributed initially to various parts of the plant, mainly in the xylem, and secondary redistribution occurs in the phloem. In this Chapter an attempt is made to explain how long-distance transport is regulated within the plant in the light of the knowledge of ion transport at the sub-cellular, cellular, tissue and organ levels as presented in earlier Chapters.

Many investigations have been made over the past hundred years of the effects of environmental factors on the distribution of materials in plants. While the results of such studies are difficult to interpret unequivocally in terms of mechanism, they must be consistent with any hypothesis that is proposed. Some of the more important features of ion distribution in the intact plant will be described and then a detailed analysis of regulation in a particular system, the germinating seedling, is presented. From this emerge some suggestions about the integration of xylem and phloem transport in relation to the demand for ions in different parts of the plant.

2. Ion Distribution and Transpiration

One of the earliest proposals about the uptake and transport of ions in plants was that put forward by SACHS (1887). He conjectured that mineral salts are taken up from the soil with the water that enters a plant and are carried into the leaves in the transpiration stream. There the water evaporates, leaving the salts behind to accumulate. This simple mechanism, although true in part, does not adequately account for the selectivity with which ions are absorbed and distributed within the plant. COLLANDER (1941) was among the first to demonstrate that different plant species growing in the same nutrient solution absorb quite different amounts of particular ions and that these are distributed quite differently between roots and shoots.

Many studies have been made on the relationship between water and salt absorption. Although it has often been found that ion uptake increases with increase in transpiration, most investigators have failed to establish any quantitative relationship between the two processes (e.g. MUENSCHER, 1922; VAN DEN HONERT, 1933; BROYER and HOAGLAND, 1943; BROUWER, 1954, 1956; VAN DEN HONERT, HOOYMANS and VOLKERS, 1955; RUSSELL and SHORROCKS, 1959). Only a few researchers (e.g. SCHMIDT, 1936; HYLMÖ, 1953, 1955, 1958) have claimed to have observed an exact correlation between water and salt absorption.

It has now become evident that water absorption can influence ion transport in several ways and that the interaction between the two processes is complex. Firstly, the rate of supply of ions to the absorbing surfaces of the root may sometimes be a limiting factor, especially when plants are growing in dry soils where the resistance of the diffusional pathway is high. In this situation, mass flow of soil solution towards the root surface under the influence of transpiration will increase the supply of ions to the root in proportion to the rate of water absorption. Secondly, water movement across the root cortex affects ion fluxes through both the apoplastic and symplastic pathways (Chaps. *1, 2, 3.3, 3.4*). Finally, the concentration of ions in the xylem sap is an important factor regulating their release into the conducting elements and this is affected greatly by transpiration rate (*3.4.2.2*; see particularly Table *3.18*). RUSSELL and SHORROCKS (1959) calculated that the concentration of phosphate in the xylem sap of barley and sunflower plants was many times higher at low than at high transpiration rates. The reduction in concentration of xylem sap at high transpiration rates will promote influx of ions into the xylem whether this occurs by leakage of previously accumulated ions from cells adjoining the xylem or by the operation of an active pump. However, BOWLING and WEATHER-LEY (1965) concluded that the increased K^+ flux with increasing transpiration in *Ricinus* was not due to dilution of the xylem sap or a mass flow of nutrient solution directly to the xylem because the xylem concentration remained virtually constant at about 20 times that of the external solution at widely different water fluxes. It appears that in this case, increased transpiration causes a decline in the resistance of the root both to water and ion fluxes (cf. BROUWER, 1954).

COOIL (1974a, b) studied the relationship between exudation rate and the concentration of ions in the xylem sap of excised cucumber roots bathed in media of various ionic composition and observed that some ionic components were present in the xylem sap at concentrations nearly identical with those in the external solution. Concentrations of these components were unaffected when water flow was reduced in various ways. COOIL concluded that there must be a radial flow of solution across the cortex which is in some way controlled by ion uptake at the plasma membranes of the cortical cells.

Upward transport of ions from the roots into the shoot seems to depend mainly on the transpiration stream, despite the occasional claim that there is some upward movement in the phloem (CURTIS, 1923, 1925). STOUT and HOAGLAND (1939) showed that when ^{42}K, ^{86}Br and ^{32}P were fed to the roots of willow and geranium plants radioactivity was detected in both xylem and phloem of the shoot within a few hours. However, if the wood and bark were separated along a length of the stem by a strip of waxed paper radioactivity in this region was high in the wood and low in the bark (Fig. *10.1*). These results indicated that upward longitudinal movement occurs mainly, if not entirely, in the xylem, but that rapid lateral exchange of ions is possible between xylem and phloem where the two tissues are in contact. They also showed that there was some longitudinal transport in the phloem, especially in the downward direction, but this was evidently slow in comparison with the rate of transport upwards in the xylem in these experiments.

The speed at which radioactive isotopes are carried into the leaves of plants appears to be closely correlated with the rate of transpiration, and radioactivity has been detected in the leaves of a rapidly transpiring tomato plant within minutes of labeled phosphate being supplied to the roots (ARNON et al., 1940). On the

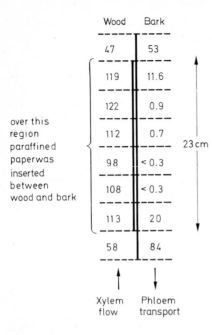

Fig. *10*.1. Results of Stout and Hoagland's (1939) experiment on distribution of ^{42}K between bark and wood in willow stems. The figures are in ppm of dry matter

other hand, if the evaporation of water from a leaf, or from part of a leaf, is prevented by enclosing the organ in a polythene bag, movement of isotopes into that leaf is drastically reduced (Sutcliffe, 1962). Closure of stomata, either as a result of water stress or artificially by treatment with abscisic acid presumably has a similar, but less marked, effect.

Although the linear rate of transport of ions in the xylem is profoundly influenced by the rate of transpiration, the *quantity* of solute transported in unit time (specific mass transfer) is largely independent of the rate of solution flow, except at very low transpiration rates. This occurs because, as already mentioned above, the concentration of the xylem sap usually increases with reduction in transpiration. Broyer and Hoagland (1943) found that there was very little difference in the amounts of ions accumulating in the shoots of barley plants when salts were provided only during the night, and when they were provided during the day, despite large differences in transpiration rates. In barley seedlings and mustard seedlings the net transport of univalent cations to the shoot from solutions containing both K^+ and Na^+ was found to be independent of transpiration rate, even though the proportions of K^+ and Na^+ taken up varied (Pitman, 1965, 1966). Not all net uptakes are independent of water flow, and often it is found that uptake of Ca^{2+} and Mg^{2+} is at least partly dependent on the rate of transpiration (see Fig. *10*.2), and so is uptake of SiO_2 (Jones and Handreck, 1965; Handreck and Jones, 1967).

The amount of solutes reaching the individual leaves *via* the xylem is a function of the volume flow of xylem sap to the leaf, and its concentration. The first is usually closely related to the transpiration rate, although in a growing leaf a significant fraction of the water taken up may be retained by expanding cells. Sap concentration is influenced (a) by the rate at which solutes are delivered to the xylem sap in the roots and (b) by the extent to which ions are removed during passage through the xylem.

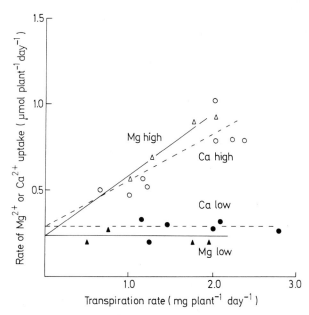

Fig. *10.2*. Relationship between rates of transport of Ca^{2+} and Mg^{2+} and water flow at high (15 mM; open symbols) and at low external concentration (0.5 mM; closed symbols). At the high concentration, uptake of univalent cation (K$^+$+Na$^+$) was unaffected by water flux. (Data from LAZAROFF and PITMAN, 1966)

Factors affecting the loading of the xylem in the root are discussed in Chap. *3.4*. Particular points are that transfer of most ions from the external solution to xylem across the root involves metabolic transport, but certain ions (e.g. NO$_3^-$, P$_i$, NH$_4^+$, K$^+$) seem more dependent on metabolism than others (e.g. Ca^{2+}, Mg^{2+}). Transport of the first group is strongly reduced by inhibitors of energy metabolism, but little affected by transpiration (see, for example, SUTCLIFFE, 1962). Where transport across the root is dependent on metabolism, there is a possibility that transfer to the shoot can be regulated by feedback of signals from the shoot (see below). The plant has less independent control over transpiration-dependent transport. (This may be important as a cause of salinity damage when Na$^+$ and Cl$^-$ are at high concentrations externally.)

During passage through the plant, the concentration of xylem sap may decrease considerably (KLEPPER and KAUFMANN, 1966). Solutes are absorbed from the transpiration stream by neighbouring cells, especially by the actively-growing cambium and xylem parenchyma, and some are transferred from xylem to phloem, perhaps *via* specialized transfer cells (GUNNING and PATE, 1968, 1974; Fig. *10.3*).

Certain ions, notably calcium, are removed from the xylem sap by absorption on to the walls of the xylem elements. BIDDULPH et al. (1961) and BELL and BIDDULPH (1963) showed that the pattern of ^{45}Ca^{2+} movement through stems of bean plants includes a reversible exchange phase and an irreversible adsorption phase. JACOBY (1967) demonstrated that calcium moved more readily through bean stems when it was supplied together with EDTA than in the absence of the chelator. FERGUSON

(1972) found that the rate of movement of Ca^{2+} through woody stems was increased by addition of chelating agents, such as EDTA or by presence of a competing cation such as Sr^{2+}. This was taken to indicate that Ca^{2+} becomes bound to negative exchange sites in the xylem. WALLACE (1963) has shown similarly that ethylene diamine di(o-hydroxyphenylacetate) (EDDHA) facilitates the transfer of iron from roots to shoots of bush beans. Immobility of ionic iron in plant tissues and particular-

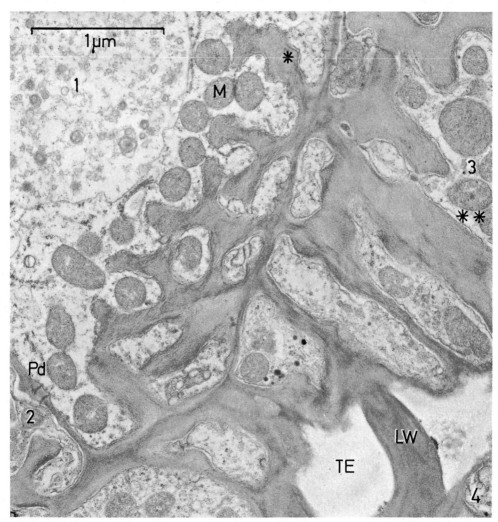

Fig. *10.3.* Transfer cells of xylem parenchyma from the vascular tissue at the base of the ovary in a wheat floret, i.e. from where the ovary is attached to the rachilla. The micrograph shows flange-type wall ingrowths in various views and also the abundance of mitochondria in the vicinity of the ingrowths. The empty space crossed by one bar of thickened wall (*LW*) at the lower right is the lumen of a tracheary element (*TE*) that these cells abut. Parts of 3 transfer cells (numbers *1–3*) occupy most of the field, and a small part of a fourth one (No. *4*) is lying to the lower right. *M* mitochondrion, *TE* lumen of tracheary element, *LW* lignified wall of tracheary element, *Pd* plasmodesmata, * flange-type cell wall ingrowths or thickenings in transverse section, ** flange-type thickenings in longitudinal section. (Original electron-micrograph by courtesy of Drs. S.-Y. ZEE and T.P. O'BRIEN, Monash University, Melbourne)

ly in the xylem appear to be related to the level of phosphate which leads to deposition of iron as insoluble iron phosphates (cf. BECKETT and ANDERSON, 1973).

The possibility of interchange of ions between xylem and phloem may lead to a circulation of ions in the plant. MASON and MASKELL (1931) suggested that ions such as P_i and K^+ may sometimes be transported through the xylem into the leaves of plants more quickly than they are utilized there and the excess ions are exported elsewhere in the phloem. Some may travel back to the roots and there re-enter the xylem for re-circulation, probably to another part of the shoot. BIDDULPH et al. (1958) demonstrated such a circulation of ^{32}P in bean (*Phaseolus vulgaris* L.) plants. Circulation of some other ions, e.g. SO_4^{2-}, was rapidly curtailed as a result of incorporation into insoluble compounds, while Ca^{2+} did not circulate at all, presumably because it is immobile in the phloem (see below).

3. Ion Distribution and Plant Development

3.1 Ion Transport and Growth

Early investigators of ion transport in whole plants soon discovered that absorption and distribution were closely related to growth of the plant and its individual parts. MUENSCHER (1922), for example, found that salt uptake by intact barley plants was more exactly correlated with the intensity of photosynthesis and growth than with water absorption. The importance of growth in regulating ion accumulation in *Nitella* can be inferred from early experiments of HOAGLAND and his collaborators (see STEWARD and SUTCLIFFE, 1959 for references). Since these pioneering studies it has often been reported that the uptake of ions under a variety of environmental conditions is closely related with growth of plants. GREENWAY (1965) and, more recently, PITMAN (1972) established a correlation between relative growth rates and the transport of K^+ into the shoots of barley plants.

Comparison of fluxes of K^+ in roots and shoots shows that the correlation between growth and transport arises in the roots and not in the leaves. Fluxes of K^+ into the root were reduced at low relative growth rates but fluxes in cells of the leaves were the same at high and low relative growth rate (PITMAN et al., 1974, see also *3.3.8*, Table *3.15*).

An implication of the correlation between uptake and relative growth rate is that the concentration in the shoot is independent of growth rate. It is found too that the concentration of nutrients in the shoot remains generally independent of the external concentration once it has reached a certain level. Table *10.1* shows the effect of varying the concentration of K^+ and P_i in flowing culture solution on the concentrations in the shoots of several plant species. Various factors determine the concentration in the shoot, as is discussed below, but the overall impression is that for K^+ and P_i there is some orderly relation between growth and nutrient absorption determined by the plant and not by its environment (see also *3.3.7* and Fig. *3.23*).

Table *10*.1. Content of K^+ and P in shoots of plants relative to fresh weight (μmol g_{FW}^{-1}) as a function of external concentration

	K^+				
Concentration in solution (μM)	1	8	24	95	1,000
Relative content of shoots (μmol g_{FW}^{-1})	22	110	144	162	190
	P				
Concentration in solution (μM)	0.04	0.2	1	5	25
Relative content of shoots (μmol g_{FW}^{-1})	3.8	10.5	23	25	27

K^+ data: 14 species (from ASHER and OZANNE, 1967).
P data: 7 species (from ASHER and LONERAGAN, 1967).

3.2 Redistribution during Growth

MASON et al. (1936) studied the movement of carbohydrates and other solutes in the phloem of cotton plants and from their observations developed the concept of sources and sinks. Sinks are essentially growing regions of the plant, e.g. meristems, young leaves and developing storage organs towards which materials move from sources of supply elsewhere in the plant.

STEWARD and his collaborators (see STEWARD and SUTCLIFFE, 1959, for references) established the importance of growth in regulating the absorption of ions in storage tissue slices (Part A, Chap. *8*) and went on to study the accumulation patterns along the axis of intact roots and in the whole plant in relation to growth (STEWARD and MILLAR, 1954). The ability of growing cells to accumulate ions by incorporating them into insoluble cell constituents or by sequestration in vacuoles is an important factor regulating the pattern of distribution in the plant as a whole.

The supply of inorganic solutes to growing regions comes partly from the external medium *via* the roots and partly from reserves within the plant itself. WILLIAMS (1955) calculated that over 90% of the total nitrogen and phosphorus taken up by a cereal plant may be accumulated by the time 25% of the dry matter is synthesized and later growth is mainly at the expense of the accumulated reserves. ARNON and HOAGLAND (1943) showed that if tomato plants are grown in a complete nutrient medium until the flowering stage and then transferred to one lacking P_i, some P moves into the developing fruits from leaves and stems. Plants which were deflowered upon transfer to the P-free medium made more vegetative growth than those which were allowed to fruit, presumably because more P was available. WILLIAMS (1948) found that when oat plants were grown under conditions of low P_i supply, the P required for development of inflorescences was obtained mainly from the medium, but when they were grown in the presence of abundant P_i a larger proportion was supplied by depletion of leaves and stems (Fig. *10*.4). There is a relatively greater export of N from leaves into the developing grains of wheat if the N supply to the roots is cut off at anthesis than if N is supplied continuously (NEALES et al., 1963). KISSEL and RAGLAND (1967) deprived *Zea mays* plants of an external supply of N-compounds, P_i, K^+, Mg^{2+} and Ca^{2+} immediately after silking and showed that there was increased movement of all ions except for Ca^{2+} from the leaves into the developing fruits.

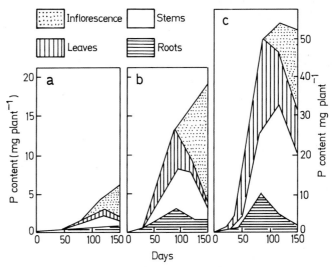

Fig. 10.4a–c. The pattern of distribution of phosphorus in oat plants with varying phosphate supply (a) low phosphate, (b) medium phosphate, (c) high phosphate (re-drawn from WILLIAMS, 1948)

Shoots can deplete roots of previously accumulated P if the external supply is cut off (LOUGHMAN and RUSSELL, 1957). However, CROSSETT and LOUGHMAN (1966) showed that only a small proportion of the total orthophosphate of barley is normally available for transport to the shoot. It is postulated that there is a small cytoplasmic P_i fraction available for transport and this is maintained at a constant level by regulation of the processes of absorption and transport (LOUGHMAN, 1969).

Ion retranslocation to developing organs is often extremely selective even compared with transport to other parts of the shoot. Table 10.2 shows that high salinity has a much greater effect on K^+, Na^+ and Cl^- levels in leaves of barley than in the developing grain.

Because fruits and other storage organs transpire relatively slowly, the bulk of the ions they receive comes *via* the phloem. It is not surprising, therefore, that the Ca^{2+} content of fruits is relatively low and Ca^{2+} deficiency symptoms often appear first in fruits, e.g. blossom-end rot of tomatoes. It has been demonstrated

Table 10.2. Response of barley grains and leaves to different salinites. Data from GREENWAY et al. (1965)

	Leaves (mmol g_{DW}^{-1})			Grain (mmol g_{DW}^{-1})		
	K^+	Na^+	Cl^-	K^+	Na^+	Cl^-
Low salinity[a]	1.3	0.1	0.2	0.29	0.02	0.05
High salinity[b]	0.5	1.0	1.0	0.25	0.04	0.09

[a] 1 mM NaCl; [b] 50 mM NaCl; K^+ was 6 mM in both cases.

that there is very little transport of Ca^{2+} from the parent plant into the fruits of peanut (*Arachis hypogaea* L.) which normally develop underground, unless they are allowed to transpire (SKELTON and SHEAR, 1971). Some calcium is taken up directly from the soil by buried fruits (BLEDSOE et al., 1949) and this appears to be the main source of Ca^{2+} under natural conditions.

3.3 Regulation of Ion Supply to Leaves

When a leaf is young it receives nutrients both *via* the phloem and the xylem. As the leaf expands and the transpiration rate increases, the xylem becomes a more important channel of supply than the phloem. GREENWAY and PITMAN (1965) estimated the amounts of K^+ moving through the xylem and phloem into barley leaves of different ages and the data they obtained are presented in Table *10*.3.

The reduction in the intake of nutrients through the phloem by a leaf as it matures is accompanied by development of photosynthetic activity and an associated reversal in the direction of net flow of carbohydrates. Observations on a number of dicotyledonous plants suggest that a leaf often begins to produce assimilates surplus to its requirement by the time it is 25% of its final area, and export attains a maximum at about the same time as the leaf reaches its full size (THROWER, 1967). In contrast, net intake of phosphorus *via* the phloem appears to continue until the leaf reaches maturity and only during senescence does net export occur (HOPKINSON, 1964). The likelihood that a leaf at a certain stage of its development may act as a source for one substance (e.g. assimilate) and a sink for another (e.g. phosphorus) must be taken into account when discussing the control of distribution (see below). It has been observed that the phloem remains functional and transports large amounts of various substances from a leaf after it has stopped transporting significant amounts of assimilates. Old leaves are therefore of greater significance as sources of nitrogen and phosphorus than of assimilates (HOPKINSON, 1966).

The situation in monocot leaves is somewhat different from that in dicots because of the presence of an intercalary meristem and the consequent existence of cells at various stages of development in the same lamina. However, the general picture appears to be the same. Leaves are net importers of K^+ and P_i for a sustained period and net export only occurs in the later stages of leaf development (Table

Table *10*.3. Rates (μmol d^{-1}) of intake of potassium ions through the xylem and phloem in barley leaves of different ages (GREENWAY and PITMAN, 1965)

	Age of leaf		
	Young	Inter- mediate	Old
Intake *via* xylem	2.0	2.7	1.9
Intake *via* phloem	1.3	0.7	-1.6
Total intake	3.3	3.4	0.3

10.3). There is a possibility of significant redistribution of ions within a monocot leaf as it grows. GREENWAY and GUNN (1966) showed that when ^{32}P was applied to the tip of a recently-emerged barley leaf, some of the radioactivity moved into the basal part of the same leaf but not into the other parts of the plant. Transport of phosphate into a mature leaf in the xylem was reduced when the concentration of phosphate in the nutrient medium was lowered and the rate of export in the phloem also declined. On the other hand, the supply of phosphate to young growing tissues was similar at both high and low phosphate levels.

LINCK, 1955 (see SWANSON, 1959) made a comprehensive study of the pattern distribution of ^{32}P fed to individual leaves of pea (*Pisum sativum* L.) and some of his results are presented diagrammatically in Fig. *10*.5. As the Figure shows, movement from the lower leaves (L5 and L7) was predominantly towards the roots, while that from higher leaves (L9 and L10) was mainly towards the apex. A large fraction of the translocate from leaf 9 moved into the developing pods and most of it went into the pod at node 11 (P$_2$). On the other hand, ^{32}P applied to leaf 10 was exported mainly into the pod developing in the axil of this leaf (P$_1$).

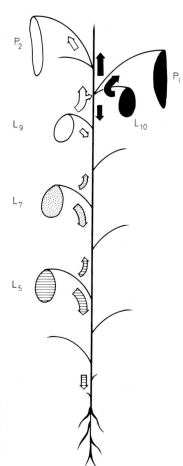

Fig. *10*.5. Diagram of a pea (*Pisum sativum*) plant showing distribution of labeled phosphate from individual leaves as a function of their position on the stem. The size of the arrow indicates the amount of radioactivity moving in a particular direction from the fed leaves (L5–10) (re-drawn from LINCK, 1955)

4. Ion Distribution in Germinating Seeds

4.1 Mineral Content of Seeds

During the maturation of seeds, organic substances and inorganic ions are transported from the parent plant and stored either in the cotyledons of the young embryo, or in extra-embryonic tissues such as endosperm. The mineral elements are stored either as free ions or in organic complexes involving such substances as RNA, protein and phytin (Mg^{2+} and Ca^{2+} salts of myo-inositol hexaphosphate). These materials are mobilized and utilized to support growth of the axis in the early stages of germination. Some plants, e.g. tomato, have relatively small seed reserves and need to start absorbing ions from the medium at an early stage if normal growth is to be maintained. Others, notably crop plants such as legumes which have been bred for maximum size and nutritive content of the seed have sufficient reserves to grow reasonably well for several weeks without an external supply of nutrients. The micronutrient content of many seeds is more than sufficient to supply the needs of one generation and a large proportion of an element such as zinc is transferred from one crop of seeds to the next (Sudia and Green, 1972).

A study of the factors which regulate the movement of endogenous substances from the storage tissues of a germinating seed into the developing axis might help to clarify some of the problems of ion distribution in plants generally. Early investigators (e.g. Le Clerk and Breazeale, 1911; Buckner, 1915, 1921) established a correlation between the rate of depletion of cotyledons or endosperm and growth of the axis but were unable to decide whether it was the supply of reserve materials that controlled growth or *vice versa*.

4.2 Transport in Cereal Grains

Analyses have shown that although movement of mineral elements from endosperm parallels in general that of dry matter, different elements are transported at different rates and some ions, notably Ca^{2+}, are utilized much more slowly than others. Brown (1946) noted that the dry weight of attached embryos of barley seedlings increased less rapidly during the first 48 h than did nitrogen content which suggests that there was a greater export of N-compounds than dry matter from the endosperm in the early stages of germination. Mer et al. (1963) confirmed that this was also the case in dark-grown oat seedlings and supported the suggestion by Albaum et al. (1942) that growth of oat seedlings during early germination is controlled by nitrogen supply. However, it seems dangerous to conclude that supply of an N-compound controls growth on the basis of a study of the relationship between transport of N and dry matter alone. Baset and Sutcliffe (1975) studied the export of four elements, N, P, K^+ and Mg^{2+} in relation to movement of dry matter from the endosperm during germination of oat seedlings. The plants were grown in the dark in the absence of exogenous nutrients except for $CaCl_2$ which had to be added to maintain the roots in a healthy state. Under the conditions employed the endosperm was largely depleted of dry matter after seven days. As may be seen from Fig. *10*.6 the pattern of depletion was broadly similar for total dry matter and for each of the individual elements studied. There was a period of relatively

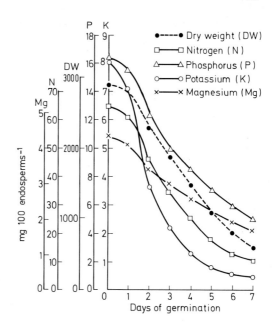

Fig. *10*.6. Changes in the composition of the endosperm of oat seedlings during germination (BASET, 1972)

slow depletion building up to a maximum rate after 2–3 days and then the rate decreased. Differences in rates of depletion of individual elements are seen more clearly when the data are plotted as in Fig. *10*.7. From this Figure it is clear that the rate of export of individual elements relative to dry matter changes markedly with time and the pattern is different for each element.

The simplest curve to analyze is that for K^+. This element is transported from the endosperm relative to dry matter most rapidly during the first day of germination and progressively more slowly thereafter. This suggests that K^+ is readily available for export immediately upon hydration of the seed and it is evident that strong sinks already exist in the growing axis. As sink capacity is likely to increase with time as the axis enters its exponential phase of growth, especially in the absence of an external supply of K^+ to the roots, the falling rate of export of K^+ from the endosperm must be related solely to diminishing supply. Supply is therefore likely to be the controlling factor over most, if not all, of the germination period.

The relationship between the export of N and dry matter is similar to that for K^+, but the initial rate of depletion of N relative to dry matter is lower and it falls less rapidly with time. This suggests that either the sinks for N in the axis are initially less powerful than those for K^+, or that N is not so immediately available for export. We cannot decide about the first possibility, but the latter proposition seems likely as about 88% of the N in the endosperm initially is in the form of protein, which presumably must be hydrolyzed before the N can be translocated. The proteolytic activity of endosperm extracts is relatively low at the beginning of germination but increases rapidly with time (SUTCLIFFE and BAṢET, 1973). They found that there is a close correlation between the *rate of increase* of protease activity in endosperm extracts and the rate of export of N. This suggests that the growing axis controls the mobilization and transport of N partly by regulating the synthesis of proteolytic enzymes in the aleurone layer. A similar correlation

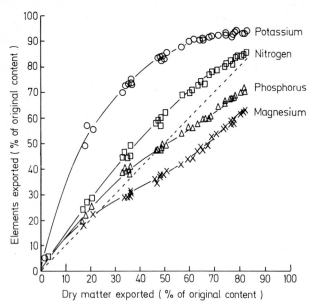

Fig. *10.7.* Export of potassium, nitrogen, phosphorus and magnesium relative to export of dry matter from the endosperm of oat seedlings during germination (Baset and Sutcliffe, 1975)

was found between the rate of depletion of starch and the rate of increase in amylase activity in endosperm extracts. The fact that proteolytic activity develops relatively more quickly than amylase activity probably accounts for the fact that N is transported from the endosperm relatively more quickly than dry matter in the early stages of germination (Fig. *10.7*).

It follows from these observations that only recently-acquired enzyme activity is involved in the breakdown of starch and protein. With time a progressively higher proportion of the total enzyme in endosperm extracts appears to be inactive *in vivo*. It was suggested that newly-synthesized enzyme becomes associated with individual starch or aleurone grains and that when each of these is depleted, the enzyme attached to it is inactivated through lack of substrate (Sutcliffe and Baset, 1973).

The curves for the depletion of Mg^{2+} and P relative to dry matter are more difficult to interpret. Both elements are transported from the endosperm relatively more quickly than dry matter during the first day of germination, but the rate does not increase as quickly thereafter, so the relative rate falls. It increases again in the later stages of germination because the rate of depletion of the two elements falls off less quickly than that of dry matter. The similarity between the two curves may be related to the fact that both Mg^{2+} and P are stored in oat endosperm mainly in the form of phytin and will be released when phytin is broken down. Discrepancies may be attributed to the fact that an appreciable amount (20–30%) of P, but not of Mg^{2+}, is stored in oat endosperm in other substances than phytin, e.g. RNA, which may be mobilized at different rates.

Alternatively, there may be a relative change during growth in the strength of the sinks in the axis for Mg^{2+} on the one hand and P on the other. This might be related to changing demands for the elements in the growing axis. It was observed that when Na^+ or K^+ phosphate was supplied to the roots, transport of P_i, but not of Mg^{2+}, from the endosperm was reduced. This is interpreted as indicating that the demand for a particular element in the growing axis, if partially satisfied from elsewhere, can exert a regulating influence on depletion of this element in the endosperm. It would be interesting to compare the rates of depletion of Mg^{2+} and P reserves in dark-grown seedlings with those of seedlings grown in the light in which chlorophyll synthesis in the later stages of germination might possibly increase the demand for Mg^{2+} relative to P. Preliminary experiments in the authors laboratory indicate that Mg^{2+} is transported from oat endosperm more rapidly relative to other elements when the seedlings are grown in the light than in the dark.

The picture that emerges from this research is that the transport of materials from the endosperm of oat seedlings during germination is controlled by growth of the axis through a "push-pull" system. The axis controls the strength of sinks for individual elements through the utilization of materials in growth or by sequestering them in vacuoles. At the same time it regulates the mobilization of reserve materials in response to demand by controlling the rate of synthesis or activation of hydrolytic enzymes in the aleurone layer presumably by changes in the rate of production of gibberellins (VARNER, 1964). Thus both utilization and mobilization are under precise control, and transport can be altered quickly in response to changing needs during early growth of the seedling.

4.3 Transport in Pea Seedlings

A similar investigation to the one described above has been made of the control of the movement of materials from the cotyledons into the developing axis of the garden pea, Pisum sativum L. (GUARDIOLA and SUTCLIFFE, 1971a, b, 1972; GUARDIOLA, 1973; GARCIA-LUIS and GUARDIOLA, 1975). It has been observed that distribution of cotyledonary reserves between the shoot and the root depends on the relative rates of growth of the two organs. Dark-grown seedlings accumulated a higher proportion of the transported reserves in the shoot than did the light grown seedlings

Table 10.4. Transport of mineral elements from pea cotyledons into the shoot during 4 weeks of growth. Data expressed as percent of the total amount transported from the cotyledons recovered in the shoot (GUARDIOLA, 1973)

	Element	Age of the seedlings (weeks)			
		1	2	3	4
Dark-grown seedlings	Nitrogen	63.3	72.6	83.3	88.8
	Phosphorus	62.2	71.9	84.1	88.6
	Potassium	56.1	69.5	77.7	81.4
	Sulfur	65.3	84.3	91.0	93.3
Light-grown seedlings	Nitrogen	43.4	57.8	73.9	73.1
	Phosphorus	44.2	54.6	73.6	78.4
	Potassium	31.3	45.8	62.3	62.6
	Sulfur	40.0	66.0	78.9	79.6

Table *10*.5. The influence of the shoot on growth and accumulation of mineral elements in the roots of pea plants grown without an external source of nutrient. Plants de-shooted when one week old. Growth parameters of the root expressed as a percentage of those from intact plants. Values marked with an asterisk (*) differ significantly from the untreated control plants (GUARDIOLA, 1973)

	Weeks after excision of the shoot		
	1	2	3
Fresh weight	77*	79*	77*
Dry weight	85*	76*	71*
Protein nitrogen	96	96	94
Total nitrogen	93	124*	126*
Phosphorus	97	133*	130*
Sulfur	101	112*	139*
Potassium	94	100	100

(Table *10*.4), and growth of the shoot also increased markedly whilst that of the root was hardly affected. Removal of the shoot from a week-old seedling did not affect movement of K^+ into the roots, suggesting that there is little or no competition between shoot and root for this nutrient (Table *10*.5). Dry matter content of the roots in de-shooted plants was lower than in intact plants but total N, P and S contents were higher. The increases are an indication that either there is competition between shoot and root for these elements or that in the intact plant there is some secondary redistribution from root to shoot.

In contrast to the situation in *Avena* the rate of transport of each individual element was related linearly with dry matter transport and the relationship was not affected when the growth rate of the axis was altered (Fig. *10*.8). There was also little change in the rates at which individual elements were transported relative to one another throughout the germination period. No correlation such as those observed in *Avena* endosperm could be found between the activity of proteolytic enzymes and phytase and the transport of nitrogen and phosphorus. It was concluded that mobilization is regulated by the rate of release of soluble substances from cotyledon cells rather than by the rate of hydrolysis of reserve materials.

Existence of a stoichiometric relationship between the movement of dry matter and individual elements from cotyledons have been observed by a number of investigators (BIDDULPH, 1951; KATSUTA, 1961; OKAMOTO, 1962; RIGA and BUKOVAC, 1961). The high rate of transport of zinc relative to phosphorus from cotyledons of *Phaseolus vulgaris* L. in the early stages of germination has been attributed to location of zinc in the cell walls (VOCHTING, 1953).

The cotyledons of dormant pea seeds contain an appreciable amount of Ca^{2+}, part of which is exported to the axis during early growth of the seedling. GUARDIOLA and SUTCLIFFE (1972) found that 26% of the cotyledonary Ca^{2+} was exported during a 4-week growth period compared with 72% of dry matter and 82% of K^+. Calcium is often said to be immobile in the phloem, but it is not likely that transport occurs out of the cotyledons in the xylem. Experiments in which the shoots of 9-day-old pea seedlings were steam-girdled showed that movement of

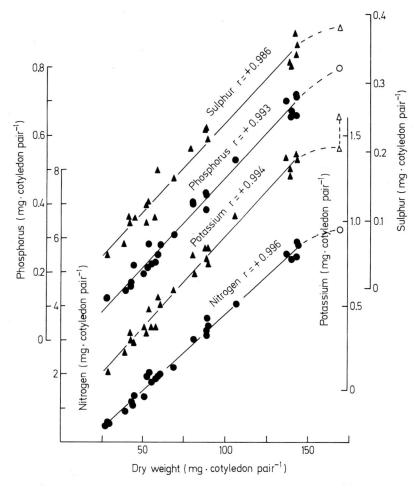

Fig. *10*.8. The relationship between dry matter content and the amounts of individual elements in the cotyledons of *Pisum sativum* at different stages of growth over a period of 4 weeks. The results are from seedlings grown under several different conditions (see text). The full line has been statistically adjusted and the coefficients of correlation are given. (From GUARDIOLA and SUTCLIFFE, 1972)

Ca^{2+} as well as of other solutes, was markedly inhibited and it was concluded that export occurs in the phloem (GUARDIOLA and SUTCLIFFE, 1972).

Up to three-quarters of all the Ca^{2+} in pea cotyledons remains at the time of abscission and attempts to increase the amount of Ca^{2+} mobilized have been unsuccessful. FERGUSON (1972) found that Ca^{2+} injected into pea cotyledons was not transported to the axis and he concluded that only Ca^{2+} associated with phytin is mobilized. It appears that Ca^{2+} released from phytin is protected in some way from binding to cell walls and thus becomes available for transport in the phloem. Whether the Ca^{2+} is in ionic or in some chelated form is not known. The breakdown of phytin is achieved by sequential de-phosphorylation (COSGROVE, 1966), and it may be that calcium is released as undissociated phosphates.

5. Hormonal Control of Redistribution

As the distribution of materials in plants is profoundly affected by the pattern of growth it is not surprising that growth regulators have been found to have effects on ion transport. Mitchell and Martin (1937) showed that soluble carbohydrates, nitrogenous substances and ions accumulate in regions of a plant treated with IAA. Bode (1959) found that when decapitated tomato plants were sprayed with solutions of IAA (1 g m^{-3}) there was in increase in the total ash content of the leaves. There was a slight decrease in the proportion of K$^+$ in the total ash of the younger leaves and an increase in mature leaves. Bode suggested that auxin stimulated the formation of K$^+$ acceptors in the mature leaves. In long-term experiments such as these it is possible that the effects of IAA are the result of differential growth induced by the auxin. Even in short-term experiments, such as those of Davies and Wareing (1965) in which increased transport of ^{32}P towards the tip of a decapitated bean seedling treated with IAA was detected within a few hours the possibility of a growth effect cannot be ruled out. However, from experiments involving the use of tri-iodo benzoic acid, a known inhibitor of auxin transport, Davies and Wareing concluded that the basipetal transport of IAA was somehow involved in the stimulation of ^{32}P movement. Seth and Wareing (1967) found that kinetin and gibberellic acid had no effects on transport of ^{32}P in de-fruited peduncles of French bean plants but when these substances were applied together with IAA, the effect of the latter was increased markedly. The largest response was obtained when all three substances were applied together. Whether the effects are due to an influence on transport *per se* or to an effect on processes leading to stimulation of growth is still uncertain.

Cytokinins have the capacity to delay senescence of tissues (Mothes, 1961) and when applied to a leaf they stimulate movement of materials towards the treated region (Fig. *10*.9). Saeed (1975) has shown that application of benzyladenine delays senescence in a mature leaf of *Xanthium pennsylvanicum* and reduces the rate of export of assimilates and of ^{32}P. Growth of the next-higher leaf on the stem is somewhat reduced and smaller amounts of carbohydrate and ^{32}P are transported into it from the treated leaf. These effects of benzyladenine are thought to be related to an influence of the cytokinin on RNA and protein synthesis which affects the fluxes of ions and other materials across the plasma membranes of leaf cells. Garg and Kapoor (1972) found that ascorbic acid also retards leaf senescence and that it is effective at a lower concentration than either gibberellic acid or kinetin. Its effect on ion distribution has not been studied.

Guardiola and Sutcliffe (1972) showed that GA$_3$ stimulated shoot growth in seedlings of *Pisum sativum* cv Alaska without affecting K$^+$, P, N or S content. Garcia-Luis and Guardiola (1975) obtained a much larger effect of GA$_3$ on shoot growth of another pea variety, "Progress", and in this case the N content of the shoot was increased, while that of the root was reduced. The GA$_3$ treatment increased amylolytic activity in the cotyledons of the "Progress" seedlings and there was enhanced movement of dry matter and of nitrogen from them into the axis.

Specific effects of plant hormones on ion transport processes are discussed in Chaps. *3.4*, *7* and *8*. In general, two types of action can be distinguished, though

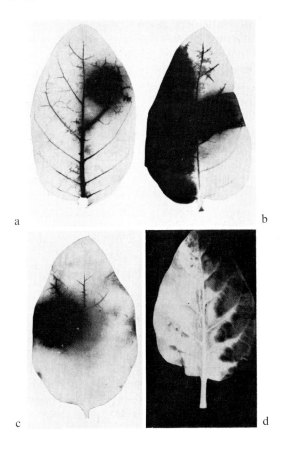

Fig. *10*.9a–d. Kinetin effect on the distribution of foliar-applied amino acid in tobacco leaves (*Nicotiana rustica*). (a) to (c) Autoradiographs: (a) ^{14}C-glycine added in a circular area on the right of the central vein. (b) ^{14}C-glycine added as in (a) but in addition the leaf blade on the left of the central vein was sprayed with kinetin solution. (c) Both kinetin and ^{14}C-glycine were applied to the left side of the leaf blade. Note that the amino acid was retained at the site of its application in the non-kinetin treated leaf (a), redistributed to the site of kinetin application in (b), and retained in (c) where the sites of the kinetin and amino acid application were the same. (d) Photograph of an isolated leaf of *N. rustica* 13 days after removal from the plant: the right side of the leaf blade was sprayed with kinetin solution and shows little senescence as compared with the untreated left side of the blade. (From MOTHES, 1961)

the response of cells and plants to particular hormones may differ from one plant or condition of growth to another. Fluxes into root cells at the plasmalemma seem to be more sensitive to auxins and analogous phytotoxins such as fusicoccin (*7.7.3*; *7.8*), particularly those processes involving H^+ secretion. Transport across the root to the xylem is specifically affected by cytokinins and by ABA (*3.4.2.5*; *7.3*; *7.5*). Since this process supplies most of the nutrients to the xylem, there is a clear possibility that cytokinins and ABA could have a role in regulating the supply of nutrients from root to shoot. Growth regulators probably play a significant role in the integration of transport in the plant as a whole.

6. Regulation of Ion Transport—A Working Hypothesis

The picture that emerges from the results presented in this Chapter is that a vascular plant functions as an integrated unit in which the distribution of solutes is controlled in such a way that the available supply is utilized most effectively.

During early growth, a seedling is more or less dependent on endogenous reserves. These are mobilized in the storage tissues at a rate determined by growth of the axis and distributed between shoot and root according to the needs of each organ *via* the phloem. Sooner or later the plant becomes dependent on inorganic solutes absorbed by the roots and carried into the shoot in the transpiration stream. The shoot exerts some control on the amounts of solutes absorbed by the roots and on the proportion transferred to the shoots. The influence of the transpiration stream in regulating ion uptake by the root and transport to the shoot has already been discussed (*10*.1). The shoot also exerts control by the provision of respiratory substances upon which absorption by the root and transport into the xylem depends. The possibility that the shoot regulates these processes by hormonal stimuli has also been considered (*10*.4).

In general, these mechanisms will result in the supply of ions to the shoot through the xylem being increased when the demand is high and reduced when the demand is low. However, the control of ion transport in the xylem does not appear to be a very effective mechanism for regulating supply to individual leaves during the course of their development. When a leaf is young, the xylem supply is evidently inadequate for the rapidly increasing needs of growing cells and it is supplemented by an additional supply through the phloem. Later when growth declines and transpiration rate reaches a maximum, supply through the xylem is more than sufficient, and excess solutes are exported in the phloem or, in some cases, excreted by salt glands (Chap. *5*.2). The evidence thus points to the phloem as having a prominent role in regulating solute distribution within the plant. It seems clear that to do this efficiently the distribution of each solute must be controlled independently and the following working hypothesis has been proposed (SUTCLIFFE, 1976).

The suggested hypothesis invokes the idea that solutes move from sources to sinks through the sieve tubes in response to gradients of electro-chemical potential ("solute potential") between the solution bathing the free space at the two ends of the system. Transport occurs from a place where the concentration of a solute is high, to one where it is lower, and each solute moves independently in response to the existing gradient (cf. MASON et al., 1936). The movement of solutes from the sources (cotyledons or endosperm) of a germinating seed into shoot and root sinks of the developing axis can be described using an electrical analogy (Fig. *10*.10).

The source is represented by a capacitor (A) to indicate that it has both "capacity", which is related to its volume, and the ability to hold a "charge", or potential. The source in a germinating seed acquires solutes (solute potential becomes negative) as a result of the enzyme breakdown of reserve materials and release of soluble

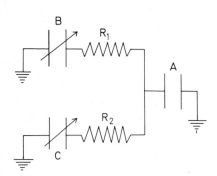

Fig. *10*.10. An electrical analogy to illustrate movement of solutes from a source (*A*) to shoot (*B*) and root (*C*) sinks in a germinating seed. (From SUTCLIFFE, 1975)

molecules or ions into the free space. When a seedling begins to take up ions through the roots sources may acquire additional solutes via the xylem. The sinks (B and C) likewise have both capacity and potential. The capacity is assumed to increase during growth and is therefore represented as variable in Fig. *10.*10. Sink potential is controlled by the rate at which solutes are removed from the solution to be utilized in growth or accumulated in vacuoles, and by the rate of replenishment of the free space.

The amount of an individual solute transported in unit time from the source to an individual sink will be related to the solute potential difference between the two solutions and the resistance presented by the connecting sieve tubes (R_1 and R_2). It appears from this model that removal of one of the two sinks should not directly affect the rate of solute movement to the other and this is what was observed in the experiments with pea seedlings, at least for K^+. The rate of movement of solutes to a particular sink may be altered by a change in solute potential at either source or sink, or both, or by a change in phloem resistance. It appears that in the oat seedling, increased growth of the axis causes not only an increase in solute potential at the sinks but also a reduction in solute potential at the source by stimulating the hydrolysis of storage products.

As the plant grows and new leaves are produced, older ones begin to senesce and the pattern of solute distribution becomes more complex. According to the hypothesis, a leaf continues to act as a sink as long as the potential of the solution bathing its sieve tube terminals is sufficiently high to maintain a gradient in the right direction between it and neighbouring leaves. When growth begins to slow down, solutes start to accumulate in the solution and the gradient is reduced until finally it is reversed. The leaf then begins to act as a source, but it does not do so necessarily for all solutes simultaneously, because the rates of consumption and supply of individual substances may be different.

It has been found that in general the ability of a leaf to function as a sink or source is closely related to the intensity of RNA and protein synthesis (YAGI, 1972; SAEED, 1975). This is not surprising, because the rate of synthesis of these substances must control the rate of utilization of a variety of organic and inorganic solutes. Even the accumulation of ions in vacuoles has been shown to be dependent on protein synthesis (SUTCLIFFE, 1962, 1973). On the other hand, net protein hydrolysis as in senescing leaves leads to the accumulation of a variety of solutes in the solution bathing the sieve-tube ends. Soluble materials are released from ageing cells through an increase in the permeability of the tonoplasts and plasma membranes (cf. release of materials in cotyledons).

It has often been observed that solutes exported from individual leaves are transported mainly to the next younger leaf in the same orthostichy. This is not surprising if the resistance of the phloem is a factor in the regulation of ion distribution. When a powerful sink, such as a developing fruit, arises, solutes move into it from all over the plant but, even so, movement is mainly from the nearest mature leaf (Fig. *10.*5) presumably because the resistance is lowest.

More data on the rates of movement of individual solutes in both xylem and phloem to particular leaves during the course of their development are needed to test the solute potential gradient hypothesis.

References

ALBAUM, H.G., DONNELY, J., KORKES, S.: The growth and metabolism of oat seedlings after seed exposure to oxygen. Amer. J. Bot. **29**, 388–395 (1942).

ARNON, D.I., HOAGLAND, D.R.: Composition of the tomato plant as influenced by nutrient supply in relation to fruiting. Bot. Gaz. **104**, 576–590 (1943).

ARNON, D.I., STOUT, P.R., SIPOS, F.: Radioactive phosphorus as an indicator of phosphorus absorption of tomato plants at various stages of development. Amer. J. Bot. **27**, 791–798 (1940).

ASHER, C.J., LONERAGAN, J.F.: Responses of plants to phosphate concentration in solution culture. 1. Growth and phosphorus content. Soil Sci. **103**, 225–233 (1967).

ASHER, C.J., OZANNE, P.G.: Growth and potassium content of plants in solution cultures maintained at constant potassium concentrations. Soil Sci. **103**, 155–161 (1967).

BASET, Q.A.: Mobilization and transport of food reserves in etiolated oat seedlings. D. Phil. thesis, University of Sussex (1972)

BASET, Q.A., SUTCLIFFE, J.F.: Regulation of the export of potassium, nitrogen, phosphorus, magnesium and dry matter from the endosperm of etiolated oat seedlings (*Avena sativa* cv Victory) Ann. Bot. (London), N.S. **39**, 31–41 (1975).

BECKETT, J.T., ANDERSON, W.P.: Ferric-EDTA absorption by maize roots. In: Ion transport in plants, 630 p. (W.P. ANDERSON, ed.), p. 595–607. London-New York: Academic Press 1973.

BELL, C.W., BIDDULPH, O.: Translocation of calcium. Exchange versus mass flow. Plant Physiol. **38**, 610–614 (1963).

BIDDULPH, O.: The translocation of minerals in plants. In: Mineral nutrition of plants (E. TRUOG, ed.), p. 261–275, Madison, Wisconsin: University of Wisconsin Press, 1951.

BIDDULPH, O., BIDDULPH, S.F., CORY, R., KOONTZ, H.: Circulation patterns for P^{32}, S^{35}, and Ca^{45} in the bean plant. Plant Physiol. **33**, 293–300 (1958).

BIDDULPH, O., NAKAYAMA, F.S., CORY, R.: Transpiration stream and ascension of calcium. Plant Physiol. **36**, 429–436 (1961).

BLEDSOE, R.W., COMAR, C.L., HARRIS, H.C.: Absorption of radioactive calcium by the peanut fruit. Science **109**, 229–330 (1949).

BODE, H.R.: Über den Einfluß des Heteroauxins auf die Kationenzusammensetzung der Blattasche der Tomate. Planta **53**, 212–218 (1959).

BOWLING, D.J.F., WEATHERLEY, P.E.: The relationship between transpiration and potassium uptake in *Ricinus communis.* J. Exptl. Bot. **16**, 732–741 (1965).

BROUWER, R.: The regulating influence of transpiration and suction tension on the water and salt uptake by roots of intact *Vicia faba* plants. Acta Botan. Neerl. **3**, 264–312 (1954).

BROUWER, R.: Investigations into the occurrence of active and passive components in the ion uptake by *Vicia faba.* Acta Botan. Neerl. **5**, 287–314 (1956).

BROWN, R.: Studies on germination and seedling growth. III. Early growth in relation to certain aspects of nitrogen metabolism in the seedling of barley. Ann. Bot. (London), N.S. **10**, 73–96 (1946).

BROYER, T.C., HOAGLAND, D.R.: Metabolic activities of roots and their bearing on the relation of upward movement of salts and water in plants. Amer. J. Bot. **30**, 261–273 (1943).

BUCKNER, G.D.: Translocation of mineral constituents of seeds and tubers of certain plants during growth. J. Agr. Res. **5**, 409–453 (1915).

BUCKNER, G.D.: Comparative utilisation of the mineral constituents in the cotyledons of bean seedlings grown in soil and in distilled water. J. Agr. Res. **20**, 875–880 (1921).

COLLANDER, R.: Selective absorption of cations by higher plants. Plant Physiol. **16**, 691–720 (1941)

COOIL, B.J.: Accumulation and radial transport of ions from potassium salts by cucumber roots. Plant Physiol. **53**, 158–163 (1974a).

COOIL, B.J.: Characteristics of radial solution flow in roots of cucumber (*Cucumis sativus* L.). Ann. Bot. (London), N.S. **38**, 1053–1065 (1974b).

COSGROVE, D.J.: Chemistry and biochemistry of inositol polyphosphates. Rev. Pure Appl. Chem. **16**, 209 (1966).

CROSSETT, R.N., LOUGHMAN, B.C.: The absorption and translocation of phosphorus by seedlings of *Hordeum vulgare* L. New Phytologist **65**, 459–468 (1966).

CURTIS, O.F.: The effect of ringing a stem on the upward transfer of nitrogen and ash constituents. Amer. J. Bot. **10**, 361–382 (1923).

CURTIS, O.F.: Studies on the tissues concerned in the transfer of solutes in plants. The effect on upward transfer of solutes of cutting the xylem and compared with that of cutting the phloem. Ann. Bot. (London), **39**, 573–585 (1925)

DAVIES, C.R., WAREING, P.F.: Auxin induced transport of radio phosphorus in stems. Planta **65**, 135–156 (1965).

FERGUSON, I.B.: Calcium mobility in plants. Ph.D. Thesis, University of Auckland, New Zealand (1972).

GARCIA-LUIS, A., GUARDIOLA, J.L.: Effects of gibberellic acid on the transport of nitrogen from the cotyledons of young pea seedlings. Ann. Bot. (London), N.S. **39**, 325–330 (1975).

GARGO, O.P., KAPOOR, V.: Retardation of leaf senescence by ascorbic acid. J. Exptl. Bot. **23**, 699–703 (1972).

GREENWAY, H.: Plant responses to saline substances. VII. Growth and ion uptake throughout plant development in two varieties of *Hordeum vulgare*. Australian J. Biol. Sci. **18**, 163–181 (1965).

GREENWAY, H., GUNN, A.: Phosphorus retranslocation in *Hordeum vulgare* during early tillering. Planta **71**, 43–67 (1966).

GREENWAY, H., GUNN, A., PITMAN, M.G., THOMAS, D.A.: Plant responses to saline substrates. VI. Chloride, sodium, and potassium uptake and distribution within the plant during ontogenesis of *Hordeum vulgare*. Australian J. Biol. Sci. **18**, 525–540 (1965).

GREENWAY, H., PITMAN, M.G.: Potassium retranslocation in seedlings of *Hordeum vulgare*. Australian J. Biol. Sci. **18**, 235–247 (1965).

GUARDIOLA, J.L.: Growth and accumulation of mineral elements in the axis of young pea (*Pisum sativum*, L.) seedlings. Acta Botan. Neerl. **22**, 55–68 (1973).

GUARDIOLA, J.L., SUTCLIFFE, J.F.: Control of protein hydrolysis in the cotyledons of germinating pea (*Pisum sativum*, L.) seeds. Ann. Bot. (London), N.S. **35**, 791–807 (1971a).

GUARDIOLA, J.L., SUTCLIFFE, J.F.: Mobilisation of phosphorus in the cotyledons of young seedlings of the garden pea (*Pisum sativum*, L.). Ann. Bot. **35**, 809–823 (1971b).

GUARDIOLA, J.L., SUTCLIFFE, J.F.: Transport of materials from the cotyledons during germination of seeds of the garden pea (*Pisum sativum*, L.). J. Exptl. Bot. **23**, 322–337 (1972).

GUNNING, B.E.S., PATE, J.S.: Transfer cells. Plant cells with wall ingrowths, specialised in relation to short distance transport of solutes—their occurrence, structure and development. Protoplasma **68**, 107–133 (1968).

GUNNING, B.E.S., PATE, J.S.: Transfer cells. In: Dynamic aspects of plant ultrastructure, Chap. 13 (A.W. ROBARDS, ed.). London-New York: McGraw-Hill 1974.

HANDRECK, K.A., JONES, L.H.P.: Uptake of monosilicic acid by *Trifolium incarnatum* (L.). Australian J. Biol. Sci. **20**, 483–485 (1967).

HONERT, T.H. VAN DEN: The phosphate absorption by sugar cane. Verslag 13e. Bijeenk omst van de Vereeniging van Proefstations-Personell Buitenzorg, Java (1933).

HONERT, T.H. VAN DEN, HOOYMANS, J.J.M., VOLKERS, W.S.: Experiments on the relation between water absorption and mineral uptake by plant roots. Acta Botan. Neerl. **4**, 139–155 (1955).

HOPKINSON, J.M.: Studies on the expansion of the leaf surface. IV. The carbon and phosphorus economy of a leaf. J. Exptl. Bot. **15**, 125–137 (1964).

HOPKINSON, J.M.: Studies on the expansion of the leaf surface. VI. Senescence and the usefulness of old leaves. J. Exptl. Bot. **17**, 762–770 (1966).

HYLMÖ, B.: Transpiration and ion absorption. Physiol. Plantarum **6**, 333–405 (1953).

HYLMÖ, B.: Passive components in the ion absorption of the plant. I. The zonal ion and water absorption in Brouwer's experiments. Physiol. Plantarum **8**, 433–441 (1955).

HYLMÖ, B.: Passive components in the ion absorption of the plant. II. The zonal water flow, ion passage and pore size in roots of *Vicia faba*. Physiol. Plantarum **11**, 382–400 (1958).

JACOBY, B.: Effect of roots on calcium ascent in bean stems. Ann. Bot. (London), N.S. **31**, 725–730 (1967).

JONES, L.H.P., HANDRECK, K.A.: Studies of silica in the oat plant. III. Uptake of silica from soils by the plant. Plant Soil **23**, 79–96 (1965).

KATSUTA, M.: The breakdown of reserve protein in pine seeds during germination. J. Japan. Forestry Soc. **43**, 241–244 (1961).

KISSEL, D.E., RAGLAND, J.C.: Redistribution of nutrient elements in corn (*Zea mays* L.) I. N, P, K, Ca and Mg redistribution in the absence of nutrient accumulation after silking. Soil Sci. Soc. Am. Proc. **31**, 227–230 (1967).

KLEPPER, B., KAUFMANN, M.R.: Removal of salts from xylem sap by leaves and stems of guttating plants. Plant Physiol. **41**, 1743–1747 (1966).

Lᴀᴢᴀʀᴏꜰꜰ, N., Pɪᴛᴍᴀɴ, M.G.: Calcium and magnesium uptake by barley seedlings. Australian J. Biol. Sci. **19**, 991–1005 (1966).

Lᴇ Cʟᴇʀᴋ, J.A., Bʀᴇᴀᴢᴇᴀʟᴇ, J.F.: Translocation of plant food and elaboration of organic material in wheat seedlings. U.S. Dept. Agr. Bull. Chem. **138**, (1911).

Lɪɴᴄᴋ, A.J.: Studies on the distribution of phosphorus-32 in *Pisum sativum* in relation to fruit development. Ph.D. Dissertation. Ohio State University, Columbus, Ohio (1955).

Lᴏᴜɢʜᴍᴀɴ, B.C.: The uptake of phosphate and its transport within the plant, p. 309–322. In: Ecological aspects of the mineral nutrition of plants. (I.H. Rᴏʀɪsᴏɴ, ed.). Oxford-Edinburgh: Blackwells 1969.

Lᴏᴜɢʜᴍᴀɴ, B.C., Rᴜssᴇʟʟ, R.S.: The absorption and utilisation of phosphate by young barley plants. J. Exptl. Bot. **8**, 280–293 (1957).

Mᴀsᴏɴ, T.G., Mᴀsᴋᴇʟʟ, E.J.: Further studies on transport in the cotton plant. I. Preliminary observations on the transport of phosphorus, potassium and calcium. Ann. Bot. (London), **45**, 125–174 (1931).

Mᴀsᴏɴ, T.G., Mᴀsᴋᴇʟʟ, E.J., Pʜɪʟʟɪs, E.: Further studies on transport in the cotton plant. III. Concerning the independence of solute movement in the phloem. Ann. Bot. (London), **50**, 23–58 (1936).

Mᴇʀ, C.L., Dɪxᴏɴ, P.F., Dɪᴀᴍᴏɴᴅ, B.C., Dʀᴀᴋᴇ, C.F.: The dominant influence of nitrogen on growth correlation in etiolated oat seedlings. Ann. Bot. (London), N.S. **27**, 693–721 (1969).

Mɪᴛᴄʜᴇʟʟ, J.W., Mᴀʀᴛɪɴ, W.E.: Effects of indolyl-acetic-acid on growth and chemical composition of etiolated bean plants. Botan. Gaz. **90**, 171–183 (1937).

Mᴏᴛʜᴇs, K.: Aktiver Transport als regulatives Prinzip für gerichtete Stoffverteilung in höheren Pflanzen. In: Biochemie des aktiven Transports. Berlin-Göttingen-Heidelberg: Springer 1961.

Mᴜᴇɴsᴄʜᴇʀ, W.C.: Effect of transpiration on the absorption of salts by plants. Amer. J. Bot. **9**, 311–330 (1922).

Nᴇᴀʟᴇs, T.F., Aɴᴅᴇʀsᴏɴ, M.J., Wᴀʀᴅʟᴀᴡ, I.F.: The role of the leaves in the accumulation of nitrogen by wheat during ear development. Australian J. Agr. Res. **14**, 725–736 (1963).

Oᴋᴀᴍᴏᴛᴏ, H.: Transport of cations from cotyledons to seedling of the embryonic plants of *Vigna sesquipedalis*. Plant Cell Physiol. (Tokyo) **3**, 83–94 (1962).

Pɪᴛᴍᴀɴ, M.G.: Sodium and potassium uptake by seedlings of *Hordeum vulgare*. Australian J. Biol. Sci. **18**, 10–24 (1965).

Pɪᴛᴍᴀɴ, M.G.: Uptake of potassium and sodium by seedlings of *Sinapis alba*. Australian J. Biol. Sci. **19**, 257–269 (1966).

Pɪᴛᴍᴀɴ, M.G.: Uptake and transport of ions in barley seedlings. II. Evidence for two active stages in transport to the shoot. Australian J. Biol. Sci. **25**, 243–257 (1972).

Pɪᴛᴍᴀɴ, M.G., Lᴜᴛᴛɢᴇ, U., Lᴀᴜᴄʜʟɪ, A., Bᴀʟʟ, E.: Ion uptake to slices of barley leaves and regulation of K content in cells of the leaves. Z. Pflanzenphysiol. **72**, 75–88 (1974).

Rɪɢᴀ, A.J., Bᴜᴋᴏᴠᴀᴄ, M.J.: Distribution du ^{32}P, du ^{45}Ca et du ^{65}Zn chez le haricot (*Phaseolus vulgaris* L.) après absorption radiculaire. Redistribution de ces éléments au cours de la germination de la graine et du dévelopment de la jeune plantule. Bull. Inst. Agron. Stn. Rech. Gembloux **29**, 165–196 (1961).

Rᴜssᴇʟʟ, R.S., Sʜᴏʀʀᴏᴄᴋs, V.M.: The relationship between transpiration and the absorption of inorganic ions by intact plants. J. Exptl. Bot. **10**, 301–316 (1959).

Sᴀᴄʜs, J. ᴠᴏɴ: Lectures on the physiology of plants. English Ed. Translated by H. Marshall Ward, Oxford: Clarendon Press 1887.

Sᴀᴇᴇᴅ, A.F.H.: The distribution of mineral elements in *Xanthium pennsylvanicum*. D. Phil. Thesis, University of Sussex (1975).

Sᴄʜᴍɪᴅᴛ, O.: Die Mineralstoffaufnahme der höheren Pflanze als Funktion einer Wechselbeziehung zwischen inneren und äußeren Faktoren. Z. Bot. **30**, 289–334 (1936).

Sᴇᴛʜ, A.K., Wᴀʀᴇɪɴɢ, P.F.: Hormone-directed transport of metabolites and its possible role in plant senescence. J. Exptl. Bot. **18**, 65–77 (1967).

Sᴋᴇʟᴛᴏɴ, R.C., Sʜᴇᴀʀ, G.M.: Calcium translocation in the peanut (*Arachis hypogea* L.) Agron. J. **63**, 409–412 (1971).

Sᴛᴇᴡᴀʀᴅ, F.C., Mɪʟʟᴀʀ, F.K.: Salt accumulation in plants: A reconsideration of the role of growth and metabolism. Soc. Exptl. Biol. Symp. **8**, 367–406 (1954).

Sᴛᴇᴡᴀʀᴅ, F.C., Sᴜᴛᴄʟɪꜰꜰᴇ, J.F.: Plants in relation to inorganic salts. In: Plant physiology—a treatise (F.C. Sᴛᴇᴡᴀʀᴅ, ed.), Chap. 4, p. 253–478. New York-London: Academic Press 1959.

STOUT, P.R., HOAGLAND, D.R.: Upward and lateral movement of salt in certain plants as indicated by radioactive isotopes of potassium, sodium and phosphorus absorbed by roots. Amer. J. Bot. **26**, 320–324 (1939).

SUDIA, T.W., GREEN, D.G.: The translocation of ^{65}Zn and ^{134}Cs between seed generations in soy bean (*Glycine max* (L) Mer). Plant Soil. **37**, 695–697 (1972).

SUTCLIFFE, J.F.: Mineral salts absorption in plants. Oxford: Pergamon Press 1962.

SUTCLIFFE, J.F.: The role of protein synthesis in ion transport. In: Ion transport in plants, p. 399–406 (W.P. ANDERSON, ed.). London-New York: Academic Press 1973.

SUTCLIFFE, J.F.: Regulation of ion transport in the whole plant. Perspectives in Experimental Biology (N. SUNDERLAND, ed.), Vol. 2, Botany 542 p. I. Oxford: Pergamon Press 1976.

SUTCLIFFE, J.F., BASET, Q.A.: Control of hydrolysis of reserve materials in the endosperm of germinating oat (*Avena sativa* L.) grains. Plant Sci. Letters **1**, 15–20 (1973).

SWANSON, C.A.: Translocation of organic solutes. In: Plant physiology—a treatise (F.C. STEWARD, ed.), chap. 5, p. 481–551. New York-London: Academic Press 1959.

THROWER, S.L.: The pattern of translocation during leaf ageing. Soc. Exptl. Biol. Symp. **21**, 483–506 (1967).

VARNER, J.E.: Gibberellic-acid controlled synthesis of α-amylase in barley endosperm. Plant Physiol. **39**, 413–415 (1964).

VOCHTING, A.: Über die Zinkaufnahme von *Zea mays* L. und *Aspergillus niger* v. Tieg. in Einzelkultur und in Mischkultur. Ber. Schweiz. Bot. Ges. **63**, 103–161 (1953).

WALLACE, A.: Solute uptake by intact plants. A. Wallace, Los Angeles, California (1963).

WILLIAMS, R.F.: The effects of phosphorus supply on the rates of intake of phosphorus and nitrogen and upon certain aspects of phosphorus metabolism in graminaceous plants. Australian J. Sci. Res. (B) **1**, 333–361 (1948).

WILLIAMS, R.F.: Redistribution of mineral elements during development. Ann. Rev. Plant Physiol. **6**, 25–42 (1955).

YAGI, M.I.A.: Relationship between the distribution of mineral elements and growth of bean plants. D. Phil. Thesis, University of Sussex (1972).

Epilog: Integration of Transport in the Whole Plant

M.G. Pitman and U. Lüttge

This Section (III) has been concerned with different aspects of control and regulation in relation to transport. In Part A (Chaps. 9 and 11) it was suggested that on the cellular level there was a need to learn how to recognize the operation of a control. Difficult as it may be to distinguish control from correlations arising from common linkages to energy metabolism in the plant or the limitations imposed by physical properties of membranes, this *caveat* is even more relevant to transport in the whole plant.

Solute redistribution in the plant involves transport across cell membranes, and symplasmic transport from cell to cell (Chap. 2). In addition there may be long-distance transport in the xylem or phloem from roots to shoot, from seed to seedling, or between older and younger leaves. The particular problem we raise here is whether the plant has the means of integrating its activities to balance the demand of one organ with the supply from another.

Certain observations suggest that the plant does regulate ion uptake and redistribution. The levels of K^+ and P in the shoot (and rates of transport from root to shoot) were largely independent of external concentrations above certain quite low levels (3.3.7; 10.3.1). Further, the rate of transport from root to shoot was proportional to relative growth rate and the concentration of K^+ in the shoot was consequently very nearly independent of relative growth rate (i.e. 3.3.8; 10.3.1). There are many observations too showing that plants have a greater propensity for nutrient uptake when growing under nutrient-deficient conditions; this behavior is particularly marked for plants experiencing iron deficiency (3.3.5.4).

One view among plant physiologists of the explanation of these results has been that it is adequate to assume that operation of the supply system is in some kind of "balance" with the availability of products of metabolism. In this view the regularity of content in relation to growth is simply a reflection of the rate of supply of sugars in the phloem to the root. There is ample evidence that under certain conditions rates of uptake can be limited by supply of metabolites in the phloem (3.3.8), but other conditions can be found when availability of sugars does not seem to be the only factor limiting ion transport.

One system of integration of uptake and demand was suggested in 1971 by BENZIONI *et al.* (as discussed in Part A, Chap. 11.) Their model combines the activity of phloem and xylem in an ingenious way. Nitrate uptake was considered to be regulated by exchange for HCO_3^-; these ions were in turn provided by K^+-malate translocated from the leaf to the root. The K^+ ions were returned to the leaf in the xylem together with NO_3^- ions. A difficulty in the scheme is that the required high rates of K^+ retranslocation in the phloem do not seem to be general, though possible in particular cases.

An alternative mechanism for regulation within the whole plant involves plant hormones. Ample evidence has been obtained that plant hormones can be transported from root to shoot in xylem sap and from shoot to root in the phloem (7.5; 7.8), and that these compounds can affect operation of the xylem and phloem.

In 10.3 it was shown that the phloem can be the major pathway of redistribution in the shoot, into developing leaves and in supply of nutrients from seeds. A working hypothesis was prosposed (10.6) to show how supply and demand might be balanced during operation of phloem transport. The direction of such transport has been shown to be affected by plant hormones, providing a possible mechanism for regulation of supply and transfer of information in the plant.

In other plant systems, the supply of ions is almost entirely in the xylem, e.g. from root to shoot. In Chap. 3.4. and in 10.2 it was shown that the supply of ions to the xylem involved active transport across the root. In particular the supply of ions to the xylem could be differentiated from uptake to the root by means of the differing sensitivity of these processes to inhibitors of protein synthesis and certain plant hormones (3.4.2.5). Both xylem transport and phloem transport appear to be sensitive to plant hormones and make it possible for these compounds to transfer information linking the activities of different parts of the plant.

Integration of supply and demand of nutrients in the plant is a topic in which there are clearly many questions that need to be answered. The response of uptake in the plant to growth and external conditions poses problems both about integration and about the regulation of transport at the membrane level. This area promises to be one of the most important topics of research in the near future.

Author Index

Page numbers *in italics* refer to the bibliography.

Symbols, Units, and Abbreviations

Symbols

Some symbols are defined, and used, only in parts of particular Chapters. If no definition is given then the symbols have the following meanings.

Symbol	Description	Unit
A	area	m^2
a	activity	mM
b	partition coefficient	
C	capacity	F
c	concentrations	mM $(=mol\ m^{-3})$
$\bar{c}$	mean concentration	mM $(=mol\ m^{-3})$
D	diffusion coefficient	$m^2\ s^{-1}$
d	diameter	m
G	Gibbs free energy	J
g	gravitational force (centrifuge). Note 1,000 g often written as 1 K (Part A, Ch 10)	
g	membrane conductance	$S\ m^{-2}$
H	enthalpy	J
I	current	A
J	generalized flow $(L \cdot X)$	
J_j	net flux, $= \phi_{oi,j} - \phi_{io,j}$	flux
J_s	solute flux	e.g. mol $m^{-2}\ s^{-1}$
J_v	volume flow	$m\ s^{-1}$
K_i	inhibitor constant	mM
K_m	Michaelis-Menten constant	mM
k	rate constant	s^{-1}
L	generalized conductance coefficient, i.e. (J/X)	
L_p	hydraulic conductivity	$m\ s^{-1}\ Pa^{-1}$
l	length	m
$\ln$	logarithm to base e	
$\log$	logarithm to base 10	
n	number of moles	
P	hydrostatic pressure	Pa
P	permeability	$m\ s^{-1}$
P_K	permeability to K^+ (etc.)	
P_w	diffusion permeability to water	

Q	quantity of substance, location or ion shown by subscript	
Q^*	quantity of isotope	
q	electrical charge	C
R	electrical resistance	Ω
R	relative water content (Part A, Ch. 2) (% of content at full turgor)	
r	electrical resistance or resistivity of membrane	Ω $\Omega\,m^2$
r	radius	m
s	specific radio-activity	(counts) $mol^{-1}\,s^{-1}$
T	temperature	K or °C
t	time	s
u	electric mobility	$m^2\,V^{-1}\,s^{-1}$
$\bar{V}$	partial molar volume	$m^3\,mol^{-1}$
V	volume	
V_{max}	maximum reaction velocity; Michaelis-Menten enzyme kinetics	$mol\,s^{-1}$
v	velocity of reaction	$mol\,s^{-1}$
W	weight	
x	mole fraction	
X	generalized force	
z	valency	
α, β, γ	permeability coefficient ratios	
γ	activity coefficient	
ε	elastic coefficient (Part A, Chap. 2)	
ε	electric permittivity	
μ	chemical potential	$J\,mol^{-1}$
μ_w^o	standard state of μ_w	$J\,mol^{-1}$
$\bar{\mu}$	electrochemical potential	$J\,mol^{-1}$
$\bar{\mu}_{vo}$	difference in $\bar{\mu}$ between v and o	$J\,mol^{-1}$
π	osmotic pressure, or potential	Pa or $J\,m^{-3}$
σ	reflection coefficient	
τ	transport number	
τ	matric potential (Part A, Chap. 2)	Pa or $J\,m^{-3}$
Φ	net flux based on radioactive measurements[1] eg:—	$mol\,m^{-2}\,s^{-1}$ or $mol\,g_{FW}^{-1}\,s^{-1}$ etc.
$\Phi_{ov},$ Φ_{in}	net tracer uptake to cell or tissue relative to s_o	
$\Phi_{vo},$ Φ_{out}	net tracer efflux from cell or tissue relative to s_v	
Φ_{ox}	net tracer transport from solution to xylem, relative to s_o	

---

[1] The net flux Φ can be expressed in terms of the fluxes (ϕ) at each membrane (see Part A, Chap. 5), and Φ may thus be described as a "complex" flux in terms of the "simple" fluxes (ϕ). The usual result of experiments using tracers to measure uptake is Φ which should not be taken as equivalent to a ϕ except in certain well defined conditions.

Φ^*	net tracer uptake, or efflux ($\Phi \cdot s$)	
ϕ	unidirectional flux	mol m^{-2} s^{-1}
	or	mol g$_{FW}^{-1}$ s^{-1} etc.
ϕ_{oc}	subscripts show direction of fluxes	
$\phi_{oc,K}$	specifies the ion, and direction	
Ψ	water potential	Pa or J m^{-3}
ψ	Electrical potential difference	V
ψ_{vo}	ψ of vacuole relative to outside	
ψ_K	equilibrium potential for K$^+$ (etc.) between two phases	
ω	permeability coefficient	
[ ]	denotes concentration	
Δ	difference in	
∇	gradient operator $\left(= \dfrac{d}{dx} + \dfrac{d}{dy} + \dfrac{d}{dz} \right)$	

Subscript Labels

Subscript labels used with symbols have the following general meanings. Chemical symbols (K, Cl) when used as subscripts specify the particular ion. The order of labels shows direction, as with unidirectional fluxes.

Label	Description
a, b	two arbitrary phases in contact
c	cytoplasmic
ch	refers to a chemical reaction
d	diffusion (as in P_d)
i	inside
j	jth species
M	membrane
max	maximum level reached (eg. V_{max})
o	outside, solution
p	4th component of isotope exchange model (Part A, Chap. 5)
s,S	solute (e.g. c_s)
T	total (Part A, Chap. 5)
v	vacuole
w	water
x	xylem

Constants

Values of certain constants are:

Constant	Value
F	Faraday constant $= 96490$ C mol^{-1}

R gas constant $8.314 \, J \, mol^{-1} K^{-1}$
RT/F has the value 25.7 mV at 25° C.

Units

The International System of Units (SI) is used in this book, but with retention of the terms mM and µM for concentrations, as these have direct equivalents in the SI system (mM $=$ mol m^{-3}; µM $=$ mmol m^{-3}). The unit mol m^{-3} is used where it is more appropriate to express content relative to volume (e.g. stomata; Part A, Chap. *10*). Where possible fluxes are expressed in meter/second units or where referred to amount of tissue, relative to gram fresh weight or dry weight. Note that 1 pmol cm^{-2} s^{-1} $=$ 10 nmol m^{-2} s^{-1}; 1 mmol kg^{-1} h^{-1} $=$ 1 µmol g^{-1} h^{-1} $=$ 0.278 nmol g^{-1} s^{-1}. 1 µl $=$ 1 mm^{3} $=$ 10^{-9} m^{3}.

Prefixes:

Standard prefixes are used to all units, e.g.

M	mega	10^3
c	centi	10^{-2}
m	milli	10^{-3}
µ	micro	10^{-6}
n	nano	10^{-9}
p	pico	10^{-12}

Units:

Unit	Description
A	ampere; electrical current
bar	unit of pressure; 100 kPa or 0.987 atmospheres
C	coulomb; electrical charge
Ci	curie; radioactivity ($3.7 \cdot 10^{10}$ s^{-1})
d	day
°C	degrees Celsius; temperature
eq	equivalent; mol $\times z$
F	farad; electrical capacitance
g	gram
g_{FW}	gram fresh weight
g_{DW}	gram dry weight
Hz	hertz; frequency
h	hour
ha	hectare (10^4 m^2)
J	joule (N m)
K	kelvin (degrees)
lx	lux; illumination (lm m^{-2})
M	molar; used mainly as (mM) as this is equivalent to (mol m^{-3}), the S.I. unit

m	meter
min	minute; time
mol	mole; amount of substance, ion or compound, whose mass in grams equals the molecular weight
N	newton; force
osmol	sum of mole contribution to osmotic pressure irrespective of ionic species. Note osmol m^{-3} = mOsmolar = mOsm
Pa	pascal; pressure ($J\,m^{-3}$; $N\,m^{-2}$)
rad	absorbed dose of ionizing radiation ($10^{-2}\,J\,kg^{-1}$)
S	siemens; electrical conductance
s	second
V	volt; electrical potential difference
W	watt; power ($J\,s^{-1}$)

Abbreviations

ABA	abscisic acid
AC	alternating current
Acetyl CoA	S-acetyl coenzyme A
ACT-D	actinomycin D
ADP	adenosine 5′-diphosphate
AFS	apparent free space
AMP	adenosine 5′-monophosphate
c-AMP	cyclic AMP, adenosine 2′:3′-cyclic monophosphate
AP	action potential
APW	artificial pond water
ATP	adenosine 5′-triphosphate
ATPase	adenosine triphosphatase (Cl-ATPase, K-ATPase show ion required to stimulate activity)
BA	benzyl adenine
pBQ	p-benzoquinone
C_3	refers to photosynthetic carbon reduction cycle
C_4	refers to photosynthetic dicarboxylic acid pathway
CAM	refers to crassulacean acid metabolism
CCCP	carbonyl cyanide, m-chloro-phenyl hydrazone
CDP	cytidine 5′-diphosphate
CMP	cytidine 5′-monophosphate
CHM	cycloheximide

CMU	3′-(4-chlorophenyl)-1′, 1-dimethyl urea
CTP	cytidine 5′-triphosphate
DCCD	N,N′-dicyclohexyl carbo-diimide
DCMU	3′-(3,4 dichlorophenyl)-1′, 1-dimethyl urea
2,4-D	2,4-dichlorophenoxyacetic acid
3,5-D	3,5-dichlorophenoxyacetic acid
DFS	Donnan free space
DHAP	dihydroxyacetone phosphate
Dio-9	antibiotic of unknown structure
DMO	5,5-dimethyl-2,4-oxazolidine dione
DNA	desoxyribonucleic acid
DNP	2,4-dinitrophenol
DSPD	disalicylidenepropanediamine
DW	dry weight
EDTA	ethylenediaminetetraacetic acid
EDDHA	ethylenediamine-di(o-hydr-oxy phenylacetate)
EMF	electromotive force
ER	endoplasmic reticulum
r ER	rough ER
s ER	smooth ER
FAD	flavin adenine dinucleotide

FC	fusicoccin
FCCP	carbonyl cyanide, *p*-tri-fluoro-methoxy phenyl hydrazone
FPA	DL-*p*-fluorophenylalanine
FW	fresh weight
GA, GA$_3$	gibberellic acid
GTP	guanosine 5′-triphosphate
IAA	indole-3-acetic acid
IDP	inosine 5′-diphosphate
IMP	inosine 5′-monophosphate
ITP	inosine-5′-triphosphate
LS	low-salt (roots)
3-0-MG	3-0-methyl-glucose
α-NAA	naphthalene-1-acetic acid
β-NAA	naphthalene-2-acetic acid
NAD	nicotinamide adenine di-nucleotide;
NADH$_2$	—, reduced form
NADP	nicotinamide adenine di-nucleotide phosphate;
NADPH$_2$	—, reduced form
−ive	negative
OAA	oxaloacetic acid
∼P	"high energy" phosphate
P$_i$	inorganic phosphate
P$_r$	phytochrome (red light absorbing form = inactive form)
P$_{fr}$	phytochrome (far-red light absorbing form = active form)
PCMB	*p*-chloromercuribenzoic acid
PCMBS	*p*-chloromercuribenzene sulfonic acid

	(=*p*-chloromercuriphenyl sulfonic acid)
PD	potential difference
PEP	phosphoenolpyruvate
PEP carboxylase	phosphoenolpyruvate car-boxylase
PGA	phosphoglyceric acid
Phe	phenyl alanine
PMS	phenazine methosulfate
ppm	parts per million
PS I	photosystem I
PS II	photosystem II
+ive	positive
Q$_{10}$	temperature coefficient, ratio of rates at temp-eratures differing by 10° C
RNA	ribonucleic acid
m-RNA	messenger RNA
t-RNA	transfer RNA
RUDP	D-ribulose-1,5-diphosphate
SCC	short circuit current
SEM	standard error of the mean
SF	senescence factor
sp., spp.	species
TCA	tricarboxylic acid cycle
α-TEG	thioethyl-D-glucopyranoside
tris	tris (hydroxy methyl) amino-methane
poly-U	poly (uridine) nucleotide
UDP	uridine 5′-diphosphate
UMP	uridine 5′-monophosphate
UTP	uridine 5′-triphosphate
UV	ultraviolet light
WFS	water free space

Subject Index

Italic page numbers refer to Part A; roman page numbers refer to Part B.

Encyclopedia of Plant Physiology, New Series

Editors: A. Pirson, M.H. Zimmermann

Planta

An International Journal of Plant Biology

Editorial Board: E. Bünning, H. Grisebach, J. Heslop-Harrison, G. Jacobi, A. Lang, H.F. Linskens, H. Mohr, P. Sitte, Y. Vaadia, M.B. Wilkins, H. Ziegler

Planta publishes original articles on all branches of botany with the exception of taxonomy and floristics. Papers on cytology, genetics, and related fields are included providing they shed light on general botanical problems.

Languages used: Approximately 80% of the articles are in English; the others, in German or French, are preceded by an English summary.

1976: 3 volumes (3 issues each)

A sample copy as well as subscription and back volume information available upon request.

Please address:

Springer-Verlag
Heidelberger Platz 3
D-1000 Berlin 33
or
Springer-Verlag New York Inc.
175 Fifth Avenue
New York, NY 10010

Springer-Verlag Berlin Heidelberg New York